Hans Lewy

(Photo provided courtesy of J. Hildebrandt)

Contemporary Mathematicians

Gian-Carlo Rota†
Joseph P.S. Kung

Editors

Hans Lewy Selecta

Volume 1

David Kinderlehrer
Editor

Birkhäuser
Boston • Basel • Berlin

David Kinderlehrer
Department of Mathematical Sciences
Carnegie Mellon University
Pittsburgh, PA 15213-3890
U.S.A.

Library of Congress Cataloging-in-Publication Data

A CIP catalogue record for this book is available from the Library of Congress, Washington D.C., USA.

AMS classification codes: 00B60, 01A75, 31A05, 31B05, 32V05, 32H02, 35D05, 35J85, 35J05, 51B20, 65M06

Printed on acid-free paper

Birkhäuser

Softcover reprint of the hardcover 1st edition 2002

ISBN-13:978-1-4612-7406-3 e-ISBN-13:978-1-4612-2080-0
DOI: 10.1007/978-1-4612-2080-0

9 8 7 6 5 4 3 2 1

Hans Lewy as an infant in Germany

Hans and Helen Lewy

Summer 1973

Hans and Helen Lewy with Brigitte and Stefan Hildebrandt

Golden Gate Park, San Francisco, CA, 1979

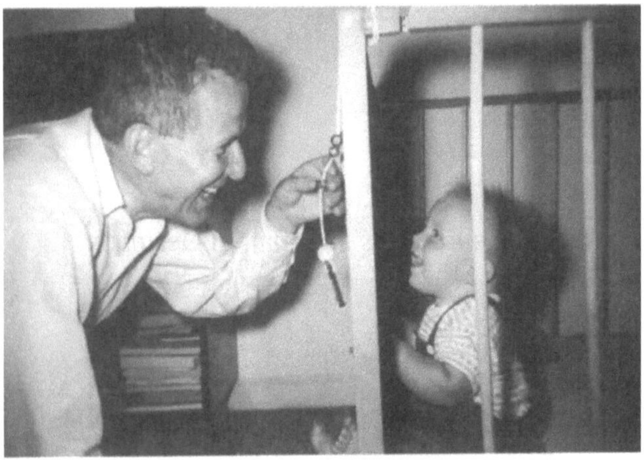

Hans Lewy

with son Michael

Hans and Helen Lewy

with son Michael

Hans Lewy with Jean and Margarite Leray

Canadian Rockies, September 1973

Contents to Volume 1

(Bracketed numbers are from the list of Publications of Hans Lewy)

Contents to Volume 2

(Bracketed numbers are from the list of Publications of Hans Lewy)

Preface

The work of Hans Lewy (1904–1988) has had a profound influence in the direction of applied mathematics and partial differential equations, in particular, from the late 1920s. We are all familiar with two of the particulars. The Courant–Friedrichs–Lewy condition (1928), or CFL condition, was devised to obtain existence and approximation results. This condition, relating the time and spatial discretizations for finite difference schemes, is now universally employed in the simulation of solutions of equations describing propagation phenomena. His example of a linear equation with no solution (1957), with its attendant consequence that most equations have no solution, was not merely an unexpected fact, but changed the viewpoint of the entire field.

Lewy made pivotal contributions in many other areas, for example, the regularity theory of elliptic equations and systems, the Monge–Ampère Equation, the Minkowski Problem, the asymptotic analysis of boundary value problems, and several complex variables. He was among the first to study variational inequalities. In much of his work, his underlying philosophy was that simple tools of function theory could help us understand the essential concepts embedded in an issue, although at a cost in generality. This was extremely successful.

In these two volumes, most all of Lewy's papers are presented, in chronological order. They are preceded by several short essays about Lewy himself, prepared by Helen Lewy, Constance Reid, and myself, and commentaries on his work by Erhard Heinz, Peter Lax, Jean Leray, Richard MacCamy, François Treves, and Louis Nirenberg. The contributions from Reid and Lax are reprinted. Finally, there are Lewy's own remarks on the occasion of his honorary degree from the University of Bonn.

David Kinderlehrer
Pittsburgh, Pennsylvania
September 20, 2001

Acknowledgments

Above all, I wish to express my gratitude to Helen Lewy and to Michael Lewy for their constant support and advice and unflagging enthusiasm at every stage in the creation of this volume. Shortly after Lewy's death, I was approached by Haïm Brezis with the idea of publishing this volume with Birkhäuser, and Edwin Beschler generously agreed to serve as the publisher's contact. In addition to them, I received constant encouragement from Stefan Hildebrandt, Mario Miranda, and Louis Nirenberg. I thank Ann Kostant for assuming the editorial and production responsibilities for the volume.

I would also like to thank the authors of the commentaries: Erhard Heinz, Peter Lax, Richard MacCamy, Louis Nirenberg, and François Treves. Jean Leray, unfortunately, passed away prior to publication. The work of Constance Reid was essential to the preparation of the short biography contained here. Special gratitude is due to Stefan Hildebrandt for his translation of Lewy's address, *Crises in Mathematics*.

Finally we should note the volume *Selecta of Hans Lewy*, prepared by Stefan Hildebrandt for the Wolf Foundation series Wolf Prize Winners, of imminent publication by World Publishing Company.

Hans Lewy Opera

(Papers with asterisks are not included in the Selecta)

[1] 1925 Über einen Ansatz zur numerischen Lösung von Randwertproblemen, Nachrichten Ges. Wiss. Göttingen, 118–121

[2]* 1927 a Verallgemeinerung der Riemannschen Methode auf mehr Dimensionen, Nachrichten Ges. Wiss. Göttingen, 118–123

[3] 1927 b Über den analytischen Charakter der Lösungen elliptischer Differentialgleichungen, Nachrichten Ges. Wiss. Göttingen, 178–186

[4] 1927 c Über die Methode der Differenzengleichungen zur Lösung von Variations- und Randwertproblemen, Mathematische Annalen, **98**, 107–124

[5] 1927 d Über das Anfangswertproblem einer hyperbolischen nichtlinearen partiellen Differentialgleichung zweiter Ordnung mit zwei unabhängigen Veränderlichen, Mathematische Annalen, **98**, 179–191

[6] 1927 e (with K. Friedrichs) Über die Eindeutigkeit und das Abhängigkeitsgebiet der Lösungen beim Anfangswertproblem linearer hyperbolischer Differentialgleichungen, Mathematische Annalen, **98**, 192–204

[7] 1928 a (with R. Courant and K. Friedrichs) Über die partiellen Differenzgleichungen der mathematischen Physik, Mathematische Annalen, **100**, 32–74

1967 On the Partial Difference Equations of Mathematical Physics, IBM Journal of Research and Development, **11**, 215–234

[8] 1928 b (with K. Friedrichs) Das Anfangswertproblem einer beliebigen nichtlinearen hyperbolischen Differentialgleichung beliebiger Ordnung in zwei Variablen. Existenz, Eindeutigkeit, und Abhängigkeitsbereich der Lösung, Mathematische Annalen, **99**, 200–221

[9] 1929 Neuer Beweis des analytischen Charakters der Lösungen elliptischer Differentialgleichungen, Mathematische Annalen, **101**, 609–619

[10] 1930 Sulla unicità della soluzione del problema di Cauchy per un'equazione ellitica del secondo ordine in due variabili, Rend, R. Accademia dei Lincei, **XI.6**, 162–165

[11] 1931 Eindeutigkeit der Lösung des Anfangsproblems einer elliptischen Differentialgleichung zweiter Ordnung in zwei Veränderlichen, Mathematische Annalen, **104**, 325–339

[12] 1932 (with K. Friedrichs) Über fortsetzbare Anfangsbedingungen bei hyperbolischen Differentialgleichungen in drei Veränderlichen, Nachrichten Ges. Wiss. Göttingen, Math-Phys. Klasse, 135–143

[13] 1935 a (with C. R. Adams) On convergence in length, Duke Mathematical Journal, **1**, 19–26

[14] 1935 b A priori limitations for solutions of the Monge-Ampère equations, Transactions of the A.M.S., **37**, 417–434

[15] 1936 a Generalized integrals and differential equations, Proc. National Academy of Sciences, **22**, 377–381

[16] 1936 b On the non-vanishing of the Jacobian in certain one-to-one mappings, Bulletin of the A.M.S., **42**, 689–692

[17] 1937 A priori limitations for solutions of the Monge-Ampère equations, II, Transactions of the A.M.S., **41**, 365–374

[18] 1938 a On the existence of a closed convex surface realizing a given Riemannian metric, Proc. National Academy of Sciences, **24**, 104–106

[19] 1938 b On differential geometry in the large, I (Minkowski's Problem), Transactions of the A.M.S., **43**, 258–270

[20] 1938 c Generalized integrals and differential equations, Transactions of the A. M. S., **43**, 437–464

[21] 1938 d A property of spherical harmonics, American Journal of Math., **LX**, 555–560

[22]* 1939 *Aspects of the Calculus of Variations*, University of California Press, 96pp, notes by J. W. Green

[23] 1946 Water waves on sloping beaches, Bulletin A.M.S., 52, 737–775

[24] 1948 a (with K. O. Friedrichs) The dock problem, C.P.A.M., **1**, 135–148

[25] 1948 b On the convergence of solutions of difference equations, Studies and Essays, Courant Anniversary Volume, 211–214

[26] 1950 a Developments at the confluence of analytic boundary conditions, Int. Cong. Mathematicians, Cambridge Mass., **1**, 601–605

[27] 1950 b Developments at the confluence of analytic boundary conditions, Univ. Calif. Publications new series, **1**, 247–280

[28] 1951 a On minimal surfaces with partially free boundary, C.P.A.M., **4**, 1–13

[29] 1951 b On the boundary behavior of minimal surfaces, Proc. Nat. Acad. Sci., **37**, 103–110

[30] 1952 a A note on harmonic functions and a hydrodynamical application, Proc. A.M.S., **3**, 111–113

[31] 1952 b A theory of terminals and the reflection laws of partial differential equations, Stanford Technical Report, No. 4

[32] 1952 c Asymptotic developments at the confluence of boundary conditions, Nat. Bureau Standards, Ser. 18, 255–256

[33] 1952 d (with P. R. Garabedian and M. Schiffer) Axially symmetric cavitational flow, Ann. Math., **56**, 560–602

[34] 1952 e On steady free surface flow in a gravity field, C. P. A. M., **V**, 413–414

[34a] 1953 An Introduction to Riemann's Work, H. Weber (ed.), Dover, New York

[35] 1955 Extension of Huyghen's principle to the ultrahyperbolic equation, Ann. Mat. Pura ed Appl., **XXIX**, 63–64

[36] 1956 a On the local character of the solutions of an atypical linear differential equation in three variables and a related theorem for regular functions of two complex variables, Ann. Math., **64**, 514–522

[37] 1956 b On the relations governing the boundary values of analytic functions of two complex variables, C. P. A. M., **9**, 295–297

[38] 1957 a An example of a smooth linear partial differential equation without solution, Ann. Math., **66**, 155–158

[39] 1957 b On linear difference-differential equations with constant coefficients, J. Math. Mech., **6**, 91–108

[40] 1959 a Composition of solutions of linear partial differential equations in two independent variables, J. Math. Mech., **8**, 185–192

[41] 1959 b On the reflection laws of second order differential equations in two independent variables, Bull. A. M. S., **65**, 37–58

[42] 1960 On hulls of holomorphy, C. P. A. M., **XIII**, 587–591

[43] 1961 a Atypical partial differential equations, Partial Differential Equations and Continuum Mechanics, (Rudolph E. Langer, ed.), U of Wisconsin Press, 171–175

[44] 1961 b (with R. S. Lehman) Uniqueness of water waves on a sloping beach, C. P. A. M., **XIV**, 521–546

[45]* 1964 a On the analytic continuation of minimal surfaces and similar problems, Proceedings of the Fourth All Union Mathematics Congress, **2**, 233–236 (in Russian)

[46]* 1964 b Sul prolungamento delle funzione armoniche, Rend. Sem. Mat. Torino, **24**, 27–29

[47] 1965 a On the definiteness of quadratic forms which obey conditions of symmetry, Annali della Scuola Normale Superiore di Pisa, **XIX**, 577–591

[48] 1965 b On the extension of harmonic functions in three variables, Journal of Math. and Mechanics, **14**, 925–928

[49] 1965 c The wave equation as limit of hyperbolic equations of higher order, C.P.A.M., **17**, 5–16

[50] 1966 Sulla riflessione delle funzioni armoniche di 3 variablili, Convegno su le Equazioni alle Derivate Parziali, Cremonese (Nervi)

[51] 1968 a On a variational problem with inequalities on the boundary, Journal of Math. and Mechanics, **17**, 861–864

[52] 1968 b On the nonvanishing of the jacobian of a homeomorphism by harmonic gradients, Annals of Math., **88**, 518–529

[53] 1969 a About the Hessian of a spherical harmonic, American J. Math., **91**, 505–507

[54] 1969 b (with Guido Stampacchia) On the regularity of the solution of a variational inequality, C.P.A.M., **22**, 153–188

[55] 1970 a On a minimum problem for superharmonic functions, Proc. Int. Conf. on Functional Analysis and Related Topics, Tokyo, Univ. of Tokyo Press, 283–284

[56] 1970 b On a refinement of Evans' law in potential theory, Rend. Accademia Nazionale dei Lincei, **48**, 1–9

[57] 1970 c On the partial regularity of certain superharmonics, Studies and Essays Presented to Yu-Why Chen on his Sixtieth Birthday, 169–171

[58] 1970 d (with Guido Stampacchia) On the smoothness of superharmonics which solve a minimum problem, Journal d'Analyse Mathématique, **23**, 227–236

[59]* 1971 a Obobscenie teoremy o prostranstvennom ugle (Generalization of the spatial angle) (translated from the English by Ju. V. Egorov) Uspekhi Mat. Nauk, **26**, 199–204

[60] 1971 b (with Guido Stampacchia) On existence and smoothness of solutions of some non-coercive variational inequalities, Arch. Rat. Mech. Anal., **41**, 241–253

[61] 1972 On the coincidence set in variational inequalities, J. Diff. Geometry, **6**, 497–501

[62] 1975 On the nature of the boundary separating two domains with different regimes, Rend. Accademia Nazionale dei Lincei, **217**, 181–188

[63] 1976 On analyticity in homogeneous first order partial differential differential equations, Annali della Scuola Normale Superiore di Pisa, **4**, 719–723

[64] 1977 a On the boundary behavior of holomorphic mappings, Accad. Nazionale dei Lincei, Contributi del Centro Linceo Interdisciplinare di Scienze Matematiche e Loro Applicazioni, No. 35, 1–8

[65] 1977 b On the minimum number of domains in which the nodal lines of spherical harmonics divide the sphere, Comm. P.D.E., **2**, 1233–1244

[66]* 1978 On the relation between analyticity in one and in several complex variables, Complex Analysis and its Applications (Russian)

[67] 1979 a An inversion of the obstacle problem and its explicit solution, Annali della Scuola Normale Superiore di Pisa, **6**, 561–571

[68] 1979 b Expansion of solutions of t'Hooft's equation. A study in the confluence of analytic boundary conditions, Manuscripta Math., **26**, 411–421

[69] 1980 On conjugate solutions of certain partial differential equations, C. P. A. M., **33**, 441–445

[70] 1982 Über die Darstellung ebener Kurven mit Doppelpunkten, Nachtrichten der Akad. der Wiss. in Göttingen, Mathematisch-Physikalische Klasse, **4**, 109–130

[71] 1986 (with Tang Zhiyuan) On free boundary problems in two dimensions, Rend. Circ. Mat. Palermo, **3**, 325–336

[72] 1987 (with U. Dierkes and S. Hildebrandt) On the analyticity of minimal surfaces at movable boundaries of prescribed length, Journal für die Reine und Angewandte Mathematik, **379**, 100–114

[73] 1989 On atypical variational problems, Partial Diff. Eqns and the Calculus of Variations, Essays in Honor of Ennio DeGiorgi, (F. Colombini, A. Mariano, L. Modica, S. Spagnolo, eds.), Birkhäuser, 697–702

HANS LEWY
A BRIEF BIOGRAPHICAL SKETCH

D. KINDERLEHRER

Hans Lewy was born October 20, 1904 in Breslau, Germany (now Wroclaw, Poland), the first son and second of three children to Max and Greta Lewy. As a child he was drawn to science and to music, cf. The Music in Hans Lewy's life.[1]

Göttingen, in 1922 when Hans Lewy matriculated, was among the premier mathematical establishments in the world. With Klein and Hilbert still in residence, as well as Runge, Prandtl, Landau, and Emmy Noether, Courant, of course, and many others {Ried, 1970} it radiated irresistible scientific excitement. We are told that Lewy's gymnasium professor suggested to his father that Breslau was too provincial to provide a good scientific education and that the young man should travel to Göttingen. In this the professor was not only prescient but, perhaps, well informed. Not only Courant, but also Born, Hellinger, and Toeplitz had come to Göttingen via Breslau.

Friedrichs also arrived in 1922 and their life-long friendship took hold immediately. Unfortunately, 1922 marked the beginning of the Weimar hyper-inflation period, which ended at the end of 1923. The situation led to considerable conflict between Lewy and his family, who understandably wanted the eighteen-year-old to return to the security of his home, and eventually led to a temporary rupture of relations. In this period he took a railroad job, doing maintenance on the tracks and ties; neither his personality nor his physique made this easy for him. Courant wrote several times to Lewy's parents expressing great satisfaction and extolling the achievements of his student, perhaps in part to relieve this family pressure.

He completed his thesis in 1926 with Courant and became, together with Friedrichs, Courant's Assistant and a Privatdozent. In the next several years he, and Friedrichs, pursued new methods for understanding elliptic and hyperbolic problems. He solved the initial value problem for general nonlinear hyperbolic equations in two independent variables [5]. He used this to give a new proof of the analyticity of solutions of elliptic equations in two variables which far exceeded the known proof in its elegance and simplicity [3], {1}. He and Friedrichs developed energy methods for wave equations [6]. Most well known, in 1928, the joint work of Courant, Friedrichs, and Lewy [7], (Lax) was published. About this, Lax writes, "... the onset of the Second World War brought an unprecedented (and unabated) pressure on the technological centers to provide numerical solutions of partial differential equations; here the ideas introduced by C-F-L—the famous stability condition, energy inequalities, and the leap frog difference method—turned out to be basic. This is an outstanding example of research undertaken for purely theoretical purposes turning out to be of immense practical importance." Lewy's work

[1] {x} a reference at the end of this biographical note,
(y) a reference to a commentary on Lewy's work, and
[z] a citation of the Lewy Bibliography.

attracted the attention of Hadamard who included a long special appendix about it in the second edition, which was the later French edition, of his book, {Hadamard}. To paraphrase Leray: he opened a world to us whose richness continues and will continue to amaze us, (Leray). These years in Göttingen also marked the period of the "boy physics."

In 1929–1930, under the sponsorship of the Rockefeller Foundation, he studied in Rome where he was especially influenced by Tullio Levi-Civita. He also met many younger Italian mathematicians, and in particular Beniamo Segre who would later have an influence on his work [52]. The fellowship then took him to Paris, 1930–1931, where he was active in Hadamard's seminar.

Lewy left Germany quite soon after Hitler assumed power in 1933. He went first to Italy and then to Paris where Hadamard had managed to obtain for him a year's support. He introduced Leray and Schauder, (Leray). Later, during the war when Leray was taken prisoner, Lewy communicated with him via the Red Cross, and after the war the families remained quite close. In Paris, on the recommendation of Hadamard, he was offered a one to two year position at Brown, funded by the Duggan Foundation. In the fall of 1933, Lewy was in Providence. He worked with C. R. Adams [13]. He completed his first work on the Monge–Ampère Equation and Minkowski's Problem [14, 17, 18, 19] (Heinz).

On the invitation of G. C. Evans, Lewy went to Berkeley in 1935. Courant, visiting Berkeley in 1932, had spoken enthusiastically of Lewy's work on that occasion. Courant himself was, in fact, offered a position at Berkeley, {Rider}. Evans arrived in Berkeley in 1934 as the central part of the plan to build up the deparment, {Rider}. C. B. Morrey arrived in 1933 and Lewy in 1935. A. P. Morse, Jerzy Neyman, and Raphael Robinson came at the end of the 1930s. Lewy's work on Minkowski's Problem was completed here [19]. Heinz (Heinz) remarks that, "His estimates are of local type ... It is remarkable that he did not make use of the Weyl estimates ... Thus he went far beyond the original goal." Although the events in Europe suggest that it was natural for America to welcome scientists from facist countries, it is well to keep in mind that the country remained in The Depression and that there was considerable concern, by all parties that emigres might be assuming positions that could be had by natives and that this could lead to a negative environment.

With the outbreak of the war, Lewy took flying lessons and obtained a solo license in the hope of offering his services, but was soon called to Aberdeen Proving Grounds as part of the University of California contingent. He also worked half-time with the Office of Naval Research in New York. Here he became interested in water waves and the Dock Problem, resuming his collaboration with Friedrichs [24]. The explicit problem in [25] also gave rise to a proof of the Law of Quadratic Reciprocity, (Nirenberg). At the war's conclusion, Lewy, then a full professor, was invited to return to Göttingen as Privatdozent!

In 1947 he and Helen Crosby, an artist, writer, and translator, were married. Their honeymoon included a trip around the world, beginning with a return to Europe. Also included was a two month stay in Chengtu, Szechuan (Chengdu, Sichuan) where Lewy gave a course on water waves and a third month visiting other institutions in China. Still under Nationalist rule, China was rarely visited by westerners in those years.

The analytic continuation of minimal surfaces bounded by analytic arcs or satisfying free boundary conditions was given in the early 1950s [28, 29], (Kinderlehrer). These and related results had a formative impact on the study of free boundary and regularity problems in variational principles. Also in this decade Lewy devoted considerable attention to the issue of the smoothness of solutions of equations and systems at the boundary [37] and their asymptotic development near change of regime boundary points [27], cf. (MacCamy).

During the Loyalty Oath controversy in the state of California, Lewy was part of the group dismissed for refusal to sign the oath. Having seen Fascism at first hand in Italy, and watched its rise in Germany, he was wary of cooperating with any totalitarian tendencies in his new country. The details of these events, insofar as he is involved, are chronicled in Constance Reid's article {Reid 1991}. He was on the faculty at Harvard in the fall of 1952 and then at Stanford in 1952–1953. This was the genesis of the work on Riabouchinsky flow with Garabedian and Shiffer [33]. In the settlement of this dispute by the courts, the professors were reinstated and Lewy returned to Berkeley. In 1955, the Lewys' son Michael was born.

Lewy's completely unexpected example of a linear equation with no solution, and with it a homogeneous equation with only the trivial zero solution, appeared in 1957, [38], "one of the major events in the modern development of PDE theory," (Nirenberg/Treves). It arose in the context of Lewy's study of Cauchy–Riemann structures in several complex variables and pseudoconvex domains, (cf. Treves) [42]. Both simple and general, this example illustrates many different aspects of differential equations. We teach, for example, that real first order linear equations admit solution by the elementary method of characteristics. But complex ones may not admit any solutions, cf. {John}. It put an end to speculation about finding a version of the fundamental solution for a general linear equation, even with analytic coefficients, although this had recently been established for equations with constant coefficients. He received the Steele Prize of the American Mathematical Society for this work in 1979.

In 1959–1960, Lewy was a Visiting Professor at the Courant Institute. In this period he was elected to the National Academy of Sciences, and the American Academy of Arts and Sciences.

In 1964–1965 Lewy accepted an invitation from the Scuola Normale Superiore and the University of Pisa. During that time, Pisa was achieving stature as a vigorous world center with Andreotti, De Giorgi, and Stampacchia among its leaders. Here, Lewy helped create the emerging area of variational inequalities with Stampacchia. Their work was fundamental to the growth of the subject, introducing new types of free boundary problems. He published in this area through the 1980s. In 1969–1970, he returned to Rome on the invitation of the Accademia dei Lincei. He was elected a Foreign Member in 1972. He also retired in 1972, continuing his research with undiminished vitality.

On retirement, he insisted that available funding should be dedicated to the support of younger scientists and declined to apply for any more grants. Also in 1973 he joined Ahlfors as one of the first two Ordway Professors at the University of Minnesota. On the initiative of Stampacchia, a volume of papers in Lewy's honor was published by the Annali della Scuola Superiore di Pisa, {Stampacchia}. A collection of fifteen papers dedicated to Hans Lewy and C.B. Morrey, Jr. appeared in *Manuscripta Mathematica* in 1978, {Manuscripta Mathematica}. Always attentive to the encouragement of young people, he returned to Minneapolis to participate in the 1984/1985 Institute of Mathematics and its Applications Program on Continuum Physics and Partial Differential Equations. In 1986 in Trento there was a meeting in his honor, {Hildebrandt et al.}.

He shared the 1984/1985 Wolf Prize with K. Kodaira "for initiating many, now classic and essential, developments in partial differential equations." Subsequently he wrote frequently to the Wolf Foundation in support of other scientists, especially De Giorgi, who received the 1990 prize.

He was awarded an honorary doctorate from Bonn University in 1986. In his acceptance speech (Hans Lewy), {12, Hildebrandt et al.} "Krisen in der Mathematik" he spoke, characteristically, about the future.

He died on August 23, 1988 in Berkeley. He and Helen had recently returned from a trip to Europe where he delivered his last paper, [73], in honor of Ennio De Giorgi.

There are many scientists whose objective is to bring order to chaos, to organize a disparate subject and bring it unity of theme and method. In an arcane way, Lewy was their antithesis: with a profoundly innovative point of view, he illustrated and revealed to us fundamental principles and ideas with the simplest of tools. Hans Lewy was a pillar of science in the twentieth century. His work touched nearly every significant area of analysis and endures in the canon that we pass on to our children.

Notes

1. Protter, M., J. L. Kelley, T. Kato, and D. H. Lehmer, Hans Lewy (UC Berkeley memorial obituary).
2. Reid, Constance (1970). *Hilbert*, Springer.
3. Reid, Constance (1976). *Courant in Göttingen and New York*, Springer.
4. Reid, Constance (1990). *Hans Lewy, More Mathematical People, Contemporary Conversations* (Donald Albers, Gerald L. Alexanderson, and Constance Reid, eds.), Boston: Harcourt, Brace, Jovanovich.
5. Reid, Constance (1991). *Hans Lewy • 1904–1988, Miscellanea Mathematica* (P. Hilton, F. Hirzebruch, and R. Remmert, eds.), Heidelberg: Springer.
6. Rider, Robin, E. (1989). An opportune time: Griffith C. Evans and Mathematics at Berkeley, in *A Century of Mathematics in America*, Part II (Peter Duran, ed.), Providence: American Mathematical Society.
7. John, F. (1982). *Partial Differential Equations*, New York: Springer.
8. Hadamard, J. (1932). *Le problème de Cauchy et les équations aux derivées partielles linéaires hyperboliques*, Paris: Hermann et cei.
9. Stampacchia, Raccolta degli scritti dedicati a Hans Lewy, Annali della Scuola Normale Superiore di Pisa, 1978.
10. *Manuscripta Mathematica*, 28.1–28.3, 1978.
11. Hildebrandt, S., Kinderlehrer, D., and Miranda, M., Calculus of Variations and Partial Differential Equations, Lecture Notes in Mathematics 1340, Springer-Verlag, 1988.
12. Hildebrandt, S., Hirzebruch, F., Leis, R., Lewy, H. and Tits, J., Mathematische Betrachtungen, Bonner Akademische Reden 68, Bouvier Verlag, Bonn, 1999.

THE MUSIC IN HANS LEWY'S LIFE

HELEN LEWY

Hans Lewy always shunned the intrusion of the personal into his professional life. His private life was his own domain. Yet he was an eminently social being with so many friends and colleagues across Europe and Asia, and in the U.S., that a great deal about his personal life—his work habits, his personality, his roots, his friends and "hang-outs"—is already well known. During his extended stay in Europe, especially in Italy, it was his custom to spend a great deal of his time socially, as well as mathematically (if the two can be separated) with the students. Like all scholars, he was always energized by the marvelous curiosity of the young.

I have been asked to write a few words about Hans Lewy for this volume. His family is, for the most part, content to let his legacy rest on his contributions to mathematical thought—for that's where it should be. I will not try to evoke the man behind the math. But, I do not feel he would object, if, to honor the students he loved to be with, I pass along some facts about the music which was always the companion of the math in his life.

The combination of mathematics and music, while not uncommon, especially in the Europe of Hans Lewy's time, was, in his case, serious enough to force him to make, in his teens, a difficult career choice. He played the violin, the viola and the piano with mastery, and, at times in his life also the clarinet and the cello. He composed many string quartets—which unfortunately were lost,—and at least one string trio discovered by Lori Courant Berkowitz after the death of her mother. (In a labor of love, in the 1990s, Lori, with fellow musicians, edited, copied and brought it to light; all three parts had been laboriously hand-copied by Nina Courant.)

According to his mother, Hans Lewy's love for music began with the piano lessons of his older sister when she was about 10 years old. At that particular time, she had little interest in learning to play the piano, but her little brother, five years her junior, did; he managed to be present during the lessons and soon he was outperforming his sister. This situation led to so much "mischief in the family," according to his mother, that a music teacher for little Hans was promptly acquired, for lessons not on the piano, but on the violin.

Hans took to his lessons with passion ("ferocity", was the word his mother used): he soon had outgrown his first two teachers, and, at length, as a pre-teen he was studying under a well-known concert violinist. By then the violin had become the mode through which his native ebullience and independence found expression. He told me, sixty years later, of his joy in playing even as a child (but added that he had not always enjoyed "being produced" for family occasions). By the time he reached his teens, he was already playing regularly with chamber music groups.

His parents would not permit the unusual skills of their son to be exploited, until, under pressure from his teacher, they agreed when he was 16 to allow him to perform in public, for one time only: the concert took place in Bautzen in July, 1920. One can sense the excitement of the young man in this partial translation of the letter he wrote to his parents two days later:

> My dear ones:
> Saturday was THE day: an orchestra, a male chorus, a woman soloist and me. I played the Chaconne by Bach and Devil's Trills by Tartini, accompanied by a very nice elderly gentleman, Prof. Crusl. During the first piece I still felt a little tight, but by the second I could hardly wait to start playing. The reviews are here today; I quote them verbatim:
>
> *Bautzner Tageblatt*: "The talented young Hans Lewy did the... "Devils Trills sonata by Tartini. The youthful musician amazed the audience with his splendid technique. Double-stopping, octaves and staccato don't seem to give him any trouble. That he also has a mature understanding of the character of the music was apparent when he played Bach's 'Chaconne' for violin only. The audience thanked him with thundering applause."
>
> *Bautzner Nachrichten*: "The youthful violinist, Hans Lewy, from Breslau, a very promising high school student there, played the 'Chaconne' by Bach and the 'Devils Trills' sonata by Tartini with admirable precision and technical expertise."

Actually the teenager was allowed another performance before the public; it took place also in Bautzen, in October of the same year. This time he was soloist for Mozart's D-sharp violin concerto. The reviews were, to say the least, enthusiastic. Here are excerpts, in translation, from two of the reviews:

> *Bautzner Tageblatt*: "The spirit and solid technique so important in Mozart's works were present. The bowing is light and elegant, and, on the other hand, does not lack energy. The left hand is excellently schooled and establishes full success in passage-work in all ranges. Nevertheless, Mozart's splendid work was interpreted with classic poise and superior precision, and the youthful violinist received rich applause that came from the heart."
>
> *The Bautzner Nachrichten* review hailed him as a "Virtuoso" and "Wunderkind with true Mozartian charm" and predicted for him a great future as a musician.

After these successes, the teenaged Lewy, of course, had to consider a career as a professional musician. The choice would have been easy if it had not been for that other passion, mathematics.

His father favored mathematics; his mother was not sure; and he, of course, wanted both.

In the end mathematics won, and he soon left home for his studies in Göttingen. There he played with the city orchestra and, during all his years as student and Privat Dozent, participated in weekly musical evenings at the home of Helene Wintgen, the wife of a professor of chemistry, and a friend and contemporary of Nina Courant.

Frau Wintgen, as an octogenarian, described for me her first meeting with my husband. The quartet needed "a violin" and a newly-arrived student had been recommended. It was raining hard that evening when the new student knocked at her door, and "there he was," she said, smiling, "the 17-year-old Hans, standing there forlorn and drenched: he wore no hat, no raincoat, and carried no umbrella—but the violin was carefully wrapped in a towel!"

From the stories that reached my ears, these musical get-togethers of Lewy's student days were very lively—with the players enjoying not only the challenges and riches of the music, but also revelling in each others' company and wit, and at times even nonsense. I remember my husband telling of one "session" that included a very accomplished Russian violist, and a good friend. They were about to begin playing, when his Russian friend put down his bow and said solemnly: "Let's try it first without sharps and flats!"

But had mathematics won? That is, had music lost? My husband said later in his life that he had never regretted his decision.

He made music almost every day of his life (exceptions were extended foreign stays when no piano was at hand). But his seminal decision had made it possible for him to make music, and live, as a free spirit—free of the constraints of a concert performer's life, especially, he told me more than once, the constant travelling, and the uprooted lifestyle...

Mathematics and Music remained the twin passions of his life, and, in the end, I believe he felt he had been lucky enough to have had the best of both worlds.

Reprint from

Miscellanea mathematica

Edited by P. Hilton, F. Hirzebruch, R. Remmert

Springer-Verlag
Berlin Heidelberg New York
London Paris Tokyo
Hong Kong Barcelona
Budapest

Constance Reid

Hans Lewy · 1904–1988

Hans Lewy was one of the first of that great group of mathematicians who, leaving Germany after Hitler came to power in 1933, substantially influenced the development of mathematics in the United States.

I became acquainted with him only some thirty years later. I had been told that in his youth he had been in Göttingen and had personally known many of the great mathematicians of the early twentieth century. It was largely as a result of my conversations with him that I wrote a life of David Hilbert, the first of my books to be published by Springer-Verlag.

Although Lewy loved to tell stories that illustrated the characters and personalities of others, he relentlessly removed any personal details from published accounts of his own life. The following is a typical Lewy entry:

Born Breslau, Ger. [now Wrocław, Poland], Oct. 20, 1904. Ph.D., U. Göttingen 1926. Married. Naturalized U.S. citizen. Priv. Doz., Göttingen 1927–33. Associate, Brown University 1933–35. Lecturer, University of California (Berkeley) 1935–37, Asst. Prof. 1937–39, Assoc. Prof. 1937–39, Prof. 1945.

No mention of honors, honorary academies or honorary degrees. He did not choose, he said, to participate in the contemporary cult of the personality, which he dismissed as a "star" system. And yet he was a star. And now that he is gone it seems that in addition to his more than six decades of influential mathematical work, something about the man and his life should also be in print.

Lewy was the first son and second child of Max and Greta (Rösel) Lewy. His father was a merchant who dealt in accessories for women's millinery, and his mother before her marriage had been a teacher of German in an expatriate enclave in Hungary. Her interest and ability in languages were inherited by her son. In addition to a thorough grounding in Greek and Latin, which he had received in his *Gymnasium* days, he was so fluent in French and Italian that he was frequently mistaken for a native on the streets of Paris and Rome. He could converse in Russian, once delivering a lecture in that language, and he actively studied Chinese up to the time of his death. On several occasions, marooned and without the solace of his piano, he set out to teach himself a new language – once it was Spanish, which he attacked by plunging into *Don Quixote.*

Both of Lewy's parents were Jewish, but only the father was religious. Lewy himself did not observe the rituals of Judaism and, as he once told a reporter for a Jewish newspaper, did not believe that the Jews were chosen by God. Needled by a friend about whether he would don a yarmulke when he was awarded the $100,000 Israeli-based Wolf Prize, which he would share with the distinguished Japanese mathematician Kunihiko Kodaira, he parried gracefully, "In matters of dress I will be guided by my co-recipient."

The boy Lewy so loved mathematics that his mother felt she had to hide his mathematics books from him, but he also loved music and was a gifted violinist. His parents, not wanting his gift to be exploited, did not permit him to perform in public until his fifteenth year, when he took part in a concert in the resort town of Bautzen. There is in existence a letter that he wrote to his family after that event: "Saturday was THE day: an orchestra, a male chorus, a woman soloist, [and] I – *Chaconne* by Bach, *Devil's Trills* by Tartini .... During the first piece I still felt a little tight, but the second I could hardly wait to start playing." One local critic praised his technical expertise and another his mature understanding of the music. A few months later, as soloist with the 46-piece Bautzen orchestra, playing Mozart's *Concerto in D Major*, he was praised for his "solid technique" and hailed as "a 'Wunderkind' with true Mozartian charm."

The choice between music and mathematics was not easy for Lewy. His parents favored mathematics as more secure, but he chose it for a different reason, seeing in the constant playing of often familiar pieces required by a musical career "a kind of enslavement." Although eventually he gave up the stringed instruments (the viola was actually his favorite), he continued to play the piano until the end of his life.

A mathematics teacher at the *Gymnasium* in Breslau prevailed upon Lewy's father to send the boy to the mathematically prestigious university in Göttingen rather than to the local university, which was "too stuffy and old-fashioned" in the teacher's opinion, although a number of distinguished mathematicians (Ernst Hellinger, Otto Toeplitz, Richard Courant) had begun their higher education there. So it happened that on the wave of the staggering inflation and hardship that followed the first World War, Lewy arrived in Göttingen. For a while he worked as a manual laborer on the railroad and (so he told me) thought more about food than about mathematics.

At eighteen he was exceptionally young for the mathematical society in which he found himself. He felt "quite overwhelmed" by the level of the lectures and the abilities of such fellow students as K.O. Friedrichs, who was a couple of years older than he. The director of the famous mathematics institute, which existed at that time only as a name on a letterhead, was Richard Courant, who had been a student of Hilbert and who had only that year (1922) taken over from Felix Klein. Hilbert, who was still lecturing, had aged prematurely as a result of an as yet undiagnosed case of pernicious anemia. In the first conversations I had with Lewy I was impressed by the fact that although he had been very young at the time he had been interested in observing the workings of the mind of this great mathematician of an earlier day, with whose disabilities other young mathematicians were more often simply annoyed.

As a student Lewy did not limit himself to mathematics. He regularly attended the lectures of Alfred Kühn, who was the head of the zoological institute – there was great interest among the Göttingen mathematicians at that time in "the new biology." He also attended the lectures of the physicist Max Born. A few years later, however, he found himself "repelled" by the quantum mechanics of Born and other physicists in Göttingen:

"Often if you would corner them, ask them to explain precisely, they would be evasive or else you would find out that they didn't quite understand what they were saying themselves. They obviously had some physical intuition which I didn't have, but their mathematics was objectionable. The type of personality that was needed at that time in physics was not to my liking. Well, what happened teaches one humility, and it showed that different situations in science require different types of personalities. So one cannot make up one's mind what is *the right way* of being for a scientist."

In Lewy's development as a mathematician both Courant and Friedrichs played significant roles, one as teacher, the other as collaborator. It was Courant who guided the younger men into analysis and – especially important in the early years – suggested that they study the application of finite difference equations in mathematical physics. In 1926, still only twenty-two, Lewy took his Ph.D. under Courant with a dissertation that used the equivalent finite difference problem to solve the question of existence and construction for the simplest variational problem. The following year Courant recommended his *Habilitation* as a *Privatdozent* to the Göttingen faculty.

The dissertation, Courant wrote, "offers something essentially new as far as methods and results are concerned, and reveals in its approach a completely original and independent mind." The success of the *Habilitation* thesis stemmed "from a novel turn that he gave to the theory of characteristics by which he completely solved a problem hitherto considered unsolvable, i.e., the reduction of integration of partial differential equations of second order to the differential equations of characteristics."

By the time Courant wrote the above, the groundbreaking Courant-Friedrichs-Lewy paper "On the partial difference equations of mathematical physics" had also been submitted for publication. It was to be a dramatic anticipation. Forty years later, "the ideas exposed still prevail[ing]," IBM was to reprint it (in English) in its entirety, describing it as "one of the most prophetically stimulating developments in numerical analysis [and] the basis of modern investigations into the numerical analysis of partial differential equations."

Although the paper was an extension of earlier publications by Courant, his young collaborators had made extensive independent contributions, using finite differences to develop an existence theory of hyperbolic equations of the first order. The result that was to be so significant for numerical analysis was Lewy's observation that a hyperbolic differential equation cannot be replaced by a difference equation in an arbitrary manner without a severe restriction on the ratio of the time difference to the space difference.

Friedrichs told me that at the time he was "absolutely amazed" by Lewy's "uncanny insights." Many mathematicians in future years were to echo Friedrichs's astonishment. Louis Nirenberg, a next-generation worker in the field, going over Lewy's papers in preparation for a talk on his work, said he found himself asking, again and again, how did he ever think of that?

"He seemed to think in a different way from most mathematicians I have met. He would have, somehow, fantastic ideas, and sometimes these ideas would come out in a paper of three or four pages. And these papers would have enormous influence on later development."

Courant wrote his letter of recommendation only a month after Lewy's twenty-third birthday; yet in it he was able to cite, in addition to the dissertation, the thesis and "C–F–L" (as it came to be known), three other significant papers, including one recently submitted for publication "of great importance" that solved "an important problem with which many mathematicians have struggled unsuccessfully" by means of "a novel and very original idea" that would "open up future investigation" of related problems.

"All in all, the obvious scientific mathematical development of this candidate presents a picture of an extremely productive and penetrating scientific mind, rich in independent ideas and possessing the technical ability to work them out."

On the personal side, Courant mentioned "a certain reticence and abruptness" in the candidate's manner, which he was sure would disappear in time since these were "associated with an inner modesty." He emphasized to the Faculty that while pursuing his mathematical studies, Lewy had also continued his music, playing quartets each week, performing with the symphony orchestra and composing chamber music: "One may indeed be amazed at the energy and ability of this man who does not appear impressive or at all robust." (This description will startle anyone who knew Lewy's expressive and interesting face in age and the physical energy that he radiated up until his death.)

As a Privatdozent, whose meager fees were paid by his students, usually few in number, Lewy jokingly reported in 1928 a "lucrative" semester: "Now when people ask why I went into mathematics I can answer, 'For the money!'"

In addition to teaching, he served with Friedrichs as one of Courant's many assistants. Although in the future Courant was often to be criticized for exploiting such younger men, Lewy always considered the time he spent as assistant exceedingly valuable.

In 1929, again on Courant's recommendation, Lewy obtained a year's fellowship from the Rockefeller Foundation. He spent the first semester in Rome – the beginning of a lifelong love affair with Italy, to which he was to return on countless occasions – and the second semester in Paris. He has described his reactions to these two centers and to the mathematicians there in an interview that will appear in the second volume of *Mathematical People* so I shall not repeat them.

If Adolf Hitler had not come to power, Lewy might well have spent the rest of his life in Göttingen; but, unlike many others, he recognized instantly that the situation in Germany could do nothing but worsen. That spring, while Courant was struggling to retain his position with letters and petitions to the Government, Lewy left Germany, passing on to a fellow Jewish student the Marks that he was not permitted to take out of the country. With support from the Duggan Foundation he obtained a position as an associate at Brown University in the fall of 1933. Two years later, when that stipend expired, Brown could not find money in its own budget to keep him. Offers came from Ohio and Kentucky and an indication of possible interest from the University of California, which had recently engaged Griffith Evans to upgrade its mathematics department. Lewy, attracted by the lure of the American west, "followed the hope" rather than the certainty – he was typically optimistic – and in the late summer of 1935 received a contract from Evans. The more than half century of his subsequent tenure at Berkeley was to span the transformation of that institution from a provincial state university to one of the great intellectual centers of the world.

Lewy grew to feel very much at home in America, becoming a citizen in 1940, but he also continued to be at home in Europe. Unlike many scholars, he took great pleasure in moving his household, even with a small child, for a number of extended stays abroad.

His mathematical work in the United States, as in Europe, was in analysis, a branch of mathematics that he once described to me as consisting of "kind of hesitant statements":

"In some ways analysis is more like life than certain other parts of mathematics ...," he said. "There are some fields in mathematics where the statements are very clear, the hypotheses are clear, the conclusions are clear. The drift of the subject is also quite clear. But not in analysis. To present analysis in this form does violence to the subject in my opinion. The conditions are or should be considered temporary, also the conclusions, sort of temporary. As soon as you try to lay down exact conditions, you artificially restrict the subject."

It was a source of regret to him that analysis was not a popular field of research in the United States.

Lewy's mathematical work was all connected in one way or another with partial differential equations. It falls into five groups, according to Nirenberg: (1) partial differential equations themselves, (2) the calculus of variations and the applications of variational methods to nonlinear partial differential equations, (3) the theory of several complex variables (a group of three famous papers for which he received the Steele Prize of the American Mathematical Society), (4) fluid dynamics, and (5) offbeat properties of solutions of certain classes of partial differential equations. Although he did some work on analytic functions themselves, in the main he used analytic function theory as his principal tool – always, according to Nirenberg, "in an extremely clever, ingenious, in fact I would say almost *devilish* way."

In spite of the playfulness that many have noted in his work, Lewy approached mathematics with a deep seriousness, seeing the "desire to fill in the details" as an ethical matter – what he described to me as "a certain fanatic adherence to the truth that is necessary in mathematics." In his foreword to the collected works of Riemann, he wrote that the mathematician must not be satisfied with the "blandishments of brilliance [but must] search for deeper truths, hidden beneath the reflection of the surface of facts."

His collected papers, which number approximately sixty over a period of more than sixty years (a relatively small number when one considers the influence they had), are being edited by David Kinderlehrer, a former student.

"Mathematics is like a Riemann surface," Lewy used to tell his students, Kinderlehrer recalls. "There are ordinary points and singular points. You have to work on the singular points."

In September 1933, as classes took up at Berkeley, Hitler invaded Poland, and the long expected Second World War began in Europe. Lewy was eager to get into combat, and the following year he began to take flying lessons. Although he was already in his mid-thirties, he hoped that by having a pilot's license he might be accepted by the Army Air Corps in spite of his small stature. Instead he spent the war years at the Aberdeen Proving Grounds, where he may have been more useful.

In his forty-third year (1947) Lewy married Helen Crosby, an artist, writer and translator who had been with the American Office of Strategic Services in Europe during the war. The couple almost immediately set off on a year-long journey around the wartorn planet, returning to Göttingen, Paris and Rome to do what they could to help old friends. The last three months they spent in the far west of China at a mathematics institute headed by a colleague from Göttingen days, Wei Si Luan. Lewy became very fond of the Chinese people, whom he liked to describe as "the Italians of the Orient."

The Lewys had one son, Michael, who is now also a mathematician.

Once he became a family man and a householder, Lewy discovered the joys of home repair, woodwork and gardening. He took pride in the fact that, unlike professors he had known in his youth, he could – and would – do things with his hands. In domestic activities he exhibited the same freedom from constraints that characterized his mathematical work. To the despair of his wife and son, he refused to read the directions that came with household appliances but insisted on discovering for himself how they worked. Because of his propensity for seeing both sides of a question, he could not make life and death decisions even for plants and left undisturbed "volunteers" that other gardeners would have eradicated. As a result the Lewy garden now contains an elm tree that first made its appearance in young Mike's sandbox.

He was often joyously unrestrained and enthusiastic. At these times he might suddenly swing up his arms and clap his hands – or even shout – in sheer delight. He was fun, his friends agree, "just enormous fun."

The abiding passion of his life was the desire for freedom that had caused him to choose a career in mathematics over one in music. While he was at the Aberdeen Proving Grounds, crowded into wartime housing, he found in sailing on Chesapeake Bay the space his soul required. But he thrilled especially to the wildness of great mountains, feeling perhaps most free there. Only a few months before his death he was hiking the Crest Trail in the Tahoe National Forest, leaping exultantly from rock to rock.

The most dramatic event of his more than fifty years at Berkeley occurred during the McCarthy era when the Regents of the University decided that members of the faculty should sign a loyalty oath – widely considered redundant because they had already sworn to uphold the Constitution and the laws of the State of California in their teaching contracts. Lewy recognized the threat to academic freedom and refused to sign, one of the three members of the Berkeley mathematics department who refused and were subsequently dismissed by the University. He was not, however, doctrinaire on the subject and counseled others, especially younger colleagues, that they must look at their weapons before they decided to fight. He himself was well armed, having resolved when he had had to leave Göttingen that he would save a year's salary as soon as possible in case he ever had to leave another job. In the tense months that followed, as the non-signers took their case to the courts, he was one of the few who managed to remain on friendly terms with those who had different views. Ultimately the California Supreme Court declared the oath unconstitutional and ordered the University to reinstate the non-signers with back pay and privileges. In spite of this difficult period in their relationship, Lewy remained grateful to the University of California "for taking him in and treating him royally" and provided in his will for a post-doctoral fellowship at Berkeley.

Lewy became emeritus in 1972, but he did not stop doing mathematics. Even in the summer of his death he gave a talk at a meeting in Cortona, Italy, on new work that involved attacking the Carathéodory conjecture from a different, variational angle. He had hoped to finish off the problem, but the complete solution was not to be granted to him. While in Europe, following a strenuous schedule in order to see as many as possible of his European friends, he caught a cold that developed into pleurisy. On his return to Berkeley he was hospitalized, fatally ill. He died on August 23, 1988, two months before his eighty-fourth birthday.

One of the last times I talked with him was in connection with the interview for *More Mathematical People* mentioned above. We were in his office in Evans Hall, surrounded by a half century's accumulation of books, journals, letters and reprints. Although he had stipulated "no personal details," recalling the great mathematical centers and the great mathematicians he had known had resulted to a certain extent in a review of his own life.

"I must say," he concluded with wry appreciation of life's ironies, "that as far as I strictly on a personal basis am concerned, I owe Hitler a much more interesting life than I would have had otherwise."

COMMENTARY ON LEWY'S PAPERS

ERHARD HEINZ

[1] A priori limitations for solutions of Monge-Ampère equations I, *Trans. Amer. Math. Soc.* **37** (1935), 417–434.
[2] A priori limitations for solutions of Monge-Ampère equations II, *Trans. Amer. Math. Soc.* **41** (1937), 365–374.
[3] On the non-vanishing of the Jacobian in certain one-to-one mappings, *Bull. Amer. Math. Soc.* **42** (1936), 689–692.
[4] On the existence of a closed surface realizing a given Riemannian metric, *Proc. Nat. Acad. Sci. USA* **24**, Nr. 2 (1938), 104–106.
[5] On differential geometry in the large I (Minkowski's Problem), *Trans. Amer. Math. Soc.* **43** (1938), 258–270.

Lewy's work on Monge-Ampère equations ([1] and [2]) was largely motivated by the problems of Weyl and Minkowski. Both problems lead to equations of Monge-Ampère type, which in terms of local parameters (u, v) take the form

$$(1)\qquad \begin{aligned}(\rho_{uu} - \Gamma^1_{11}\rho_u - \Gamma^2_{11}\rho_u - E)(\rho_{vv} - \Gamma^1_{22}\rho_u - \Gamma^2_{22}\rho_v - G)\\ - (\rho_{uv} - \Gamma^1_{12}\rho_u - \Gamma^2_{12}\rho_v - F)^2\\ = K(u,v)[2\rho^2(EG - F^2) - (G\rho_u^2 - 2F\rho_u\rho_v + E\rho_v^2)],\end{aligned}$$

and

$$(2)\qquad rt - s^2 = f(u, v).$$

respectively. (1) is the Darboux equation satisfied by the function $\rho = \frac{1}{2}|C|^2$, where C is the position vector of the surface with the given line element ds^2 of curvature K. The equations (1) and (2) are elliptic, if $K > 0$ and $f > 0$, and are particular cases of the Monge-Ampère equation

$$(3)\qquad Ar + 2Bs + Ct + rt - s^2 = E,$$

where $A, \dots, E$ are given functions of the variables x, y, z, p, q such that

$$(4)\qquad D = AC - B^2 + E > 0$$

holds on the surface $\Sigma = (x, y, z(x, y), p(x, y), q(x, y))$. In order to solve the problems of Weyl and Minkowski by a continuity method one is confronted with the problem of obtaining a priori estimates for the second and higher order derivatives of the solutions of the equations (1) and (2) in terms of bounds for the first derivatives. Lewy considered this problem in the papers [1] and [2] for the general equation (3) under the additional assumption that the coefficients $A, \dots, E$ are analytic in a complex neighborhood of Σ. His estimates are of local type and hold in arbitrary compact subsets of the parameter domain. It is remarkable that he did not make use of the Weyl estimates for the mean curvature of a closed surface in terms of the line element. Thus he went far beyond the original goal. Lewy also observed that there are singular cases of analytic Monge-Ampère equations for which interior estimates of the second derivatives cannot be obtained. The situation gets worse if one passes to higher dimensions. Thus for instance, from Pogorelow's work [C3] it follows that interior estimates for the second derivatives of the solution of the n-dimensional analogue of (2), namely

$$\det \operatorname{Hess} z = f(x_1, \dots, x_n), \tag{5}$$

where $n \geq 3$, Hess $z > 0$ and $f > 0$, do not exist even if f is analytic. Lewy's main contribution is a compactness theorem for solutions of a certain class of analytic Monge-Ampère equations. Here one only assumes that the first derivatives of the solutions are uniformly bounded. This class can be described as follows: Consider the characteristic form Φ associated with the equation (3), namely

$$\Phi \equiv (C + r)\, dx^2 - 2(B - s)\, dxdy + (A + t)\, dy^2. \tag{6}$$

Introduce new parameters (u, v) such that

$$\Phi \equiv \theta(u, v)(du^2 + dv^2) \tag{7}$$

holds. Then the functions $x(u, v)$ and $y(u, v)$ satisfy the second order sytem

$$\begin{aligned} \Delta x &= h_1(x_u^2 + x_v^2) + h_2(x_u y_u + x_v y_v) + h_3(y_u^2 + y_v^2) + h_4(x_u y_v - x_v y_u), \\ \Delta y &= \tilde{h}_1(x_u^2 + x_v^2) + \tilde{h}_2(x_u y_u + x_v y_v) + \tilde{h}_3(y_u^2 + y_v^2) + \tilde{h}_4(x_u y_v - x_v y_u), \end{aligned} \tag{8}$$

where $h_1, \dots, \tilde{h}_4$ are certain nonlinear differential expressions involving the quantities x, y, z, p, q and the coefficients of (3). The Lewy class is then characterized by the condition that the coefficients $h_1, \dots, \tilde{h}_4$ depend only on x and y. It can be shown that the equations (1) and (2) belong to this class. This fact is fundamental for Lewy's solution of the Weyl and Minkowski problem in the case of analytic data ([4] and [5]). An essential ingredient in the proof of the compactness theorem is a certain lemma on the nonvanishing of the jacobian of one-to-one mappings $(u, v) \to (x, y)$ satisfying the equations (8), where the coefficients $h_1(x, y), \dots, \tilde{h}_4(x, y)$ are analytic. This result, which was established in [3], is of considerable intrinsic interest, apart from its connection with Monge-Ampère equations. Lewy's theory was extended to the nonanalytic case by the author ([C1] and [C2]). In these investigations estimates for the jacobian of one-to-one mappings satisfying (8), play a dominant role. Recently F. Schulz [C4] improved these results by applying more refined elliptic techniques.

References

[C1] Heinz, E., On elliptic Monge-Ampère equations and Weyl's embedding problem, *J. d'Analyse Math.* **7** (1959), 1–52.

[C2] Heinz, E., Interior estimates for solutions of elliptic Monge-Ampère equations, *Proceedings of Symposia in Pure Mathematics*, **4** (1961), 149–155.

[C3] Pogorelow, A.V., *The Minkowski Multidimensional Problem*, John Wiley & Sons, New York, 1978.

[C4] Schulz, F., *Regularity Theory for Quasilinear Elliptic Systems and Monge-Ampère Equations in Two Dimensions*, Lecture Notes in Mathematics 1445, Springer-Verlag, 1990.

Peter D. Lax*

Hyperbolic Difference Equations: A Review of the Courant-Friedrichs-Lewy Paper in the Light of Recent Developments

Abstract: The portion of the Courant-Friedrichs-Lewy paper [*Math. Ann.* 100, 32 (1928)] that was devoted to hyperbolic difference equations is critically reviewed in terms of its basic contribution to the numerical solution of partial differential equations. Some subsequent developments are then discussed, including recent literature related to the von Neumann condition, some irreversible schemes, generalizations of the energy method, some new difference schemes, and mixed initial boundary value problems.

Introduction

At the time their paper was written, Courant, Friedrichs and Lewy[2] were interested in difference equations as a tool for proving the existence of solutions of partial differential equations; discretization was a way of constructing a sequence of finite dimensional problems whose solutions tended in the limit to the solution of the continuous problem.** For this reason the authors developed a method which was applicable to equations with variable coefficients; for the same reason they were not interested in studying all possible difference schemes but were satisfied when they found one that worked.

Today the methods of functional analysis—projection, mollification—are so well developed (thanks in large part to the pioneering work of Friedrichs) that descent to the finite dimensional case is not necessary for proving the existence of solutions of linear problems.† On the other hand, the onset of the Second World War brought an unprecedented (and unabated) pressure on the technological centers to provide numerical solutions of partial-differential equations; here the ideas introduced by C-F-L—the famous stability condition, energy inequalities, and the leap frog difference method—turned out to be basic. This is an outstanding instance of research undertaken for purely theoretical purposes§ turning out to be of immense practical importance.

* Courant Institute of Mathematical Sciences, New York University.

** Courant was inspired partly by Euler's use of discretization in the calculus of variations. Also, the authors were aware of the importance of difference methods for elliptic equations.

† Quite to the contrary, the difference method may be the only effective one for dealing with severely nonlinear problems, as indicated by the success of Glimm's recent work on shock waves.[6]

§ Curiously, in the late twenties the authors were interested through the work of Prandtl in hyperbolic problems of compressible gas dynamics. They considered solving these equations numerically by the method of characteristics but *not* by difference schemes.

The difference schemes of Courant-Friedrichs-Lewy

Here is a brief description of Part II of the paper, concerning hyperbolic equations: The authors start by pointing out that a scheme *cannot* be a convergent one if the ratio of the time and space mesh is so large that the domain of dependence of a point in the difference scheme does *not* contain all points of its domain of dependence in the differential equation; such a scheme ignores information which does influence the value of the solution of the differential equation. This fact is obvious to those who are used to thinking in these terms but was very much unsuspected by quite a large number of otherwise eminent scientists who, in their haste to use too large a value of Δt, fell into the trap of using difference schemes that could not converge and which in fact did diverge spectacularly.

The authors start with the centered difference scheme for the one-dimensional wave equation where $\Delta x/\Delta t$ equals the sound speed. The resulting difference equations are extremely simple and can be solved explicitly. Then the authors turn to the centered difference scheme for the wave equation on an arbitrary grid and remark that they will not bother to write down an explicit representation of the solution because "it is too complicated to yield a limiting value easily as the meshwidth tends to zero;" thereby they failed to discover the von Neumann stability criterion.*

The convergence of the difference scheme is proved by carrying over the energy method developed by Friedrichs and Lewy for the wave equation.[5] The crux of that method is a quadratic integral identity obtained as follows: multiply the wave equation by U_t, write the product as a di-

* The explicit representation was given by Lewy in a paper[16] published twenty years later.

vergence, integrate it over a lens-shaped domain, convert the volume integral into a surface integral and finally recognize that if the faces of the lens are space-like, the integrals over them are positive definite. To derive an analogous identity for their scheme, the authors multiply the difference equation by the centered approximation to U_t; they write the resulting quadratic function as a divergence sum over a tetrahedral region bounded by diagonal planes and observe that if the faces of the tetrahedron slope steeper then the characteristic cone, then the surface sums over them are positive. Although the authors do not emphasize it, the writing of the quadratic function as a divergence is much trickier for the difference scheme than for the differential operator; in the differential case it is merely integration by parts; once one has seen it he can almost certainly remember it or reconstruct it. In contrast, although I have seen the difference identity (8) [on page 63 of the original paper] a dozen times, whenever I prepare to lecture on it I waste half an hour in trying to rediscover it, then at last resort to looking it up.

In the last section of the paper the authors give an ingenious difference scheme for a certain class of second-order hyperbolic differential equations with a cross term and prove its convergence when the cross term is not too large. They observe that by a change of variables one can bring locally every hyperbolic equation into a form where the cross term is not too large. This is satisfactory if one is interested, as the authors are, in proving existence of solutions, but it may be decidedly unsatisfactory if one is interested in calculating solutions. Nevertheless, the method may be practical for equations to which it is applicable.

An interesting feature of this method (one not mentioned explicitly by the authors) is that the restriction that has to be placed on Δt to achieve convergence is more severe than merely requiring that the domain of dependence of the difference scheme contain the true domain of dependence.

Once the energy identities are derived, the authors prove convergence swiftly and elegantly. They observe that because of the positivity of the terms entering the energy identity, one obtains inequalities for the energy norm of the solution at time t in terms of the energy norm of the initial data. Analogous estimates are derived for the higher difference quotients of the solution by observing that these difference quotients satisfy an equation similar to the original one, obtained from it by differencing. With these bounds for the difference quotients, an estimate* for the maximum norm in terms of the energy norm of difference quotients is used and compactness is invoked to select a subsequence which, together with its difference quotients, converge to a limit. That limit solves the original differential equation and has the prescribed initial data; on the other hand, it follows from the energy identity for the differential equation that there can be no more than one such solution. This shows that any other convergent subsequence has the same limit and therefore the original sequence itself converges.

This indirect argument gives no error estimates; of course, using the energy inequality the authors could have estimated the difference between exact and approximate solution with the greatest of ease, had they bothered.

Some recent developments

We turn now to recent developments; these might be grouped coveniently into the following five subjects:

1) the von Neumann condition,
2) irreversible schemes,
3) generalizations of the energy method,
4) new difference schemes, and
5) mixed initial and boundary value problems.

The new developments will be illustrated rather than summarized. (For a full treatment we recommend the forthcoming second edition of Robert D. Richtmyer's book, *Initial Value Problems*).

For simplicity we shall discuss first-order systems and one-level difference operators associated with them rather than the second-order equations and second-level schemes discussed in C-F-L. In a one-level scheme the value of the approximate solution at time $t+\delta$ is obtained by applying an operator S_δ to its values at time t. S_δ is a difference operator of the form $S_\delta = \sum s_\alpha T^\alpha$, α a multi-index, T^α translation by $\delta\alpha$, and s_α a matrix-valued function. The operators S_δ act on vector-valued functions normed by the L^2 norm.

In C-F-L, the authors proved the convergence of their difference schemes by energy inequalities; in the operator language introduced here these express the uniform boundedness of powers of the operators S_δ, i.e., the existence of two constants a and b such that

$$||S_\delta^n|| \leq ae^{bt}, \quad \text{where} \quad t = n\delta, \tag{1}$$

the norm being the L^2 operator norm. A scheme satisfying (1) is called *stable*; the procedure of C-F-L demonstrates that *every stable scheme is convergent*. It turns out that something like the converse of this also is true: a scheme which converges for all square integrable data is stable in the sense of (1). Schemes which converge only for C^∞ data need be stable in a weaker sense only that these exists an integer N such that

$$||S_\delta^n|| \leq ae^{bt}c^N, \quad \text{where} \quad t = n\delta. \qquad (1)_N$$

* This is the discrete analog of the Sobelev inequality; the continuous version of that inequality, although not its sharpest form, was well known in Göttingen as an important tool in the theory of partial differential equations.

The *accuracy* of a difference scheme is measured by how closely solutions of the differential equation satisfy the difference equation. A scheme is accurate of order m if for every smooth solution u of the differential equation there is a constant k such that

$$||u(t + \delta) - S_\delta u(t)|| \leq k\delta^{m+1} \tag{2}$$

The method of C-F-L shows that if a scheme is stable in the sense of (1), and accurate of order m in the sense of (2), then for smooth data the overall error is $O(\delta^m)$:

$$||u(t) - S_\delta^n u(0)|| \leq pe^{qt}\,\delta^m, \qquad t = n\delta. \tag{3}$$

Somewhat surprisingly, Strang proved in Ref. 20 that the *same* error estimate (3) holds also for weakly stable schemes which satisfy merely $(1)_N$, provided that the solution u is sufficiently differentiable.

Weak stability is very hard to verify for operators with variable coefficients since it is an unstable concept, in the following sense: If S_δ satisfy $(1)_N$, an operator $S_\delta' = S_\delta + M$ where the $||M|| \leq$ const. $\times\ \delta$, need not satisfy $(1)_N$. In contrast, the class of strongly stable operators satisfying (1) is stable in this sense.

In Ref. 23 Thomée makes a precise and surprisingly delicate study of the degree of stability of a class of operators in the *maximum norm.*

The symbol (amplification matrix) of the operator S_δ is

$$S(x, \xi) = \sum S_\alpha(x)e^{i\alpha\xi}.$$

For operators with coefficients independent of x it follows immediately from the isometric character of Fourier transformation that necessary and sufficient for stability is the uniform boundedness of all powers of the symbol. This reduces a stability question to a pure matrix problem; this matrix problem, however, turns out to be trickier than it seems at first glance. An obvious necessary condition, the famous one due to von Neumann, is for the spectrum of such a matrix to lie in the unit disc in the complex plane. An obvious sufficient condition is for the norm of this matrix to be ≤ 1. A non-obvious sufficient condition is that the numerical range of the matrix be in the unit disc.[1,3,4] Necessary and sufficient conditions were given by Kreiss[8] and Buchanan;[1] see also Morton[17] and Morton and Schechter.[18]

If the difference scheme is reversible, i.e., can be used to compute solutions both forward and backward in time, then both positive and negative powers of the symbol have to be bounded. In this case the von Neumann condition requires that the spectrum be located on the unit circle. One of the new developments has been the observation that although hyperbolic problems are reversible, irreversible difference schemes for them can be quite useful. Such a difference scheme was introduced by Friedrichs in Ref. 4. In some nonlinear cases the slight dissipation which makes the difference scheme irreversible is even indispensable.

It is not hard to show that the von Neumann condition is necessary for the stability of schemes with variable coefficients as well. Von Neumann conjectured that it is also sufficient; something like this has been demonstrated by a surprisingly elaborate extension of the energy method.

In the hands of Courant, Friedrichs, and Lewy, energy inequalities were derived from energy identities, obtained from difference analogs of formulas for the differentiation of products. This has been generalized and systematized by Lees[15] who has devised a large number of inequalities in this way. It was observed in Ref. 11 that all such energy inequalities can be obtained by the purely algebraic method of expressing certain non-negative symbols as sums of squares; according to Hilbert's theory of higher order positive forms, this is not always possible and so the classical energy method has inherent limitations. Nonetheless, it is possible to derive energy inequalities without energy identities as was done, for instance, in Ref. 13 with the aid of a certain amount of dissipation, and in Ref. 12 with the aid of nothing at all except sufficient differentiability of the coefficients. The theorem of Lax and Nirenberg reads: If the symbol is a sufficiently differentiable function of x, ξ and if $|S(x, \xi)| \leq 1$ for all x and ξ, then the operator S_δ satisfies the inequality.

$$||S_\delta|| \leq 1 + K\delta. \tag{4}$$

Obviously the expression of (4) implies stability in any finite time interval. The more delicate problem of proving stability when the symbol is not bounded by inequality (1) in norm but merely satisfies the von Neumann condition has been handled by Kreiss.[9] He found that if in addition to the von Neumann condition one requires dissipation in a definite sense then it is possible to introduce a new norm equivalent to the L^2 norm for which inequality (4) is true. The proof of the surprisingly delicate matrix theorems needed to do this have been simplified by Parlett.[19]

As already mentioned, earlier practical needs have called for a whole army of novel difference schemes. Among these one might single out the crude but useful method introduced by Friedrichs,[4] the more accurate methods devised by Du Fort-Frankel,[3] by Lax-Wendroff,[14] and the high-order schemes studied by Strang,[21] and many others. As more and more complicated schemes are being tried for more and more complicated systems of equations in more and more variables, the verification of the von Neumann condition, although merely algebraic, becomes quite difficult and is often done numerically on the machine.

Almost all problems of practical importance involve boundary conditions as well as initial conditions. There is a large body of practical experience with these and a modest body of theory. An analog of the von Neumann condition has been given by Godunov and Ryabenkii[7] using some observations of Gelfand. In spite of interesting beginnings, such as the work of Strang[22] who has found a relation to the theory of Wiener-Hopf equations, and some recent work of Kreiss[10] using high-order extrapolation, there is as yet no general theory for problems with variable coefficients.

Bibliography

1. M. L. Buchanan, "A necessary and sufficient condition for stability of difference schemes for initial value problems," *J. Soc. Indust. Appl. Math.* 11, 919–935 (1963).
2. R. Courant, K. Friedrichs, and H. Lewy, "On Partial Difference Equations of Mathematical Physics," *IBM Journal* 11, 215–234 (1967). (English translation of the original work, "Über die Partiellen Differenzengleichungen der Mathematischen Physik," *Math. Ann.* 100, 32–74 (1928).
3. E. C. Du Fort and S. P. Frankel, "Stability conditions in the numerical treatment of parabolic differential equations," *Math. Tables and Other Aids to Computation.* 7, 135–152 (1953).
4. K. O. Friedrichs, "Symmetric hyperbolic linear differential equations," *Comm. Pure and Applied Math.*, 7, 345–392 (1954).
5. K. O. Friedrichs and H. Lewy, "Über die Eindeutigkeit und das Abhängigkeitsgebiet der Lösungen beim Anfangswertproblem linearer hyperbolischer Differentialgleichungen," *Math. Ann.* 98, 192–204 (1928).
6. J. Glimm, "Solutions in the large for nonlinear hyperbolic systems of equations," *Comm. Pure and Applied Math.*, 18, 697–715 (1965).
7. Godunov and Ryabenkii, "Spectral criteria for stability of boundary problems for non-self-adjoint difference equations," *Uspekii Math. Nauk* 18, 3 (1963).
8. H. O. Kreiss, "Über die Stabilitätsdefinition für Differenzengleichungen die partielle Differentialgliechungen approximieren" (English Summary), *Nordisk Tidskr. Informations Behandlung* 2, 153–181 (1962).
9. H. O. Kreiss, "On difference approximations of the dissipative type for hyperbolic differential equations," *Comm. Pure and Applied Math.*, 17, No. 3, 335–353 (1964).
10. H. O. Kreiss, "Difference approximations for hyperbolic differential equations," *Numerical Solution of Partial Differential Equations.* Edited by James H. Bramble Academic Press, N. Y., 1966, pp. 125–152.
11. P. D. Lax, "The scope of the energy method," *Bull. Amer. Math. Soc.*, 66, 32–35 (1960).
12. P. D. Lax and L. Nirenberg, "On stability for difference schemes," *Comm. Pure and Applied Math.*, 19, 4, 473–492 (1966).
13. P. D. Lax and B. Wendroff, "On the stability of difference schemes: A Sharp Form of Gårding's Inequality," *Comm. Pure and Applied Math.* 15, 363–371 (1962).
14. P. D. Lax and B. Wendroff, "Difference schemes for hyperbolic equations with high order accuracy," *Comm. Pure and Applied Math.* 17, No. 3, 381–398 (1964).
15. M. Lees, "A priori estimates for the solutions of difference approximations to parabolic partial differential equations," *Duke Math. J.*, 27, 297–312 (1960).
16. H. Lewy, "On the convergence of solutions of difference equations," *Studies and Essays Presented to R. Courant on his 60th Birthday*, Jan. 8, 1948, Interscience Publishers Inc., N. Y., 1948, pp. 211–214.
17. K. W. Morton, "On a matrix theorem due to H. O. Kreiss," *Comm. Pure and Applied Math.* 17, 375–379 (1964).
18. K. W. Morton and S. Schechter, "On the stability of finite difference matrices," *J. Soc. Indust. and Applied Math. Ser. B Numer. Anal.* 2, 119–128 (1965).
19. B. Parlett, "Accuracy and dissipation in difference schemes," *Comm. Pure and Applied Math.*, 19, No. 1, 111 (1966).
20. W. Gilbert Strang, "Difference methods for mixed problems," *Duke Math. Journal*, 27, No. 2, 221–232 (June 1960).
21. W. Gilbert Strang, "Trigonometric polynomials and difference methods of maximum accuracy," *Journal of Math. and Physics*, 41, No. 2, 147–54 (1962).
22. W. Gilbert Strang, "Wiener-Hopf difference equations," *Journal of Math. and Mech.* 13, 85–96 (1964).
23. V. Thomée, "On maximum-norm stable difference operators," *Numerical Solution of Partial Differential Equations*, Edited by James H. Bramble. Academic Press, N. Y., 1966, pp. 125–152.

Received August 16, 1966

LES PREMIERS TRAVAUX DE HANS LEWY

JEAN LERAY

De 1925 à 1931 Hans Lewy, qui naquit en 1904, publia en langue allemande dix articles; ils concernent les équations aux dérivées partielles, tout spécialement les équations à deux variables indépendantes. Ces articles lui valurent, très précocement une réputation mondiale: ils ouvraient un domaine dont la richesse continue et continuera certainement de nous étonner. Comparés à ce qui les précédait, ces articles étaient des révélations; comparés à ce qui les suivit, ils préfiguraient l'avenir. L'illustre Université de Göttingen apprécia Hans Lewy: il en devint Dozent.

Au Collège de France il participa souvent au fameux Séminaire de Jacques Hadamard; il parlait un français impeccable; Jacques Hadamard était radieux d'accueillir ce brillant mathématicien, déjà célèbre, si jeune et sympathique.

Mais, en 1933, la catastrophe nazie déferla, dispersant l'école de Göttingen. Hans Lewy eut la sagesse de venir aussitôt à Paris rechercher calmement, outre-Atlantique, une nouvelle patrie et une place sûre.

Il m'avait ébloui au Séminaire de Jacques Hadamard; je n'étais qu'un débutant; il fut cordial envers moi; il me fit rencontrer Jules Schauder, à Paris, au printemps 1933; c'est avec une vive émotion que j'évoque les amitiés qui naquirent ainsi.

Ce n'est pas à moi de décrire l'influence qu'eut Hans Lewy aux Etats-Unis et aussi en Italie et même en Chine. Il fut profondément heureux de revenir, pour une visite, à Göttingen et d'y publier à nouveau, en 1982, un article en langue allemande: c'était, au bout d'un demi-siècle, le dénouement du grand drame qu'il avait vécu.

Jean Leray

CONFLUENCE OF BOUNDARY CONDITIONS

RICHARD MACCAMY

Immediately after World War II there was a considerable amount of work on water wave problems, an outgrowth of war sponsored research. One aspect of this was the study of waves on sloping beaches. Lewy considered this problem, in a simplified form, in reference 23. The problem consists of Laplace's equation in a wedge shaped region. One has a Neumann condition on one edge and a Robin type condition onto the other edge. There is then a confluence of the two conditions at the vertex. For wedge angles $p\pi/2q$ Lewy constructed an explicit (although complicated) solution by a very clever use of analytic continuation. Lewy was quite fascinated by the dependence of the solution on the angle of the wedge and he observed that continuity with respect to the angles was related to a result in pure number theory.

Once Lewy started on the study of confluence of boundary conditions he became very interested in the subject. His next effort was done in conjunction with K. Friedrichs. They considered the problem of a semi-infinite dock, that is, the wedge problem with angle π, and they again obtained an explicit solution (reference 24). (For many years after this work Lewy worked on the problem of scattering from a dock of finite width. This problem, like that of scattering of sound waves from a finite strip, is a very well known one. Despite the efforts of Lewy, and many others, it remains unsolved.)

Lewy continued his work on confluence of conditions by undertaking a study of the local behavior near the point of joining. This work started in references 26 and 27 in which he considered the dock problem. Here there is no angular dependence and what he found was that just the two different boundary conditions produced a very complicated asymptotic expansion of the solution in terms of z and $z \log z$ where $z = x + iy$ with the point of confluence at $(0,0)$. These developments were later used in the study of singular integral equations. Lewy's student, R. S. Lehman, extended the local behavior studies to the case of a wedge and those results proved very valuable in the study of finite difference methods in polygonal domains.

Lewy continued these studies on a more and more abstract level (see for instance, reference 39). A unifying theme to all of this work was the idea that appeared very early in Lewy's work that many results for elliptic equations can be obtained from using ideas of reflection and analytic continuation results from complex variable theory.

COMMENTS ON SOME OF HANS LEWY'S WORK

LOUIS NIRENBERG

Hans Lewy's work has had much impact on the theory of partial differential equations, especially on elliptic equations, and free boundary problems.

The subjects in which he worked may be grouped into several categories. Almost all involved partial differential equations—which he usually assumed to be analytic. A recurrent theme in all of them in fact, is the use of analytic function theory—often in truly ingenious, almost devilish, ways:

1. Papers devoted purely to partial differential equations—some with applications to geometry—involving existence and regularity theory for elliptic and hyperbolic equations; also difference schemes for numerical approximation of solutions.
2. Existence and regularity of solutions of variational problems, and of variational inequalities, including the study of free boundaries. Papers on minimal surface theory. This is a large and important part of his work.
3. Partial differential equations connected with several complex variables. This includes his famous papers 36, 38 for which the American Mathematical Society awarded him the Steele Prize.
4. Partial differential equations connected with water waves, and fluid dynamics.
5. Offbeat properties of solutions of partial differential equations.

I would add a sixth category, which overlaps with 3, 4 and 5.

6. Extension of solutions of some partial differential equations across the boundary, by some general "reflection" method.

Lewy's work was extremely original and inventive; from his early work to the latest. Many of his ideas opened up new fields of research which were then taken up by others. These ideas often occurred when Lewy was considering a very particular problem, sometimes in a paper of only a few pages. On reading his papers, time after time, one is struck by some new ingenious, but simple looking, idea or technique.

Here I will discuss a few representative results—except from categories 2 and 3, which are fully described by others.

Category 1. Beginning with his earliest papers, Lewy was interested in difference methods for finding approximate solutions of boundary value problems, and with variational problems; see papers 1 and 4. In addition, he began his work on analyticity of solutions of nonlinear partial differential and variational equations in two dimensions; this was to remain a constant interest. In paper 3, he proved the analyticity of a solution u of $\Delta u = f(x, y, u, u_x, u_y)$, while in 9, he treated a fully nonlinear second order elliptic equation in the plane, assuming $u \in C^2$. His method of proof is very novel, even today. He extends the solution to complex values of x, by solving an initial value problem for a particular *hyperbolic* system. This involves using existence results for such systems, and he makes use of characteristics; the argument is restricted to two dimensions. During this period Lewy wrote a series of basic papers on solving the initial value problem for hyperbolic partial differential equations, 5, 6, 8, and: "Über fortsetzbare Anfangsbedingungen bei hyperbolischen Differentialgleichungen in drei Veränderlichen," Ges. Wiss. Göttingen, Math Physik Klasse (1932), pp. 135–143. All but the first are joint with K. O. Friedrichs. In 6 they use the energy method to prove uniqueness, and to obtain bounds on domains of dependence. They treat second order hyperbolic equations in n dimensional and special systems. Paper 8 contains similar results for higher order equations in two dimensions. In the last paper above, for the wave equation in two space dimensions, they use energy estimates and inequalities which we now call "of Sobolev-type" to treat the initial value problem. They approximate the problem via difference schemes, use related energy estimates, etc., and prove the convergence of the difference scheme solutions to the one for the differential equation. Furthermore they obtain solutions with square integrable derivatives of higher order, and infer, via Sobolev-type estimates, that the solutions are in C^2. They use energy estimates as in paper 7.

Paper 7 is a fundamental paper on difference methods. Its aim was to prove solvability of partial differential equations by difference scheme approximations; the paper has had a great influence in numerical analysis. For a hyperbolic problem in one space dimension, it introduced energy inequalities for difference schemes, and the fundamental stability condition. For stability of the scheme, it is necessary to build in the analogue of finite propagation speed of characteristics, that is, in the difference scheme, the ratio of the time step to each step in the space variable should be bounded by some fixed number. For a more complete description of 7 and for further developments up to 1967, I recommend the comments by P. D. Lax, included here on page 49 of Vol. 1 of Selected Works of Friedrichs published by Birkhäuser in 1986.

Returning to elliptic equations, in 10, he adapted and generalized the method used in 9, involving going into the complex and using hyperbolic systems, to prove uniqueness for the Cauchy problem for solutions of a fully nonlinear elliptic equation, which is analytic, in the plane. The solutions are supposed to be in C^4 and to agree, together with their first derivatives, on a C^2 curve. About the same time, Carlemann proved unique continuation for the Cauchy problem for second order elliptic partial differential equations in two dimensions, requiring only some smoothness, not analyticity, of the equations. He introduced the method of establishing integral inequalities involving powers of a suitable weight function. "Estimates of Carlemann type" have become the standard way to approach such problems for various types of equations.

Finally, in this category, we mention briefly the papers 14, 17, 18 and 19, which are concerned with estimates for solutions of analytic Monge-Ampère equations in two dimensions, and their application to two problems in differential geometry. These are discussed in greater detail by E. Heinz. In 14, he shows that if one has bounds on the C^2-norm of a solution, then one can derive bounds for further derivatives, and then, using 9, infer that the solution is analytic. In 17 he shows, for an analytic solution, how to obtain a bound on the C^2-norm, knowing a bound on the C^1-norm. The proof involves an interesting contact transformation, by which solutions are transformed to solutions of a related equation whose second derivatives are automatically bounded. The estimates of 12 and 15 are then applied in the geometric problems in 18 and 19.

In 18 Lewy solves the Weyl problem, which is to embed isometrically, the two dimensional sphere S^2, with a given metric having positive Gauss curvature, as a closed convex surface in $\mathbb{R}^3$. Lewy assumes the metric is analytic, and uses the continuity method as outlined by Weyl. Surprisingly, he does not use the estimates for the second derivatives derived by Weyl—valid also for nonanalytic metrics. Paper 19 solves the Minkowski problem: Given a positive function K on S^2 satisfying the (vector) condition

$$\int_{S^2} K(x)x\,dA = 0$$

(here x is the unit vector in $\mathbb{R}^3$), there exists a closed convex surface having K as a function of its normal, or spherical image. Again he works with analytic K and obtains an analytic surface. Only in the late 1940's and early 1950's were both problems solved in the nonanalytic category.

Category 4. Shortly after the Second World War, a number of people studied water waves in two dimensions in an angular sector corresponding to a sloping beach or a dock. Lewy's papers 21, 22, 25, 29, 30, 31, the Stanford Technical Report (ONR) No. 31: The development of functions associated with surface waves over an inclined bottom (1954), and paper 44, with several different coauthors, were devoted to this subject. Some papers involved studying harmonic functions u in a sector Σ in the complex z plane: $0 > \text{Arg } z > -2\pi\alpha$, $0 < \alpha \leq 1$, with u subject to certain boundary conditions. The holomorphic function f, with u as real part, was then extended across the boundary to a Riemann surface, on which it then satisfied some difference scheme. Paper 35 and a later paper 64 dealt with such difference schemes.

In several of these papers, determination of rather precise behaviour of the wave near the origin and near infinity, plays a crucial role. The behavior involves expansions in powers of z and $z\log z$. Paper 23 deals with the case that $\alpha = \pi p/4q$, p, q integers, p odd, and constructs standing waves. In paper 24 the dock problem is treated: $\alpha = 1$. Quite different methods from those used in 23 are employed, methods related to Laplace transform. I will discuss, mainly, the interesting paper 44, with R. S. Lehman, which settles a uniqueness question concerning a mixed boundary value problem in Σ, with $\alpha < 1$. Consider a real bounded function u satisfying

$$\Delta u - k^2 u = 0 \ \text{ in } \ \Sigma$$

$$u_y = u \ \text{ on } \{\theta = 0\}, \quad u_\theta = 0 \ \text{ on } \ \{\theta = -2\pi\alpha\}.$$

Here $\theta = \arg z$. For some k, a bounded, nontrivial solution was known, and the problem was, in general, to determine if the space of bounded solutions is at most one dimensional.

In fact, for $0 \leq k < 1$, two solutions Φ and Ψ were already known: Φ is continuous in $\overline{\Sigma}$ with $\Phi(0) = 1$, and is bounded. Ψ is bounded near infinity, but $\Psi - \log|z| \to$ constant as $z \to 0$. In 44 it is proved that any bounded solution is a multiple of Φ (for $k < 1$). Furthermore, for $k \geq 1$, two solutions Φ, Ψ with the properties described above cannot both exist.

The paper is tricky and highly unusual. It makes use of paper 40 which contains a curious result concerning solutions of a homogeneous linear partial differential equation with constant coefficients in the plane: $Lu = 0$. Given two solutions, in a simply connected domain $\Omega \subset \mathbb{R}^2$, it constructs a new solution in Ω. The construction relies on a path integral, and makes ingenious use of Green's Theorem. It only works in two dimensions. (Further work on the operator $\Delta + \lambda$ may be found in paper 41.)

Uniqueness questions similar to the one treated in 44 arise often in studying solutions of boundary value problems in unbounded domains. Many authors have worked on such problems but to my knowledge, no one has followed up the unusual methods of this paper.

Lewy had one more paper in fluid dynamics; the well-known paper 33, with P. Garabedian and M. Schiffer. It is concerned with axially symmetric cavitational flow in three dimensions. Such flow occurs when a body moves so quickly in water that the fluid pressure falls to vapor pressure, and a cavity of stream develops behind the body. If the body is axially symmetric, and moves in the direction of the axis, and if the fluid (and cavity) fill all of $\mathbb{R}^3$, then the flow is also symmetric about the axis. (For plane cavitational irrotational flow in two dimensions, there had been much earlier work.)

This basic paper is the first to provide existence of axially symmetric cavitational flow. The problem is treated as a variational one, in which the boundary of the cavity is a free—a priori unknown—boundary surface. This is Lewy's second paper, after 30, concerned with free boundaries. Later, he had many papers in the subject. 33 combines a number of basic arguments and techniques (some of which had been used earlier): (i) Symmetrization; (ii) Interior variations, i.e., a variation of the independent variables—here via explicit conformal mappings; (iii) It proves that the free boundary is analytic, by extending the stream function analytically to complexified values of the independent variables, as in paper 9. A functional equation is then obtained, in the complex domain, related to the geometry of the free boundary. Its solution yields the analyticity of the boundary. (Similar arguments for analyticity of free boundaries occur in papers 28 and 29.) (iv) Uniqueness, and continuous dependence on the "cavitation constant," are proved using a comparison lemma similar to Schwarz' lemma. In fact this lemma follows easily from the Hopf boundary lemma for elliptic equations; this was also published in 1952.

Category 5. Lewy wrote several papers, 16, 21, 52 and 53 concerned with special properties of a local map near the origin in $\mathbb{R}^2$ or $\mathbb{R}^3$, related to solutions of certain elliptic equations. In 16, he shows that for a 1–1 map given by two functions in $\mathbb{R}^2$ near the origin, and satisfying a particular quasilinear second order system of elliptic partial differential equations, the Jacobian of the map is nonzero. This result is used in his papers 18 and 19 in geometry, mentioned earlier.

Papers 19, 49 and 50 are concerned with maps near the origin of $\mathbb{R}^3$, into $\mathbb{R}^3$, given by the gradient of a harmonic function u. One of the results in 21 is that if u is a homogeneous polynomial of degree > 2, then the Jacobian of the gradient map of u, i.e., the determinant of the Hessian H of u, must vanish at some point other than the origin. In 1953, B. Segre gave a different proof and showed, furthermore, that either H vanishes identically or changes sign. In 53, Lewy gives a new proof of Segre's result by extending the argument of 21. Finally in 52 he proves the same conclusion in case u is merely harmonic (homogeneity is dropped). As a corollary, he obtains: If u is harmonic and the gradient map is 1–1, then $\det H \neq 0$, everywhere. No extension was made to higher dimensions until very recently. In 1991, S. Gleason and T. H. Wolff published "Lewy's harmonic gradient maps in higher dimensions," Comm. P.D.E., 12, pp. 1925–1968. One of their results is: If u is harmonic in a domain in $\mathbb{R}^n$, $n \geq 3$, such that at every point x, the Hessian matrix H has at most one negative eigenvalue, then either $\det H$ never vanishes, or it vanishes identically.

Category 6. Lewy was clearly fascinated by the Cauchy reflection principle for harmonic or holomorphic functions. It played an essential role in his paper 64 on boundary behaviour of holomorphic mappings.

Paper 41 is a very interesting one based on a lecture he gave. It contains a number of results:

a) He studies reflection of harmonic functions in a domain Ω in two dimensions, across a straight bounding segment—under rather general boundary conditions. He obtains extension of the functions to the full reflection of the domain in the segment. Furthermore, he shows that in three dimensions a similar result need not hold, i.e., the harmonic functions extend only locally across the boundary segment.

b) He treats more general linear elliptic equations in Ω, having Laplacian as leading part, and with analytic coefficients in the other terms. He proves that the solutions extend to the full mirror image of Ω. The proof is most ingenious. He extends each coordinate into the complex, and then works in a region in $\mathbb{C}^2$.

c) An interesting consequence of b) is the following: In the half disc D: $\{|z| < 1, \text{Im } z > 0\}$, let u and $v \in C^2(\overline{D})$ satisfy $\Delta u = 0 = \Delta v + \lambda v$, $\lambda = \text{constant} \neq 0$. Suppose u and v have the same Cauchy data on $\text{Im } z = 0$, i.e., $\nabla u = \nabla v$ there. Then u and v are real analytic up to that boundary: $\text{Im } z = 0$. (This was extended to higher dimensions by D. Kinderlehrer, L. Nirenberg and J. Spruck in 1978.)

d) The paper also describes some of the results of his paper: "A theory of terminals and the reflection laws of partial differential equations," Stanford Technical Report (ONR), No. 4, 1952, 31. In his work on reflection of solutions, in a simply connected domain in $\mathbb{C}$, Lewy often considers integrals along an arc p from some fixed point to a variable point z, of some holomorphic function. In the report, he introduces a more general abstract concept, which he calls a terminal. It is a functional defined on functions f in the domain in $\mathbb{C}$, and which depends on the path, $\Phi(f, p)$. If f is holomorphic, then Φ is to depend only on f and the endpoint z. Some other properties of terminals are required. They are then used to prove reflection laws for partial differential equations. I believe that no one else has taken up further study of terminals.

Lewy's ideas have inspired many mathematicians, and some of his ideas are still waiting to be developed further.

THREE DISCOVERIES AND ONE QUESTION OF HANS LEWY IN CR ANALYSIS

FRANÇOIS TREVES

Today CR (Cauchy–Riemann) analysis and geometry are well-established mathematical fields and the focus of very active research. But in the first half of the century, up to the 1950s, when Lewy made his first discoveries, they hardly existed as an independent area of study—they were peripheral aspects of several complex variable theory. I do not think that the notion of an abstract CR manifold had been defined (the adjective CR was first introduced in [Greenfield, 1968]). In fact Hans Lewy was the first to ask the question, alluded to in the title of this note, as to whether the homogeneous equation $\mathrm{L}h = 0$ in Euclidean three-dimensional space admits two "independent" solutions, under the hypothesis that the smooth complex vector field L, its complex conjugate $\overline{\mathrm{L}}$ and the commutation bracket $[\mathrm{L}, \overline{\mathrm{L}}]$ are linearly independent. In a different language, Lewy was asking whether any abstract three-dimensional strongly pseudoconvex CR manifold can be locally embedded in $\mathbb{C}^2$ (as a real hypersurface). The answer, negative, was provided by the example in [Nirenberg, 1974].

Naturally, mathematicians had been doing CR prose for some time without being aware of it. Some of the basic discoveries in the subject date back to the beginning of the century—to the work of H. Poincaré and E. Cartan on the biholomorphic invariants of a real hypersurface in multidimensional complex space, and to the fundamental discovery by E. Levi of the quadratic form that bears his name, and of the attendant concept of (Levi) pseudoconvexity ([Levi, 1910]). These were contributions to the theory of holomorphic functions of several complex variables, which was gaining momentum. A powerful effort, begun in the 1930s, aimed at characterizing the domains of holomorphy, and culminated in their identification with the pseudoconvex domains. The names of K. Oka, H. Cartan, H. Grauert, dominate those developments. In the 1950s a different approach was initiated by D. Spencer, based on the deformation theory of complex structures and centered around the solution of the complex-analysis analogue of the Neumann problem, familiar to potential theorists. As a boundary value problem the $\overline{\partial}$-Neumann problem gives a prominent role to the boundary (the equation in the interior is the familiar Laplace equation; the complex structure is reflected in the boundary conditions). The existence and regularity (up to the boundary) of solutions to the $\overline{\partial}$-Neumann problem was established by J. J. Kohn in the middle Sixties (see [Folland–Kohn, 1972]). Central to the sheaf-theoretical approach (Oka–Cartan theory) as well as to the carrying out of the Spencer program, the study of the Dolbeault complex ([Dolbeault, 1956]) on a complex manifold with boundary led, rather naturally, to that of the differential complex induced on the boundary. The concept was formalized in [Kohn–Rossi, 1965] and given the name of "boundary complex" (with boundary operator $\overline{\partial}_b$).

Circa 1956 complex variables and related PDE problems were in the air at Berkeley. C. B. Morrey was devoting much thought to the smoothness up to the boundary of the solutions of the $\overline{\partial}$-Neumann problem (whose formulation was still somewhat murky). If I am not wrong, Hans Bremerman was about to join the faculty. It is not sure that all this had much influence on Hans Lewy; Lewy was one of those mathematicians who followed a relatively eccentric trajectory of their own, impervious to outside attractions or repulsions. The fact remains that 1956 and 1957 were the years, and Berkeley was the place, where he made two of his most important discoveries, and that both had to do with CR structures.

I shall now attempt to describe the mathematics, and to give complete, essentially self-contained proofs of Lewy's results, and of Nirenberg's answer to Lewy's question.

A hypersurface Σ in $\mathbb{R}^4$ (for us, a smooth submanifold of codimension one) acquires an interesting new structure when $\mathbb{R}^4$ acquires the complex structure that identifies it to $\mathbb{C}^2$. Call $z = x + iy$ and $w = u + iv$ the complex coordinates ($i = \sqrt{-1}$). The holomorphic functions of (z, w) in some open subset $\mathcal{O}$ of $\mathbb{C}^2$ are the $\mathcal{C}^1$ functions (or the distribution) in $\mathcal{O}$ that are annihilated by the *Cauchy–Riemann vector fields*

$$\partial/\partial\overline{z} = \frac{1}{2}(\partial/\partial x + i\partial/\partial y), \quad \partial/\partial\overline{w} = \frac{1}{2}(\partial/\partial u + i\partial/\partial v).$$

At an arbitrary point of Σ there is one (and only one, up to a scalar multiple) linear combination, L, of the two Cauchy–Riemann vector fields that is tangent to Σ. Let $\rho(x, y, u, v) = 0$ be an equation defining Σ in a neighborhood U of the point under consideration; ρ is a real-valued $\mathcal{C}^\infty$ function, and $d\rho$ does not vanish. As the *tangential Cauchy–Riemann* vector field L one may select any scalar multiple of the vector field

$$(\partial\rho/\partial\overline{w})\partial/\partial\overline{z} - (\partial\rho/\partial\overline{z})\partial/\partial\overline{w}.$$

It might help to keep in mind three prototypical examples: a flat manifold, the hyperplane $v = 0$, with the obvious selection $\mathrm{L} = \partial/\partial\overline{z}$; two convex boundaries: the unit sphere $\mathbf{S}^3$, with $\mathrm{L} = w\partial/\partial\overline{z} - z\partial/\partial\overline{w}$, and the hyperquadratic $\Sigma_0 = \{(z, w) \in \mathbb{C}^2;\ v = z\overline{z}\}$, with $\mathrm{L} = \partial/\partial\overline{z} = 2iz\partial/\partial\overline{w}$.

A fundamental concept of several complex variable theory is that of a strongly (or strictly) pseudoconvex domain. The geometrical meaning is that the domain is biholomorphically equivalent, near any one of its boundary points, to a convex region $|z|^2 + |w|^2 < 1 + 0(|z|^3 + |w - i|^3)$. A consequence of this property is that, on the hypersurface $\partial\Omega$, the three vector fields L, $\overline{\mathrm{L}}$ and $(2i)^{-1}[\overline{\mathrm{L}}, \mathrm{L}]$ are linearly independent at every point (L is the tangential Cauchy–Riemann vector field on $\partial\Omega$). Conversely, if L, $\overline{\mathrm{L}}$ and $(2i)^{-1}[\overline{\mathrm{L}}, \mathrm{L}]$ are everywhere linearly independent on a real hypersurface Σ, Σ is locally the boundary of a strongly pseudoconvex domain (the opposite sign is strongly pseudoconcave). One says, somewhat improperly, that the hypersurface Σ is *strongly pseudoconvex.* Of the three prototypes above, $\mathbf{S}^3$ and Σ_0 are strongly pseudoconvex; a real hyperplane is not (it is weakly pseudoconvex).

It is natural to inquire about the properties of the functions whose domain of definition lies in Σ and which are, in that set, solutions of the tangential Cauchy–Riemann equation $\mathrm{L}f = 0$. In modern terminology these are called *CR functions*, and the structure that Σ inherits from $\mathbb{C}^2$ is called a CR structure.

Such an inquiry could be, and perhaps was, the starting point of an analytic theory of CR structures. But other approaches are conceivable, approaches that are more geometric or more algebraic. And it seems that the geometric approach predates all others. CR theory was born shortly after the emergence of several complex variable theory. It is not unreasonable to date its birth to 1907, when Poincaré observed that two real hypersurfaces in $\mathbb{C}^2$ cannot, in general, be transformed into each other by means of a biholomorphism (i.e., a diffeomorphism between open sets containing the hypersurfaces, which is holomorphic and whose inverse is holomorphic). Such an impossibility might not seem so striking were it not that it is thoroughly local, inherent to the germs of the hypersurfaces at one of their points. It begs the question of finding what are the local invariants (under biholomorphisms) of hypersurfaces in complex space. The answer was provided, for strongly pseudoconvex hypersurfaces, in the work of E. Cartan and later, through a different route, by J. Moser. The results were extended to nondegenerate hypersurfaces in $\mathbb{C}^n$ with $n > 2$, in [Tanaka, 1962, 1976] and, independently, in [Chern–Moser, 1974] (on all this, see [Jacobowitz, 1990]).

The algebraic ramifications of CR theory stem from the commutation properties of the vector field L and of its adjoint, $\overline{\mathrm{L}}$. This is most easily seen in the case of the hyperquadratic Σ_0. If the coordinates x, y, u are used on Σ_0, the expression of L is $\partial/\partial\bar{z} - iz\partial/\partial u$. Writing $X = \mathit{Re}\,\mathrm{L} = \frac{1}{2}\partial/\partial x + y\partial/\partial u$, $Y = \mathit{Im}\,\mathrm{L} = \frac{1}{2}\partial/\partial y - x\partial/\partial u$ and $\mathrm{Z} = -\partial/\partial u$, the commutation relations read

$$[\mathrm{X},\mathrm{Y}] = (2i)^{-1}[\overline{\mathrm{L}},\mathrm{L}] = \mathrm{Z}, \quad [\mathrm{X},\mathrm{Z}] = [\mathrm{Y},\mathrm{Z}] = 0.$$

In other words, the Lie algebra generated by X, Y and Z is a copy of the Lie algebra $\mathfrak{h}$ of the Heisenberg group H, which leads one to expect that there be a local isomorphism between Σ_0 and H. This type of correspondence, between pseudoconvex hypersurfaces and nilpotent groups, has been exploited to great length (see [Folland–Stein, 1974]).

The one-forms dz and dw span the orthogonal of L in the complexified cotangent bundle $\mathbb{C}\mathrm{T}^*\Sigma_0$; this is a vector bundle of (complex) rank 2 which we shall denote by T'. Note that T' intersects the real cotangent bundle $\mathrm{T}^*\Sigma_0$ along a line bundle. The intersection is the characteristic set of the CR structure of Σ_0, *Char* Σ_0, the zero set of the *symbol* of the vector field L, $\sigma(\mathrm{L}) = \frac{1}{2}(\xi + i\eta) - iz\lambda$ (ξ, η, λ denote the dual variables associated to x, y, u respectively). *Char* Σ_0 is spanned by $\theta = \mathrm{d}u + i(z\,\mathrm{d}\bar{z} - \bar{z}\,\mathrm{d}z) = \mathrm{d}w - 2i\bar{z}\,\mathrm{d}z$. The one-form θ is not exact, a reflection of the fact that the subbundle of $\mathrm{T}\Sigma_0$ spanned by X and Y is not integrable, i.e., does not satisfy the Frobenius condition (which, here, means that X and Y do not commute). Note that $\mathrm{d}\theta = 2i\,\mathrm{d}z \wedge \mathrm{d}\bar{z}$ and $\theta \wedge \mathrm{d}\theta = 2i\mathrm{d}z \wedge \mathrm{d}\bar{z} \wedge \mathrm{d}u$ is proportional to the volume form of Σ_0 (for the Riemannian metric inherited from the ambient space); in other words, θ is a contact form on Σ_0. This exemplifies a general property of strongly pseudoconvex hypersurfaces in complex space: their characteristic set is spanned by a contact form θ. The relation between CR and conformal geometry is also a fertile ground to till (see, e.g. [Jerison–Lee, 1985]).

The preceding "hints" are meant to suggest the richness of the CR concept and some (only some) of the avenues that might be explored. Now, to return to the properties of CR functions on a strongly pseudoconvex hypersurface, such as $\mathbf{S}^3$ or Σ_0, the first shocking news is that most of them do not very much resemble holomorphic functions: no smoothness, no unique continuation, no maximum principle preside over their behaviour. Example: for any odd integer $k \geq 1$ the CR function in Σ_0, $w^{\frac{1}{2}k} = (u + i|z|^2)^{\frac{1}{2}k}$ is of class $\mathcal{C}^{\frac{1}{2}(k-1)}$, but it is not $\mathcal{C}^\infty$ in any open set containing the origin (if $k \geq 3$, it satisfies the tangential CR equation in the classical sense); the function $\exp(-(i/w)^{\frac{1}{2}})$

vanishes to infinite order at the origin, and does not vanish anywhere else (thus its absolute value has a strict minimum at 0).

Yet CR functions on Σ_0 have one strong link with holomorphic functions: precisely, they are the boundary values of the functions defined and holomorphic on the convex side of Σ_0, $\{(z,w) \in \mathbb{C}^2;\ v > |z|^2\}$. This is the first one of the three major discoveries of Hans Lewy concerning the solutions of the tangential Cauchy–Riemann equations. The property is very general; it is valid for the boundary of any strongly pseudoconvex domain. It ought not to be confused with *Hartog's principle*, according to which any holomorphic function in the complement of a ball in $\mathbb{C}^n$ ($n \geq 2$) extends as a holomorphic function to its interior (hence to the whole of $\mathbb{C}^n$). The Hartog principle is global and bears little relation to pseudoconvexity. The extension, to the pseudoconvex side, of a CR function h defined in a subset of a pseudoconvex hypersurface is a purely local phenomenon, and need not occur if the subset lies in a flat portion of the boundary. Example: The CR functions on the hyperplane $v = 0$ are simply the functions $h(z,s)$ which are holomorphic with respect to z and arbitrary with respect to s. Surely such a function need not be the boundary value of a holomorphic function $\tilde{h}(z,w)$ in the region $v > 0$. The hyperplane $v = 0$ is *Levi flat*, meaning that it is foliated by complex curves: the vector fields $\mathrm{X} = \partial/\partial x$ and $\mathrm{Y} = \partial/\partial y$ do commute, and define the foliation whose leaves are the z-planes $u =$ *const.*

The extendability of CR functions in a strongly pseudoconvex hypersurface Σ is a direct consequence of the existence of *analytic disks* in the pseudoconvex side of Σ, *with boundary on* Σ. Let Δ denote the open unit disk in $\mathbb{C}$ and let φ be a continuous map $\overline{\Delta} \to \mathbb{C}^2$ which is holomorphic in the interior Δ; $\varphi(\Delta)$ is an analytic disk with boundary on Σ if $\varphi(\partial\Delta) \subset \Sigma$. That such analytic disks fill one side of a convex hypersurface is most readily seen when the hypersurface is the unit sphere $\mathbf{S}^3$ or the hyperquadric Σ_0. Every affine complex line intersects the unit ball $\mathbf{B}^4$ along an analytic disk. And each point (z_0, w_0) such that $v_0 > |z_0|^2$ belongs to an analytic disk with boundary on $\partial\Sigma_0$, for instance to the disk $\zeta \to \left(v_0^{\frac{1}{2}}\zeta, w_0\right)$.

The idea underlying the original proof in [Lewy, 1956] is not hard to grasp. Let h be a CR function in Σ_0, endowed with some degree of smoothness. One can check directly, by differentiation under the integral sign, that the Cauchy integrals

$$I_h(z,w) = (2\pi i)^{-1} \oint_{|\zeta|=\sqrt{v}} (\zeta - z)^{-1} h(\zeta, u)\, \mathrm{d}\zeta$$

define a holomorphic function of (z,w) in the region $|z|^2 < v$. What must be further checked is that the boundary value of that holomorphic function at a point (z,w) such that $|z|^2 = v$ is equal to $h(z,u)$. The key to the latter property is that the tangential Cauchy–Riemann equations $\mathrm{L}h = 0$ precisely ensure that the function $z \to h(z,u)$ is the boundary value of a holomorphic function in the whole disk $|z| < v^{\frac{1}{2}}$.

There is a more recent, and shorter, proof of Lewy's result, based on the property, not difficult to establish (cf. [Baouendi–Treves, 1981]), that a continuous CR function on Σ_0 is the uniform limit, on compact subsets, of a sequence of holomorphic polynomials $P_j(z,w)$. By the maximum principle the polynomials $P_j(z,w)$, if they converge on the union of the circumferences $|z|^2 = v \leq R$, must also converge uniformly on the set $|z|^2 \leq v \leq R$, to a continuous function that is holomorphic in the interior $|z|^2 < v < R$.

The existence of analytic disks with boundaries on Σ_0 is also the key to the second major discovery of Hans Lewy—undoubtedly his most striking: that *the inhomogeneous tangential CR equation on Σ_0 is not locally solvable.* This means the following: given any open subset $\mathcal{O}$ of Σ_0 there is a $\mathcal{C}^\infty$ function f in Σ_0 such that no $\mathcal{C}^1$-function (and, as is now known, no distribution) g in $\mathcal{O}$ satisfies, in that set, the differential equation $\mathrm{L}g = f$. Once this is established, a fairly easy Baire category argument shows that the same is true for all $\mathcal{C}^\infty$ functions f except those that belong to a certain set of the second category (for the standard $\mathcal{C}^\infty$ topology; see [Lewy, 1957]).

There have been many proofs of Lewy's result since the original one. The proof in [Hörmander, 1960], based on Functional Analysis, is very different from Lewy's; it was the starting point of the general theory of local solvability for linear PDE. It has the advantage of directly proving that the Lewy equation does not have *distribution* solutions (for "most" smooth right-hand sides). That it does not have *hyperfunction* solutions was first shown by Schapira (see [Schapira, 1969]). Precise descriptions of the range of the Lewy operator were obtained at the microlocal level in [Sato–Kawai–Kashi–Wara, 1971] and at the local and global level in [Greiner–Kohn–Stein, 1975]). Later proofs (e.g., [Ehrenpreis, 1986, 1988]) relate the Lewy equation to Fourier Analysis on nilpotent Lie groups, or to various consequences of strong pseudoconvexity in several complex variable theory (existence of peak functions on the hyperquadric, Hartog's extension). Like all great mathematical discoveries, Lewy's example seems to have an inexhaustible supply of facets that can be explored.

We present here an elementary and self-contained proof, inspired by an argument of [Grushin, 1971]. As a matter of fact, our proof is close to Lewy's, and to the proof above, of the holomorphic extension of CR functions to the convex side. Like those it also uses analytic disks with boundary on Σ_0. It will prove the nonsolvability of the equation $\mathrm{L}g = f$ in an arbitrarily small open neighborhood $\mathcal{O}$ of the origin, for a certain smooth right-hand side f with compact support in $\mathcal{O}$. Nonsolvability at the origin suffices, as one can see by a variety of arguments: e.g., a "translation" in the Heisenberg group will extend the result to any other point; or by noting that the hyperquadric is biholomorphically equivalent, locally, to the 3-sphere, whose points are indistinguishable.

Let $g \in \mathcal{C}^1(\mathcal{O})$ be arbitrary (not CR) and consider the integral

$$G(u,v) = (2\pi i)^{-1} \oint_{|\zeta|=\sqrt{v}} g(\zeta,u)\,\mathrm{d}\zeta.$$

We have, if $v = r^2 > 0$,

$$\begin{aligned}(\partial/\partial v - i\partial/\partial u)G &= (2\pi)^{-1}\int_0^{2\pi} r^{-1}\left(\frac{1}{2}\partial/\partial r - ir\partial/\partial u\right)(g\zeta)\,\mathrm{d}\theta \\ &= (2\pi)^{-1}\int_0^{2\pi} r^{-1}\left[\frac{1}{2}(\partial/\partial r + ir^{-1}\partial/\partial\theta) - ir\partial/\partial u\right](g\zeta)\,\mathrm{d}\theta \\ &= (2\pi)^{-1}\int_0^{2\pi} r^{-1}(\partial/\partial\bar{\zeta} - i\zeta\partial/\partial u)(g\zeta)\,\mathrm{d}\theta = (2\pi)^{-1}\oint_{|\zeta|=\sqrt{v}} \mathrm{L}g\, e^{i\theta}\,\mathrm{d}\theta.\end{aligned}$$

Suppose

(1) $$\mathrm{L}g = f_0(|z|,u)/z,$$

with $f_0(\sqrt{v}, u) \in \mathcal{C}^\infty(\mathbb{R}^2)$ vanishing identically in the complement of the compact set $\mathrm{K}_\epsilon = \{(u, v) \in \mathbb{R}^2;\ u^2 + (\sqrt{v} - i\epsilon)^2 \le \frac{1}{4}\epsilon^2\}$, and $f_0(\sqrt{v}, u) > 0$ in the interior of K_ϵ (note that $f_0(|z|, u) \neq 0 \Rightarrow |z| > \frac{1}{2}\epsilon$). We get

$$(\partial/\partial v - i\partial/\partial u)G = f_0(\sqrt{v}, u)/\sqrt{v}.$$

In particular, G is a holomorphic function of w in $\mathbb{C} \setminus \mathrm{K}_\epsilon$. But by its very definition, $v = 0 \Rightarrow G = 0$. Unique continuation demands that $G \equiv 0$ in $\mathbb{C} \setminus \mathrm{K}_\epsilon$. In turn, this implies, thanks to Stokes' theorem,

$$0 = \oint_{\partial \mathrm{K}_\epsilon} G\, \mathrm{d}w = \iint_{\mathrm{K}_\epsilon} (\partial G/\partial \overline{w})\, \mathrm{d}\overline{w} \wedge \mathrm{d}w = -2 \iint f_0(r, u)\, \mathrm{d}r\, \mathrm{d}u$$

which is absurd since $f_0 > 0$ in the interior of its support. We have reached a contradiction and we conclude that Eq. (1) does not have any $\mathcal{C}^1$ solution.

Lewy's discovery of his unsolvable equation is one of the major events in the modern development of PDE theory. It has motivated the whole effort at characterizing those linear PDE with smooth coefficients that can be solved in distributions. The characterization of all solvable PDEs with simple real characteristics was completed in the early 1970s. The proof of the final result, equating local solvability with the so-called Property $(\mathcal{P})$, has served as a testing ground for the primary tools of Microlocal Analysis, Pseudodifferential and Fourier Integral Operators (for a thorough exposition see [Hörmander, 1983], Ch. XXVI). In a somewhat different direction the Lewy example and its generalizations (as in [Andreotti–Hill, 1976]) can be interpreted as stating that the cohomology of the tangential Cauchy–Riemann differential complex, associated with the CR structure of the manifold, is not finite in certain dimensions. In such a light Lewy's theorem provides us with a vital piece of information if we are to build a theory of differential complexes more general than the De Rham and Dolbeault ones.

A minor modification of the preceding proof enables us to answer, in the negative, the question of Hans Lewy referred to in the title and produces the central portion of the result in [Nirenberg, 1972]; namely, that there are smooth perturbations of the Lewy operator,

$$\mathrm{L}^{\#} = \mathrm{L} - A_0 \partial/\partial z - A_1 \partial/\partial u,$$

such that the differential of any solution h of the *homogeneous* equation

(2) $$\mathrm{L}^{\#} h = 0,$$

in a neighborhood of the origin, must vanish at the origin. Indeed, (2) can be reinterpreted as an inhomogeneous Lewy equation

(3) $$\mathrm{L}h = A_0 h_z + A_1 h_u.$$

Take then $A_0 = z^{-1} \sum_k a_{2k}(|z|, u)$, $A_1 = z^{-1} \sum_k a_{2k+1}(|z|, u)$ to be series that converge in $\mathcal{C}^\infty(\mathbb{R}^3)$, with $a_k \in \mathcal{C}^\infty(\mathbb{R}^2)$, $a_k \equiv 0$ in $\mathbb{C} \setminus \tilde{K}_k$, $a_k > 0$ in the interior of $\tilde{K}_k = \{(r, u) \in \mathbb{C}^2;\ (r^2, u) \in K_\epsilon\}$ for $\epsilon = \epsilon_k \searrow 0$; the numbers ϵ_k must also be such that the sets $\tilde{K}_k$ do not intersect one another. The argument in the proof of Lewy's result shows that, for all $k = 1, 2, \ldots,$

(4) $$\iint_{\tilde{K}_{2k}} a_{2k}(r, u) h_z^{\#}(r, u)\, \mathrm{d}r\, \mathrm{d}u = \iint_{\tilde{K}_{2k+1}} a_{2k+1}(r, u) h_u^{\#}(r, u)\, \mathrm{d}r\, \mathrm{d}u = 0$$

(with the notation $\phi^{\#}(r,u) = (2\pi)^{-1}\int_0^{2\pi}\phi(r,\theta,u)\,d\theta$). If either $h_z(0,0,0) \neq 0$ or $h_u(0,0,0) \neq 0$, at least one of the two equations (4) cannot be true for k sufficiently large. But if $h_z(0,0,0) = h_u(0,0,0) = 0$, then (3) $\Rightarrow h_{\bar{z}}(0,0,0) = 0$, i.e., $dh\big|_0 = 0$.

In the preceding construction, the coefficients A_0, A_1 vanish to infinite order at the origin. An adaptation of the Baire category argument in [Lewy, 1957] enables one to deduce the following from the existence of nonintegrable flat perturbations of L such as $L^{\#}$ ([Jacobowitz–Treves, 1981]). Call *aberrant* any complex vector field ϑ in $\mathbb{R}^3$ endowed with the property that, in any open subset $\mathcal{O}$ of $\mathbb{R}^3$, every $\mathcal{C}^{1+\sigma}$ ($\sigma > 0$) solution h in $\mathcal{O}$ of the homogeneous equation $\vartheta h = 0$ is constant. Then, in the space (isomorphic to $(\mathcal{C}^\infty(\mathbb{R}^3))^3$) of all smooth complex vector fields in $\mathbb{R}^3$, the set of the vector fields that are *not* aberrant is of the second category!

In the wake of Nirenberg's finding much effort has been expended to show that embedding in Euclidean complex space is feasible provided the strongly pseudoconvex "hypersurface" has dimension ≥ 5 (CR structures of the hypersurface type make sense only on manifolds of odd dimension). A proof of the result in dimension ≥ 9 was given in [Kuranishi, 1982]; the seven-dimensional case was settled in [Akahori, 1987]. These proofs are very much in the spirit of Spencer's program; they are based on solving a $\overline{\partial}_b$- Neumann problem akin to the $\overline{\partial}$-Neumann problem. A different, and shorter, proof of the same embedding theorems can be found in [Webster, 1989]. Webster's proof exploits the Henkin–Ramirez homotopy formulas. From the *global* standpoint, it should be mentioned that the sphere $\mathbf{S}^2$ equipped with the CR structure defined by the vector field $L_\epsilon = z\partial/\partial\overline{w} - w\partial/\partial\overline{z} + \epsilon(\overline{z}\partial/\partial w - \overline{w}\partial/\partial z)$, with $0 < \epsilon < \frac{1}{2}$, cannot be realized as a hypersurface in $\mathbb{C}^2$ ([Rossi, 1964]). But any compact strongly pseudoconvex manifold of dimension $2n-1 \geq 5$ can be embedded in $\mathbb{C}^n$ ([Boutet de Monvel, 1975]).

We come now to the third and last of the discoveries of Hans Lewy about CR functions to which this note is devoted (see [Lewy, 1977]). Lewy shares the credit for it with Pinčuk who proved it independently ([Pinčuk, 1975]). The Lewy–Pinčuk theorem has to do with CR diffeomorphisms of one strongly pseudoconvex, real-analytic hypersurface Σ onto another one, Σ', both contained in $\mathbb{C}^n$. For the sake of simplicity in the notation we shall limit ourselves to the case n = 2, but both the statement and its proof are the same in higher dimensions (the CR function denoted by f in what follows becomes a CR map when n $\geq$ 2). A map $(f,g)\colon \Sigma \to \mathbb{C}^2$, say of class $\mathcal{C}^1$, is said to be CR if both its components are CR functions. Suppose the image of Σ under the map (f,g) is contained in Σ'. Then to say that the map is CR is the same as saying that the push-forward of a tangential Cauchy–Riemann vector field L on Σ is a Cauchy–Riemann vector field L$'$ tangent to Σ', or that the pull-back of a holomorphic function in $\mathbb{C}^2$, or that of a CR function on Σ', is a CR function on Σ.

The property about to be established is purely local; it applies to the germs of the hypersurfaces at one of their points. We may as well assume that those points are the origin. By a rotation we may also assume that both tangent hyperplanes to Σ and to Σ' at the origin are the hyperplane $v = 0$. This means that defining equations of Σ and Σ' in a neighborhood of 0 can be taken to be, respectively,

$$v = \phi(z,\overline{z},u), \quad v = \psi(z,\overline{z},u),$$

with ϕ, ψ real-valued and ϕ, ψ, $d\phi$, $d\psi$ all vanishing at 0. The hypothesis is that both ϕ, ψ are of class $\mathcal{C}^\omega$. On the other hand, the functions f and g are CR functions in Σ, of class $\mathcal{C}^1$ (what really matters here are the germs of ϕ, ψ, f, g at 0). We must have

$$g - \overline{g} = 2i\psi\left(f, \overline{f}, \frac{1}{2}(g + \overline{g})\right). \tag{5}$$

We avail ourselves of the analyticity of ψ and of the fact that $d\psi\big|_0 = 0$, and apply the implicit function theorem with respect to $\overline{g}$, getting

$$\overline{g} = \mathcal{Q}(f, \overline{f}, g) \tag{6}$$

with $\mathcal{Q}$ holomorphic in some neighborhood of 0 in $\mathbb{C}^3$. Now let

$$\mathrm{L} = \partial/\partial\overline{z} - i(1 + i\phi_u)^{-1}\phi_{\overline{z}}\partial/\partial u$$

denote the tangential Cauchy–Riemann vector field on Σ. Note in passing that $\mathrm{L}\big|_0 = \partial/\partial\overline{z}$. By hypothesis we have $\mathrm{L}f = \mathrm{L}g = 0$. Part of the hypothesis is also that (f, g) is a diffeomorphism. Since (6) demands that dg be real at 0 we must have $\mathrm{L}\overline{f} \neq 0$. By differentiating both sides of (6) we get

$$\mathrm{L}\overline{g}/\mathrm{L}\overline{f} = \mathcal{Q}_{\overline{z}}(f, \overline{f}, g). \tag{7}$$

Note that (6) is equivalent to $g = \overline{\mathcal{Q}}(\overline{f}, f, \overline{g})$ and therefore (7) is equivalent to

$$\mathrm{L}\overline{g}/\mathrm{L}\overline{f} = \mathcal{Q}_{\overline{z}}(f, \overline{f}, \overline{\mathcal{Q}}(\overline{f}, f, \overline{g})). \tag{8}$$

So far the strong pseudoconvexity hypothesis has not been exploited. In fact we shall avail ourselves only of the pseudoconvexity of Σ' (but if Σ is CR isomorphic to Σ' it will also be strongly pseudoconvex). Let

$$\mathrm{L}' = \partial/\partial\overline{z} - i(1 + i\psi_u)^{-1}\psi_{\overline{z}}\partial/\partial u$$

denote the tangential CR vector field on Σ'. The hypothesis that the commutator $[\mathrm{L}', \overline{\mathrm{L}}']$ is linearly independent, at 0, of L' and $\overline{\mathrm{L}}'$, is equivalent to the fact that $\psi_{z\overline{z}}(0,0,0) \neq 0$, in turn equivalent to $\mathcal{Q}_{z\overline{z}}(0,0,0) \neq 0$. The latter allows us to apply the implicit function theorem to (8) and solve with respect to f:

$$f = \mathcal{P}(\overline{f}, \overline{g}, \mathrm{L}\overline{g}/\mathrm{L}\overline{f}),$$

with $\mathcal{P}$ holomorphic in a neighborhood of 0 in $\mathbb{C}^3$. By the Lewy extension theorem we know that f and g extend holomorphically to the *convex* side of Σ. As a consequence $\overline{f}$ and $\overline{g}$ extend holomorphically to the *concave* side of Σ, and so do $\mathrm{L}\overline{f}$ and $\mathrm{L}\overline{g}$, since the coefficients of L extend to both sides. We conclude that $\mathcal{P}(\overline{f}, \overline{g}, \mathrm{L}\overline{g}/\mathrm{L}\overline{f})$ extends holomorphically to the concave side of Σ. But since it is equal to f it also extends to the opposite side, and because f is of class $\mathcal{C}^1$, applying the Cauchy–Riemann operators $\partial/\partial\overline{z}$ and $\partial/\partial\overline{w}$ to its extensions leads to the conclusion that f is the restriction to Σ of a holomorphic function in a full neighborhood of 0. But then $g = \overline{\mathcal{Q}}(\overline{f}, f, \overline{g})$ extends to the concave side, hence also to a full neighborhood of 0. We have proved the Lewy–Pinčuk theorem: *The germ of a CR diffeomorphism of class $\mathcal{C}^1$ of the germ of a strongly pseudoconvex real-analytic hypersurface in $\mathbb{C}^2$ onto the germ of another such hypersurface is real-analytic.*

The Lewy–Pinčuk theorem has been generalized in a variety of directions. We content ourselves with mentioning a few and refer for more details to [Fostnerič, 1989]. A $\mathcal{C}^\infty$ version can be found in [Nirenberg–Webster–Yang, 1980]. Extension to $\mathbb{C}^n$ of the analytic result, under more general hypotheses than strong pseudoconvexity, is carried out in [Baouendi–Jacobowitz–Treves, 1985]. The generalization of the Lewy–Pinčuk theorem has been pushed the farthest when the hypersurfaces are embedded in $\mathbb{C}^2$, as shown in [Baouendi–Rothschild, 1989]: *in order that any proper CR map* (f, g), *assumed to be of class* $\mathcal{C}^\infty$ *and such that* $\det \partial(f,g)/\partial(z,w)$ *does not vanish to infinite order at the origin, of a* $\mathcal{C}^\omega$ *hypersurface* $\Sigma \subset \mathbb{C}^2$ *into another one,* Σ', *be analytic, it is necessary and sufficient that* Σ *not be Levi flat* (i.e., biholomorphically equivalent to the hyperplane $v = 0$).

I would like to conclude on a personal note. In the summer of 1958 I arrived in Berkeley as a beginning assistant professor, having just defended my Ph.D. thesis at the University of Paris. A couple of years earlier I had worked on a project suggested by my thesis supervisor, Laurent Schwartz: to prove that every linear PDE, with $\mathcal{C}^\infty$ coefficients that did not all vanish simultaneously at any point and with any $\mathcal{C}^\infty$ right-hand side, could be solved in distributions. Thinking about that problem turned out to be rather discouraging (after a totally unfruitful year I changed problems and managed to complete my Ph.D. work). Soon after my arrival in Berkeley I met Hans Lewy and had the opportunity of hearing him lecture. But I do not believe that, during the two years of my stay at Berkeley, he and I ever had an in-depth mathematical conversation. We had conversations in French, at the small gatherings organized by M. Loève at the Faculty Club. We got along well and, of course, I had much admiration and respect for Lewy (not least for what I had heard of his courage during the McCarthy era). Not once did we discuss the local solvability of linear partial differential equations or the holomorphic extension of their solutions. This is the more strange, and perhaps regrettable, considering that I had been preoccupied by some of those same questions in the recent past, and that the near totality of my research, for the next thirty years and more, would revolve around them.

I saw Hans Lewy for the last time during a visit I made to Berkeley in May 1988. I was amazed, and very happy, to find the 83 year old gentleman in such fantastic shape, mentally and physically. Not for one moment did I suspect that he was gravely ill. It was heartwarming to see again the fine features, the youthful smile. The subtle curtain that had hung between us, in matters mathematical, seemed finally to have lifted. He kept coming up with questions about CR functions, good questions, of course, but a bit out of the beaten path, as his questions sometimes tended to be. Retropsectively I am glad that, on those last occasions, I directed some questions to him, not solely about math, but also about more mundane matters: about his last few months in Germany, on the eve of the accession of power of Hitler, and about his coming to America. He reminisced willingly, spoke of the antisemitism, then widespread among German students, of Teichmüller attending his course with the Nazi insignia on his lapel. He described his arrival in the U.S. and how, after a short period of unemployment, he had gotten a job offer on the West Coast and driven across the continent to California, to spend the rest of his life there.

Back home from Berkeley, I sent him photocopies of some sections of the book I was writing, sections devoted to the special CR structures that I had decided, with every justification, I believe, to name after him. He wrote back that he could not understand the terminology. Shortly thereafter, I came upon the few lines in the *New York Times* that reported his death and alluded to some of his contributions; among these, explicitly mentioned, were supercomputers. In the midst of the sadness there was the certainty that, in one form or another, what Hans Lewy had found and created would last a long time.

REFERENCES

Akahori, T., A new approach to the local embedding theorem of CR structures for $n \geq 4$ (the local solvability for the operator $\overline{\partial}_b$ in the abstract sense), *Memoirs of the A.M.S.* **67**, No. 366, May, 1987.

Andreotti, A. and Hill, C. D., E. E. Levy convexity and the Hans Lewy problem, I and II, *Ann. Scuola Norm. Sup. Pisa Sci. Fis. Mat.* **26** (1972), 325–363, 747–806.

Baouendi, M. S. and Rothschild, L. P., A general reflection principle in $\mathbb{C}^2$, *J. Funct. Anal.*

Baouendi, M. S. and Treves, F., A property of the functions and distributions annihilated by a locally integrable system of complex vector fields, *Ann. of Math.* **113** (1981), 387–421.

Boutet de Monvel, L., *Intégration des équations de Cauchy–Riemann induites formelles*, Sémin. Goulouic–Lions–Schwartz 1974–1975, Exposé IX.

Dolbeault, P., Formes différentielles et cohomologie sur une variété analytique complexe, I, *Ann. Math.* **64** (1956), 83–130; *Ann. Math.* **65** (1957), 282–330.

Ehrenpreis, L., Lewy's operator and its ramifications, *J. Funct. Analysis* **68** (1986), 329–366.

Ehrenpreis, L., *Lewy Unsolvability and Several Complex Variables*, 1988.

Folland, G. B. and Kohn, J. J., *The Neumann Problem for the Cauchy-Riemann Complex*, Annals of Math. Studies **75** (1972), Princeton University Press.

Folland, G. B. and Stein, E. M., Estimates for the $\overline{\partial}_b$-complex and analysis on the Heisenberg group, *Comm. Pure Applied Math.* **27** (1974), 429–522.

Forstnerič, F., Proper holomorphic mappings: A survey, Preprint Series, Dept. Math., University E. K. Ljubljana **27** (1989), 1–47.

Greiner, P. C., Kohn, J. J., and Stein, E. M., Necessary and sufficient conditions for solvability of the Lewy equation, *Proceedings N.A.S.* **72** (1975), 3287–3289.

Greenfield, S. T., Cauchy–Riemann equations in several variables, *Ann. Scuola Norm Sup. Pisa, Sci. Fis. Mat.* **22** (1968), 275–314.

Grushin, V. V., A certain example of a differential equation without solutions, *Mat. Zametki* **10** (1971), 125–128. (English translation in *Math. Notes* **10** (1971), 499–501.)

Hörmander, L., Differential operators of principal type, *Math. Ann.* **140** (1960), 124–146.

Hörmander, L., *The Analysis of Linear Partial Differential Operators*, Springer-Verlag, Berlin, 1983.

Jacobowitz, H., *An Introduction to CR Structures*, Math. Surveys and Monographs, Amer. Math. Soc., Providence, RI, 1990.

Jacobowitz, H. and Treves, F., Nowhere solvable homogeneous partial differential equations, *Bull. Amer. Math. Soc.* **8** (1983), 467–469.

Jerison, D. and Lee, J. M., Intrinsic CR normal coordinates and the CR Yamabe problem, *J. Diff. Geometry*, 1989.

Kohn, J. J. and Rossi, H., On the extension of holomorphic functions from the boundary of a complex manifold, *Ann. of Math.* **81** (1965), 451–472.

Kuranishi, M., Strongly pseudoconvex CR structures over small balls, I, *Ann. of Math.* **115** (1982), 451–500; II, *ibid.* **116** (1982), 1–64; III, *ibid.* **116** (1982), 249–330.

Levi, E., Studii sui punti singolari essenziali delle funzioni analitiche di due o più variabili complesse, *Ann. Mat. Pura Ed. Appl.* **17** (1910), 61–87.

Lewy, H., On the local character of the solution of an atypical differential equation in three variables and a related problem for regular functions of two complex variables, *Ann. of Math.* **64** (1956), 514–522.

Lewy, H., An example of a smooth linear partial differential equation without solution, *Ann. of Math.* **66** (1957), 155–158.

Lewy, H., On the boundary behavior of holomorphic mappings, *Accad. Nazion. Lincei* **35** (1977), 1–5.

Nirenberg, L., On a question of Hans Lewy, *Russian Math. Surveys* **29** (1974), 251–262.

Nirenberg, L., Webster, S., and Yang, P., Local boundary regularity of holomorphic mappings, *Comm. Pure Applied Math.* **33** (1980), 305–338.

Pinčuk, S. I., On the analytic continuation of biholomorphic mappings, (Russian) *Mat. Sb.* **98**(18) (1975), 416–435. (English transl. in *Math. USSR Sb.* **27** (1975), 375–392.)

Rossi, H., Attaching analytic spaces to an analytic space along a pseudoconcave boundary, *Proc. Conf. Complex Analysis Minneapolis, 1964*, Springer, New York, 1965, pp. 242–256.

Sato, M., Kawai, T., and Kashiwara, M., Microfuntions and pseudo-differential equations, in *Hyperfunction and Pseudodifferential Equations*, Springer Lecture Notes No. 187, 1971.

Schapira, P., Solutions hyperfunctions des équation aux dérivées partielles du premier ordre, *Bull. Soc. Math. France* **97** (1969), 665–667.

Tanaka, N., On pseudo-conformal geometry of hypersurfaces of the space of n complex variables, *J. Math. Soc. Japan* **14** (1962), 397–429.

Tanaka, N., On non-degenerate real hypersurfaces, graded Lie algebras and Cartan connections, *Japanese J. Math.* **2** (1976), 131–190.

Webster, S. M., On the proof of Kuranishi's embedding theorem, *Ann. Inst. Poincaré* **6** (1989b), 183–207.

Department of Mathematics
Rutgers University
New Brunswick, NJ 08903

CRISES IN MATHEMATICS

HANS LEWY

Your Magnificence, Colleagues of the Science Faculty, Ladies and Gentlemen,

I am very grateful to you for the honour you have conferred to me.

The title of my lecture is "Crises in Mathematics". I will not define the word "crises"; I hope to make clear what is the point in question.

Our century has seen scientific discoveries in abundance, and in many cases mathematics provided essential assistance. Often these discoveries have overthrown older notions and conceptions and replaced them by new notions, expressing them in a mathematical way in order to formulte them as incontestable as possible. For, one may think, mathematics expresses eternal truths since it only states the logical consequences of a few principles, and these principles or axioms are of such a simplicity and immediacy that to deny them would only lead to fruitless sophistry.

But is mathematics truly always the "resting pole amidst the whirl of phenomena", or have there been bitter differences of opinion also in mathematics? Do they exist even today? Are there, possibly unspoken, opinions underlying our mathematical thinking, views which cannot be defended by logic alone?

The first attempt towards an axiomatization of mathematics goes back to the ancient Greeks and is presented in Euclid's books, which were thought to be basic for almost two thousand years. Why? It is known that many mathematicians of the previous generations had studied in Babylon where they had learned their algebra. In Babylon one considered as numbers the so-called natural numbers: $1, 2, 3$, etc., and the fractions of these as quantities which we nowadays denote as positive rational numbers. But then came the discovery that there are geometric quantities which are not fractions of integers. Namely, in a rectangular triangle with two sides of length one, the third side has a length which is not a fraction of integers. For this fact we say today that the square root of two is irrational. How exciting this discovery must have been for the natural philosophers of that time who had learned to consider the natural numbers as a very successful foundation for all contemporary descriptions of nature! The echo of this crisis is reflected in Plato's utterance that a human being who is not shaken in his heart by learning that the square root of two is not a rational number resembles a stupid beast. And undoubtedly the geometric expression of all axioms of Euclid's books can be attributed to the desire to give these axioms a generality which could not be achieved by the algebra of the time.

Since Descartes the modern conception takes the opposite way and reduces all geometric notions to algebraic ones.

But the endeavour to formulate a sufficiently general concept of quantities algebraically requires the notion of the so-called actual infinity, thereby generating a deeper rooted crisis the offshoots of which have had an influence up to the recent past. Today we learn in school that each quantity is an infinite decimal fraction, say 1.2357... ad infinitum. This concept of quantities requires the specification of infinitely many numbers from 0 to 9. Such a definition would have been unbearable for Euclid. If he speaks of the infinity of the set of all prime numbers, that is, of all those numbers having no divisors except themselves and the unit, he expresses and proves this by: *There exist more prime numbers than in any given set.*

The need to include infinity into mathematics became of immediate interest only with the invention of the infinitesimal calculus. The success of the infinitesimal calculus and its overwhelming fertility in solving old problems and creating new questions led in the 18th century to the well-known slogan: "En avant, toujours en avant, et la foi vous viendra." "Forwards, always forwards, and belief will follow." But was it really easy to believe? For example, the infinite series $1 - 1/2 + 1/3 - 1/4 + \ldots$ converges to natural logarithm of 2. However, if one orders the terms of this series in a different way, then one can obtain any positive or negative number as sum, or we can let the series diverge, just as we wish. This example must have made a devastating impression upon the contemporaries. In fact, the foundation of the infinitesimal calculus was even in the 19th century a topic of vivid discussion, and this not only in mathematical circles. So for instance Karl Marx wrote an essay about the foundation of the infinitesimal calculus, and his concept of this field has been taught in the schools of the Soviet Union and of China up to our time. I myself witnessed the lecture of a Russian mathematician entitled "Marx's foundation of the calculus" held in the course of the International Conference of Mathematics in Zurich, 1932. The inclusion of this lecture of 45 minutes in the program of the meeting was a precondition for the participation of the Russian colleagues in the Congress. The speaker characterized Weierstrass's $\varepsilon - \delta$-definition of convergence as a static and bourgeois concept, whereas Marx's definition was seen to be something fluent and vivid.

Weierstrass's concept of convergence rested on preparatory work by Cauchy and Abel and allowed a clarification of all seeming contradictions of the infinitesimal calculus. But this long desired goal had scarcely been attained when in the second half of the 19th century there arose a completely new conception of infinity in the set theory of Georg Cantor. Cantor introduced the notion of denumerability of a set and showed that there are infinite sets which are not denumerable, i.e., sets where it is impossible to furnish each of its elements with a natural number in such a way that two different elements obtain different numbers. The simplest example is the set of all decimals between zero and one. For us, having grown up with the notions of denumerability and nondenumerability of infinite sets, it is hard to understand any longer why so many outstanding mathematicians of those days fought with such hostility against these new concepts. Weierstrass's point of view is interesting: The famous Göttingen mathematician David Hilbert, in that time a young man, visited Weierstrass in Berlin and asked him for his opinion about the new set theory. Weierstrass told him that he would accept this theory as soon as it would be able to prove the existence of transcendental numbers, that is, the existence of numbers which do not satisfy any algebraic equation with integers as coefficients. However, this was proved by Cantor in the following year, actually by verifying a strong result: The set of transcendental numbers is not only infinite, but even nondenumerable. Further successful applications of set theory to other branches of mathematics which previously seemed to have nothing to do with it secured a firm place for set theory in mathematics. But with the new attitude against the actual infinity new clouds arose in the firmament of mathematics. Certain paradoxes such as the set of all sets containing themselves showed that even the axioms of logic have to be revised if infinite sets are taken into consideration. In particular the so-called axiom of choice by Zermelo in its unrestricted application to infinite sets made many mathematicians feel ill at ease.

A radical cure for all doubts was suggested by so-called intuitionism. Essentially, intuitionism expelled all statements from mathematics containing notions which cannot be defined in a finite number of steps. Particularly, for intuitionists the principle of *tertium non datur* is not applicable to infinite sets. For example, let us take the statement: In the decimal representation of $\pi = 3.1415\ldots$ there is an uninterrupted sequence of ten Nines. The intuitionist says: As long as there is no finite proof of this statement or of its reverse, one is not allowed to say that this statement is either correct or false. This was the example cited by the well-known topologist L.E.J. Brouwer in a great lecture at Göttingen, about 1930. According to Brouwer most of the mathematics taught at the time was false. In the discussion following his lecture he was asked as to whether this statement should also be extended to his own papers that had made him famous. Yes, he answered, it's all false.

The majority of mathematicians rejected intuitionism, mainly for the reason that intuitionism was not only extensively restricting the area of allowed questions, but would also make all proofs uncommonly complicated, even in those cases where it led to the same results as ordinary mathematics. Still one has to admit that the problem of consistency within mathematics is, in the broadest sense, unsolved. Intuitionists would cripple mathematics in order to make sure that there is no contradiction. In any case, logicians had step by step erected a building of mathematics, beginning with the natural numbers, when in the thirties Gödel's discovery struck like a lightening bolt that, expressed briefly, for any given finite system of axioms there is always an assertion formulable by means of the notions of this system whose correctness cannot be decided by these axioms alone, and also that the consistency cannot be proved with this sytem of axioms alone.

Now I turn to another crisis in which present-day mathematics finds itself. We older mathematicians and the majority of our younger colleagues believe it to be absolutely unavoidable to convince ourselves personally of the correctness of the results of others if we want to use them. But today there exist published papers based on results whose examination would require the study of more than 5000 printed pages, a task which is admittedly beyond human capability. In all experimental sciences it is of course impossible to check personally the validity of all experiments preceding one's own research and forming its foundation. But should we mathematicians, so far having been proud to recognize no other authority than our own intellect, also resign ourselves to accept upon trust the printed words as the truth? And here is another source of discomfort—the reliance placed on the computer. A celebrated, more than a hundred-year-old problem is the so-called four-colour-problem. This is the question of whether it is always possible to paint a globe with many countries by only four colours in such a way that any two countries with a common border line have different colours. Two American mathematicians have reduced this problem to another one, approachable by the computer, and after the work of many hours the computer had decided: Yes, it is possible. But do I have to accept it thereby to be proved that four colours suffice?

What is a mathematical proof? The answer to this question contains a personal element, even if one rejects placing trust in the computer and the acceptance of unchecked results. I know of a paper which was published prior to the invention of the computer, and whose aim was to prove a certain conjecture by Leonhard Euler. This paper makes a distinction of several hundred cases and investigates each case separately, thereby arriving at the desired result. Even if I had succeeded in examining all these cases, I would not take this as a proof of the correctness of this assertion, but only as a confirmation of my inability to discover a mistake. It seems to me that one has to ask more of a proof, namely, an understanding of the connections which underlie the truth of an assertion.

The nonmathematician envies us for the incontestability of our results but the questions of conscience may well escape his notice that, as in the past, so also today have contributed to a crisis in mathematics.

(This lecture was held at Bonn University on the occasion of the bestowal of the honorary doctorate of the Science Faculty to Hans Lewy, November 28, 1986.)

Über einen Ansatz zur numerischen Lösung von Randwertproblemen.

Von

Hans Lewy.

Aus den Nachrichten der Gesellschaft der Wissenschaften zu Göttingen.
Mathematisch-physikalische Klasse. 1925.

Über einen Ansatz zur numerischen Lösung von Randwertproblemen.

Von

Hans Lewy in Göttingen.

Vorgelegt von R. Courant in der Sitzung vom 18. Dezember 1925.

In einer Note in den Göttinger Nachrichten [1]) hat Herr Courant vor kurzem den Nachweis erbracht, daß die Lösung der Randwertaufgabe der Potentialtheorie gleichmäßig approximiert wird durch die Lösungen einer entsprechenden Aufgabe, in der die Ableitungen der auftretenden Funktion durch die Differenzenquotienten von Funktionswerten in diskreten Punkten ersetzt worden sind. Herr Courant hat diesen Gedanken weiter verfolgt für aus sehr allgemeinen Variationsproblemen entspringende lineare Differentialgleichungen, und es gelang ihm, die entsprechenden Randwertprobleme zu lösen. Ich habe nun, ermutigt durch den Erfolg dieser Differenzenmethode und angeregt durch persönliche Gespräche, in meiner Doktordissertation den Nachweis erbracht, daß in der Tat unter gewissen sehr einfachen und wenig einschränkenden Voraussetzungen die Lösungen der entsprechenden Differenzenaufgaben gegen die Lösung des Variationsproblems

$$(1) \qquad \int_0^1 F(x, y, y')\,dx = \text{Min.}, \quad y(0) = a_0, \quad y(1) = a_1$$

konvergieren, und daß sogar die ersten Differenzenquotienten gleichmäßig die entsprechenden Differentialquotienten approximieren. Es wird also ausreichen, wenn man zur numerischen Berechnung der Lösung von (1) nur die „Polygonlösungen“ der Aufgabe

$$(2) \qquad l \sum_{\nu=1}^{n} F\left(x_\nu, y_\nu, \frac{y_\nu - y_{\nu-1}}{l}\right) = \text{Min.}$$

berechnet. Hierbei ist $l = \frac{1}{n}$, $x_\nu = \frac{\nu}{n}$; $y_0 = a_0$, $y_n = a_1$ und n

1) 23. 10. 1925.

selbst ein ganzzahliger positiver Parameter, der gegen ∞ streben wird.

Der Ausgangspunkt der numerischen Methode, die ich nun auseinandersetzen will, ist, wie ich nachträglich bemerkt habe, mit andrer Tendenz von Hadamard in seinem „Mémoire sur les plaques encastrés“ [1]) benutzt worden, und ist folgender:

Ich frage nach der bestmöglichen Verbesserung von vorgelegten Werten $\overset{0}{y}_\nu$ $(\nu = 1, 2, \ldots, n-1)$ im Sinne unseres Differenzenproblemes (2). Dabei soll die Verbesserung selbst infinitesimal klein und die Summe der Quadrate der Verbesserungen eine feste Zahl sein. Das ergibt zur Bestimmung der Korrektion $\varepsilon\overset{0}{\eta}_\nu$ das Problem:

$$l\sum_{\nu=1}^{n} F\left(x_\nu, \overset{0}{y}_\nu, \frac{\overset{0}{y}_\nu - \overset{0}{y}_{\nu-1}}{l}\right) - l\sum_{\nu=1}^{n} F\left(x_\nu, \overset{0}{y}_\nu + \varepsilon\overset{0}{\eta}_\nu, \frac{\overset{0}{y}_\nu - \overset{0}{y}_{\nu-1}}{l} + \varepsilon\frac{\overset{0}{\eta}_\nu - \overset{0}{\eta}_{\nu-1}}{l}\right) = \text{Max.},$$

$$\sum_{1}^{n-1} \overset{0}{\eta}{}_\nu^2 = c = \text{konst.}$$

für „sehr kleines“ ε. Die erste Forderung geht durch Ausrechnung über in

$$-\varepsilon\sum_{\nu=1}^{n-1} \overset{0}{\eta}_\nu\left(l\frac{\partial\overset{0}{F}_\nu}{\partial y} + \frac{\partial\overset{0}{F}_\nu}{\partial y'} - \frac{\partial\overset{0}{F}_{\nu-1}}{\partial y'}\right) = \text{Max.}$$

Hier bedeutet $\frac{\partial\overset{0}{F}_\nu}{\partial y}$ nichts anderes als $\frac{\partial F}{\partial y}\left(x_\nu, \overset{0}{y}_\nu, \frac{\overset{0}{y}_\nu - \overset{0}{y}_{\nu-1}}{l}\right)$ etc. etc. Halten wir ε fest und setzen, worauf es nicht ankommt, die Konstante

$$\sum_{\nu=1}^{n-1} \overset{0}{\eta}{}_\nu^2 = c = \sum_{\nu=1}^{n-1} E_\nu^2(\overset{0}{F}) = \sum_{\nu=1}^{n-1}\left(l\frac{\partial\overset{0}{F}_\nu}{\partial y} + \frac{\partial\overset{0}{F}_\nu}{\partial y'} - \frac{\partial\overset{0}{F}_{\nu-1}}{\partial y'}\right)^2$$

$$\left(E_\nu(\overset{0}{F}) = l\frac{\partial\overset{0}{F}_\nu}{\partial y} + \frac{\partial\overset{0}{F}_\nu}{\partial y'} - \frac{\partial\overset{0}{F}_{\nu-1}}{\partial y'}\right)$$

so folgt aus der Schwarzschen Ungleichung, daß das Maximum von $|\sum \overset{0}{\eta}_\nu E_\nu(\overset{0}{F})|^2$ angenommen wird und gleich $\sum \overset{0}{\eta}{}^2 \sum E_\nu^2(\overset{0}{F})$ ist, falls $\overset{0}{\eta}_\nu$ proportional zu $E_\nu(\overset{0}{F})$ ist. Wir haben durch unsere Bestimmung von c bewirkt, daß gerade

$$\overset{0}{\eta}_\nu = E_\nu(\overset{0}{F})$$

ist.

Wir schreiten nunmehr zur Bestimmung eines passenden Wertes des vorhin erwähnten ε. Da wir möglichst schnell unsere Summe

1) Paris 1908; Mémoires de divers savants.

(2) erniedrigen wollen, so wird es ratsam sein, dieses ε so groß zu wählen, daß man gerade die stärkste Abnahme der Summe (2) erhält, wenn man $\overset{0}{y}_\nu$ durch $\overset{0}{y}_\nu + \varepsilon \overset{0}{\eta}_\nu = \overset{0}{y} + \varepsilon E_\nu(\overset{0}{F})$ ersetzt; d. h. man bestimme $\varepsilon = \overset{0}{\varepsilon}$ aus der Forderung, daß die Verbesserung für $\varepsilon = \overset{0}{\varepsilon}$ stationär ist; das ergibt die Bestimmungsgleichung für $\overset{0}{\varepsilon}$

$$\sum_{\nu=1}^{n-1} E_\nu(\overset{0}{F}) E_\nu(\overset{1}{F}) = 0;$$

$$\overset{1}{F}_\nu = F\left(x_\nu, \overset{0}{y}_\nu + \overset{0}{\varepsilon}\overset{0}{\eta}_\nu, \frac{\overset{0}{y}_\nu - \overset{0}{y}_{\nu+1}}{l} + \overset{0}{\varepsilon}\frac{\overset{0}{\eta}_\nu - \overset{0}{\eta}_{\nu-1}}{l}\right).$$

Über das Vorzeichen von $\overset{0}{\varepsilon}$ läßt sich im Voraus einsehen, daß es negativ ist; eine Folge der Definition von $\overset{0}{\varepsilon}$. Setzt man voraus, daß die Summe (2) gegen ∞ strebt, wenn mindestens ein $|y_\nu|$ über alle Grenzen wächst, so folgt, daß $\overset{0}{\varepsilon}$ einen endlichen Wert besitzt, sobald mindestens ein $E_\nu(\overset{0}{F}) \neq 0$ ist. Man setze nun

$$\overset{0}{y}_\nu + \overset{0}{\varepsilon}\overset{0}{\eta}_\nu = \overset{1}{y}_\nu$$

und wende auf $\overset{1}{y}_\nu$ dasselbe Verfahren wie vorher auf $\overset{0}{y}_\nu$ an. D. h. man bestimme $E_\nu(\overset{1}{F}) = \overset{1}{\eta}_\nu$ und $\overset{1}{\varepsilon}$ und setze $\overset{2}{y}_\nu = \overset{1}{y}_\nu + \overset{1}{\varepsilon}\overset{1}{\eta}_\nu$. Auf $\overset{2}{y}_\nu$ wende man dasselbe Verfahren an u. s. f. Wir behaupten, daß das Verfahren die Werte von $\overset{m}{y}_\nu$ gegen die Lösung der Eulerschen Differenzengleichung

$$E_\nu(F) = 0; \quad y_0 = a_0; \quad y_n = a_1$$

konvergieren läßt, wenn m gegen ∞ strebt. Nach der gemachten Voraussetzung über das Verhalten der Summe (2) bei großen $|y_\nu|$ sind alle $|y_\nu|$ beschränkt. Es existiert also eine Häufungspolygonfunktion h_ν. Wenn für diese nicht die Beziehung

$$E_\nu(F(h)) = l\frac{\partial F}{\partial y}\left(x_\nu, h_\nu, \frac{h_\nu - h_{\nu-1}}{l}\right) + \frac{\partial F}{\partial y'}(\cdots)_\nu - \frac{\partial F}{\partial y'}(\cdots)_{\nu-1} = 0$$

bestände, so gäbe es zwei endliche Größen $\varkappa > 0$ und $\tau > 0$, sodaß die Verbesserung $\bar{h}_\nu = h_\nu - \varkappa E_\nu(F(h))$, die durch unser Verfahren geliefert wird, die Summe (2) um mehr als $\tau.\varkappa \sum E_\nu^2(F(h))$ verkleinert. Für eine h_ν hinreichend benachbarte Polygonfunktion h'_ν gibt es entsprechende τ' und $\varkappa'$, die gegen τ und $\varkappa$ konvergieren, wenn h'_ν gegen h_ν strebt. Es konvergiert aber

$$l\sum F\left(x_\nu, h'_\nu, \frac{h'_\nu - h'_{\nu-1}}{l}\right) \text{ gegen } l\sum F\left(x_\nu, h_\nu, \frac{h_\nu - h_{\nu-1}}{l}\right)$$

und

$$\sum E_\nu^2(F(h')) \text{ gegen } \sum E_\nu^2(F(h)).$$

Das bedeutet einen Widerspruch, denn die Verbesserung von (2) kann nicht gleichzeitig gegen 0 und gegen eine positive Zahl konvergieren. Unser Verfahren konvergiert also gegen die Lösung der Eulerschen Differenzengleichung

$$E_\nu(F) = 0.$$

Die Größen $E_\nu(F)$ liefern — wenigstens prinzipiell — eine Abschätzung, wie weit unsere m-te Näherung $\overset{m}{y}_\nu$ von dem wahren Werte h_ν noch entfernt ist.

Selbstverständlich ist es nicht etwa nötig, den vorhin mit $\overset{0}{\varepsilon}, \overset{1}{\varepsilon}, \ldots$ bezeichneten Proportionalitätsfaktor der Verbesserung genau in der angegebenen Weise zu wählen. Es wird z. B. genügen, ihn seinem Betrage nach nicht größer als $|\overset{0}{\varepsilon}|, |\overset{1}{\varepsilon}|, \ldots$ und nicht kleiner als $\left|\frac{\overset{0}{\varepsilon}}{2}\right|, \left|\frac{\overset{1}{\varepsilon}}{2}\right|, \ldots$ zu machen.

Hat man nun angenähert das lösende Polygon mit $n+1$ Ecken bestimmt, so kann man zum lösenden Polygon mit $2n+1$ Ecken aufsteigen, indem man die Polygonlösung mit $n+1$ Ecken als erste Näherung des lösenden Polygons mit $2n+1$ Ecken auffaßt.

Die vorhergehenden Angaben stellen die ersten rohen Ansätze eines numerischen Verfahrens dar, das nicht bloß auf Probleme der Form (1) und verwandte beschränkt zu sein scheint. Ich bin mir bewußt, daß seine Vorzüge und Fehler erst durch die Praxis werden erschlossen werden können; darum halte ich es auch noch für verfrüht, genauere Angaben und eventuelle Kunstgriffe mitzuteilen, die die Rechnung abzukürzen und die Schnelligkeit der Konvergenz zu erhöhen geeignet sein könnten.

Nachrichten der Gesellschaft der Wissenschaften zu Göttingen
1927, pp. 178–186

Über den analytischen Charakter der Lösungen elliptischer Differentialgleichungen.

Von

Hans Lewy in Göttingen.

Vorgelegt von R. Courant in der Sitzung vom 28. Oktober 1927.

Im folgenden zeige ich an dem speziellen Beispiel der Differentialgleichung

$$(1) \qquad \Delta u = \frac{\partial^2 u}{\partial x^2} + \frac{\partial^2 u}{\partial y^2} = f(x, y, u, u_x, u_y)$$

für die Funktion $u(x, y)$ der Variablen x und y, daß — unter gewissen Differenzierbarkeits-Voraussetzungen — u eine analytische Funktion von x und y ist, wenn f analytisch von seinen Argumenten abhängt. Einen allgemeinen Beweis, daß die Lösung einer beliebigen elliptischen Differentialgleichung beliebiger Ordnung in zwei Variablen analytisch von diesen abhängt, falls die Differentialgleichung selbst analytisch ist, werde ich in kurzem an anderem Orte ausführen.

Bisher liegen in der Literatur für Gleichungen zweiter Ordnung zwei Beweise vor, von denen der erste von S. Bernstein (Math. Ann. 1904), der zweite von Gevrey (Ann. de l'École normale 1918) stammt. Der Bernsteinsche Beweis ist kürzlich von Radó neu dargestellt worden (Math. Zeitschr. 25). Am Anfang gibt der Verfasser einen kritischen Überblick über die genannten Arbeiten, auf den ich hier verweisen möchte.

Während nun die zitierten Arbeiten das eine gemeinsam haben, daß sie nämlich Reihenentwicklungen für die zu betrachtenden Funktionen herstellen und die Koeffizienten, die dabei auftreten, abschätzen, geht der vorliegende Beweis einen grundsätzlich anderen Weg. Seine Idee ist, vermöge der *Permanenz der Differentialgleichung* (1) *die analytische Fortsetzung der Lösung für komplexe Werte der unabhängigen Variablen* zu ermitteln. Denkt man sich nämlich beispielsweise x auf reelle Werte beschränkt, ersetzt dagegen y durch einen komplexen Wert $y_1 + iy_2$, versucht man also, die vorgelegte reelle Lösung von (1) in eine $x y_2$-Ebene fortzusetzen, so wird sie in eine komplexwertige Funktion von x und y_2 übergehen, welche noch von dem Parameter y_1 abhängt. Die Differentialgleichung (1) wird formal in die nunmehr *hyperbolische* Differentialgleichung

$$u_{xx} - u_{y_2 y_2} = f(x, y_1 + i y_2, u, u_x, -i u_{y_2})$$

überführt, und die gewünschte analytische Fortsetzung verlangt, die Lösung dieser Differentialgleichung so zu bestimmen, daß sie für $y_2 = 0$ in die vorgegebenen *Anfangswerte* $u(x, y)$ übergeht. Das ist ein *hyperbolisches Anfangswertproblem* der Art, wie ich es in einer Arbeit „Über das Anfangswertproblem einer beliebigen hyperbolischen Differentialgleichung zweiter Ordnung“ Math. Ann. 98 gelöst habe. Mit Hilfe eines Eindeutigkeitssatzes für die Lösungen gewisser partieller Differentialgleichungen läßt sich zeigen, daß die so fortgesetzte Funktion $u(x, y)$, angesehen als Funktion des komplexen Arguments $y = y_1 + i y_2$ den *Cauchy-Riemannschen Differentialgleichungen* bezüglich y_1 und y_2 genügt, und zwar in einem die Anfangswerte von y im Inneren enthaltenden Gebiete einer $y_1 y_2$-Ebene. Damit ist aber bewiesen, daß die Funktion u schon für $y_2 = 0$, wo sie mit dem vorgelegten $u(x, y)$ übereinstimmt, analytisch von $y = y_1$ abhängt, weil nur solche Funktionen die Fortsetzung ins Komplexe im genannten Sinne erlauben.

Da jedoch in der genannten Abhandlung über das hyperbolische Anfangsproblem alle Resultate unter Beschränkung auf reelle Größen abgeleitet wurden, werde ich den Beweis hier nicht unter Berufung auf jene Resultate durchführen, sondern auf die charakteristischen Differentialgleichungen unserer partiellen Differentialgleichung zurückgehen.

Als einen Vorzug meines Beweises glaube ich es ansehen zu dürfen, daß er trotz seines elementaren Charakters fast frei von Abschätzungen ist, und daß diese Gedankengänge wohl zum ersten Male die mehr als formalen Zusammenhänge aufdecken, die zwischen den von Dubois-Reymond als „himmelweit verschieden“ bezeichneten Differentialgleichungen vom elliptischen und hyperbolischen Typus vorhanden sind. Übrigens erschöpft sich die Verwendbarkeit dieses Zusammenhangs nicht im Beweise des obigen allgemeinen Satzes, sondern er gibt z. B. auch leicht einen Beweis des bekannten Satzes von Cauchy-Kowalewsky für Differentialgleichungen in zwei unabhängigen Veränderlichen, worauf ich hier jedoch nicht näher eingehen will.

§ 1.

Um die folgenden Entwicklungen für sich möglichst vollständig zu machen, leiten wir für unsere Differentialgleichung

$$\Delta u - f(x, y, u, p, q) = 0, \qquad p = u_x, \; q = u_y \tag{1}$$

zunächst die charakteristischen Differentialgleichungen ab [1].

Führt man nämlich die Differentiationssymbole ein

$$\text{I} \qquad \begin{aligned} \frac{\partial}{\partial x} + i\frac{\partial}{\partial y} &= {}', \\ \frac{\partial}{\partial x} - i\frac{\partial}{\partial y} &= {}^{\backprime}, \end{aligned}$$

so folgt aus den für eine Lösung von (1) gültigen Gleichungen

$$\begin{aligned} p_x + i q_x \qquad\quad - p' &= 0 \\ q_x + i q_y - q' &= 0 \\ p_x \qquad\quad + q_y - f &= 0 \end{aligned}$$

durch Multiplikation bzw. mit $-1, i, 1$ und nachheriger Addition der entstehenden Gleichungen

$$p' - iq' - f^{\backprime} = 0. \tag{2}$$

Ganz ähnlich erhält man die Gleichung

$$p^{\backprime} + iq^{\backprime} - f = 0. \tag{3}$$

Diese Gleichungen zusammen mit

$$u' - p - iq = 0 \tag{4}$$

sind natürlich nur eine Schreibweise gewisser leicht aus ihnen herzustellender reeller Gleichungen, die man durch Aufspaltung in Real- und Imaginärteil erhält.

§ 2.

Die Frage, die wir uns jetzt stellen, ist also folgende: Gibt es *komplexe* Funktionen u, p, q des reellen Arguments $x_1 = x$ und des komplexen Arguments $y = y_1 + iy_2$, für welche die Gleichungen (2), (3), (4) erfüllt sind, wenn bei der partiellen Differentiation nach y der Realteil y_1 konstant gehalten wird, und die überdies für reelles y mit den Funktionen u, p, q, der vorgelegten Lösung von (1) übereinstimmen?

---

1) Vgl. hierzu K. Friedrichs und H. Lewy, „Lösung des Anfangswertproblems einer beliebigen etc.", Math. Ann. 98.

Um uns solche Funktionen herzustellen, begeben wir uns in eine Ebene mit den rechtwinkligen Koordinaten $x_1 = x$ und y_2. Wir nehmen wieder die Gleichungen (2), (3), (4), geben ihnen aber jetzt einen völlig anderen Sinn, indem wir die Striche als die folgenden Abkürzungen

$$\text{II} \qquad \begin{aligned} &= \frac{\partial}{\partial x_1} + \frac{\partial}{\partial y_2}, \\ &= \frac{\partial}{\partial x_1} - \frac{\partial}{\partial y_2} \end{aligned}$$

und die Größen $y, u, p, q,$ als Abkürzungen für die Ausdrücke

$$\begin{aligned} u &= u_1 + i u_2 \\ p &= p_1 + i p_2 \\ q &= q_1 + i q_2 \\ y &= y_1 + i y_2 \end{aligned}$$

auffassen, in denen $u_1, \ldots, q_2$ reelle Funktionen der beiden reellen Argumente x_1 und y_2 sein sollen. Da f analytisch von seinen Argumenten abhängt, ist hierdurch die Bedeutung von f als analytischer Funktion für komplexe Argumente in einer gewissen Umgebung der Anfangswerte festgelegt.

Die Gleichungen (2), (3), (4) in der neuen Bedeutung können wir durch Aufspaltung nach Real- und Imaginärteil durch die folgenden sechs ersetzen:

$$(*) \qquad \begin{aligned} p_1' + q_2' - \Re(f) &= 0 \\ p_2' - q_1' - \Im(f) &= 0 \\ p_1^{\backprime} - q_2^{\backprime} - \Re(f) &= 0 \\ p_2^{\backprime} + q_1^{\backprime} - \Im(f) &= 0 \\ u_1' - p_1 + q_2 &= 0 \\ u_2' - p_2 - q_1 &= 0. \end{aligned}$$

Die Größe y_1 spielt in diesen Gleichungen die Rolle eines unmittelbar nicht interessierenden Parameters. Für die obigen Gleichungen (*) stellen wir nun ein Anfangswertproblem, indem wir auf der Geraden $y_2 = 0$ die folgenden Werte vorgeben, wobei wir unter $u(x, y)$, $p(x, y)$, $q(x, y)$ die vorgelegte Lösung von (1) verstehen:

$$\begin{aligned} &u_2(x_1, 0) = 0; && p_2(x_1, 0) = 0; && q_2(x_1, 0) = 0. \\ &u_1(x_1, 0) = u(x_1, y_1); && p_1(x_1, 0) = p(x_1, y_1); && q_1(x_1, 0) = q(x_1, y_1). \end{aligned}$$

Besitzt die vorgelegte Lösung $u(x, y)$ von (1) stetige Ableitungen vierter Ordnung, so sind alle unsere Anfangswerte offenbar dreimal stetig in Richtung der Anfangsgeraden differenzierbar; $\mathfrak{R}(f)$ und $\mathfrak{J}(f)$ sind in einer Umgebung der Anfangswerte beliebig oft nach den Argumenten $x_1, y_1, y_2, \ldots, q_2$ partiell differenzierbar, da sie ja nach unserer Voraussetzung, daß f analytisch von seinen Argumenten abhängt, sogar analytisch von $x_1, y_1, \ldots, q_2$ abhängen.

Ein solches Anfangsproblem besitzt nun eine Lösung, wie ich in einer Abhandlung über das hyperbolische Anfangswertproblem [1]) gezeigt habe, weil die Determinante des Systems (*) nicht verschwindet. Es ist dies nämlich die Determinante

$$\begin{vmatrix} 1 & 0 & 0 & 1 & 0 & 0 \\ 0 & 1 & -1 & 0 & 0 & 0 \\ 1 & 0 & 0 & -1 & 0 & 0 \\ 0 & 1 & 1 & 0 & 0 & 0 \\ 0 & 0 & 0 & 0 & 1 & 0 \\ 0 & 0 & 0 & 0 & 0 & 1 \end{vmatrix} = 4.$$

Diese Lösung $u_1, u_2, \ldots, q_2$ ist mit stetigen Ableitungen nach x_1 und y_2 bis zur zweiten Ordnung versehen. Sie ist aber auch nach dem Parameter y_1 differenzierbar [2]) und besitzt auch stetige Ableitungen $\frac{\partial^2}{\partial x_1 \partial y_1}$ und $\frac{\partial^2}{\partial y_2 \partial y_1}$.

---

1) Vgl. Hans Lewy, l. c., Der dort bewiesene Satz enthält die oben ausgesprochene Behauptung unmittelbar nur dann, wenn f von x_1, y_2 nicht abhängt. Führt man aber zwei neue Funktionen $\bar{x}_1$ und $\bar{y}_2$ von x_1 und y_2 ein, so hat das Problem

$$(\bar{x}_1 - \bar{y}_2)' = 0, \quad (\bar{x}_1 + \bar{y}_2)^{\backprime} = 0,$$
$$\bar{x}_1(x_1, 0) = x_1, \quad \bar{y}_2(x_1, 0) = 0$$

offenbar die einzige Lösung

$$\bar{x}_1 \equiv x_1; \quad \bar{y}_2 \equiv y_2;$$

es wird

$$\bar{x}_1' = 1, \quad x_1^{\backprime} = 1$$

und das nach Real- und Imaginärteil gespaltene System

$$\begin{aligned} p' - iq' - f(\bar{x}_1, y_1 + i\bar{y}_2, u_1 + iu_2, \ldots, q_1 + iq_2)\,\bar{x}_1' &= 0 \\ p^{\backprime} - iq^{\backprime} - f(\cdot \;\cdot\;\cdot\;\cdot\;\cdot\;\cdot\;\cdot\;\cdot\;\cdot\;\cdot\;\cdot\;\cdot\;\cdot\;\cdot)\,\bar{x}_1^{\backprime} &= 0 \\ u' - (p + iq)\,\bar{x}_1' &= 0 \\ (\bar{x}_1 - \bar{y}_2)' = 0, \quad (\bar{x}_1 + \bar{y}_2)^{\backprime} &= 0 \end{aligned}$$

ist offenbar mit dem System (*) gleichwertig, und die zugehörigen Anfangsprobleme sind äquivalent, wenn man dieselben Anfangswerte nimmt. Das neue System erfüllt aber die Voraussetzungen der zitierten Arbeit.

2) Diese Tatsache der Differenzierbarkeit nach einem Parameter ist in der zitierten Arbeit über das hyperbolische Problem nicht ausdrücklich hervorgehoben,

§ 3.

Nachdem wir im vorigen Paragraphen die Funktionen $u_1, \ldots, q_1$ in eine gewisse dreidimensionale Umgebung eines Gebietes der Ebene $y_1 = 0$ des $x_1 y_1 y_2$-Raumes fortgesetzt haben, indem wir ihre Werte in der Ebene $y_2 = 0$ mit den Werten der vorgelegten Lösung der Differentialgleichung (1) identifizierten, werden wir jetzt in zwei Schritten das Erfülltsein der Cauchy-Riemannschen Differentialgleichungen für die Funktionen

$$u = u_1 + i u_2, \quad \ldots, \quad q = q_1 + i q_2$$

hinsichtlich der Variablen $y_1 + i y_2$ nachweisen.

Eine Funktion v der komplexen Veränderlichen $y_1 + i y_2$ hängt offenbar analytisch von dieser ab, wenn die Beziehung

$$\nabla v = \left(\frac{\partial}{\partial y_1} + i \frac{\partial}{\partial y_2}\right) v = 0$$

in dem in Betracht kommenden Gebiete der $y_1 y_2$-Ebene erfüllt ist.

Wir weisen nun erstens die Gleichungen

$$\nabla u = 0, \quad \nabla p = 0, \quad \nabla q = 0 \tag{5}$$

in der Anfangsebene $y_2 = 0$ nach. Berücksichtigt man nämlich, daß für die Anfangsebene die Gleichungen (2), (3), (4) in doppelter Weise Gültigkeit besitzen, nämlich einmal mit der Bedeutung I des Striches bezw. , zweitens aber auch in der Bedeutung II von S. 181, so folgt vermöge der Identität

$$\left(\frac{\partial}{\partial x_1} \pm i \frac{\partial}{\partial y_1}\right) - \left(\frac{\partial}{\partial x_1} \pm \frac{\partial}{\partial y_2}\right) = \pm i \left(\frac{\partial}{\partial y_1} + i \frac{\partial}{\partial y_2}\right)$$

durch Subtraktion der entsprechenden Gleichungen, weil für $y_2 = 0$ ja $x_1 = x$ und $y_1 = y$ ist,

$$\begin{aligned} \nabla p - i \nabla q &= 0 \\ \nabla p + i \nabla q &= 0 \\ \nabla u &= 0, \end{aligned}$$

woraus sich unmittelbar die Behauptung ergibt.

Um nun die Gültigkeit der Gleichungen (5) auch für von null verschiedene Werte von y_2 nachzuweisen, wenden wir den Operator $\nabla = \frac{\partial}{\partial y_1} + i \frac{\partial}{\partial y_2}$ auf die Gleichungen (2), (3), (4) mit der Bedeutung II des Striches an. Da nach der Bemerkung über die

---

weil sie für die dort behandelte Frage irrelevant ist. Sie ist aber unschwer aus dem Beweis des dortigen Hauptsatzes abzulesen.

Existenz von stetigen zweiten Ableitungen der in Betracht kommenden Funktionen (§ 2 Ende) der Operator ∇ mit den Differentiationen $\frac{\partial}{\partial x_1}$ und $\frac{\partial}{\partial y_2}$ vertauschbar ist, so folgen die Gleichungen

$$
(6) \qquad \begin{aligned}
(\nabla p)' - i(\nabla q)' - \nabla f &= 0 \\
(\nabla p)^{\backprime} + i(\nabla q)^{\backprime} - \nabla f &= 0 \\
(\nabla u)' - \nabla p - i\nabla q &= 0.
\end{aligned}
$$

Nun ist aber f als analytische Funktion seiner komplexen Argumente nach diesen differenzierbar, so daß wir erhalten

$$\nabla f = \frac{\partial f}{\partial u}\nabla u + \frac{\partial f}{\partial p}\nabla p + \frac{\partial f}{\partial q}\nabla q + \frac{\partial f}{\partial y}\nabla y;$$

weiter ist $\nabla y = 0$. Die Gleichungen (6) gehen also in die folgenden drei komplexen Differentialgleichungen über, die mit sechs reellen äquivalent sind:

$$
(7) \qquad \begin{aligned}
(\nabla p)' - i(\nabla q)' - \frac{\partial f}{\partial u}\nabla u - \frac{\partial f}{\partial p}\nabla p - \frac{\partial f}{\partial q}\nabla q &= 0 \\
(\nabla p)^{\backprime} + i(\nabla q)^{\backprime} - \frac{\partial f}{\partial u}\nabla u - \frac{\partial f}{\partial p}\nabla p - \frac{\partial f}{\partial q}\nabla q &= 0 \\
(\nabla u)' - \nabla p - i\nabla q &= 0.
\end{aligned}
$$

Dies sind drei homogene Differentialgleichungen mit nicht verschwindender Determinante für die Funktionen ∇p, ∇q, ∇u mit den Anfangswerten $\nabla p = \nabla q = \nabla u = 0$ für $y_2 = 0$.

Eine Lösung ist $\nabla p \equiv \nabla q \equiv \nabla u \equiv 0$. Daß es nur diese Lösung gibt, dafür wollen wir einen einfachen Beweis geben[1]).

Zur Abkürzung setzen wir

$$
\begin{aligned}
\nabla p - i\nabla q &= A_1 + iA_2 \qquad (A_1 \text{ und } A_2 \text{ reell}) \\
\nabla p + i\nabla q &= B_1 + iB_2 \qquad (B_1 \text{ und } B_2 \text{ reell}) \\
\nabla u &= C_1 + iC_2 \qquad (C_1 \text{ und } C_2 \text{ reell}).
\end{aligned}
$$

Dann haben wir anstatt (7) die sechs reellen Gleichungen

$$
(8) \qquad \begin{aligned}
A_1' &= \cdots \\
A_2' &= \cdots \\
C_1' &= \cdots \\
C_2' &= \cdots \\
B_1^{\backprime} &= \cdots \\
B_2^{\backprime} &= \cdots
\end{aligned}
$$

1) Vgl. zu dieser Tatsache die schwächere Formulierung eines allgemeineren Satzes, Hans Lewy, l. c. Im obigen Beweise wird nur die Existenz der in (7) auftretenden Ableitungen von ∇p, ∇q, ∇u vorausgesetzt.

Die rechten Seiten sind hierbei Linearformen in den $A_1\, A_2\, B_1\, B_2\, C_1\, C_2$ mit stetig von x_1 und y_2 abhängigen Koeffizienten, die somit in einer Umgebung der Anfangsebene $y_2 = 0$ beschränkt sind. Wir multiplizieren die Gleichungen (8) der Reihe nach mit $A_1\, A_2\, C_1\, C_2\, B_1\, B_2$ und können dann die folgenden Ungleichungen aufschreiben:

$$(D^2)' \equiv (A_1^2+A_2^2+C_1^2+C_2^2)' \leqq c(A_1^2+A_2^2+C_1^2+C_2^2+B_1^2+B_2^2) \leqq c(D^2+E^2)$$
$$(E^2)' \equiv \quad (B_1^2+B_2^2)' \quad \leqq c(A_1^2+A_2^2+C_1^2+C_2^2+B_1^2+B_2^2) \leqq c(D^2+E^2)$$

wo c eine passende Konstante ist. Jetzt beachten wir, daß D^2 und E^2 die Anfangswerte null besitzen, und können daher die Ungleichung

$$(9) \qquad (D^2+E^2)_P \leqq c\sqrt{2}\int_1 (D^2+E^2)\,dy_2 + c\sqrt{2}\int_2 (D^2+E^2)\,dy_2$$

ableiten, in der das erste Integral über die Seite 1 und das zweite über die Seite 2 des in der Figur gezeichneten Dreieckes zu erstrecken ist. Das Dreieck selbst ist in einer Ebene $y_1 = \text{const}$ des $x_1\, y_1\, y_2$-Raumes liegend zu denken. Da der Ausdruck $D^2 + E^2$ als stetige Funktion in einer Umgebung der Anfangsgeraden $y_2 = 0$ beschränkt ist, folgt dort aus (9) die Abschätzung

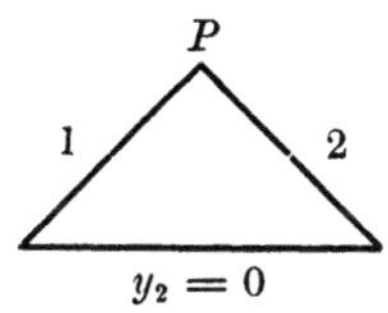

$$D^2 + E^2 \leqq c_1\, y_2$$

mit passenden konstantem c_1. Führen wir sie in die Ungleichung (9) ein, so kommt die neue Abschätzung

$$D^2 + E^2 \leqq c\sqrt{2}\, c_1 \frac{y_2^2}{2} \cdot 2 = c c_1 \sqrt{2}\, y_2^2.$$

Verfahren wir entsprechend weiter, so erhalten wir schließlich für beliebige y_2 in unserer Umgebung der Anfangsgeraden

$$D^2 + E^2 \leqq 2c\sqrt{2}\, c c_1 \sqrt{2}\, \frac{y_2^3}{3}, \quad \cdots,$$

also

$$D^2 + E^2 \leqq c_1 (c\sqrt{2})^{n-1} y_2^n$$

was für große n und hinreichend kleine y_2 nur möglich ist, wenn dort $D^2 + E^2 \equiv 0$ ist. Es gibt also eine Umgebung der Anfangsgeraden $y_2 = 0$, in der $A_1 = A_2 = B_1 = B_2 = C_1 = C_2 = 0$ ist, d. h. die Cauchy-Riemannschen Differentialgleichungen (5) gelten.

Wir haben somit das Ergebnis: *eine viermal* [1]) *stetig diffe-*

1) Da die Voraussetzungen des allgemeinen Existenzsatzes, auf dem der Beweis beruht, gegenüber ihrer Fassung (l. c.) gemildert werden können, so könnte

renzierbare Lösung u der Gleichung

$$\Delta u = f(x, y, u, p, q) \tag{1}$$

in der f eine analytische Funktion der Argumente $x, y, u, p = u_x$, $q = u_y$ *ist, ist eine analytische Funktion von y.*

Wegen der Drehungsinvarianz von Δu ist aber die Lösung auch eine analytische Funktion irgend einer linearen Verbindung $ax + by$ mit konstantem a und b. *Es ist also u eine analytische Funktion von x und y.*

---

auch unsere Behauptung unter geringeren Voraussetzungen bewiesen werden. Man vgl. die scharfe Formulierung bei Gevrey, Bull. des Sciences Math. 50 (1926).

Mathematische Annalen
98 (1927), pp. 107–124

Über die Methode der Differenzengleichungen zur Lösung von Variations- und Randwertproblemen.

Von

Hans Lewy in Göttingen.

In einer Abhandlung mit dem Titel „Über direkte Methoden bei Variations- und Randwertproblemen", welche Herr Courant in den Jahresberichten der Deutschen Mathematikervereinigung **34**, 1925 veröffentlicht hat[1]), sind — zum Teil in programmatischer Form — eine Reihe von Gedanken zu den direkten Methoden der Variationsrechnung entwickelt worden. Unter anderem findet sich dort der Hinweis, man müsse die Differentialgleichungen der Variationsrechnung als Grenzfälle entsprechender Differenzengleichungen[2]) auffassen und durch die Durchführung dieses Grenzüberganges zu den Lösungen der Variationsprobleme gelangen können. Herr Courant hat diesen Gedanken in einer Arbeit in den Göttinger Nachrichten[3]), sowie in einer jüngst in den Acta Mathematica **49** erschienenen Arbeit für den Fall gewöhnlicher und partieller linearer Differentialgleichungen und gewisser Klassen anderer sehr allgemeiner Funktionalgleichungsprobleme linearen Charakters weiter ausgeführt.

Angeregt durch diese Untersuchungen sowie durch persönliche Gespräche habe ich versucht, allgemeine Variationsprobleme, welche nicht mehr auf lineare Differentialgleichungen führen, mit der Methode der Differenzengleichungen zu behandeln. Das erste Ergebnis dieser Untersuchungen möchte ich hier vorlegen.

In den vier Paragraphen des ersten Teils dieser Arbeit betrachte ich Variationsprobleme, bei denen es sich um die Bestimmung einer Funktion y

---

[1]) Mit einigen Änderungen und Zusätzen in den Math. Annalen 97 abgedruckt unter dem Titel „Über direkte Methoden in der Variationsrechnung und verwandte Fragen".

[2]) Eine derartige Auffassung bildet übrigens schon für Euler den Ausgangspunkt der Untersuchung. Seine Ansätze sind neuerdings von A. Kneser aufgegriffen und mit den modernen Begriffsbildungen in Einklang gebracht worden.

[3]) „Über die Theorie der linearen partiellen Differenzengleichungen." 23. 10. 1925.

einer unabhängigen Veränderlichen x handelt und bei denen in dem zu variierenden Integrale nur die erste Ableitung y' von y nach x außer x und y selbst auftritt. Für den Integranden machen wir dabei als wesentliche Einschränkung nur die, daß er die Legendresche Bedingung in der starken Form erfüllen soll.

Es ergibt sich, daß dann in einem beliebigen konvexen (x, y)-Bereich es *immer eine mit stetiger Tangente versehene Lösung des Variationsproblems gibt, wenn man den Begriff der Lösung geeignet erweitert.*

Zur Behandlung des Variationsproblems bilde ich nun eine spezielle Minimalfolge. Ich teile das Intervall $0 \leqq x \leqq 1$ in n gleiche Teile und ersetze das zum Mimimum zu machende Integral durch eine entsprechend für diese Teilpunkte gebildete Summe, wobei an die Stelle der Differentialquotienten der Funktionen entsprechende Differenzenquotienten treten. Legen wir, was in § 1 geschieht, den Differentialquotienten bzw. Differenzenquotienten die Beschränkung auf, absolut genommen nicht größer als eine feste Schranke M zu werden, so führt das Minimumproblem auf eine Eulersche Differenzengleichung bzw. an den Stellen, wo in der Beschränkungsbedingung das Gleichheitszeichen erreicht wird, eine Eulersche Differenzenungleichung. Die Lösungen dieser Eulerschen Relationen werden durch eine Anzahl diskreter Werte dargestellt, die den Ordinaten der Eckpunkte einer „Polygonfunktion" entsprechen. Es wird sodann in § 1 der Grenzübergang für n gegen ∞ ausgeführt und dadurch die Lösung des entsprechenden Variationsproblems für Funktionen mit beschränkter Ableitung erzielt.

Nach einigen Hilfsbetrachtungen in § 2 wird sodann in § 3 die Einschränkung

$$|y'| \leqq \mathsf{M}$$

aufgegeben, indem ein Grenzübergang für wachsendes M ausgeführt wird. Beschränkt man, wie schon angedeutet, die zugelassenen Vergleichsfunktionen auf einen konvexen abgeschlossenen Bereich, so zeigt es sich, daß es mindestens eine Grenzkurve gibt, die als Lösung des Variationsproblems *ohne* Ableitungsbeschränkung aufzufassen ist.

Merkwürdigerweise braucht das Vorhandensein einer unteren Grenze des Integrals nicht vorausgesetzt zu werden. Auf Grund der Existenz der Lösung läßt sich nun aber auch die Existenz einer unteren Grenze des Variationsintegrales erschließen.

Im zweiten Teil entwickle ich den Grundgedanken einer numerischen Methode, welche keineswegs darauf beschränkt ist, daß in dem Variationsproblem nur eine unahängige Veränderliche auftritt, oder darauf, daß höhere als erste Ableitungen fehlen. Diese Methode, welche, wie ich nachträglich bemerkt habe, sich in ihrem Ausgangspunkt eng mit einer von

Hadamard zum Existenzbeweise für die Lösung gewisser Variationsprobleme verwandten Idee berührt[4]), ist mannigfacher Ausgestaltung in theoretischer Hinsicht fähig[5]) und in praktischer Hinsicht bedürftig. Der ihr zugrunde liegende allgemeine Gedanke läßt sich auch auf die numerische Behandlung von Problemen anwenden, die der Variationsrechnung nicht mehr angehören.

I. Teil.

§ 1.

Lösung des Variationsproblems bei Beschränkung der Ableitung.

Wir behandeln im folgenden Variationsprobleme des einfachsten Typus mit einer unabhängigen Veränderlichen und einer gesuchten Funktion, d. h. Variationsprobleme der Form

$$J[y] = \int_0^1 F(x, y, y')\,dx = \text{Min.}, \tag{1}$$

wo außerdem für den Rand noch gewisse Bedingungen vorgeschrieben sein können. Wir wollen uns zunächst auf den speziellen Fall beschränken, daß die unbekannte Funktion am Rande fest vorgegeben ist:

$$y(0) = c_0; \qquad y(1) = c_1. \tag{2}$$

Von F setze ich voraus, daß es samt seinen Ableitungen bis zur zweiten Ordnung für die zu betrachtenden (in § 1 abgeschlossenen) Bereiche seiner Argumente x, y, y' stetig ist.

Wir schicken zur Orientierung folgende, später aber nicht benutzte Bemerkung voraus: Das Problem (1) hat sicher eine Lösung, wenn man von den ersten Ableitungen $y'(x)$ der zur Konkurrenz zugelassenen Funktionen $y(x)$ verlangt, daß sie bei festen $K > 0$ und beliebigem h der Relation genügen

$$\left| \frac{y'(x+h) - y'(x)}{h} \right| \leqq K.$$

Denn in jeder zugehörigen Minimalfolge

$$\varphi_1, \varphi_2, \ldots$$

sind außer den Funktionen φ_n auch ihre ersten Ableitungen gleichmäßig

---

[4]) Hadamard, Mémoire sur les plaques encastrées. Mémoires de divers savants présentés à l'Académie des Sciences Paris **33**, 1908.

[5]) Vgl. Göttinger Nachrichten 18. Dezember 1925: H. Lewy, Über einen Ansatz zur numerischen Lösung usw. und R. Courant, Bemerkungen zur Frage der numerischen Auflösung usw.

stetig und beschränkt, also existiert in der Folge der Ableitungen nach einem bekannten Satze[6]) eine geeignet zu wählende Teilfolge

$$\varphi'_{i_1}, \varphi'_{i_2}, \ldots$$

mit $\lim\limits_{k\to\infty} \varphi'_{i_k} \Rightarrow u'$; hierbei ist u' die stetige Ableitung einer stetigen Funktion u, gegen die die Funktionen φ_{i_k} selbst gleichmäßig konvergieren. Es folgt dann

$$\lim_{k\to\infty} \int_0^1 F(x, \varphi_{i_k}, \varphi'_{i_k})\,dx = \int_0^1 F(x, \lim_{k\to\infty} \varphi_{i_k}, \lim_{k\to\infty} \varphi'_{i_k})\,dx = \int_0^1 F(x, u, u')\,dx,$$

womit der Beweis für die Behauptung geführt ist[7]).

Anders schon werden die Verhältnisse, wenn wir von den Funktionen $y(x)$, die zur Konkurrenz in dem Variationsproblem (1) zugelassen werden, nur die Beschränktheit der Beträge der ersten Ableitungen voraussetzen. Wir wollen von vornherein annehmen, daß die Schranke $\mathsf{M} > 0$ für die Beträge der Ableitungen so groß ist, daß jedenfalls mindestens eine Kurve vorhanden ist, für die die Ableitung zwar durch M beschränkt ist, die aber die Punkte $(0, c_0)$ und $(1, c_1)$ verbindet[8]). Ist $\mathsf{M} > 0$ demgemäß gewählt, so ist von selbst auch $y(x)$ auf einen gewissen abgeschlossenen Bereich beschränkt. Es gilt nun folgender Satz:

Unter allen Funktionen $y(x)$, für die $y(0) = c_0$, $y(1) = c_1$ ist und deren erste Ableitungen stückweise stetig und ihrem Betrage nach nicht größer als M sind, existiert eine Funktion $u(x)$, die das Variationsintegral zum Minimum macht, falls für alle in Frage kommenden Werte x, y, y' die Legendresche Bedingung

$$(3) \qquad F_{y'y'}(x, y, y') > 0$$

erfüllt ist. Die Funktion u besitzt eine überall stetige erste Ableitung[9]).

Beweis. Infolge der Abgeschlossenheit des durch

$$0 \leqq x \leqq 1; \quad |y'| \leqq \mathsf{M}; \quad y(0) = c_0, \; y(1) = c_1$$

definierten Bereiches von (x, y, y') und der vorausgesetzten Stetigkeit von $F_{y'y'}(x, y, y')$ folgert man aus (3)

$$(4) \qquad F_{y'y'} \geqq \alpha > 0,$$

wobei α eine positive von M abhängige Konstante bedeutet. Der Mittel-

---

[6]) Vgl. Courant-Hilbert, Methoden der mathemat. Physik **1** (Berlin 1924), S. 40.

[7]) Das Zeichen $\Rightarrow$ soll bedeuten: „konvergiert gleichmäßig".

[8]) Ich habe mir die Freiheit genommen, je nach Belieben Funktion oder Kurve zu sagen.

[9]) Es ist leicht, diesen Satz und seinen Beweis auf Variationsprobleme mit mehreren gesuchten Funktionen $y, z, \ldots$ einer Variablen x zu übertragen.

wertsatz der Differentialrechnung liefert dann für $|a| \leqq M$, $|b| \leqq M$, $a \neq b$ und zulässige x, y

$$(5) \qquad \frac{F_{y'}(x, y, a) - F_{y'}(x, y, b)}{a - b} \geqq \alpha.$$

Wir wollen jetzt das Intervall $0 \leqq x \leqq 1$ durch $n+1$ äquidistante Punkte $0, \frac{1}{n}, \frac{2}{n}, \ldots, \frac{n}{n} = 1$ teilen und das korrespondierende n-te Differenzenproblem — um ein Wort zu haben — lösen:

$$(6) \quad l \sum_{\nu=1}^{n} F\left(x_\nu, y_\nu, \frac{y_\nu - y_{\nu-1}}{l}\right) = \text{Min.}, \quad y(0) = y_0 = c_0;\ y(1) = y_n = c_1.$$

$$(7) \qquad \left|\frac{y_\mu - y_{\mu-1}}{l}\right| \leqq M.$$

Wir bedienen uns hierbei und im folgenden der Abkürzungen

$$l = \frac{1}{n}; \quad x_\mu = \frac{\mu}{n}; \quad F_\nu = F\left(x_\nu, y_\nu, \frac{y_\nu - y_{\nu-1}}{l}\right);$$

$$\frac{\partial F_\nu}{\partial y} = \frac{\partial F}{\partial y}\left(x_\nu, y_\nu, \frac{y_\nu - y_{\nu-1}}{l}\right) \quad \text{usw.}$$

Die Wertepaare (x_μ, y_μ) für $\mu = 0, 1, 2, \ldots, n$ bestimmen die Ecken einer Polygonfunktion. Die Existenz der das Problem (6) unter der Einschränkung (7) lösenden, durch die $n+1$ Werte y_μ bestimmten Polygonfunktion ist durch den Weierstraßschen Satz über das Minimum einer stetigen Funktion im abgeschlossenen Bereich gesichert.

Greifen wir nunmehr drei sukzessive Punkte $(x_{\nu-1}, y_{\nu-1})$, (x_ν, y_ν), $(x_{\nu+1}, y_{\nu+1})$ heraus. Wenn wir die Werte von $y_{\nu-1}$ und $y_{\nu+1}$ festhalten, so findet man y_ν aus der Forderung

$$(*) \qquad l F\left(x_\nu, y_\nu, \frac{y_\nu - y_{\nu-1}}{l}\right) + l F\left(x_{\nu+1}, y_{\nu+1}, \frac{y_{\nu+1} - y_\nu}{l}\right) = \text{Min.}$$

Aus dieser Forderung leitet man für y_ν eine Eulersche Gleichung oder Ungleichung her, je nachdem, ob für dieses optimale y_ν die Beträge von

$$\left|\frac{y_\nu - y_{\nu-1}}{l}\right| \quad \text{und} \quad \left|\frac{y_{\nu+1} - y_\nu}{l}\right|$$

beide kleiner sind als M oder nicht. Gilt nämlich in den Beziehungen (7), angewandt auf $\mu = \nu$ und $\mu = \nu + 1$, beide Male das Zeichen der Ungleichheit, so muß sich y_ν extremal verhalten gegenüber einer freien Variation, d. h. einer hinreichend kleinen Vergrößerung wie Verkleinerung. Nullsetzen des Differentialquotienten der Summe (*) ergibt für das extremale y_ν

$$0 = l \frac{\partial F_\nu}{\partial y} + \frac{\partial F_\nu}{\partial y'} - \frac{\partial F_{\nu+1}}{\partial y'}.$$

Gilt jedoch für das optimale y_ν in den Beziehungen (7) für $\mu = \nu$ und $\mu = \nu + 1$ mindestens einmal das Gleichheitszeichen, so müssen wir drei verschiedene Fälle unterscheiden:

1. Eine Variation des optimalen y_ν im positiven Sinne verstößt gegen (7), im negativen aber nicht. Dann folgt durch Variation im negativen Sinne, wenn $\varepsilon > 0$ und hinreichend klein ist:

$$0 \geqq \frac{1}{\varepsilon}\Big\{l F_\nu + l F_{\nu+1} - l F\Big(x_\nu, y_\nu - \varepsilon, \frac{y_\nu - \varepsilon - y_{\nu-1}}{l}\Big) \\ - l F\Big(x_{\nu+1}, y_{\nu+1}, \frac{y_{\nu+1} - y_\nu + \varepsilon}{l}\Big)\Big\},$$

also durch Grenzübergang für $\varepsilon \to 0$

$$l \frac{\partial F_\nu}{\partial y} + \frac{\partial F_\nu}{\partial y'} - \frac{\partial F_{\nu+1}}{\partial y'} \leqq 0. \tag{8a}$$

2. Eine Variation des optimalen y_ν im negativen Sinne verstößt gegen (7), jedoch nicht im positiven Sinne; dann folgt entsprechend

$$-l \frac{\partial F_\nu}{\partial y} - \frac{\partial F_\nu}{\partial y'} + \frac{\partial F_{\nu+1}}{\partial y'} \leqq 0. \tag{8b}$$

Im ersten Fall gilt die Beziehung

$$\frac{y_\nu - y_{\nu-1}}{l} > \frac{y_{\nu+1} - y_\nu}{l},$$

im zweiten die entsprechende Annahme

$$\frac{y_{\nu+1} - y_\nu}{l} > \frac{y_\nu - y_{\nu-1}}{l}.$$

3. Der dritte Fall, der eintritt, wenn sowohl positive wie negative Variation unzulässig ist, wird durch die Beziehung

$$\frac{y_\nu - y_{\nu-1}}{l} = \frac{y_{\nu+1} - y_\nu}{l}$$

repräsentiert.

Die im ersten Falle gültige Formel (8a) schreiben wir jetzt so:

$$\frac{\partial F}{\partial y'}\Big(x_\nu, y_\nu, \frac{y_\nu - y_{\nu-1}}{l}\Big) - \frac{\partial F}{\partial y'}\Big(x_\nu, y_\nu, \frac{y_{\nu+1} - y_\nu}{l}\Big) \\ \leqq -l \frac{\partial F}{\partial y}\Big(x_\nu, y_\nu, \frac{y_\nu - y_{\nu-1}}{l}\Big) + \frac{\partial F}{\partial y'}\Big(x_\nu + l, y_\nu + l\frac{y_{\nu+1} - y_\nu}{l}, \frac{y_{\nu+1} - y_\nu}{l}\Big) \\ - \frac{\partial F}{\partial y'}\Big(x_\nu, y_\nu, \frac{y_{\nu+1} - y_\nu}{l}\Big). \tag{**}$$

Auf die beiden letzten Glieder der rechten Seite kann man den Mittelwertsatz anwenden und findet so, daß wegen der Beschränktheit von $\frac{y_{\mu+1} - y_\mu}{l}$ und der Stetigkeit der ersten und zweiten Ableitungen von F

die rechte Seite kleiner ist als eine vom Index und von l unabhängige Zahl, multipliziert mit l; die linke Seite ist gemäß (5) größer als $\alpha\left(\frac{y_\nu - y_{\nu-1}}{l} - \frac{y_{\nu+1} - y_\nu}{l}\right)$. Es gibt also eine von l unabhängige Zahl $K > 0$, so daß sicher

$$0 < \frac{y_\nu - y_{\nu-1}}{l} - \frac{y_{\nu+1} - y_\nu}{l} \leqq Kl,$$

falls $\frac{y_\nu - y_{\nu-1}}{l} > \frac{y_{\nu+1} - y_\nu}{l}$ (erster Fall).

Entsprechend liefert (8 b)

$$0 < \frac{y_{\nu+1} - y_\nu}{l} - \frac{y_\nu - y_{\nu-1}}{l} \leqq Kl,$$

falls $\frac{y_{\nu+1} - y_\nu}{l} > \frac{y_\nu - y_{\nu-1}}{l}$ (zweiter Fall).

Die Formel (8) liefert dasselbe wie (8a) oder (8b), je nachdem, ob $\frac{y_\nu - y_{\nu-1}}{l}$ größer oder kleiner als $\frac{y_{\nu+1} - y_\nu}{l}$ ist. Zusammenfassend können wir also die Beziehung

$$(9) \qquad \left|\frac{y_{\nu+1} - y_\nu}{l} - \frac{y_\nu - y_{\nu-1}}{l}\right| \leqq Kl$$

konstatieren, der sich offenbar auch der letzte Fall 3 unterordnet.

Bezeichnen wir die in den Punkten $l, 2l, \ldots, nl = 1$ durch die Werte $\frac{y_1 - y_0}{l}, \frac{y_2 - y_1}{l}, \ldots$ definierte Funktion mit $\varphi_n'(x)$. Wir können sie zu einer im ganzen Intervall $0 \leqq x \leqq 1$ stetigen Funktion *ergänzen*, indem wir sie innerhalb der einzelnen Teilintervalle der Länge l linear interpolieren und ihr für $0 \leqq x \leqq l$ den Wert $\frac{y_1 - y_0}{l}$ zuschreiben.

Es gilt dann für beliebiges positives h die Ungleichung

$$(9') \qquad |\varphi_n'(x+h) - \varphi_n'(x)| \leqq Kh.$$

Lassen wir jetzt n unbegrenzt wachsen. (9') verbürgt uns die gleichmäßige Stetigkeit der φ_n'. Da die $|\varphi_n'|$ durch M beschränkt sind, folgt die Existenz einer gleichmäßig gegen eine stetige Grenzfunktion $u'(x)$ konvergierenden Teilfolge:

$$\lim_{k\to\infty} \varphi_{i_k}'(x) \Rightarrow u'(x).$$

Setzt man $\varphi_{i_k}(\nu l)$ gleich demjenigen y_ν, das zur Intervalleinteilung i_k gehört, so wird φ_{i_k} eine Summe über $l\varphi_{i_k}'$, genommen in den Punkten μl, die vor νl liegen (abgesehen von der additiven Konstanten c_0). Bezeichnen wir mit $u(x)$ den Ausdruck $c_0 + \int_0^x u'(x)\,dx$, so wird $u(0) = c_0$, und wegen der gleichmäßigen Konvergenz auch

$$u(1) = c_0 + \int_0^1 u'(x)\,dx = c_0 + \lim_{k\to\infty} l \sum_{\mu=0}^{i_k} \varphi'_{i_k}(\mu l),$$
$$u(1) = c_1.$$

Es sei jetzt $\psi(x)$ irgendeine Funktion, für die ψ' stückweise stetig und dem Betrage nach nicht größer als M ist, und für die $\psi(0) = c_0$, $\psi(1) = c_1$. Nun ist, wenn $\bar\psi$ die durch die Werte $\psi\left(\frac{\nu}{m}\right)$ gelieferte Polygonfunktion und $\bar\psi'\left(\frac{\nu}{m}\right)$ ihren Differenzenquotienten bedeutet, nach der Bedeutung von φ_m und φ'_m

$$\frac{1}{m}\sum_1^m F\left(\frac{\nu}{m}, \bar\psi\left(\frac{\nu}{m}\right), \bar\psi'\left(\frac{\nu}{m}\right)\right) \geqq \frac{1}{m}\sum_1^m F\left(\frac{\nu}{m}, \varphi_m\left(\frac{\nu}{m}\right), \varphi'_m\left(\frac{\nu}{m}\right)\right),$$

und schließlich infolge der gleichmäßigen Konvergenz für $m = i_1, i_2, \ldots$

$$\int_0^1 F(x, \psi(x), \psi'(x))\,dx \geqq \int_0^1 F(x, u, u')\,dx,$$

d. h. u löst das gestellte Minimumproblem.

In einem Intervall von x, in welchem $|u'(x)| < \mathsf{M}$ ist, kann man die du Bois-Reymondsche Beziehung und die aus ihr folgende Eulersche Differentialgleichung herleiten. Für Polygone (aus unserer Teilfolge) mit genügend vielen Ecken muß in diesem Intervall natürlich auch der Differenzenquotient dem Betrage nach kleiner als M sein. Es gilt also die Differenzen*gleichung* (8), aus der durch Summation die Relation

$$l \sum_{\nu=\nu_0}^{\nu_1} \frac{\partial F_\nu}{\partial y} = \frac{\partial F_{\nu_1}}{\partial y'} - \frac{\partial F_{\nu_0}}{\partial y'}$$

folgt. Bei unserem Grenzübergang geht sie über in die klassische Relation

$$\int_{x_0}^{x_1} \frac{\partial F}{\partial y}\,dx = \frac{\partial F}{\partial y'}\Bigg|_{x_0}^{x_1} \qquad \text{(du Bois-Reymond).}$$

Durch Differentiation nach x_1 kommt

$$\frac{\partial F}{\partial y} = \left(\frac{\partial F}{\partial y'}\right)' \qquad \text{(Euler).}$$

§ 2.

Bemerkungen zu dem vorangehenden Beweis.

I. Wenn wir dasselbe Problem wie eben stellen, jedoch die Punkte $x, y(x)$ noch auf einen Bereich B einschränken, der von zwei zueinander konkaven Kurven mit stetiger erster Ableitung sowie ev. noch zwei der

y-Achse parallelen Geradenstücken begrenzt wird (s. Figur), so kommen wir zu demselben Existenztheorem wie eben. Man folgert nämlich wie auf S. 111 u. 112 aus der Unzulässigkeit einer Vergrößerung von y_ν die Beziehung $\frac{y_\nu - y_{\nu-1}}{l} \geqq \frac{y_{\nu+1} - y_\nu}{l}$, aus der Unzulässigkeit einer Verkleinerung von y_ν hingegen $\frac{y_\nu - y_{\nu-1}}{l} \leqq \frac{y_{\nu+1} - y_\nu}{l}$. Man schließt die Formel (8) im Falle der freien, die Formeln (8a) oder (8b) im Falle der einseitigen Variation, und alles übrige geht ebenso vor sich wie oben.

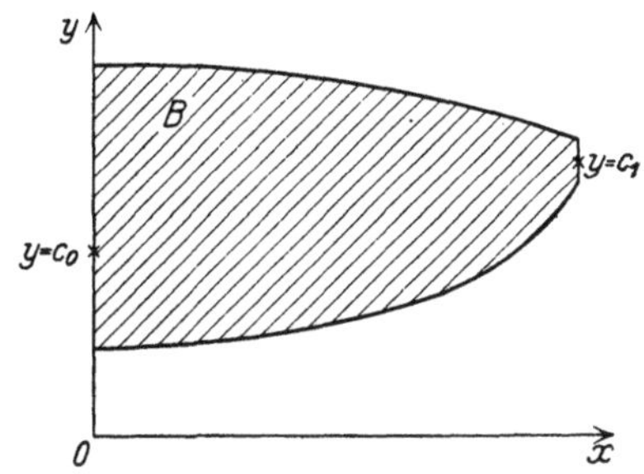

II. Da wir später den Grenzübergang $\mathsf{M} \to \infty$ ausführen werden, ist es wichtig festzustellen, daß zwei Werte von x, etwa ζ und $\bar{\xi}$, nicht beliebig nah aneinander liegen können, wenn die zugehörigen Werte der Funktion $u'(x)$ zwei feste Zahlen a bzw. $\bar{a} \neq a$ sind, und zwar werde ich zeigen, daß ihr Abstand größer als eine feste positive Zahl bleibt, *unabhängig davon, bei welchem* M *die Funktion* u *Lösung des Problems* (1) *war*. Ein solches Intervall $[\zeta, \bar{\xi}]$ können wir nämlich von vornherein als ein solches annehmen, in dessen Innerm die Funktion $u'(x)$ nur Werte *zwischen* a und $\bar{a}$ annimmt; denn jedes Intervall, an dessen Endpunkten die Funktion u' die Werte a und $\bar{a}$ annimmt, enthält wegen der Stetigkeit von u' eins der bezeichneten Art. Nun war ja unsere Funktion u' als der gleichmäßige Limes von gewissen Polygonfunktionen gewonnen. Für diese gewannen wir aus der Ungleichung (**) eine Schranke des zweiten Differenzenquotienten. In § 1 brauchten wir nur eine rohe Abschätzung für diese Schranke, da wir versichert waren, daß der erste Differenzenquotient seinem Betrage nach unterhalb M lag; wir können aber dieselbe Abschätzung in für unser jetziges Ziel schärferer Form aussprechen, indem wir uns auf ihre Entstehung besinnen (Anwendung des Mittelwertsatzes) und sie gleich in der „integrierten“ Form aufschreiben:

Sind a und $\bar{a} \neq a$ zwei Werte, die unsere Polygonfunktion in zwei Punkten ζ und $\bar{\xi}$ des Intervalles $0 \leqq x \leqq 1$ annimmt, so haben diese Punkte ζ und $\bar{\xi}$ eine Entfernung, die größer als $\frac{|a - \bar{a}|}{\beta(a, \bar{a})}$ ist, wo $\beta(a, \bar{a})$ eine von a und $\bar{a}$ abhängige positive Zahl bedeutet, die jedoch von M nicht abhängt.

Diese Beziehung bleibt natürlich beim Grenzübergang zu verschwindender Maschenweite erhalten, so daß wir also sagen können:

Es sei $y_{\mathsf{M}}(x)$ unsere Lösungsfunktion des Problems (1), wenn zur Konkurrenz alle Funktionen mit stückweise stetiger erster Ableitung des

8*

Betrages $\leqq \mathsf{M}$ zugelassen werden, für die der Wert im Punkte 0 gleich c_0, im Punkte 1 gleich c_1 ist, und die so beschaffen sind, daß die Kurve $(x, y(x))$ ganz dem Bereiche B angehört; dann haben zwei Punkte ζ und $\bar{\xi}$, für die $y'_{\mathsf{M}}(\zeta) = a$, $y'_{\mathsf{M}}(\bar{\xi}) = \bar{a} + a$ ist, einen Abstand, der *größer oder gleich der positiven Zahl* $\frac{|a - \bar{a}|}{\beta(a, \bar{a})}$ ist, die *nicht von* M *abhängt*. Unsere Schlußweise setzt auch unmittelbar die folgende Tatsache in Evidenz:

Wenn von den Zahlen a und $\bar{a}$ mindestens eine absolut kleiner als eine feste Konstante H ist und sie so variieren, daß ihr Abstand größer als eine feste positive Zahl bleibt, so bleibt auch $|\zeta - \bar{\xi}|$ größer als eine feste positive Zahl.

§ 3.

Grenzübergang zu $\mathsf{M} = \infty$; allgemeiner Existenzsatz.

Um den in der Überschrift angedeuteten Grenzübergang auszuführen, zeigen wir erstens, daß die sämtlichen Funktionen $y_{\mathsf{M}}(x)$ eine beschränkte Bogenlänge besitzen, und zweitens, daß ihre Ableitungen nach der Bogenlänge $\frac{dy_{\mathsf{M}}}{ds}$ als Funktionen derselben gleichmäßig stetig sind. Es ist nun leicht nachzuweisen, daß für unsere oben definierten Lösungsfunktionen $y_{\mathsf{M}}(x)$ bei beliebigem $\mathsf{M} > 0$ der Wert des Integrals der Bogenlänge $\int_0^1 \sqrt{1 + y'_{\mathsf{M}}(x)^2}\, dx$ oder (was damit gleichbedeutend ist wegen $\int_0^1 \sqrt{1 + y'^2}\, dx \leqq \int_0^1 (1 + |y'|)\, dx$) der Wert des Integrales $\int_0^1 |y'_{\mathsf{M}}(x)|\, dx$ unter einer von M unabhängigen Schranke liegt. Bezeichnet man nämlich mit $A > 0$ den Maximalabstand der Ordinaten irgend zweier Punkte aus B (vgl § 2), so wäre das Gegenteil der Behauptung, daß bei geeignetem M das Integral $\int_0^1 |y'_{\mathsf{M}}(x)|\, dx$ größer wäre als $2A$, multipliziert mit einer beliebig groß wählbaren natürlichen positiven Zahl n. Teilt man aber das Intervall in n gleiche Teile, so gibt es sicher einen, in dem $\int |y'_{\mathsf{M}}(x)|\, dx \geqq 2A$ ist. Weil aber für beliebiges x_1 und x_2 aus $0 \leqq x \leqq 1$ die Relation $\left|\int_{x_1}^{x_2} y'_{\mathsf{M}}(x)\, dx\right| \leqq A$ gilt, so kann y'_{M} in dem vorher bezeichneten n-ten Teilintervall nicht ständig größer als Null und nicht ständig kleiner als Null sein. Es ist also dort mindestens einmal $y'_{\mathsf{M}} = 0$ und andererseits wegen $\int |y'_{\mathsf{M}}|\, dx \geqq 2A$ auch sicher $|y'_{\mathsf{M}}| = An$ in mindestens einem Punkte; also ist $|y'_{\mathsf{M}}|$ sicher einmal auch gleich z. B. der Zahl 1. Nun kann aber der Abstand zweier Punkte, in deren einem $y'_{\mathsf{M}} = 0$, deren anderem $y'_M = \pm 1$ ist, nicht beliebig klein sein, denn er ist größer als die kleinere der beiden Zahlen $\frac{1}{\beta(1,0)}$ und $\frac{1}{\beta(0,-1)}$. Er ist aber nach der Annahme kleiner als $\frac{1}{n}$, wo n beliebig groß ist, wenn nur M geeignet gewählt wurde, und damit ist ein Widerspruch hergeleitet.

Fassen wir nunmehr x und $y_M(x)$ als Funktionen der Bogenlänge s auf, so wollen wir beweisen, daß $\frac{dy_M}{ds} = \dot{y}_M(s)$ gleichmäßig stetig und beschränkt ist. Die Beschränktheit folgt aus der Beziehung $|\dot{y}_M(s)|^2 \leqq 1$. Wären nun die $\dot{y}_M(s)$ nicht gleichmäßig stetig in M, so gäbe es also eine Folge von Zahlen M und zugehöriger s_1 und s_2, so daß $|s_1 - s_2| \to 0$, und trotzdem wäre $|\dot{y}_M(s_1) - \dot{y}_M(s_2)| \geqq \varepsilon > 0$, wo ε eine feste positive Zahl ist. Nehmen wir ohne Beschränkung der Allgemeinheit an, daß $\dot{y}_M(s_1) > \dot{y}_M(s_2)$. Es folgt

$$\frac{\left|\frac{\dot{y}_M}{\sqrt{1-\dot{y}_M^2}}(s_1) - \frac{\dot{y}_M}{\sqrt{1-\dot{y}_M^2}}(s_2)\right|}{|\dot{y}_M(s_1) - \dot{y}_M(s_2)|} = \frac{1}{\left(\sqrt{1-z^2}\right)^3} \geqq 1$$

aus dem Mittelwertsatz der Differentialrechnung; z ist eine bestimmte Zahl aus $\dot{y}_M(s_2) < z < \dot{y}_M(s_1)$. Das ergibt also wegen $\frac{\dot{y}}{\sqrt{1-\dot{y}^2}} = \frac{dy}{dx} = y'$ die Beziehung

$$\left|\frac{dy_M}{dx}(s_1) - \frac{dy_M}{dx}(s_2)\right| \geqq |\dot{y}_M(s_1) - \dot{y}_M(s_2)| \geqq \varepsilon.$$

Außerdem gilt

$$|x(s_1) - x(s_2)| \leqq |s_1 - s_2|;$$

daher kommt $|x(s_1) - x(s_2)| \to 0$.

Nach den vorhin gemachten Bemerkungen wird nun, wenn $|x(s_1) - x(s_2)|$ hinreichend klein wird, $\left|\frac{dy_M}{dx}(s_1)\right|$ und $\left|\frac{dy_M}{dx}(s_2)\right|$ für fast alle M oberhalb einer beliebigen festen positiven ganzen Zahl H liegen, weil eben andernfalls $|x(s_1) - x(s_2)|$ größer bleiben müßte als eine von H und ε abhängige Zahl. Nun können aber $\frac{dy_M}{dx}(s_1)$ und $\frac{dy_M}{dx}(s_2)$ nicht verschiedenes Vorzeichen haben, weil wenigstens eine von beiden Größen sich dann von Null um mehr als $\frac{\varepsilon}{2}$ unterscheiden müßte und $|x(s_1) - x(s_2)|$ dann immer größer als $\frac{\varepsilon}{2\beta\left(\frac{\varepsilon}{2}, 0\right)}$ oder $\frac{\varepsilon}{2\beta\left(0, \frac{-\varepsilon}{2}\right)}$ sein müßte. Es folgt also wegen

$$\dot{y}_M(s_1) - \dot{y}_M(s_2) = \frac{dy_M}{dx} \cdot \frac{1}{\sqrt{1+\left(\frac{dy_M}{dx}\right)^2}}(s_1) - \frac{dy_M}{dx} \cdot \frac{1}{\sqrt{1+\left(\frac{dy_M}{dx}\right)^2}}(s_2)$$

für hinreichend kleines $|x(s_1) - x(s_2)|$

$$|\dot{y}_{\mathsf{M}}(s_1) - \dot{y}_{\mathsf{M}}(s_2)| < \left|1 - \frac{H}{\sqrt{1+H^2}}\right|,$$

und der letzte Ausdruck ist bei genügend kleinem $|x(s_1) - x(s_2)|$, d. h. bei hinreichend großem H, sicher kleiner als ε. Damit ist die Voraussetzung der ungleichmäßigen Stetigkeit von $\dot{y}_{\mathsf{M}}(s)$ als widerspruchsvoll aufgezeigt.

Durch eine einfache Betrachtung[10]) sieht man, daß sich aus den Funktionen $\dot{y}_{\mathsf{M}}(s)$ bei $\mathsf{M} \to \infty$ eine Teilfolge auswählen läßt, die gleichmäßig gegen eine stetige Grenzfunktion $\dot{\varphi}(s)$ strebt, woraus wegen $y_{\mathsf{M}}(0) = c_0$ für alle M auch die gleichmäßige Konvergenz der entsprechenden Teilfolge der $y_{\mathsf{M}}(s)$ folgt. Es wird $\varphi(0) = c_0$ und φ (Endpunkt) $= c_1$.

Im allgemeinen wird sich die Funktion $\varphi(s)$ nicht als Funktion des zugehörigen $x(s)$ schreiben lassen. Sie wird nämlich möglicherweise Stücke besitzen, die zur x-Achse senkrecht verlaufen[11]).

Machen wir nun über unseren Integranden, den wir auch als Funktion von $x(s)$, $y(s)$, $\dot{x}(s)$, $\dot{y}(s)$ schreiben können,

$$F(x, y, y') = G(x, y, \dot{y}, \dot{x}); \quad \frac{dx}{ds} = \dot{x}, \quad \frac{dy}{ds} = \dot{y}, \quad \dot{x}^2 + \dot{y}^2 = 1$$

die Voraussetzung: Es seien $K_1, K_2, \ldots$ eine beliebige Folge von Kurven, die *samt ihrer als Funktion der Bogenlänge stetigen Tangente gleichmäßig* gegen eine ebensolche Grenzkurve konvergieren; dann möge das Integral $\int G(x, y, \dot{y}, \dot{x})\dot{x}\,ds$ gegen einen endlichen oder unendlichen Grenzwert streben, der von der Gestalt der Grenzkurve allein abhängig ist und den wir als Wert des Integrals über die Grenzkurve ansehen wollen.

Ist $\mathsf{M}_1 > \mathsf{M}_2$, so gilt wegen der Minimaleigenschaft von y_{M}

$$J[y_{\mathsf{M}_1}] \leqq J[y_{\mathsf{M}_2}].$$

Aus der oben nachgewiesenen gleichmäßigen Konvergenz der $\dot{y}_{\mathsf{M}}(s)$, $\dot{x}_{\mathsf{M}}(s) = \sqrt{1 - \dot{y}^2_{\mathsf{M}}(s)}$ folgt also offenbar der Satz:

Stellt man das Problem (1), *in dem man zur Konkurrenz alle die Punkte* $(0, c_0)$ *und* $(1, c_1)$ *verbindenden Kurven zuläßt, die eine stückweise*

---

[10]) Man führe etwa die neue unabhängige Variable $\bar{s} = \frac{s}{L_{\mathsf{M}}}$ ein. Hierbei ist $L_{\mathsf{M}} = \int_0^1 \sqrt{1 + y'_{\mathsf{M}}(x)^2}\,dx$. Aus den Größen L_{M} läßt sich wegen ihrer Beschränktheit eine gegen eine positive Zahl L konvergierende Teilfolge auswählen, wenn $\mathsf{M} \to \infty$ strebt. Auch als Funktionen von $\bar{s}$ betrachtet, sind aber unsere $\dot{y}_{\mathsf{M}}$ gleichmäßig stetig und beschränkt.

[11]) Vgl. Courant, Jahresbericht der deutschen Mathematikervereinigung loc. cit.

stetig von s abhängende Tangente besitzen und die ganz im Innern oder auf dem Rande von B liegen und für die $\dot{x}(s)$ *immer größer oder gleich Null ist, so hat das Problem* (1) *eine Lösung mit stetiger Tangente. — D.h. es ist möglich, durch diese Erweiterung des Begriffes der zur Konkurrenz zugelassenen Funktionen im wesentlichen lediglich aus der Legendreschen Bedingung* (3) *die Existenz einer solchen Kurve nachzuweisen, für die das Integral* (1) *nicht größer ist als das Integral über irgendeine andere solche Kurve.*

Von Interesse scheint mir die Bemerkung, daß nirgends von der Existenz der unteren Grenze des Integrals Gebrauch gemacht ist. Man kann jedoch nunmehr ihre Existenz erschließen. Denn das Integral J, genommen für eine Lösung des Variationsproblems, ist endlich. Dies folgt, wenn man die vorausgesetzte Bedingung $F_{y'y'} > 0$ beachtet. Sie besagt nämlich, daß für festes x, y aus B und variables y' die Ungleichung gilt

$$F(x, y, y') \geqq F(x, y, 0) + \frac{\partial F}{\partial y'}(x, y, 0)\, y',$$

aus der für beliebiges x, y, y' wegen der Stetigkeit von F und $\frac{\partial F}{\partial y'}$

$$F(x, y, y') \geqq P - Q\,|y'|$$

mit konstantem P, Q folgt. Das Integral

$$\int (P - Q\,|y'|)\, dx$$

über eine feste Kurve $x(s)$, $y(s)$, nämlich unsere Lösung, bleibt aber endlich.

§ 4.

Eigenschaften der Lösung.

Die in der klassischen Variationsrechnung behandelten „hinreichenden Kriterien“ für die Existenz einer Lösung setzen im allgemeinen nicht den skizzierten erweiterten Begriff der „Lösung“ voraus und sind deshalb Kriterien dafür, daß die Lösung in unserem Sinne keine zur X-Achse senkrechten Stücke aufweist. Wir können nun aber mit einem Schlage eine Fülle von Fällen angeben, in denen offenkundig unsere Lösung u keine zur X-Achse senkrechten Stücke aufweisen wird: nämlich wenn das Integral über eine jede ein solches Stück enthaltende Kurve gegen unendlich divergiert.

Ein einfaches Beispiel liefert das Integral

$$J = \int_0^1 p(x, y)\, y'^2\, dx; \quad y(0) = 0; \quad y(1) = 1; \quad p(x, y) \geqq 1.$$

Eine zweite ebenso einfache Bemerkung bezieht sich auf die Frage, in welchen Fällen unsere Lösung überall der Eulerschen Differentialgleichung genügen wird. Nehmen wir an, daß wir bereits irgendwie bewiesen haben, daß zur x-Achse senkrechte Stücke in unserer Lösung nicht vorkommen. Ist nun der Bereich B so groß, daß man nachweisen kann, daß das Integral J für eine beliebige den Rand von B außer in $(0, c_0)$ und $(1, c_1)$ in mindestens noch *einem* Punkte treffende Funktion von x größer ist als das für eine feste, sonst *ganz im Innern von* B gelegene Funktion genommene Integral, so ist damit gezeigt, daß die *Lösung* von (1) im Innern von B gelegen und daher der Eulerschen Differentialgleichung im ganzen Intervall $0 \leqq x \leqq 1$ genügt, weil sie dann ein Extremum bietet gegenüber einer freien Variation.

Auch hierfür liefert das obige Integral ein Beispiel. Als Bereich B wähle man etwa

$$0 \leqq x \leqq 1; \qquad -\int_0^1 p(x, x)\, dx \leqq y \leqq \int_0^1 p(x, x)\, dx,$$

und vergleiche das Integral J über eine beliebige den Rand treffende Kurve mit dem Integral J über die geradlinige Verbindung von $(0, 0)$ und $(1, 1)$, d. i. mit

$$\int_0^1 (p(x, x)\, dx.$$

Zum Schluß dieser Ausführungen wollen wir noch einige Erweiterungen besprechen, die der Hauptsatz des § 3 zuläßt. Sie beziehen sich sowohl auf die durch (2) geforderten sogenannten künstlichen Randbedingungen wie auf das zum Minimum zu machende Funktional (1).

Gibt man nämlich die Forderung (2) auf, die den zugelassenen Funktionen die Randwerte vorschreibt, d. h. stellt man die Aufgabe, $J[y]$ zum Minimum zu machen gegenüber *allen* Funktionen y, die mit ihren ersten Ableitungen stückweise stetig und in B enthalten sind, so existiert auch in diesem Falle eine mit stetiger Tangente versehene Lösung des Problems. Dasselbe gilt, wenn wir einen Randwert, etwa $y(0) = c_0$, festhalten, für den anderen aber keine Bedingung stellen. Unter Lösung verstehen wir dabei immer den in § 3 besprochenen Begriff.

Ähnliche Resultate erhält man, wenn man an Stelle von $J[y] = \text{Min.}$ die Aufgabe stellt

$$J[y] + \sum_k Z_k(y(\zeta_k)) = \text{Min.}$$

Die Punkte ζ_k sind hierbei gewisse feste Punkte des Intervalls $0 \leqq \zeta_k \leqq 1$ und die Z_k irgendwelche mit ihrer Ableitung nach $y(\zeta_k)$ stetige Funktionen, sogenannte Zusatzglieder. Auch diese Probleme haben eine Lösung; jedoch wird im allgemeinen die Lösung in den Punkten ζ_k keine stetige Tangente mehr besitzen.

In der Literatur sind Fragen der Existenz von Lösungen in bekannten Arbeiten von Hilbert, Carathéodory und anderen behandelt worden. Vgl. Bolzas zusammenfassenden und ergänzenden Bericht: Bolza, Lehrbuch der Variationsrechnung, Leipzig und Berlin 1909. Für die Variationsprobleme, die Hilbert behandelt, ist charakteristisch, daß auch Kurven $x(s)$, $y(s)$ in die Konkurrenz eintreten, für die die Bedingung von S. 119 $\frac{dx}{ds} \geqq 0$ nicht mehr überall erfüllt ist. Aber schon das triviale Problem

$$\int_0^1 y'^2 |dx| = \text{Min.}, \quad y(0) = 0, \quad y(1) = 1$$

ist in diesem Sinne nicht mehr lösbar[12]) und fällt daher nicht unter die von Hilbert behandelten Fälle. Andererseits folgt der Hilbertsche Satz sowie seine Verallgemeinerung auf mehrere gesuchte Funktionen aufs einfachste aus dem Satz des § 1 bzw. dessen Verallgemeinerung. Jedoch möge der Beweis hierfür einer zweiten Abhandlung vorbehalten bleiben.

II. Teil.

Ansätze zu einer numerischen Methode der Lösung von Minimumproblemen[13]).

Wir geben nunmehr einen Ansatz zur numerischen Lösung von Randwertproblemen. Ich habe nachträglich bemerkt, daß er sich in seinem Ausgangspunkt eng mit einem schon bei Hadamard vorkommenden Gedanken berührt.

Die Erörterungen des ersten Teiles zeigten, daß das Problem der Lösung von (1) im wesentlichen auf die Auffindung der Polygonlösung der entsprechenden Differenzenaufgabe (6) für großes n hinausläuft.

Unser Verfahren ist eine sukzessive Approximation einer Lösung y_ν der Eulerschen Differenzengleichung

$$E_\nu(F) = l\frac{\partial F_\nu}{\partial y} + \frac{\partial F_\nu}{\partial y'} - \frac{\partial F_{\nu+1}}{\partial y'} = 0. \tag{8}$$

Wir machen die Voraussetzung, daß der Ausdruck

$$Q(y_1, \ldots, y_{n-1}) = l\sum_{\nu=1}^{n} F\left(x_\nu, y_\nu, \frac{y_\nu - y_{\nu-1}}{l}\right) \tag{10}$$

gegen unendlich divergiert, wenn mindestens ein $|y_\mu|$ $(\mu = 1, \ldots, n-1)$ gegen Unendlich strebt.

[12]) Man kann das Integral beliebig klein machen, indem man als Kurve einen Streckenzug wählt, der, von $(0, 0)$ ausgehend, mit hinreichend geringer Steigung bis zur Abszisse 1 führt, von dort wieder mit nur geringer Steigung zurück zur y-Achse und so im Zickzack weiter, um schließlich im Punkte $(1, 1)$ zu enden.

[13]) Vgl. Göttinger Nachrichten 18. Dez. 1925, loc. cit.

Wir bezeichnen mit y_ν eine beliebige Polygonfunktion, die den Randbedingungen genügt. Denken wir uns $y_1, y_2, \ldots, y_{n-1}$ als Komponenten eines Vektors im $(n-1)$-dimensionalen Raum. Durch

$$Q = \text{konst.}$$

werden gewisse Hyperflächen des $(n-1)$-dimensionalen Raumes charakterisiert. Ändert man den Vektor $(y_1, \ldots, y_{n-1})$ durch Addition eines unendlich kleinen Vektors vom Betrage $|ds|$, so gibt es eine Richtung $dy_1, \ldots, dy_{n-1}$ der stärksten Abnahme von Q, nämlich die des Gradienten von Q im Punkte $y_1, \ldots, y_{n-1}$. Seine Richtung wird durch die Formeln

$$(11) \qquad \frac{dy_1}{-\frac{\partial Q}{\partial y_1}} = \frac{dy_2}{-\frac{\partial Q}{\partial y_2}} = \ldots = ds$$

mit

$$\frac{\partial Q}{\partial y_1} = E_1(F), \qquad \frac{\partial Q}{\partial y_2} = E_2(F), \ldots$$

gegeben. Es ist $|ds| = \sqrt{dy_1^2 + \ldots + dy_{n-1}^2}$. s spielt die Rolle eines unwesentlichen Parameters. Wir haben das Vorzeichen von ds so eingerichtet, daß einer Zunahme von s eine Abnahme von Q entspricht:

$$(12) \qquad \frac{dQ}{ds} = \frac{\partial Q}{\partial y_1}\frac{dy_1}{ds} + \ldots + \frac{\partial Q}{\partial y_{n-1}}\frac{dy_{n-1}}{ds} = -\sum_{i=1}^{n-1}\left(\frac{\partial Q}{\partial y_i}\right)^2.$$

Wir behaupten, daß die Lösungen $y_\nu(s)$ dieser Differentialgleichung bei beliebigen Anfangswerten $y_{0,1}, \ldots, y_{0,n-1}$ gegen die Lösungen y_ν der Gleichungen

$$E_\nu(F) = 0 \qquad (\nu = 1, \ldots, n-1)$$

konvergieren, wenn s zunimmt. In der Tat, nach Voraussetzung sind Q und seine partiellen Ableitungen nach $y_1, \ldots, y_{n-1}$ stetige Funktionen. Q strebt gegen Unendlich, wenn mindestens ein $|y_\nu|$ über alle Grenzen wächst. Q besitzt also ein Minimum und die Vektorlösung der Differentialgleichungen (11) bewegt sich in einem festen abgeschlossenen Gebiete des $(n-1)$-dimensionalen Raumes, das nur von den Anfangswerten $y_{0,1}, \ldots, y_{0,n-1}$ abhängt. Es muß nun $-\frac{dQ}{ds}$ kleiner als jede positive Zahl werden, wenn s wächst. Solange nämlich

$$-\left(\frac{dQ}{ds}\right)_{s=s_1} = \sum_i \left(\frac{\partial Q}{\partial y_i}\right)^2 = \alpha > 0,$$

gilt wegen der Stetigkeit der $\frac{\partial Q}{\partial y_1}, \ldots, \frac{\partial Q}{\partial y_{n-1}}$ auch für ein endliches Intervall $s_2 \geqq s \geqq s_1$

$$-\frac{dQ}{ds} > \frac{\alpha}{2}.$$

Die Länge dieses Intervalls hängt nur von α ab. Q selbst wird sich also zwischen s_1 und s_2 um mindestens $|s_2 - s_1|\frac{\alpha}{2}$ verkleinern. Es folgt nun leicht unsere Behauptung, daß $-\frac{dQ}{ds}$ nach Null strebt, wenn s wächst. Also strebt wegen (12) $\frac{\partial Q}{\partial y_i}$ $(i = 1, \ldots, n-1)$ gegen Null, was wir beweisen wollten.

Um nun dieses Ergebnis zur numerischen Rechnung zu verwerten, werden wir versuchen, die Integration der Differentialgleichungen (11) durch eine ähnliche, aber rechnerisch leichtere Operation zu ersetzen. Dies ist um so eher möglich, als uns ja nicht die Differentialgleichungen (11) interessieren, sondern die *Grenzwerte* y_ν ihrer Lösungen, wenn s wächst, bei beliebigen Ausgangswerten $y_{0\nu}$.

Zu dem Zweck setzen wir in Anlehnung an unsere Differentialgleichungen (11)

$$y_{1\nu} = y_{0\nu} - \varepsilon\,\eta_{0\nu}.$$

$\eta_{0\nu}$ ist gleich dem Eulerschen Ausdruck im Punkte $(x_\nu, y_{0\nu})$

$$\eta_{0\nu} = E_\nu(F_0).$$

ε ist eine positive Zahl, die wir sogleich passend bestimmen werden. Für sehr kleine ε ist gemäß (12) die Abnahme von Q proportional $\varepsilon\sum_{\nu=1}^{n-1}\left(\frac{\partial Q}{\partial y_\nu}\right)^2$ also eine monotone Funktion von ε. Da Q gegen Unendlich divergiert, wenn mindestens ein $|y_\mu|$ gegen Unendlich strebt, so muß aber der Ausdruck

$$Q(y_{01} - \varepsilon\,\eta_{01},\; y_{02} - \varepsilon\,\eta_{02}, \ldots)$$

bei wachsendem ε wieder zunehmen. Es muß also einen ersten Wert ε_0 geben, für den $Q(y_0 - \varepsilon\,\eta_0)$ stationär ist. Er entsteht durch Differentiation von $Q(y_0 - \varepsilon\,\eta_0)$ nach ε. Setzen wir

$$y_{1\nu} = y_{0\nu} - \varepsilon_0\,\eta_{0\nu}, \tag{13}$$

so genügt ε_0 der Gleichung

$$\sum_{\nu=1}^{n-1} E_\nu(F_0)\,E_\nu(F_1) = 0. \tag{14}$$

Hier haben wir F_0, F_1 gesetzt als Abkürzung für die Funktion F, genommen für die Werte $y_{0\nu}$, $y_{1\nu}$ der Unbekannten y_ν.

Auf $y_{1\nu}$ wende man dasselbe Verfahren wie auf $y_{0\nu}$ an. Man bestimme also $\eta_{1\nu}$ und ε_1 und setze

$$y_{2\nu} = y_{1\nu} - \varepsilon_1\,\eta_{1\nu}. \tag{13}$$

Allgemein bilde man sukzessive für $m = 1, 2, \ldots$

(13) $$y_{m,\nu} = y_{m-1,\nu} - \varepsilon_{m-1}\,\eta_{m-1,\nu}.$$

Daß die $y_{m,\nu}$ mit wachsendem m, d. h. bei Wiederholung des Verfahrens gegen die Lösungen y_ν der Eulerschen Differenzengleichungen konvergieren, erkennt man durch eine ähnliche Betrachtung, wie wir sie soeben für die Lösungen der Differentialgleichungen (11) durchgeführt haben.

Man sieht ohne Mühe, daß es für das numerische Verfahren ausreicht, an Stelle des durch (14) bestimmten ε_m in der Formel (13) irgendein ε'_m einzuführen, wenn man nur voraussetzen darf, daß z. B.

$$\varepsilon_m > \varepsilon'_m > \frac{\varepsilon_m}{2}$$

ist.

Hat man die lösende Polygonfunktion mit $n+1$ Eckpunkten approximiert, so kann man dieselbe als Polygonfunktion mit $2n+1$ Eckpunkten ansehen und sie als Ausgangsfunktion eines neuen Verfahrens benutzen, das die lösende Polygonfunktion mit $2n+1$ Ecken liefern soll.

Grundsätzlich wenigstens geben die Werte der $E_\nu(F)$ ein Maß für die Güte der Approximation an.

(Eingegangen am 1. 8. 1926.)

Mathematische Annalen
98 (1927), pp. 179–191

Über das Anfangswertproblem einer hyperbolischen nichtlinearen partiellen Differentialgleichung zweiter Ordnung mit zwei unabhängigen Veränderlichen.

Von

Hans Lewy in Göttingen.

Einleitung.

Die vorliegende Arbeit handelt von dem Cauchyschen Anfangswertproblem für partielle Differentialgleichungen höherer Ordnung. Schon vor vielen Jahren wurde es von Frau Kowalewski in einer bekannten Arbeit gelöst, unter der Voraussetzung, daß die Anfangswerte analytisch sind. Jedoch scheint diese Lösung nur eine mehr formale Bedeutung zu haben, denn unter Voraussetzung analytischer Anfangswerte treten noch nicht einmal die Unterschiede zwischen hyperbolischen und elliptischen Differentialgleichungen zutage. Es gibt eben immer in einem gewissen Bereich der unabhängigen Veränderlichen eine Lösung, die selbst analytisch ist. Feinere und charakteristischere Fragestellungen treten erst auf, wenn man nicht mehr die Anfangswerte als analytisch voraussetzt. Man merkt alsbald, daß für elliptische lineare .Differentialgleichungen zweiter Ordnung das Anfangswertproblem im allgemeinen unlösbar ist, während es für hyperbolische lineare Differentialgleichungen sehr wohl eine Lösung besitzt. Die Form der Abhängigkeit der Lösung von den Anfangswerten ist bei den hyperbolischen Differentialgleichungen äußerst merkwürdig. Es reicht aus, nur einen Teil der Anfangswerte zu kennen, um die Lösung in einem gewissen Gebiet zu bestimmen. Diese Tatsache ist bisher meines Wissens nur für lineare partielle Differentialgleichungen bewiesen worden. Ich habe mir nun die Frage gestellt, ob alle hyperbolischen partiellen Differentialgleichungen zweiter Ordnung diese Eigenschaft besitzen[1]), d. h. speziell, ob für hyperbolische Differentialgleichungen zweiter Ordnung

[1]) Der hyperbolische Charakter der Differentialgleichung braucht nur für die Anfangswerte gefordert zu werden; vgl. § 3.

1. *das Anfangswertproblem bei beliebigen, auch nicht analytischen Vorgaben lösbar ist,*

2. *die Lösung in jedem Punkte nur von einem Teil der Anfangswerte abhängig,*

3. *die Lösung durch die Anfangswerte eindeutig bestimmt ist.*

In dieser Arbeit werden alle diese drei Fragen mit *Ja* beantwortet, wenigstens für den allein ins Auge gefaßten Fall zweier unabhängiger Variablen x, y. Die Punkte 1. und 2. geben den Grund für die Bedeutung, die die hyperbolischen Differentialgleichungen in der Theorie irgendwelcher Ausbreitungsvorgänge spielen, die mit endlicher Geschwindigkeit vor sich gehen.

Der Umstand, daß die Lösung einer hyperbolischen Differentialgleichung zweiter Ordnung in jedem Punkte zwar nur von einem Teil der vorgeschriebenen Anfangswerte, aber im allgemeinen jedenfalls von *sämtlichen* Anfangswerten auf einem bestimmten Kurvenstück abhängt, gibt den wesentlichen Unterschied zwischen dem Anfangswertproblem bei einer hyperbolischen Differentialgleichung zweiter Ordnung und bei einer partiellen Differentialgleichung erster Ordnung. Bekanntlich genügt es bei einer partiellen Differentialgleichung erster Ordnung ein Flächenelement zu kennen, um sofort einen ganzen zur Lösung gehörigen Streifen zu bestimmen. In diesem Sinne ist also hier die Lösung in irgendeinem Punkte des Streifens durch die Kenntnis des einen Flächenelements bestimmt. Anders bei einer hyperbolischen Differentialgleichung zweiter Ordnung. Hier ist die Lösung in einem Punkte nicht bloß durch die Angabe endlich vieler Anfangswerte bestimmbar; sie ist ein Funktional von unendlich vielen Anfangswerten. Setzt man daher die charakteristischen Differentialgleichungen an, so kann man nach der im allgemeinen üblichen Auffassung derselben als eines Systems gewöhnlicher Differentialgleichungen durch ihre Integration möglicherweise zu einem intermediären Integral, nicht aber zur Lösung selbst gelangen. Genauer: Die charakteristischen Gleichungen, aufgefaßt als ein System gewöhnlicher Differentialgleichungen auf einer Charakteristik, sind unterbestimmt.

In dieser Arbeit wird nun von der Tatsache ausgegangen, daß es zwei Scharen von Charakteristiken und entsprechend *zwei* Systeme von charakteristischen Differentialgleichungen gibt, deren simultane Behandlung allein zur Lösung führen kann. Es mußten zu dem Zweck diese beiden Systeme als *partielle* Differentialgleichungen aufgefaßt werden, in denen die unabhängigen Variablen zwei passend gewählte Kurvenparameter auf den Charakteristiken sind.

Die *beiden* Systeme von charakteristischen Differentialgleichungen *zusammen* bilden nun ein überbestimmtes System von *partiellen* Differentialgleichungen erster Ordnung für mehrere gesuchte Funktionen. Die Aufgabe, gewisse der charakteristischen Differentialgleichungen als Folge der übrigen darzustellen, bildet den einen Teil unserer Betrachtungen; daß das restliche System von charakteristischen Differentialgleichungen wirklich zur Bestimmung der Lösung der partiellen Differentialgleichung zweiter Ordnung ausreicht, wobei beliebige zweimal stetig differenzierbare Anfangswerte gegeben sein können, bildet den anderen Teil. Zu diesem Zwecke wird das System der charakteristischen Differentialgleichungen durch ein entsprechendes *Differenzen*-Gleichungssystem ersetzt. Da in ihm die Differenzenquotienten nur linear auftreten, gelingt es leicht nachzuweisen, daß die Differenzengleichungen bzw. ihre Lösungen bei Verengung der Maschenweite des zugrunde gelegten Gitters gegen die entsprechenden Differentialgleichungen und deren Lösungen konvergieren. Gleichzeitig wird so der oben erwähnte Punkt 2. mit erledigt, indem die darin ausgesprochene Behauptung gewissermaßen schon im Ansatz der Differenzengleichungen antizipiert wird. Vermöge der Bemerkung, daß jede Lösung der charakteristischen Differentialgleichungen wiederum die Lösung der entsprechenden Differenzengleichungen approximiert, die letztere aber infolge der Linearität leicht als eindeutig erkennbar ist, wird schließlich auch die Eindeutigkeit der Lösung der partiellen Differentialgleichung nachgewiesen[2]).

§ 1.

Ein spezielles System von Differentialgleichungen.

Um den Gedanken der folgenden Untersuchungen in möglichst klarem Lichte zu zeigen, erscheint es ratsam, sich vorerst mit einem ganz speziellen Problem zu beschäftigen.

In einer Ebene mit den rechtwinkligen Koordinaten α, β seien auf der Geraden $\alpha + \beta = 0$ oder besser nur auf einem Teil von ihr, sagen wir für

$$\alpha \leqq \alpha_0, \quad \beta \leqq \beta_0, \quad \alpha_0 + \beta_0 > 0,$$

die Funktionen p und q als stetig differenzierbare Funktionen der Bogenlänge gegeben (siehe Figur).

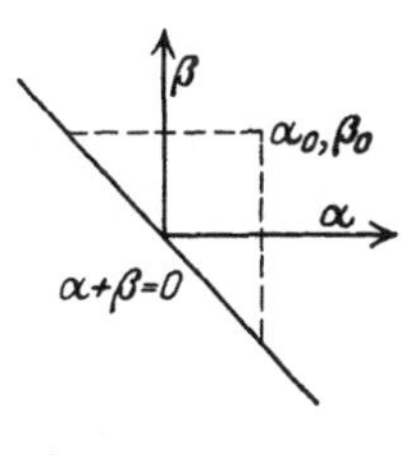

Indem wir unter dem links gerichteten Akzent die partielle Ableitung nach α, dem rechts gerichteten diejenige nach β verstehen, fragen wir, ob es möglich ist, die Funktionen p, q in der α, β-

---

[2]) In einer gemeinsamen Arbeit ist es Herrn K. Friedrichs und mir gelungen, die Ergebnisse der vorliegenden Abhandlung auf den Fall einer hyperbolischen Differentialgleichung *beliebiger* Ordnung in zwei Variablen zu verallgemeinern. Vgl. Math. Annalen 98 „Lösung des Anfangswertproblems einer belieb. hyperb. Diff.-Gl. usw."

Ebene stetig an ihre Werte auf der Geraden $\alpha+\beta=0$ anzuschließen, so daß dabei das folgende System von Differentialgleichungen erfüllt ist:

$$
(1)\qquad \begin{aligned} a(p,q)p' + b(p,q)q' &= f(p,q),\\ c(p,q)p' + d(p,q)q' &= g(p,q). \end{aligned}
$$

Von den a, b, c, d, f, g verlangen wir, daß sie stetige erste Ableitungen nach p und q besitzen, solange p und q im Innern von gewissen später genauer zu definierenden Bereichen liegen. Weiter verlangen wir, daß die Determinante

$$
\Delta = \begin{vmatrix} a & b \\ c & d \end{vmatrix}
$$

für die Werte von p, q auf dem betrachteten Stück der Anfangsgeraden $\alpha+\beta=0$ nicht verschwindet. Es kann dann offenbar für die Anfangswerte die Voraussetzung

$$
\Delta > \delta > 0
$$

gemacht werden, wo δ eine feste positive Zahl bedeutet.

Wir legen in die α, β-Ebene ein quadratisches Gitter der Maschenweite h, welches aus den Punkten $(\mu h, \nu h)$ bestehen möge $(\mu, \nu = \ldots -1, 0, 1, \ldots)$. Wir ersetzen die Gleichungen (1) für $\mu+\nu \geqq 0$ durch das entsprechende Differenzengleichungssystem

$$
(\mathrm{I})\qquad \begin{aligned} a_{\mu\nu}(p_{\mu+1,\nu} - p_{\mu\nu}) + b_{\mu\nu}(q_{\mu+1,\nu} - q_{\mu\nu}) &= h f_{\mu\nu}\\ c_{\mu\nu}(p_{\mu,\nu+1} - p_{\mu\nu}) + d_{\mu\nu}(q_{\mu,\nu+1} - q_{\mu\nu}) &= h g_{\mu\nu}, \end{aligned}
$$

wo

$$
p_{\mu\nu} = p(\mu h, \nu h), \qquad q_{\mu\nu} = q(\mu h, \nu h)
$$

und

$$
a_{\mu\nu} = a(p_{\mu\nu}, q_{\mu\nu}), \qquad b_{\mu\nu} = b(p_{\mu\nu}, q_{\mu\nu}), \ldots
$$

bedeuten. Da auf der Geraden $\alpha+\beta=0$ liegende Gitterpunkte die Koordinaten $\alpha=\mu h, \beta=-\mu h$ besitzen, so sind uns die Werte $p_{\mu,-\mu}$, $q_{\mu,-\mu}$ (jedenfalls für gewisse μ) bekannt.

Greifen wir die drei Punkte $[\mu h, (\nu+1)h]$, $[(\mu+1)h, \nu h]$ und $[(\mu+1)h, (\nu+1)h]$ heraus. Sie bilden ein rechtwinkliges „Elementardreieck". Kennt man $p_{\mu,\nu+1}$, $q_{\mu,\nu+1}$, $p_{\mu+1,\nu}$, $q_{\mu+1,\nu}$, so bestimmen die Gleichungen (I) die Werte $p_{\mu+1,\nu+1}$ und $q_{\mu+1,\nu+1}$, vorausgesetzt, daß

$$
\bar{\Delta}_{\mu+1,\nu+1} = \begin{vmatrix} a_{\mu,\nu+1} & b_{\mu,\nu+1} \\ c_{\mu+1,\nu} & d_{\mu+1,\nu} \end{vmatrix} \neq 0.
$$

Wir können also sofort vermöge der vier Anfangswerte $p_{\mu,-\mu}$, $q_{\mu,-\mu}$, $p_{\mu-1,-\mu+1}$, $q_{\mu-1,-\mu+1}$ die Werte $p_{\mu,-\mu+1}$, $q_{\mu,-\mu+1}$ bestimmen, vorausgesetzt, daß die nur von Anfangswerten abhängige Determinante

$$\bar{\varDelta}_{\mu,\,-\mu+1} \neq 0$$

ist. Wenn aber h genügend klein gewählt ist, so unterscheidet sich $\bar{\varDelta}$ genügend wenig von $\varDelta$ und daher folgt $\bar{\varDelta} \neq 0$. So werden die Werte auf der ersten Parallelreihe zur Geraden $\alpha + \beta = 0$ bestimmt. Von dieser ausgehend kann man nunmehr die auf der zweiten Parallelreihe liegenden Werte p, q ermitteln usw., solange die zugehörige Determinante $\bar{\varDelta}$ nicht verschwindet. Dieser Prozeß werde wiederholt für kleines und kleineres h. Wir wollen den Nachweis führen, daß die so erhaltenen Funktionen p und q und ihre Differenzenquotienten $p', q', p^{\backprime}, q^{\backprime}$ bei Verkleinerung der Maschenwerte h in einem festen Gebiete der α, β-Ebene gleichartig stetig bleiben.

1. Eine erste wichtige Bemerkung müssen wir über die Konvergenz der Differenzenquotienten $p', q', p^{\backprime}, q^{\backprime}$ auf der Anfangsgeraden $\alpha + \beta = 0$ machen. Der Übersichtlichkeit wegen wollen wir für die folgenden Betrachtungen die etwas umständliche Bezeichnung der Punkte durch Doppelindizes aufgeben. Es seien P und Q zwei aufeinander folgende, auf der Anfangsgeraden gelegene Punkte unseres Gitters; R der dritte Eckpunkt des „Elementardreiecks" PQR

$$\begin{matrix} P & \cdot R \\ & \cdot Q \end{matrix}$$

Wir hatten die Gleichungen

$$\begin{aligned} a_P(p_R - p_P) + b_P(q_R - q_P) &= h f_P \\ c_Q(p_R - p_Q) + d_Q(q_R - q_Q) &= h g_Q. \end{aligned}$$

Die zweite schreibt sich auch in der Form

$$c_Q(p_R - p_P) + d_Q(q_R - q_P) = h g_Q + c_Q(p_Q - p_P) + d_Q(q_Q - q_P).$$

Da aber die Anfangswerte der p und q stetig differenzierbar gegeben waren, also $\frac{p_Q - p_P}{h}$ und $\frac{q_Q - q_P}{h}$ gleichmäßig konvergieren, wenn h gegen null strebt, so folgt die gleichmäßige Konvergenz der Differenzenquotienten p', q' und entsprechend die der $p^{\backprime}, q^{\backprime}$ auf der Anfangsgeraden gegen stetige Grenzfunktionen.

2. Wir betrachten die vier Ecken $ABCD$ eines Gitterquadrates der Länge h (s. Figur)

$$\begin{matrix} \dot{E} & \cdot A & \cdot B \\ & \cdot D & \cdot C \\ & & \cdot F. \end{matrix}$$

Man erhält durch Bildung von Differenzen aus (I)

$$\begin{aligned} & a_A(p_B - p_A) + b_A(q_B - q_A) \\ & - a_D(p_C - p_D) - b_D(q_C - q_D) = h(f_A - f_D), \\ & c_C(p_B - p_C) + d_C(q_B - q_C) \\ & - c_D(p_A - p_D) - d_D(q_A - q_D) = h(g_C - g_D). \end{aligned}$$

Durch elementare Umformung kommt

$$\begin{aligned} & a_A(p_B - p_A - p_C + p_D) + b_A(q_B - q_A - q_C + q_D) \\ & = h(f_A - f_D) + (p_C - p_D)(a_D - a_A) + (q_C - q_D)(b_D - b_A), \\ & c_C(p_B - p_C - p_A + p_D) + d_C(q_B - q_C - q_A + q_D) \\ & = h(g_C - g_D) + (p_A - p_D)(c_D - c_C) + (q_A - q_D)(d_D - d_C). \end{aligned}$$

Hieraus läßt sich $\frac{p_B - p_A - p_C + p_D}{h^2} = p''$ und $\frac{q_B - q_A - q_C + q_D}{h^2} = q''$ berechnen, wenn

$$\Delta_{AC} = \begin{vmatrix} a_A & b_A \\ c_C & d_C \end{vmatrix} \neq 0.$$

Weiter entsteht infolge der Voraussetzung über $a, b, \ldots$ durch Anwendung des Mittelwertsatzes der Differentialrechnung

$$\left|\frac{\bar{\Delta}_{AC} - \bar{\Delta}_{ED}}{h}\right| \leqq H\left(\frac{|p_A - p_E|}{h} + \frac{|q_A - q_E|}{h} + \frac{|p_C - p_D|}{h} + \frac{|q_C - q_D|}{h}\right),$$

wo $H > 0$ eine von h unabhängige Konstante bezeichnet. Es wird also, falls $\bar{\Delta}_{ED} > 0$ ist,

$$\bar{\Delta}_{AC} \geqq \bar{\Delta}_{ED}(1 - hHK),$$

wo $K > 0$ ebenfalls von h unabhängig ist (s. u.).

Nun folgen aber die von h unabhängigen Beschränkungen

$$\left|\frac{p_B - p_A - p_C + p_D}{h^2}\right| \leqq \frac{L}{\bar{\Delta}_{AC}} \quad \text{und} \quad \left|\frac{q_B - q_A - q_C + q_D}{h^2}\right| \leqq \frac{L}{\bar{\Delta}_{AC}} \quad (L > 0),$$

d. h.

$$(2) \qquad |p''| = \frac{1}{h}\left|\frac{p_B - p_A}{h} \quad \frac{p_C - p_D}{h}\right| \leqq \frac{L}{(1 - hKH)\bar{\Delta}_{ED}}.$$

Die Konstanten H, K, L werden durch Abschätzungen gewonnen, unter der Voraussetzung, daß die p, q und die Differenzenquotienten $p', q', p^{\backprime}, q^{\backprime}$ im Innern von gewissen Bereichen liegen; diese Bereiche wiederum mögen so gewählt sein, daß sie die Anfangswerte der $p, q, p', q', p^{\backprime}, q^{\backprime}$ im Innern enthalten. Nehmen wir an, daß die Bereiche heißen

$$p_0 < p < p_1, \quad q_0 < q < q_1, \quad p'_0 < p' < p'_1, \ldots .$$

Liegen EDF auf der Anfangsgeraden, so ist sicher für hinreichend kleines h

$$\bar{\Delta}_{ED} > 0.$$

Infolge unserer Abschätzung (2) ändern sich beim Übergang zur ersten Parallelreihe die $p', q', \ldots$ um weniger als $\frac{hL}{(1-hKH)\bar{\Delta}_{ED}}$; ist h genügend klein, so gehören die neuen Werte der $p, q, p', \ldots$ wieder dem Bereich

$$p_0 < p < p_1, \qquad q_0 < q < q_1; \qquad p_0' < p' < p_1', \ldots$$

an und infolgedessen läßt sich die obige Abschätzung (2) mit denselben Konstanten H, K, L und mit $\bar{\Delta}_{AC}$ noch einmal anwenden.

Man findet durch Fortsetzung dieses Verfahrens z. B. für den Differenzenquotienten $p^{\prime\backprime}$ auf der k-ten Parallelreihe

$$|p^{\prime\backprime}| \leqq M e^{kh\gamma}$$

mit von h unabhängigen Konstanten $\gamma > 0$, $M > 0$ derselben Natur wie H, K, L (s. o.). Andrerseits folgt hieraus, daß man ein festes an die Anfangsgerade angrenzendes Flächenstück der α, β-Ebene angeben kann, in dessen Punkten die $p, q, p', \ldots$ sicher nicht aus ihren obigen Bereichen heraustreten können, wie klein auch h sein mag; denn für einen festen Punkt der α, β-Ebene strebt kh gegen einen endlichen Grenzwert.

Setzt man nun die Anfangswerte zweimal stetig differenzierbar voraus, was wir, um Weiterungen zu vermeiden, tun wollen, und verlangt, daß auch die a, b, c, d, f, g zweimal stetig nach den Argumenten differenzierbar sind, so leitet man in ganz gleicher Weise die Beschränktheit von $p^{\prime\prime\backprime}$, $p^{\prime\backprime\backprime}$ bzw. $q^{\prime\prime\backprime}$, $q^{\prime\backprime\backprime}$ ab. Nach einem bekannten Satze läßt sich nun aus der Funktionenfolge $p^{\prime\backprime}$, die man erhält, wenn man h monoton gegen Null fallen läßt und die zugehörigen Funktionen $p^{\prime\backprime}$ durch den Differenzenprozeß (I) bildet, eine gleichmäßig gegen eine stetige Funktion $\bar{p}^{\prime\backprime}$ konvergente Teilfolge auswählen. Die Differenzenquotienten $p', p^{\backprime}$ konvergieren dann natürlich gleichmäßig gegen die entsprechenden Integrale über die Grenzfunktion $\bar{p}^{\prime\backprime}$. Schließlich konvergieren auch die p, und die Grenzfunktion $\bar{p}$ hat die richtigen Anfangswerte. Dasselbe gilt entsprechend von $\bar{q}^{\prime\backprime}, \bar{q}', \bar{q}^{\backprime}, \bar{q}$. Die Grenzfunktionen $\bar{p}, \bar{q}$ genügen den Differentialgleichungen (1).

Wir wollen jetzt die Eindeutigkeit einer Lösung $\bar{p}, \bar{q}$ der Differentialgleichungen bei den vorgegebenen Anfangswerten zeigen, falls man nur Lösungen p, q mit stetigen gemischten Differentialquotienten $p^{\prime\backprime}, q^{\prime\backprime}$ zuläßt. Legt man ein Gitter $(\mu h, \nu h)$ der Maschenweite h in die α, β-Ebene, so gehorchen irgendwelche Lösungen $\bar{p}, \bar{q}$ der Differentialgleichungen (1) den Differenzengleichungen

$$(3) \qquad \begin{aligned} a\bar{p}^{\prime\backprime} + b\bar{q}^{\prime\backprime} &= (f(\bar{p},\bar{q}))^{\backprime} - a^{\backprime}\bar{p}' - b^{\backprime}\bar{q}' + \varepsilon_h, \\ c\bar{p}^{\prime\backprime} + d\bar{q}^{\prime\backprime} &= (g(\bar{p},\bar{q}))' - c'\bar{p}^{\backprime} - d'\bar{q}^{\backprime} + \varepsilon_h^*, \end{aligned}$$

wo ε_h und ε_h^* gewisse Größen bezeichnen, die mit h gleichmäßig im ganzen Gebiete gegen null streben. Wir behaupten nun, daß der Unterschied der Lösung $\bar{p}, \bar{q}$ dieser Differenzengleichungen (3) von der Lösung p, q der Differenzengleichungen (I) bei gleichen Anfangswerten gegen null strebt, wenn h gegen null abnimmt. Der Beweis ergibt sich aufs einfachste dergestalt, daß sogar die Differenzen der ersten *Differenzenquotienten* $|\bar{p}' - p'|$, $|\bar{p}^{`} - q^{`}|, \ldots$ auf der k-ten Parallelreihe von der Größenordnung $k \cdot h(\varepsilon + h)$ und daher auch die Differenzen $|a(\bar{p}, \bar{q}) - a(p, q)|$, $|a'(\bar{p}, \bar{q}) - a'(p, q)|, \ldots$ von derselben Größenordnung sind, also mit h gegen null streben. Damit ist nachgewiesen, daß die Lösung $\bar{p}, \bar{q}$ der Differentialgleichungen der *Limes der (eindeutig bestimmten) Lösungen* p, q *der Differenzengleichungen* (I) bei verschwindender Maschenweite ist. Die Lösung $\bar{p}, \bar{q}$ der Differentialgleichungen (1) mit den vorgeschriebenen Anfangswerten ist also eindeutig.

Der Wert der $\bar{p}, \bar{q}$ in einem Punkte $\alpha_1, \beta_1, \alpha_1 + \beta_1 > 0$ hängt, wie jede einzelne Lösung der Differenzengleichungen, nur ab von den Anfangswerten, die auf demjenigen Teil der Geraden $\alpha + \beta = 0$ liegen, der von den Geraden $\alpha = \alpha_1$, $\beta = \beta_1$ ausgeschnitten wird. Die Voraussetzung $\alpha_1 + \beta_1 > 0$ ist übrigens natürlich unwesentlich für die vorangehenden Überlegungen. Man würde nur zweckmäßig für $\alpha_1 + \beta_1 < 0$ die Differenzengleichungen entsprechend, d. h. unter Umkehrung gewisser Vorzeichen anzusetzen haben.

§ 2.

Ein allgemeiner Existenzsatz.

Das Beispiel des § 1 ist typisch für einen allgemeinen Satz, dessen Beweis auch dem Sinne nach ohne jede Modifikation genau wie der des vorigen Paragraphen zu führen ist. Er lautet:

Wenn auf der Strecke $\alpha + \beta = 0$, $\alpha \leqq \alpha_0$, $\beta \leqq \beta_0$, $\alpha_0 + \beta_0 > 0$ die Funktionen $\varphi_1, \varphi_2, \ldots, \varphi_n$ zweimal stetig nach der Bogenlänge differenzierbar vorgegeben sind, so lassen sie sich in eine gewisse Umgebung der Strecke derart fortsetzen, daß die n partiellen Differentialgleichungen erfüllt sind:

$$\sum_{k=1}^{n} a_{ik} \varphi_k' = b_i \qquad (i = 1, 2, \ldots, m < n),$$

$$\sum_{k=1}^{n} a_{ik} \varphi_k^{`} = b_i \qquad (i = m + 1, \ldots, n)$$

vorausgesetzt, daß die Determinante $|a_{ik}|$ für die Anfangswerte von null verschieden ist, und daß die a_{ik} und b_i zweimal stetig nach ihren Argumenten, den $\varphi_1, \ldots, \varphi_n$, differenzierbar sind. Die φ_i besitzen die stetige zweite Ableitung φ_i''. Es gibt nur ein einziges System von Funktionen φ_i mit stetiger gemischter zweiter Ableitung φ_i'', das diese Differentialgleichungen erfüllt und die vorgegebenen Anfangswerte annimmt.

§ 3.

Das Anfangsproblem von $F = 0$.

Mit Hilfe der charakteristischen Gleichungen ist es nun möglich, das Anfangswertproblem der für die Anfangswerte hyperbolischen Differentialgleichung

$$F(x, y, u, p, q, r, s, t) = 0 \tag{4}$$

auf den Satz des § 2 zurückzuführen. In der Gleichung (4) bedeuten x und y die beiden unabhängigen Variablen, u die gesuchte Funktion und p, q, r, s, t Abkürzungen für bzw. $u_x, u_y, u_{xx}, u_{xy}, u_{yy}$. Wir setzen F als dreimal stetig differenzierbare Funktion ihrer Argumente voraus.

Das Anfangswertproblem läßt sich dann so formulieren: In einer α, β-Ebene seien auf der Strecke

$$\alpha + \beta = 0; \quad \alpha \leqq \alpha_0, \quad \beta \leqq \beta_0, \quad \alpha_0 + \beta_0 > 0$$

die acht der Gleichung (4) gehorchenden Funktionen x, y, u, p, q, r, s, t als zweimal stetig nach der Bogenlänge differenzierbare Funktionen gegeben. Bezeichnet man die Differentiation nach der Bogenlänge auf der Geraden $\alpha + \beta = 0$ durch einen Stern, so mögen die genannten Funktionen die Gleichungen

$$u^* = p x^* + q y^*,$$
$$p^* = r x^* + s y^*,$$
$$q^* = s x^* + t y^*,$$

und die Ungleichung

$$\begin{vmatrix} F_r & F_s & F_t \\ x^* & y^* & 0 \\ 0 & x^* & y^* \end{vmatrix} \neq 0 \tag{5*}$$

erfüllen; die letzte besagt, daß $\alpha + \beta = 0$ nicht charakteristisch ist. Gefragt wird nach den Funktionen $x, y, u, p, \ldots, t$ der beiden Argumente α und β, die die Gleichung (4) erfüllen (wobei p, q, r, s, t die obige Bedeutung haben) und die auf der Geraden $\alpha + \beta = 0$ die angegebenen Anfangswerte annehmen.

Für das Folgende ist es praktisch, F_r und F_t ungleich null vorauszusetzen, eine Forderung, die durch Drehung des Koordinatensystems der x, y zu erfüllen ist und der Vermeidung lästiger Fallunterscheidungen dient.

Die charakteristische Gleichung

$$(5)\qquad \begin{vmatrix} F_r & F_s & F_t \\ \dot{x} & \dot{y} & 0 \\ 0 & \dot{x} & \dot{y} \end{vmatrix} = F_r \dot{y}^2 - F_s \dot{x}\dot{y} + F_t \dot{x}^2 = 0,$$

in der die Punkte Ableitungen nach einem Parameter darstellen, läßt sich unter der für die Anfangswerte z. B. sicher erfüllten Voraussetzung $F_s^2 - 4F_r F_t > 0$ (die besagt, daß die Differentialgleichung dort hyperbolisch ist), in die reellen linearen Faktoren zerfällen:

$$(\varrho_1 \dot{x} - \sigma_1 \dot{y})(\varrho_2 \dot{x} - \sigma_2 \dot{y}) = 0.$$

Die Funktionen $\varrho_1, \sigma_1, \varrho_2, \sigma_2$ genügen den Beziehungen

$$\varrho_1 \varrho_2 = F_t, \qquad \sigma_1 \sigma_2 = F_r, \qquad \varrho_1 \sigma_2 + \varrho_2 \sigma_1 = F_s.$$

Aus ihnen folgt

$$(\varrho_1 \sigma_2 - \varrho_2 \sigma_1)^2 = F_s^2 - 4F_r F_t.$$

Wir wählen etwa $\sigma_1 = 1$; dann ergibt sich

$$\sigma_2 = F_r \neq 0,$$

$$\varrho_2 = \frac{1}{2}\left(F_s - \sqrt{F_s^2 - 4F_r F_t}\right) \neq 0,$$

$$\varrho_1 = \frac{1}{2F_r}\left(F_s + \sqrt{F_s^2 - 4F_r F_t}\right) \neq 0.$$

Wir machen nunmehr die Linien $\alpha =$ konst. und $\beta =$ konst. zu Charakteristiken, indem wir in den Bezeichnungen der vorigen Paragraphen schreiben:

$$(6)\qquad \varrho_1 x' - \sigma_1 y' = 0,$$

$$(7)\qquad \varrho_2 x^{\backprime} - \sigma_2 y^{\backprime} = 0.$$

Wir benutzen außerdem die in bekannter Weise zu findenden weiteren charakterischen Gleichungen[3]):

$$(8)\qquad \varrho_1 F_r r' + \sigma_1 F_t s' + \sigma_1 (F_p r + F_q s + F_u p + F_x) y' = 0,$$

$$(9)\qquad \varrho_1 F_r s' + \sigma_1 F_t t' + \sigma_1 (F_p s + F_q t + F_u q + F_y) y' = 0,$$

$$(10)\qquad u' - p x' - q y' = 0,$$

$$(11)\qquad p' - r x' - s y' = 0,$$

$$(12)\qquad q' - s x' - t y' = 0,$$

$$(13)\qquad \varrho_2 F_r r^{\backprime} + \sigma_2 F_t s^{\backprime} + \sigma_2 (F_p r + F_q s + F_u p + F_x) y^{\backprime} = 0.$$

---

[3]) Vgl. z. B. K. Friedrichs und H. Lewy, „Lösung des Anfangswertproblems einer beliebigen hyperbolischen Differentialgleichung usw.", Math. Annalen **98**, wo eine einfache Ableitung gegeben wird.

Die Determinante der Gleichungen (6) bis (13) ist bis aufs Vorzeichen

$$\begin{vmatrix} \varrho_1 & -\sigma_1 & 0 & 0 & 0 & 0 & 0 & 0 \\ \varrho_2 & -\sigma_2 & 0 & 0 & 0 & 0 & 0 & 0 \\ 0 & * & \varrho_1 F_r & \sigma_1 F_t & 0 & 0 & 0 & 0 \\ 0 & * & 0 & \varrho_1 F_r & \sigma_1 F_t & 0 & 0 & 0 \\ 0 & * & \varrho_2 F_r & \sigma_2 F_t & 0 & 0 & 0 & 0 \\ * & * & 0 & 0 & 0 & 1 & 0 & 0 \\ * & * & 0 & 0 & 0 & 0 & 1 & 0 \\ * & * & 0 & 0 & 0 & 0 & 0 & 1 \end{vmatrix} = (\varrho_1 \sigma_2 - \sigma_1 \varrho_2)^2 \sigma_1 F_r F_t^2$$

$$= (F_s^2 - 4 F_r F_t) F_r F_t^2 \neq 0 .$$

Die Sterne vertreten gewisse Größen, die für den Wert der Determinante belanglos sind. Wir lösen nun die Differentialgleichungen (6) bis (13) für die Funktionen $x(\alpha, \beta)$, $y(\alpha, \beta)$, $u(\alpha, \beta), \ldots, t(\alpha, \beta)$, indem wir von der ursprünglichen *Bedeutung* von $p, q, \ldots$ *abstrahieren*, d. h. diesen Funktionen sonst keinerlei Beschränkung auferlegen. Nach dem Satze des vorigen Paragraphen gibt es in einer Umgebung der Geraden $\alpha + \beta = 0$ ein eindeutiges Lösungssystem der Gleichungen (6) bis (13) mit den vorgeschriebenen Anfangswerten. Man kann nunmehr wegen der Ungleichung

$$\frac{y'}{x'} - \frac{y^{\backprime}}{x^{\backprime}} = \frac{\varrho_1}{\sigma_1} - \frac{\varrho_2}{\sigma_2} \neq 0$$

auch wieder x und y als unabhängige Variable anstelle von α und β einführen und, wenn man will, also $u, \ldots$ als Funktionen von x, y auffassen.

Wir haben jetzt erstens zu beweisen, daß für die gefundenen Lösungsfunktionen $x(\alpha, \beta), \ldots, t(\alpha, \beta)$ die Gleichung

$$F = 0 \tag{4}$$

und zweitens die Beziehungen

$$U \equiv u^{\backprime} - p x - q y^{\backprime} = 0, \tag{14}$$

$$P \equiv p^{\backprime} - r x^{\backprime} - s y^{\backprime} = 0, \tag{15}$$

$$Q \equiv q^{\backprime} - s x^{\backprime} - t y = 0 \tag{16}$$

gelten. Erst dann können wir wegen der Gleichungen (10) bis (12) behaupten, daß wir eine *Lösung der Differentialgleichung* (4) konstruiert haben, weil dann tatsächlich die p, q, r, s, t die entsprechenden Ableitungen von u nach x und y sind.

Beweis. Addiert man zu der Gleichung (8) die mit ϱ_1 multiplizierte Gleichung (9), so erhält man bei Benutzung der Relationen

$$F_r \varrho_1^2 - F_s \varrho_1 \sigma_1 + F_t \sigma_1^2 = 0,$$
$$\sigma_1 = 1, \qquad \varrho_1 \neq 0$$

die Gleichung $F' = 0$. Da F die Anfangswerte Null besitzt, verschwindet es also identisch.

Um ferner $P = 0$ zu konstatieren, schreiben wir die Gleichungen (8) und (13) in der Form

(17) $$\frac{F_r}{x'} r' + \frac{F_t}{y'} s' + F_p r + F_q s + F_u p + F_x = 0,$$

(18) $$\frac{F_r}{x^{\backprime}} r^{\backprime} + \frac{F_t}{y^{\backprime}} s^{\backprime} + F_p r + F_q s + F_u p + F_x = 0.$$

Hieraus folgt
$$\frac{F_r}{x'} r' + \frac{F_t}{y'} s' = \frac{F_r}{x^{\backprime}} r^{\backprime} + \frac{F_t}{y^{\backprime}} s^{\backprime}.$$

Diese Gleichung nimmt nach Einführung von x, y als unabhängigen Variablen in Anbetracht der Ungleichung $\frac{y^{\backprime}}{x^{\backprime}} \neq \frac{y'}{x'}$ die Form an

$$F_r r_y \left(\frac{y'}{x'} - \frac{y^{\backprime}}{x^{\backprime}}\right) + F_t s_x \left(\frac{x'}{y'} - \frac{x^{\backprime}}{y^{\backprime}}\right) = 0.$$

Berücksichtigt man gemäß Gleichung (5), daß
$$\frac{F_r}{F_t} = \frac{x' x^{\backprime}}{y' y^{\backprime}},$$
so folgt

(19) $$r_y = s_x.$$

Für die Anfangswerte gelten nun die Gleichungen
$$p^* = r x^* + s y^*$$
und
$$p' = r x' + s y'.$$

Da $x^* : y^*$ wegen (5*) von $x' : y'$ verschieden ist, haben wir am Anfang $p_x = r$, $p_y = s$ oder $P = 0$. Nun gilt
$$P' = p'^{\backprime} - r' x^{\backprime} - s' y^{\backprime} - r x'^{\backprime} - s y'^{\backprime}$$
und die abgeleitete Gleichung (11)
$$0 = p'^{\backprime} - r^{\backprime} x' - s^{\backprime} y' - r x'^{\backprime} - s' y'^{\backprime}.$$
Es folgt daraus
$$P' = r^{\backprime} x' + s^{\backprime} y' - r' x^{\backprime} - s' y^{\backprime}$$
und unter Benutzung von (19)
$$P' = 0,$$
d. h. also jetzt
$$P \equiv 0; \qquad p_x \equiv r, \qquad p_y \equiv s.$$

Für U' und Q' findet man durch Anwendung dieser letzten Gleichung sowie von (12) die Ausdrücke

$$\begin{aligned} U' &= -p'x^{\backprime} - q'y^{\backprime} + p^{\backprime}x' + q^{\backprime}y' \\ &= (q^{\backprime} - s x^{\backprime} - t y^{\backprime})y' - (q' - s x' - t y')y^{\backprime} \\ &= Q y', \\ Q' &= -s'x^{\backprime} - t'y^{\backprime} + s^{\backprime}x' + t^{\backprime}y' \\ &= (t_x - s_y)(x^{\backprime}y' - x'y^{\backprime}). \end{aligned}$$

Subtrahiert man von (8) die Gleichung $\frac{dF}{dx} = 0$ und von (9) die Gleichung $\frac{dF}{dy} = 0$ und addiert die Resultate nach Multiplikation mit $x^{\backprime}$ bzw. $y^{\backprime}$, so erhält man unter Verwendung von (19) die Relation

$$\begin{aligned} &F_t(t_x - s_y)x^{\backprime} + F_q(q_x - s)x^{\backprime} + F_u(u_x - p)x^{\backprime} \\ -\,&F_t\frac{x'}{y'}(t_x - s_y)y^{\backprime} + F_q(q_y - t)\,y^{\backprime} + F_u(u_y - q)\,y^{\backprime} = 0. \end{aligned}$$

Man hat also die beiden *gewöhnlichen* linearen Differentialgleichungen für U und Q

$$\begin{aligned} \frac{F_t}{y'}Q' + F_q Q + F_u U &= 0, \\ U' &= Q y', \end{aligned}$$

aus denen sich infolge der Anfangswerte $U = 0$, $Q = 0$ die Relationen $U \equiv 0$, $Q \equiv 0$ leicht ableiten lassen.

Damit haben wir nachgewiesen, daß es in einem gewissen Gebiet zu willkürlichen zweimal stetig differenzierbaren Anfangswerten eine Lösung der Differentialgleichung (4) gibt. Die Lösung in einem Punkte hängt nur von einem Teile der Anfangswerte ab; nämlich nur von den Werten auf dem ausgeschnittenen Stück der Anfangslinie, das man erhält, wenn man, von diesem Punkte ausgehend, die Charakteristiken zieht, bis sie die Anfangskurve schneiden.

Da wir wissen, daß jede Lösung der Differentialgleichung (4) notwendigerweise die charakteristischen Gleichungen (6) bis (13) erfüllt, und da die Eindeutigkeit der Lösung der letzteren feststeht, so folgt, daß die Lösung der Differentialgleichung (4) eindeutig durch die Anfangswerte bestimmt ist[4]).

---

[4]) Allerdings unter der Voraussetzung, daß man nur Lösungen mit *stetigen* zweiten Differentialquotienten u', $p'^{\backprime}$, ... zuläßt.

(Eingegangen am 28. 10. 1926.)

Mathematische Annalen
98 (1927), pp. 192–204

Über die Eindeutigkeit und das Abhängigkeitsgebiet der Lösungen beim Anfangswertproblem linearer hyperbolischer Differentialgleichungen.

Von

Kurt Friedrichs und Hans Lewy in Göttingen.

Die folgenden Betrachtungen handeln von der Eindeutigkeit der Lösung des Anfangswertproblems für hyperbolische lineare partielle Differentialgleichungen. Bei den üblichen Beweisen werden gewisse Voraussetzungen über das Verhalten der Anfangswerte und der Lösung im *unendlichen* gemacht. Andrerseits zeigt aber die in gewissen Spezialfällen explizit bekannte Form der Lösung, daß ihr Wert in irgendeinem Punkt nur von den Vorgaben auf einem im *endlichen* liegenden Stück der Anfangslinie abhängt, daß also das Verhalten der Lösung im unendlichen ganz belanglos ist. Es ist daher zu erwarten, daß man den oben erwähnten Beweis für die Eindeutigkeit so umgestalten kann, daß er von dieser Tatsache Rechenschaft gibt [1]). In der neuen Form läßt sich der Beweis auf die allgemeine hyperbolische lineare Differentialgleichung zweiter Ordnung (auch mit nichtkonstanten Koeffizienten) übertragen; desgleichen auf gewisse Systeme von solchen Differentialgleichungen mit mehreren gesuchten Funktionen, wie sie z. B. in der Kristalloptik vorkommen. Damit ergibt sich auch für diese Anfangswertprobleme eine Antwort auf die Frage, von welchem Teil der Anfangswerte die Lösung in irgendeinem Punkte allein abhängen kann.

---

[1]) Nach Einlieferung dieser Arbeit haben wir erfahren, daß A. Rubinowicz auf Grund derselben Überlegung, die wir hier angestellt haben, die Eindeutigkeit der Lösungen der Maxwellschen Gleichungen nachgewiesen hat. Phys. Zeitschr. **27** (1926), S. 707. Vgl. auch Math. Annalen **96** (1927).

1.

Um uns zu orientieren, gehen wir aus von der einfachsten hyperbolischen Differentialgleichung zweiter Ordnung

$$u_{tt} - u_{xx} = 0 \tag{1}$$

für die Funktion u der beiden unabhängigen Variablen t und x. Wir stellen für sie das Anfangswertproblem, indem wir die Werte der Funktion u und ihrer ersten Ableitungen auf einer mit stetiger Tangente versehenen Kurve der (x, t)-Ebene vorschreiben und nach ihrer Fortsetzung in die (x, t)-Ebene gemäß der Differentialgleichung (1) fragen. Um zu zeigen, daß es höchstens *eine* Lösung geben kann, genügt es bekanntlich wegen der Linearität, nachzuweisen, daß die Lösung überall verschwindet, wenn die Anfangswerte verschwinden.

Wir betrachten das Integral

$$2 \iint_G u_t (u_{tt} - u_{xx})\, dx\, dt \tag{2}$$

über ein Gebiet G der (x, t)-Ebene, welches für eine Lösung der Differentialgleichung (1) offenbar verschwindet. Dieses Integral können wir folgendermaßen umformen:

$$\iint_G \{(u_t^2)_t + (u_x^2)_t - 2(u_x u_t)_x\}\, dx\, dt,$$

und dies läßt sich wiederum als folgendes Randintegral schreiben:

$$\int_\Gamma \left\{(u_t)^2 \frac{\partial t}{\partial n} + (u_x)^2 \frac{\partial t}{\partial n} - 2 u_x u_t \frac{\partial x}{\partial n}\right\} ds, \tag{3}$$

wobei s die Bogenlänge auf der Randkurve Γ und $\frac{\partial t}{\partial n}$ und $\frac{\partial x}{\partial n}$ die Ableitungen von t und x in Richtung der äußeren Normalen bedeuten. Der Integrand stellt eine quadratische Form in den Variablen u_x und u_t dar. Die Bedingung dafür, daß diese Form definit ist, lautet:

$$\begin{vmatrix} \frac{\partial t}{\partial n} & -\frac{\partial x}{\partial n} \\ -\frac{\partial x}{\partial n} & \frac{\partial t}{\partial n} \end{vmatrix} > 0. \tag{4}$$

Wir verlangen nun von unserer Anfangskurve A, daß sie dieser Ungleichungsbedingung genügt und betrachten ein einfach zusammenhängendes Gebiet G (vgl. Fig. 1), das außer von einem Teil der Anfangskurve von einer anderen Kurve B begrenzt wird, die gleichfalls der Bedingung (4) genügt, einer, wie wir sagen wollen, raumartigen Kurve. Da auf der Anfangskurve u_x und u_t verschwinden, reduziert sich unser Rand-

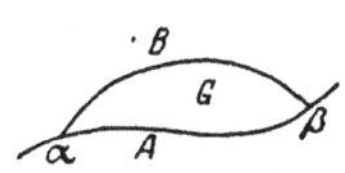

Fig. 1.

integral (3) auf ein Integral, das lediglich längs B zu erstrecken ist. Da der Integrand eine definite Form in u_x und u_t darstellt, der Wert des Integrals aber null ist, müssen auf der Linie B u_x und u_t verschwinden. Da wir die Linie B stetig in die Linie A so überführen können, daß in den Zwischenstadien stets die Ungleichung (4) erfüllt ist, so folgt, daß in dem ganzen Gebiet G u_x und u_t verschwinden. Daraus ergibt sich, daß auch die Funktion u selber in G identisch verschwindet.

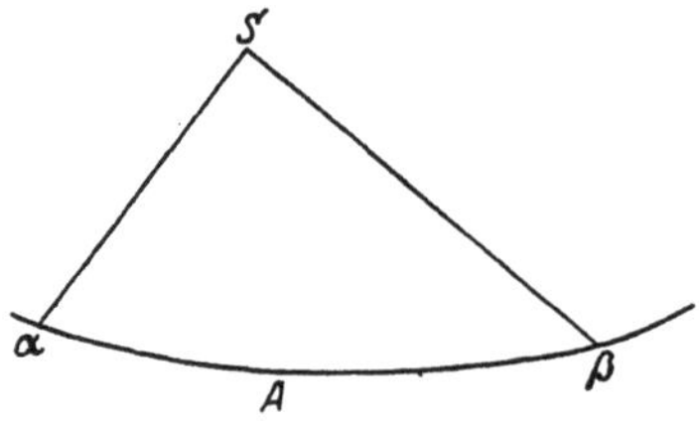

Fig. 2.

Den Schluß, daß u_x und u_t verschwinden, können wir offenbar in derselben Weise durchführen für das ganze Innere des krummlinigen Dreiecks (vgl. Fig. 2), das sich ergibt, wenn man von den Endpunkten α und β des Stückes A die beiden — offenbar geraden — Linien zieht, die der Gleichung

$$\left(\frac{\partial t}{\partial n}\right)^2 - \left(\frac{\partial x}{\partial n}\right)^2 = \begin{vmatrix} \frac{\partial t}{\partial n} & -\frac{\partial x}{\partial n} \\ -\frac{\partial x}{\partial n} & \frac{\partial t}{\partial n} \end{vmatrix} = 0$$

genügen und sich in einem Punkte S treffen (siehe Figur). Dies Dreieck können wir in der Grenze dadurch entstanden denken, daß wir die Linie B so weit wie möglich „aufblasen", ohne die Bedingung (4) zu verletzen.

Daß die Lösung u in der Spitze S verschwindet, können wir auch dadurch schließen, daß wir von vornherein dies Gebiet zugrunde legen. Dann reduziert sich das Integral (3) auf die Summe der beiden folgenden Integrale über die beiden „Charakteristiken" βS und $S\alpha$:

$$\sqrt{2}\left(\int\limits_{\beta S}\left(\frac{\partial u}{\partial s}\right)^2 ds + \int\limits_{S\alpha}\left(\frac{\partial u}{\partial s}\right)^2 ds\right) = 0.$$

Hieraus folgt $\frac{\partial u}{\partial s} = 0$, wobei s die Bogenlänge auf einer Charakteristik bedeutet; das heißt $u \equiv 0$ auf den beiden Charakteristiken und insbesondere in der Spitze S.

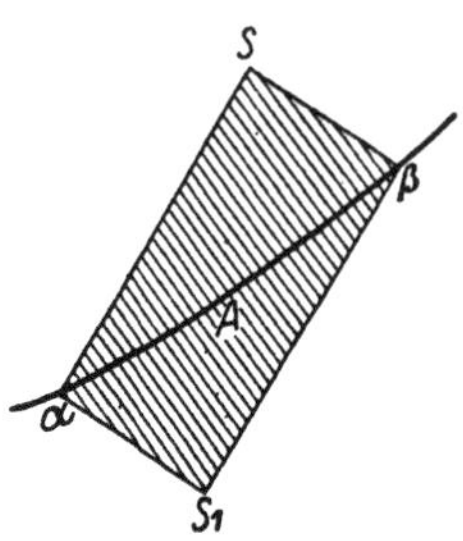

Fig. 3.

Unsere Betrachtungen lehren uns, daß zwei Lösungen der Differentialgleichung (1), deren Anfangswerte auf dem Bogen $\alpha\beta$ der Anfangslinie übereinstimmen, auch im ganzen Innern des „Charakteristikendreiecks" $S\alpha\beta$ übereinstimmen. Dasselbe gilt natürlich für das Charakteristikendreieck $S_1\alpha\beta$ (vgl. Fig. 3), das durch Ziehen der Charakteristiken durch α und β nach *unten* entsteht. Mit anderen Worten:

Ändern wir die Anfangswerte außerhalb von $\alpha\beta$, so ändert sich die Lösung nicht innerhalb des zu $\alpha\beta$ gehörigen Charakteristikendreiecks.

Wollen wir jetzt umgekehrt von irgendeinem Punkte S feststellen, von welchem Teil der Anfangskurve der Wert der Lösung in S höchstens abhängt, so brauchen wir also bloß durch diesen Punkt die beiden Charakteristiken zu ziehen, bis sie die Anfangskurve schneiden; der ausgeschnittene Teil ist das gesuchte Gebiet der Abhängigkeit.

Das soeben geschilderte Verhalten der Lösung u der Differentialgleichung (1) erweist sich in der Folge als typisch für die ins Auge gefaßten Verallgemeinerungen.

2.

Wir betrachten zunächst solche Differentialgleichungen, die durch Nullsetzen der Eulerschen Ableitung einer quadratischen Form F in den ersten Ableitungen zweier Funktionen u und v der unabhängigen Variablen $x_1, \dots, x_m, t$ entstehen. Wir beschränken uns der Einfachheit halber[2]) auf eine Form von der Gestalt

$$F = a_{ik} u_i u_k + b_{ik} u_i v_k + c_{ik} v_i v_k - a u_t^2 - c v_t^2 \;{}^{3})$$

mit konstanten Koeffizienten $a_{ik}, \dots, a, c$.

Hier bedeutet

$$u_i = \frac{\partial u}{\partial x_i}, \quad u_t = \frac{\partial u}{\partial t}; \quad v_i = \frac{\partial v}{\partial x_i}, \quad v_t = \frac{\partial v}{\partial t}.$$

Wir verlangen für a_{ik} und c_{ik} die Gleichungen

$$a_{ik} = a_{ki}, \quad c_{ik} = c_{ki},$$

was keine Beschränkung der Allgemeinheit bedeutet, und fordern, daß die Form

$$\bar{F}(u_i, v_i) = a_{ik} u_i u_k + b_{ik} u_i v_k + c_{ik} v_i v_k$$

eine nicht ausgeartete positiv definite Form in den Variablen $u_1, \dots, u_m, v_1, \dots, v_m$ ist. Die Bedeutung dieser Voraussetzungen, die unter anderem den hyperbolischen Charakter der aus der Form F entstehenden Eulerschen Differentialgleichungen garantieren, wird im Laufe dieses Paragraphen von selbst klar werden. Diese Differentialgleichungen lauten:

$$(5) \qquad \begin{aligned} 0 &= -[F]_u = 2 a_{ik} u_{ik} + b_{ik} v_{ik} - 2 a u_{tt} \\ 0 &= -[F]_v = 2 c_{ik} v_{ik} + b_{ik} u_{ik} - 2 c v_{tt}, \end{aligned}$$

wobei $u_{ik} = \dfrac{\partial^2 u}{\partial x_i \partial x_k}$, $u_{tt} = \dfrac{\partial^2 u}{\partial t^2}, \dots$ gesetzt ist.

---

[2]) Man beachte das Fehlen von Gliedern der Form $u_i u_t, \dots, u_t v_t$.

[3]) Hier und im folgenden ist wie üblich über doppelt auftretende Indizes von 1 bis m zu summieren.

Wir denken uns nunmehr auf einem mit stetigen Normalen versehenen n-dimensionalen Flächenstück A des $m+1$-dimensionalen (x_i, t)-Raumes, über dessen Natur wir später noch genauere Voraussetzungen machen werden, die Werte der Funktionen u und v und ihrer ersten Ableitungen vorgegeben, und fragen nach ihrer Fortsetzung in den $(x_i\, t)$-Raum gemäß den Differentialgleichungen (5). Es genügt aus dem in § 1 erörterten Grunde, die Anfangswerte als null anzunehmen.

Nunmehr bilden wir für die Lösung unseres Anfangswertproblems unter Beachtung der Identität

$$u_{ik} v_t + v_{ik} u_t = (u_i v_t)_k + (v_k u_t)_i - (u_i v_k)_t$$

die folgende Gleichung

$$\begin{aligned} 0 &= u_t [F]_u + v_t [F]_v \\ &= (a_{ik} u_i u_k + b_{ik} u_i v_k + c_{ik} v_i v_k + a u_t^2 + c v_t^2)_t \\ &- a_{ik}(u_i u_t)_k - a_{ik}(u_k u_t)_i - c_{ik}(v_i v_t)_k - c_{ik}(v_k v_t)_i \\ &- b_{ik}(u_i v_t)_k - b_{ik}(v_k u_t)_i \end{aligned}$$

und integrieren die rechte Seite über ein einfach zusammenhängendes Gebiet G des (x_i, t)-Raumes, das von zwei Flächen mit stetigen Normalen berandet wird. Dies Integral läßt sich offenbar in folgendes Randintegral überführen:

$$\int\limits_\Gamma J\,d\omega = \int\limits_\Gamma \left[\begin{array}{l} (a_{ik} u_i u_k + b_{ik} u_i v_k + c_{ik} v_i v_k + a u_t^2 + c v_t^2) \dfrac{\partial t}{\partial n} \\ - a_{ik}(u_i u_t) \dfrac{\partial x_k}{\partial n} - a_{ik}(u_k u_t) \dfrac{\partial x_i}{\partial n} - c_{ik}(v_i v_t) \dfrac{\partial x_k}{\partial n} \\ - c_{ik}(v_k v_t) \dfrac{\partial x_i}{\partial n} - b_{ik}(u_i v_t) \dfrac{\partial x_k}{\partial n} - b_{ik}(v_k u_t) \dfrac{\partial x_i}{\partial n} \end{array} \right] d\omega,$$

wobei $d\omega$ das Oberflächenelement und $\dfrac{\partial x_i}{\partial n}, \dfrac{\partial t}{\partial n}$ die Ableitungen von x_i und t in Richtung der äußeren Normalen der Oberfläche Γ bedeuten.

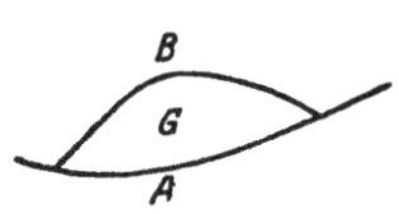

Fig. 4.

Wählen wir jetzt als die eine der beiden Flächen des Randes Γ ein einfach zusammenhängendes Stück der Anfangsfläche A, so reduziert sich das Randintegral auf ein solches, das nur über die andere Fläche B zu erstrecken ist, da der Integrand auf A verschwindet.

Wir können jetzt die Fläche B so wählen, daß der Integrand J definit wird und also verschwindet, da das ganze Integral verschwindet. Um dies einzusehen, schreiben wir den Integranden in folgender Form:

$$
(6)\qquad
\begin{aligned}
J = {} & \frac{1}{\frac{\partial t}{\partial n}} \overline{F}\left(u_i \frac{\partial t}{\partial n} - u_t \frac{\partial x_i}{\partial n},\ v_i \frac{\partial t}{\partial n} - v_t \frac{\partial x_i}{\partial n}\right) \\
& + \frac{1}{\frac{\partial t}{\partial n}} \left(a\left(\frac{\partial t}{\partial n}\right)^2 - a_{ik} \frac{\partial x_i}{\partial n} \frac{\partial x_k}{\partial n}\right) u_t^2 \\
& - \frac{1}{\frac{\partial t}{\partial n}} b_{ik} \frac{\partial x_i}{\partial n} \frac{\partial x_k}{\partial n} u_t v_t \\
& + \frac{1}{\frac{\partial t}{\partial n}} \left(c\left(\frac{\partial t}{\partial n}\right)^2 - c_{ik} \frac{\partial x_i}{\partial n} \frac{\partial x_k}{\partial n}\right) v_t^2 .
\end{aligned}
$$

Verlangen wir nun von den Normalen der Fläche B und der Einfachheit halber auch von denen der Fläche A, daß die Bedingungen der „Raumartigkeit"

$$
(7)\qquad
\begin{aligned}
&1.\ D = \begin{vmatrix} a\left(\frac{\partial t}{\partial n}\right)^2 - a_{ik} \frac{\partial x_i}{\partial n} \frac{\partial x_k}{\partial n} & -\frac{b_{ik}}{2} \frac{\partial x_i}{\partial n} \frac{\partial x_k}{\partial n} \\ -\frac{b_{ik}}{2} \frac{\partial x_i}{\partial n} \frac{\partial x_k}{\partial n} & c\left(\frac{\partial t}{\partial n}\right)^2 - c_{ik} \frac{\partial x_i}{\partial n} \frac{\partial x_k}{\partial n} \end{vmatrix} > 0 \\
&2.\ E = a\left(\frac{\partial t}{\partial n}\right)^2 - a_{ik} \frac{\partial x_i}{\partial n} \frac{\partial x_k}{\partial n} + c\left(\frac{\partial t}{\partial n}\right)^2 - c_{ik} \frac{\partial x_i}{\partial n} \frac{\partial x_k}{\partial n} > 0
\end{aligned}
$$

erfüllt sind, so ergibt sich, daß J eine nicht ausgeartete positiv definite quadratische Form der Variablen $u_i \frac{\partial t}{\partial n} - u_t \frac{\partial x_i}{\partial n}$, $v_i \frac{\partial t}{\partial n} - v_t \frac{\partial x_i}{\partial n}$, u_t, v_t ist, woraus folgt, daß alle diese Ausdrücke verschwinden, daß also, da wegen $(7)_2$ $\frac{\partial t}{\partial n} \neq 0$ sein muß, sämtliche Ableitungen von u und v auf B null sind.

Dasselbe kann man ebenso wie in § 1 nunmehr für das ganze Innere des von A und B eingeschlossenen Gebietes G beweisen. Und somit folgt aus dem Verschwinden von u und v auf der Fläche A das Verschwinden im ganzen Gebiet G.

Wenn wir jetzt umgekehrt fragen, von welchem Teil der Anfangswerte der Wert der Lösungen u, v in einem Punkte S abhängt, brauchen wir nur durch den Punkt S als Spitze den „charakteristischen" Kegel, d. h. denjenigen Kegel zu ziehen, dessen Gleichung in „Ebenen"koordinaten $D = 0$ lautet. Dieser Kegel schneidet aus der Anfangsfläche A, da diese raumartig angenommen ist, ein im Endlichen liegendes Stück C heraus. Nimmt man alle die Mantellinien, die den Punkt S mit der äußeren Berandung von C verbinden, so bilden sie zugleich mit dem Stück C ein kegelartiges Gebiet G. Dieses Gebiet [4]) besteht aus endlich vielen der

[4]) Über mögliche Gestalten des charakteristischen Kegels vgl. G. Herglotz, Sitzungsberichte der sächs. Akademie d. Wissensch. 1926.

oben von uns zugrunde gelegten Art, nur daß nicht auf dem Kegelmantel $D > 0$ ist, sondern $D = 0$, und daß anstatt $E > 0$ eventuell $E = 0$ ist. Infolgedessen können wir nicht mehr schließen, daß auf der Mantelfläche B alle Ableitungen von u und v verschwinden; jedenfalls aber sind wir sicher, daß die Variablen, die in $\overline{F}$ in (6) auftreten, nämlich $u_i \frac{\partial t}{\partial n} - u_t \frac{\partial x_i}{\partial n}$, $v_i \frac{\partial t}{\partial n} - v_t \frac{\partial x_i}{\partial n}$ einzeln verschwinden. Diese Ausdrücke sind aber unter der Bedingung $D = 0$ je m unabhängige Ableitungen der Funktionen u und v nach Richtungen, die nur auf der m-dimensionalen Mantelfläche $D = 0$ liegen[5]). Daraus folgt wiederum, daß auf der ganzen Mantelfläche und insbesondere in der Spitze S u und v verschwinden. Das heißt, der Wert der Lösung des Anfangswertproblems unserer Differentialgleichungen (5) im Punkte S hängt höchstens von den Vorgaben auf dem Stück C der Anfangsfläche A ab und ist eindeutig durch sie bestimmt.

Daß die Lösung wirklich im ganzen Innern des charakteristischen Kegels verschwindet, wenn sie auf C mit ersten Ableitungen Null ist, ließe sich wieder dadurch zeigen, daß man das Stück C durch raumartige Flächen so in den äußeren Kegelmantel überführt, daß dadurch das ganze Kegelinnere überstrichen wird.

Wir bemerken noch zum Schluß dieses Paragraphen, daß unser Anfangswertproblem für die Differentialgleichungen (5) gleichwertig ist mit einem Anfangswertproblem für eine Differentialgleichung vierter Ordnung in *einer* Funktion, entweder u oder v. Die betreffende Differentialgleichung, die durch Elimination aus den Gleichungen (5) entsteht, lautet

$$(8) \qquad \begin{vmatrix} 2 a_{ik} \frac{\partial^2}{\partial x_i \partial x_k} - 2a \frac{\partial^2}{\partial t^2} & b_{ik} \frac{\partial^2}{\partial x_i \partial x_k} \\ b_{ik} \frac{\partial^2}{\partial x_i \partial x_k} & 2 c_{ik} \frac{\partial^2}{\partial x_i \partial x_k} - 2c \frac{\partial^2}{\partial t^2} \end{vmatrix} \cdot u = 0,$$

wobei wir eine symbolische Schreibweise verwandt haben, deren Bedeutung auf der Hand liegt. Die zugehörige charakteristische Gleichung ist, wie man sieht, unsere Gleichung $D = 0$ [vgl. $(7)_1$]. Damit ist jedenfalls für solche Typen von Differentialgleichungen vierter Ordnung (bzw. höherer gerader Ordnung entsprechend mehr gesuchten Funktionen in den ursprünglichen Gleichungen) nachgewiesen, daß der Wert der Lösung in einem Punkte S eindeutig bestimmt ist durch die Vorgaben auf demjenigen Stück der Anfangsfläche, das von dem charakteristischen Kegel durch S als Spitze ausgeschnitten wird.

---

[5]) Der Ausdruck $u_i \frac{\partial t}{\partial n} - u_t \frac{\partial x_i}{\partial n}$ stellt die Ableitung von u nach einer Richtung dar, deren Projektion in den $(x_1, \ldots, x_m)$-„Raum" parallel zur x_i-Achse ist; daraus folgt sofort die Unabhängigkeit dieser Richtungen.

Der Fall einer Differentialgleichung für *eine* Funktion ist offenbar als Grenzfall in unserem Ansatz enthalten, während andererseits die Übertragung auf den Fall von mehr Funktionen sich ohne weiteres durchführen läßt.

3.

Ein Beispiel zu den vorhergehenden Überlegungen bieten die *Differentialgleichungen der Kristalloptik* [6]). Es handelt sich hier um eine quadratische Form in den ersten Ableitungen der drei Funktionen u, v, w der Variablen x, y, z, t, den Komponenten des elektrischen Vektors,

$$F \equiv \alpha_1 (u_y - v_x)^2 + \alpha_2 (v_z - w_y)^2 + \alpha_3 (w_x - u_z)^2 - u_t^2 - v_t^2 - w_t^2,$$
$$\alpha_1, \alpha_2, \alpha_3 > 0.$$

Wir stellen in entsprechender Weise das Anfangswertproblem und können alle Schlüsse wie vorher ausführen, mit Ausnahme desjenigen, daß auf der hier dreidimensionalen Fläche B alle Ableitungen erster Ordnung von u, v, w verschwinden, da die Form $\overline{F}$ ausgeartet ist. Wir können aber jedenfalls schließen, daß die zeitlichen Ableitungen u_t, v_t, w_t auf B null sind. Die Anfangsfläche und die Fläche B ist hier eingeschränkt durch die folgenden, den Relationen (7) entsprechenden Bedingungen für ihre Normalen:

$$(9)\quad 1.\quad 0 < D \equiv \begin{vmatrix} \left(\frac{\partial t}{\partial n}\right)^2 - \alpha_1 \left(\frac{\partial y}{\partial n}\right)^2 - \alpha_3 \left(\frac{\partial z}{\partial n}\right)^2 & \alpha_1 \frac{\partial y}{\partial n}\cdot\frac{\partial x}{\partial n} & \alpha_3 \frac{\partial z}{\partial n}\cdot\frac{\partial x}{\partial n} \\ \alpha_1 \frac{\partial x}{\partial n}\cdot\frac{\partial y}{\partial n} & \left(\frac{\partial t}{\partial n}\right)^2 - \alpha_2 \left(\frac{\partial z}{\partial n}\right)^2 - \alpha_1 \left(\frac{\partial x}{\partial n}\right)^2 & \alpha_2 \frac{\partial z}{\partial n}\cdot\frac{\partial y}{\partial n} \\ \alpha_3 \frac{\partial x}{\partial n}\cdot\frac{\partial z}{\partial n} & \alpha_2 \frac{\partial y}{\partial n}\cdot\frac{\partial z}{\partial n} & \left(\frac{\partial t}{\partial n}\right)^2 - \alpha_3 \left(\frac{\partial x}{\partial n}\right)^2 - \alpha_2 \left(\frac{\partial y}{\partial n}\right)^2 \end{vmatrix}$$

2. Die Summe der zweireihigen Hauptminoren ist > 0.
3. Die Summe der einreihigen Hauptminoren ist > 0.

Die „Ebene" $t = 0$ z. B. genügt diesen Bedingungen.

Die Gleichung des charakteristischen Kegels, den wir durch einen Punkt S ziehen, lautet in Ebenenkoordinaten $D = 0$. Die Determinante D zerfällt in zwei Faktoren

$$D = \left(\frac{\partial t}{\partial n}\right)^2 \cdot D_1 .$$

Infolgedessen enthält der charakteristische Kegel $D = 0$ noch die Schar der Ebenen, die der t-Achse parallel sind. Der andere Anteil, der Kegel mit der Gleichung $D_1 = 0$, zerfällt bekanntlich in zwei sich berührende

[6]) Vgl. hierzu z. B. Enzyklopädie der math. Wissensch. IV 4, 30, Artikel von E. Hellinger über Mechanik der Kontinua, Nr. 13.

Kegel. Für den äußeren Mantel gelten die Ungleichungen $(9)_2$ und $(9)_3$ (eventuell mit dem Gleichheitszeichen).

Wir betrachten jetzt das Stück C, das der äußere Kegelmantel aus der Anfangsfläche herausschneidet, und das Gebiet G, das von dem äußeren Mantel und dem Stück C umschlossen wird. Da jeder Punkt des Inneren von G Punkt einer Fläche B sein kann, verschwinden u_t, v_t, w_t überall im Inneren von G. Zieht man nun im Punkte S die Parallele zur t-Achse, so trifft sie, wie man leicht sieht, das Stück C, ohne den Kegelmantel zu durchsetzen. Hieraus ergibt sich, daß u_t, v_t, w_t in der Spitze S verschwinden, da sie auf dem Stück C der Anfangsfläche verschwinden. Das heißt, der Wert irgendeiner Lösung des Anfangswertproblems der Differentialgleichungen der Kristalloptik hängt höchstens von den Vorgaben auf dem Stück C der Anfangsfläche A ab und ist eindeutig durch sie bestimmt.

4.

Wir zeigen nunmehr, daß die Überlegungen der vorigen Paragraphen sich auch durchführen lassen, wenn die Differentialgleichungen

1. noch *lineare Zusatzglieder* in der Funktion u und ihren ersten Ableitungen enthalten;

2. wenn außerdem die Koeffizienten *nicht mehr konstant* sind, sondern Funktionen der unabhängigen Veränderlichen sind.

Wir beschränken uns dabei auf einen Spezialfall mit typischem Verhalten. Es sei eine hyperbolische Differentialgleichung

$$(10) \qquad L[u] \equiv a\,u_{tt} - b\,u_{xx} + c\,u_t + d\,u_x + e\,u = 0$$

gegeben, in der also $a > 0$ und $b > 0$ ist; wir setzen dabei voraus, daß die Koeffizienten a, b, c, d, e innerhalb eines gewissen Bereiches der Variablen x, t — auf den wir uns ein für allemal beschränken — mit ihren ersten Ableitungen stetig sind.

Wir nehmen an, daß auf einer raumartigen Anfangskurve A, das heißt auf einer solchen, für deren Normalenkomponenten $\frac{\partial t}{\partial n}$, $\frac{\partial x}{\partial n}$ die charakteristische Form

$$(10)_a \qquad a\left(\frac{\partial t}{\partial n}\right)^2 - b\left(\frac{\partial x}{\partial n}\right)^2 > 0$$

ist, die Werte einer Lösung u unserer Differentialgleichung (10) null sind mitsamt ihren ersten Ableitungen. Wir betrachten nun eine weitere raumartige Kurve B, die die Anfangskurve A in den Punkten α und β schneidet (vgl. Fig. 1), und ziehen wieder das über das eingeschlossene Gebiet G erstreckte Integral

$$2\iint_G \dot{u}\,L[u]\,dx\,dt$$

heran, wobei $\dot{u}$ die Ableitung von u in einer zeitartigen Richtung ist, d. h. in einer Richtung, die zu einer raumartigen Kurve (bzw. bei mehr Variablen zu einer raumartigen Fläche) normal ist. Man sieht, daß in unserem Spezialfall jede Parallele zur t-Achse zeitartig ist; daher können wir $\dot{u} = u_t$ wählen. Wir formen nun das Integral

$$0 = 2\iint_G u_t L[u]\,dx\,dt$$

wie früher in § 1 um und erhalten

$$\begin{aligned} 0 = &\int_\Gamma \left[a u_t^2 \frac{\partial t}{\partial n} + b u_x^2 \frac{\partial t}{\partial n} - 2b\, u_x u_t \frac{\partial x}{\partial n}\right] ds \\ (11) \qquad &- \iint_G [a_t u_t^2 + b_t u_x^2 - 2b_x u_x u_t]\,dx\,dt \\ &+ 2\iint_G [c u_t^2 + d u_x u_t + e u u_t]\,dx\,dt, \end{aligned}$$

wobei Γ wieder der Rand des Gebiets G ist Der Anteil des Randintegrals, der von dem Teil $(\alpha\beta)$ der Anfangskurve A herrührt, verschwindet wegen der Anfangsbedingungen. Den Anteil, der von der Kurve B herrührt, können wir offenbar unabhängig von der Funktion u nach unten durch

$$M\int_B (u_t^2 + u_x^2)\,ds$$

abschätzen, da der Integrand, als Form in u_x und u_t betrachtet, nicht ausgeartet positiv definit ist. Die Konstante $M > 0$ hängt nur von den Werten von $a, b, \frac{\partial t}{\partial n}, \frac{\partial x}{\partial n}$ auf B ab. Die Gebietsintegrale in (11) können wir offenbar nach der Schwarzschen Ungleichung nach oben durch das Integral

$$N\iint_G (u_t^2 + u_x^2 + u^2)\,dt\,dx$$

abschätzen, wobei $N > 0$ (ebenso wie alle folgenden Konstanten) nicht von der Funktion u abhängen. Da aber die Funktion u auf A verschwindet, gilt die leicht zu beweisende[7]) Ungleichung

$$\iint_G u^2\,dx\,dt \leqq P\iint_G (u_x^2 + u_t^2)\,dx\,dt \qquad (P > 0),$$

so daß wir schließlich aus unserer Gleichung (11) die Ungleichung

$$(12) \qquad M\int_B (u_t^2 + u_x^2)\,ds \leqq Q\iint_G (u_t^2 + u_x^2)\,dt\,dx \qquad (Q > 0)$$

erhalten.

---

[7]) Vgl. z. B. Friedrichs, Randwertprobleme der elastischen Platten I Nr. 2. 2., dieses Heft, S. 212.

Wir denken uns nun das Stück $(\alpha\beta)$ der Anfangskurve A durch eine Schar von *raumartigen* Kurven C mit stetiger Normalen, die sämtlich durch die Punkte α und β gehen und sich im übrigen nicht schneiden, mit ihren Normalen stetig in die Kurve B übergeführt. Jede Kurve C der Schar schließt mit dem Stück $(\alpha\beta)$ ein Gebiet D ein, für das wiederum eine Ungleichung der Form (12) mit gewissen anderen Konstanten M und Q gilt. Aus Stetigkeitsgründen erkennt man aber leicht, daß für alle Gebiete D die Ungleichung

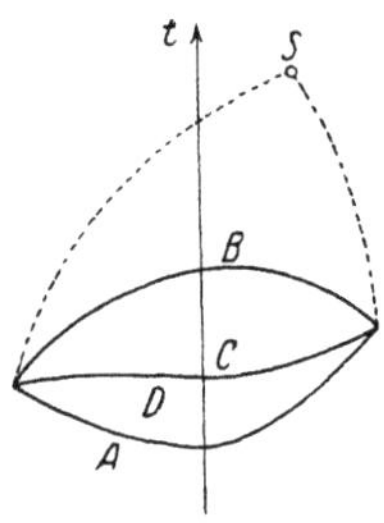

Fig. 5.

$$(13)\qquad \bar{M}\int_C (u_t^2+u_x^2)\,ds \leqq \bar{Q}\iint_D (u_t^2+u_x^2)\,dt\,dx$$

mit *denselben* festen positiven Konstanten $\bar{M}$, $\bar{Q}$ gilt.

Aus dem Bestehen dieser Ungleichungen folgt nunmehr, daß u_t und u_x im Gebiet G verschwinden. Um das einzusehen, ziehen wir etwa durch den Mittelpunkt des Stückes (α, β) die Parallele zur t-Achse und führen als neue Veränderliche erstens die Bogenlänge auf der Kurve C und zweitens die Werte von t im Schnittpunkt mit der eben beschriebenen Geraden ein. Da die Kurven C sämtlich raumartig sind, ist der Winkel ihrer Tangente mit der t-Richtung größer als ein gewisser fester positiver Winkel. Integrieren wir nun das Integral $\int_C (u_t^2+u_x^2)\,ds$ nach der so definierten Variablen t von der Anfangskurve A bis zur Kurve B, so läßt sich infolgedessen das so entstehende Doppelintegral nach unten durch $R\iint_G (u_t^2+u_x^2)\,dt\,dx$ abschätzen, wo $R>0$ ist und nicht vom Gebiet G abhängt. Integrieren wir aber das Integral $\iint_D (u_t^2+u_x^2)\,dt\,dx$ nach der neuen Variablen t, so können wir den so entstehenden Ausdruck durch $(t_B-t_A)\iint_G (u_t^2+u_x^2)\,dt\,dx$ nach oben abschätzen, so daß wir schließlich aus unserer Ungleichung (13) die Ungleichung

$$(14)\qquad \iint_G (u_t^2+u_x^2)\,dt\,dx \leqq \frac{\bar{Q}\,|t_B-t_A|}{\bar{M}\cdot R}\iint_G (u_t^2+u_x^2)\,dt\,dx$$

erhalten. Wählen wir nun die Kurve B so nahe an A, daß $|t_B-t_A| < \frac{\bar{M}\cdot R}{\bar{Q}}$ ist, so kann unsere Ungleichung offenbar nur bestehen, wenn das Integral über G selbst verschwindet, d. h. wenn u_t und u_x gleich null sind, woraus dann auch $u=0$ folgt.

Denken wir uns nun durch einen Punkt S die beiden Charakteristiken, d. h. die Linien mit der Gleichung $a\left(\frac{\partial t}{\partial n}\right)^2 - b\left(\frac{\partial x}{\partial n}\right)^2 = 0$, gezogen, bis sie die Anfangskurve A in den Punkten α und β schneiden, so wollen wir das Stück $(\alpha\beta)$ der Anfangskurve A durch eine Schar von raumartigen sich außer in den Punkten α und β nicht schneidenden Kurven C stetig[8]) in die beiden Charakteristiken S_α und S_β überführen. Nach demselben Verfahren wie oben schreiben wir jeder Kurve C eindeutig einen Wert t_C zu. Dann können wir zu jeder Kurve C eine andere Kurve $B(C)$ unserer Schar mit $t_B > t_C$ angeben, so daß in dem von C und $B(C)$ eingeschlossenen Gebiet die Funktion u verschwindet, sobald sie mit ihren ersten Ableitungen auf C verschwindet. Nach einem bekannten Schlußverfahren folgt daraus, daß u im ganzen Inneren des Dreiecks $S\alpha\beta$ verschwindet, wenn u mit ersten Ableitungen auf A verschwindet.

Im Falle der allgemeinsten linearen hyperbolischen Differentialgleichung zweiter Ordnung liegen die Verhältnisse ähnlich, so daß die Beweise ohne irgendwelche wesentliche Änderung sinngemäß übertragen werden können. Dasselbe gilt für Systeme von Differentialgleichungen zweiter Ordnung, sobald man verlangt, daß die Glieder in den Ableitungen zweiter Ordnung zu einem Variationsproblem gehören (was gewisse Symmetriebedingungen in diesen Gliedern bedeutet), und wenn man geeignete Voraussetzungen über die Natur der charakteristischen Form macht, die den hyperbolischen Charakter des Systems ausdrücken (entsprechend den in § 2 gestellten Definitheitsforderungen).

Damit ist aber der Beweis geführt, daß die Lösungen solcher Differentialgleichungen in bezug auf ihre Abhängigkeit und Eindeutigkeit von den Anfangswerten dasselbe Verhalten zeigen wie die Lösungen der Differentialgleichungen mit konstanten Koeffizienten ohne Zusatzglieder.

5.

Schlußbemerkung. Die Ergebnisse dieser Arbeit sind von Nutzen, wenn man die Lösung des Anfangswertproblems durch Integration über Einzellösungen explizite herzuleiten sucht (wie das z. B. durch die Methode des Fourierschen Integrals oder durch die damit eng zusammenhängende durch G. Herglotz[9]) jüngst angewandte Methode geschieht). Man weiß nämlich jetzt von vornherein, daß es genügt, das Integral über die Einzellösungen nur über das Abhängigkeitsgebiet zu erstrecken.

Als einfachstes Beispiel erwähnen wir das Problem der Differentialgleichung $u_{tt} - u_{xx} = 0$ mit den Anfangsbedingnngen $u(x,0) = 0$, $u_t(x,0) = g(x)$. Die Methode des Fourierschen Integrals führt zu der Auflösungsformel

$$u(x,t) = \frac{1}{2\pi}\int_{-\infty}^{+\infty} g(x+q)\,dq \cdot \int_{-\infty}^{+\infty} e^{-i\xi q}\frac{\sin t\xi}{\xi}\,d\xi.$$

---

[8]) Bei diesem Übergang sollen sich nicht nur die Kurven, sondern auch die Normalen stetig ändern, natürlich außer an der Spitze S.

[9]) A. a. O.

Aus unseren Ergebnissen entnehmen wir nunmehr sofort, daß es genügt, sich bei der Integration nach der Variablen q auf das Gebiet $-t \leqq q \leqq t$ zu beschränken, mit anderen Worten, es folgt ganz von selbst, daß der Ausdruck

$$\int_{-\infty}^{+\infty} e^{-i\xi q} \frac{\sin t\xi}{\xi}\, d\xi$$

für $q > t$ verschwindet.

In diesem Falle ist das Ergebnis natürlich auch sehr leicht direkt zu gewinnen. Anders aber liegt es in komplizierteren Fällen, vor allem bei mehr unabhängigen Veränderlichen, z. B. im Falle der Kristalloptik, wo es sich um mehrfache Fouriersche Integrale handelt, die zunächst über den ganzen unendlichen Raum zu erstrecken sind. Der Nachweis, daß diese Integrale außerhalb des Abhängigkeitsgebietes verschwinden, bildet gerade eine charakteristische Schwierigkeit bei Verwendung der Methode des Fourierschen Integrals; unsere Ergebnisse erlauben uns sofort, diesen Schluß zu ziehen.

(Eingegangen am 1. 12. 1926.)

R. Courant*
K. Friedrichs*
H. Lewy†

On the Partial Difference Equations of Mathematical Physics

Editor's note: This paper, which originally appeared in *Mathematische Annalen* 100, 32–74 (1928), is republished by permission of the authors. We are also grateful to the Atomic Energy Commission for permission to republish this translation, which had appeared as AEC Report NYO-7689, and to Phyllis Fox, the translator, who did the work at the AEC Computing Facility at New York University under AEC Contract No. AT(30-1)-1480. Professor Eugene Isaacson had made suggestions on this translation.

Introduction

Problems involving the classical linear partial differential equations of mathematical physics can be reduced to algebraic ones of a very much simpler structure by replacing the differentials by difference quotients on some (say rectilinear) mesh. This paper will undertake an elementary discussion of these algebraic problems, in particular of the behavior of the solution as the mesh width tends to zero. For present purposes we limit ourselves mainly to simple but typical cases, and treat them in such a way that the applicability of the method to more general difference equations and to those with arbitrarily many independent variables is made clear.

Corresponding to the correctly posed problems for partial differential equations we will treat boundary value and eigenvalue problems for elliptic difference equations, and initial value problems for the hyperbolic or parabolic cases. We will show by typical examples that the passage to the limit is indeed possible, i.e., that the solution of the difference equation converges to the solution of the corresponding differential equation; in fact we will find that for elliptic equations in general a difference quotient of arbitrarily high order tends to the corresponding derivative. Nowhere do we assume the existence of the solution to the differential equation problem—on the contrary, we obtain a simple existence proof by using the limiting process.[1] For the case of elliptic equations convergence is obtained independently of the choice of mesh, but we will find that for the case of the initial value problem for hyperbolic equations, convergence is obtained only if the ratio of the mesh widths in different directions satisfies certain inequalities which in turn depend on the position of the characteristics relative to the mesh.

We take as a typical case the boundary value problem of potential theory. Its solution and its relation to the solution of the corresponding difference equation has been extensively treated during the past few years.[2] However in contrast to the present paper, the previous work has involved the use of quite special characteristics of the potential equation so that the applicability of the method used there to other problems has not been immediately evident.

In addition to the main part of the paper, we append an elementary algebraic discussion of the connection of the boundary value problem of elliptic equations with the random walk problem arising in statistics.

---

[1] Our method of proof may be extended without difficulty to cover boundary value and eigenvalue problems for arbitrary linear elliptic differential equations and initial value problems for arbitrary linear hyperbolic differential equations.

* Now at Courant Institute of Mathematical Sciences, New York University.

† Now at University of California, Berkeley.

[2] J. le Roux, "Sur le problem de Dirichlet", *Journ. de mathém. pur. et appl.* (6) **10**, 189 (1914). R. G. D. Richardson, "A new method in boundary problems for differential equations", *Trans. of the Am. Math. Soc.* **18**, p. 489 ff, (1917). H. B. Philips and N. Wiener, *Nets and the Dirichlet Problem*, Publ. of M.I.T. (1925). Unfortunately these papers were not known by the first of the three authors when he prepared his note "On the theory of partial difference equations," *Gött. Nachr.* 23, X, 1925, on which the present work depends. See also L. Lusternik, "On an application of the direct method in variation calculus," Recueil de la Société Mathém. de Moscou, 1926. G. Bouligand, "Sur le probléme de Dirichlet," *Ann. de la soc. polon. de mathém.* **4**, Cracow (1926). On the meaning of the difference expressions and on further applications of them, see R. Courant, "Über direkte Methoden in der Variationsrechnung," *Math. Ann.* **97**, p. 711, and the references given therein.

I. The elliptic case

Section 1. Preliminary remarks

• 1. *Definitions*

Consider a rectangular array of points in the (x, y)-plane, such that for mesh width $h > 0$ the points of the lattice are given by

$$\left.\begin{aligned} x &= nh \\ y &= mh \end{aligned}\right\} \quad m, n = 0, \pm 1, \pm 2, \cdots .$$

Let G be a region of the plane bounded by a continuous closed curve which has no double points. Then the corresponding mesh region, G_h—which is uniquely determined for sufficiently small mesh width—consists of all those mesh points lying in G which can be connected to any other given point in G by a connected chain of mesh points. By a connected chain of mesh points we mean a sequence of points such that each point follows in the sequence one of its four neighboring points. We denote as a *boundary point* of G_h a point whose four neighboring points do not all belong to G_h. All other points of G_h we call *interior points*.

We shall consider functions $u, v, \cdots$ of position on the grid, i.e., functions which are defined only for grid points, but we shall denote them as $u(x, y), v(x, y), \cdots$. For their forward and backward difference quotients we employ the following notation,

$$u_x = \frac{1}{h}[u(x+h, y) - u(x, y)],$$

$$u_y = \frac{1}{h}[u(x, y+h) - u(x, y)].$$

$$u_{\bar{x}} = \frac{1}{h}[u(x, y) - u(x-h, y)],$$

$$u_{\bar{y}} = \frac{1}{h}[u(x, y) - u(x, y-h)].$$

In the same way the difference quotients of higher order are formed, e.g.,

$$(u_x)_{\bar{x}} = u_{x\bar{x}} = u_{\bar{x}x}$$
$$= \frac{1}{h^2}[u(x+h, y) - 2u(x, y) + u(x-h, y)], \text{ etc.}$$

• 2. *Difference expressions and Green's function*

In order to study linear difference expressions of second order, we form (using as a model the theory of partial differential equations), a bilinear expression from the forward difference quotients of two functions, u and v,

$$B(u, v) = au_xv_x + bu_xv_y + cu_yv_x + du_yv_y + \alpha u_xv$$
$$+ \beta u_yv + \gamma uv_x + \delta uv_y + guv,$$

where $a = a(x, y), \cdots, \alpha = \alpha(x, y), \cdots, g = g(x, y)$ are functions on the mesh.

From the bilinear expression of first order we derive a difference expression of second order in the following way: we form the sum

$$h^2 \sum_{G_h}\sum B(u, v)$$

over all points of a region G_h in the mesh, where in $B(u, v)$ the difference quotients between boundary points and points not belonging to the mesh are to be set equal to zero. We now transform the sum through partial summation, i.e., we arrange the sum according to v, and split it up into a sum over the set of interior points, G_h' and a sum over the set of boundary points, Γ_h. Thus we obtain:

$$h^2 \sum_{G_h}\sum B(u, v)$$
$$= -h^2 \sum_{G_h'}\sum vL(u) - h\sum_{\Gamma_h} v\Re(u). \qquad (1)$$

$L(u)$ is a linear difference expression of second order defined for all interior points of G_h:

$$L(u) = (au_x)_{\bar{x}} + (bu_x)_{\bar{y}} + (cu_y)_{\bar{x}} + (du_y)_{\bar{y}}$$
$$- \alpha u_x - \beta u_y + (\gamma u)_{\bar{x}} + (\delta u)_{\bar{y}} - gu.$$

$\Re(u)$ is, for every boundary point, a linear difference expression whose exact form will not be given here.

If we arrange $\sum_{G_h}\sum B(u, v)$ according to u, we find

$$h^2 \sum_{G_h}\sum B(u, v)$$
$$= -h^2 \sum_{G_h'}\sum uM(v) - h\sum_{\Gamma_h'} u\mathcal{S}(v). \qquad (2)$$

$M(v)$ is called the adjoint difference expression of $L(u)$ and has the form

$$M(v) = (av_x)_{\bar{x}} + (bv_y)_{\bar{x}} + (cv_x)_{\bar{y}} + (dv_y)_{\bar{y}}$$
$$+ (\alpha v)_{\bar{x}} + (\beta v)_{\bar{y}} - \gamma v_x - \delta v_y - gv,$$

while $\mathcal{S}(v)$ is a difference expression corresponding to $\Re(u)$ for the boundary.

Formulas (1) and (2) give

$$h\sum_{G_h'}\sum [vL(u) - uM(v)]$$
$$+ h\sum_{\Gamma_h'} [v\Re(u) - u\mathcal{S}(v)] = 0. \qquad (3)$$

Formulas (1), (2), and (3) are called *Green's formula*.

The simplest and most important case results if the bilinear form is symmetric, i.e., if the relations $b = c$, $\alpha = \gamma$, $\beta = \delta$ hold. In this case $L(u)$ is identical with $M(u)$—the self-adjoint case—and it can be derived from the quadratic expression

$$B(u, u) = au_x^2 + 2bu_xu_y + du_y^2$$
$$+ 2\alpha u_xu + 2\beta u_yu + gu^2.$$

In the following we shall limit ourselves mainly to expressions $L(u)$ which are self-adjoint. The character of the difference expression $L(u)$ depends principally on the nature of those terms in the quadratic form $B(u, u)$ which are quadratic in the first difference quotients. We call this part of $B(u, u)$ the *characteristic form*:

$$P(u, u) = au_x^2 + 2bu_xu_y + du_y^2.$$

We call the corresponding difference expression $L(u)$ elliptic or hyperbolic, depending on whether the function $P(u, u)$ of the difference quotients is (positive) definite or indefinite.

The difference expression $\Delta u = u_{x\bar{x}} + u_{y\bar{y}}$ with which we shall concern ourselves in the following paragraph is elliptic, i.e., it comes from the quadratic expression

$$B(u, u) = u_x^2 + u_y^2 \quad \text{or} \quad u_{\bar{x}}^2 + u_{\bar{y}}^2.$$

The corresponding Green's formulas are

$$h^2 \sum_{G_h} \sum (u_x^2 + u_y^2) = -h^2 \sum_{G_h'} \sum u\Delta u - h \sum_{\Gamma_h} u\mathfrak{R}(u) \qquad \text{[Note 3]} \tag{4}$$

$$h^2 \sum_{G_h'} \sum (v\Delta u - u\Delta v) + h \sum_{\Gamma_h} [v\mathfrak{R}(u) - u\mathfrak{R}(v)] = 0. \tag{5}$$

The difference expression, $\Delta u = u_{x\bar{x}} + u_{y\bar{y}}$, is obviously the analogue of the differential expression $(\partial^2 u/\partial x^2) + (\partial^2 u/\partial y^2)$ for a function $u(x, y)$ of the continuous variables x and y. Written out explicitly the difference expression is

$$\Delta u = \frac{1}{h^2} [u(x + h, y) + u(x, y + h) + u(x - h, y) + u(x, y - h) - 4u(x, y)].$$

Therefore $(h^2/4)\Delta u$ is the excess of the arithmetic mean of the functional values at the four neighborhood points over the functional value at the point in question.

Completely similar considerations lead to linear difference expressions of the fourth order and corresponding Green's formula, provided one starts from bilinear difference expressions which are formed from the difference quotients of second order. Consider for example the difference expression

$$\Delta\Delta u = u_{xx\bar{x}\bar{x}} + 2u_{x\bar{x}y\bar{y}} + u_{yy\bar{y}\bar{y}}.$$

This corresponds to the quadratic expression

$$B(u, u) = (u_{x\bar{x}} + u_{y\bar{y}})^2 = (\Delta u)^2,$$

provided one orders the sum

$$h^2 \sum_{G_h'} \sum \Delta u \Delta v$$

according to v, or equivalently replaces u by Δu in Eq. (5). One must notice however that in the expression $\Delta\Delta u$, the functional value at a point is connected with the values at its neighboring points and at their neighboring points, and accordingly is defined only for such points of the region G_h as are also interior points of the region G_h' (See Section 5). The entirety of such points we designate as G_h''. We obtain then Green's formula

$$h^2 \sum_{G_h'} \sum \Delta u \cdot \Delta v = h^2 \sum_{G_h''} \sum v \cdot \Delta\Delta u + h \sum_{\Gamma_h + \Gamma_h'} v \cdot \mathfrak{R}(u), \tag{6}$$

where $\mathfrak{R}(u)$ is a definable linear difference expression for each point of the boundary strip $\Gamma_h + \Gamma_h'$. Γ_h' indicates the set of boundary points of G_h'.

Section 2. Boundary value and eigenvalue problems

• 1. *The theory of boundary value problems*

The boundary value problem for linear elliptic homogeneous difference equations of second order, which corresponds to the classical boundary value problem for partial differential equations, can be formulated in the following way.

Let there be given a self-adjoint elliptic linear difference expression of second order, $L(u)$, in a mesh region, G_h. $L(u)$ results from a quadratic expression $B(u, u)$ which is positive definite in the sense that it cannot vanish if u_x and u_y do not themselves vanish.

A function, u, is to be determined satisfying in G_h the difference equation $L(u) = 0$, and assuming prescribed values at the boundary points.

Under these requirements there will be exactly as many linear equations as there are interior points of the mesh at which the function u is to be determined.[4] Some of these equations which involve only mesh points whose neighbors also lie in the interior of the region are homogeneous; others which involve boundary points of the mesh region are inhomogeneous. If the right-hand side of the inhomogeneous equations is set equal to zero, that is if $u = 0$ on the boundary, then it follows from Green's formula (1), by setting $u = v$ that $B(u, u)$ vanishes, and further, from the definiteness of $B(u, u)$, that u_x and u_y vanish, and hence that u itself vanishes. Thus the difference equation for zero boundary values has the solution $u = 0$, or in other words the solution is uniquely determined since the difference of two solutions with the

[3] The boundary expression $\mathfrak{R}(u)$ may be written as follows: Let $u_0, u_1, \ldots, u_\nu$ be values of the function at a boundary point and at its ν neighboring points ($\nu \leq 3$), then

$$\mathfrak{R}(u) = \frac{1}{h} (u_1 + \cdots + u_\nu - \nu u_0).$$

[4] If the matrix of the linear system of equations corresponding to an arbitrary difference equation of second order, $L(u) = 0$, is transposed, then the transposed set of equations corresponds to the adjoint difference equation $M(v) = 0$. Thus the above self-adjoint system gives rise to a set of linear equations with symmetric coefficients.

same boundary value must vanish. Further, if a linear system of equations with as many unknowns as equations is such that for vanishing right-hand side the unknowns must vanish, then the fundamental theorem of the theory of equations asserts that for an arbitrary right-hand side exactly one solution must exist. In our case there follows at once the existence of a solution of the boundary value problem.

Therefore we see that for elliptic difference equations the uniqueness and existence of the solution of the boundary value problem are related to each other through the fundamental theorem of the theory of linear equations, whereas for partial differential equations both facts must be proved by quite different methods. The basis for this difficulty in the latter case is to be found in the fact that the differential equations are no longer equivalent to a finite number of equations, and so one can no longer depend on the equality of the number of unknowns and equations.

Since the difference expression $\Delta u = 0$ can be derived from the positive definite quadratic expression

$$h^2 \sum_{G_h} \sum (u_x^2 + u_y^2),$$

the boundary value problem for the difference expression is uniquely solvable.

The theory for difference equations of higher order is developed in exactly the same way as that for difference equations of second order; for example one can treat the fourth order difference equation $\Delta\Delta u = 0$. In this case the values of the function must be prescribed on the boundary strip $\Gamma_h + \Gamma_h'$. Evidently here also the difference equation yields just as many linear equations as there are unknown functional values at the points of G_h''. In order to demonstrate the existence and uniqueness of the solution one needs only to show that a solution which has the value zero in the boundary strip $\Gamma_h + \Gamma_h'$ necessarily vanishes identically. To this end we note that the sum over the corresponding quadratic expression

$$h^2 \sum_{G_h''} \sum (\Delta u)^2 \tag{7}$$

for such a function vanishes, as can be seen by transforming the sum according to Green's formula (6). The vanishing of the sum (7) however implies that Δu vanishes at all points of G_h', and according to the above proof this can only happen for vanishing boundary values if the function u assumes the value zero throughout the region. Thus our assertion is proved, and both the existence and uniqueness of the solution to the boundary value problem for the difference equations are guaranteed.[5]

[5] For the actual process of carrying through the solution of the boundary value problem by an iterative method, see among others the treatment: "Über Randwertaufgaben bei partiellen Differenzengleichungen" by R. Courant, *Zeitschr. f. angew. Mathematik u. Mechanik* **6**, 322–325 (1925). Also there is a report by H. Henky, in *Zeitschr. f. angew. Math. u. Mech.* **2**, 58 ff (1922).

• 2. *Relation to the minimum problem*

The above boundary value problem is related to the following minimum problem: among all functions $\varphi(x, y)$ defined in the mesh region G_h and assuming given values at the boundary points, that function $\varphi = u(x, y)$ is to be found for which the sum

$$h^2 \sum_{G_h} \sum B(\varphi, \varphi)$$

over the mesh region assumes the least possible value. We assume that the quadratic difference expression of first order, $B(u, u)$ is positive definite in the sense described in Section 1, Part 2. One can show that the difference equation $L(\varphi) = 0$ results from this minimum requirement on the solution $\varphi = u(x, y)$, where $L(\varphi)$ is the difference expression of second order derived previously from $B(\varphi, \varphi)$. In fact this can be seen either by applying the rules of differential calculus to the sums $h^2 \sum_{G_h} \sum B(\varphi, \varphi)$ as a function of a finite number of values of φ at the grid points, or by employing the usual methods from the calculus of variations.

By way of example, solving the boundary value problem of finding the solution to the equation $\Delta\varphi = 0$ which assumes given boundary values, is equivalent to minimizing the sum $h^2 \sum_{G_h} \sum (\varphi_x^2 + \varphi_y^2)$ over all functions which take on the boundary values.

There is a similar correspondence for the fourth-order difference equations, where we limit ourselves to the example $\Delta\Delta\varphi = 0$. The boundary value problem corresponding to this difference equation is equivalent to the problem of minimizing the sum

$$h^2 \sum_{G_h'} \sum (\Delta\varphi)^2$$

for functions that take on given values on the boundary strip Γ_h'. Besides this sum there are yet other quadratic expressions in the second derivatives which give rise to the equation $\Delta\Delta u = 0$ under the process of being minimized. For example this is true in G_h for the sum

$$h^2 \sum_{G_h'} \sum (u_{xx}^2 + 2u_{xy}^2 + u_{yy}^2).$$

That the minimum problem posed above always has a solution follows from the theorem that a continuous function of a finite number of variables (the functional values of φ at the grid points) always has a minimum if it is bounded from below and if it tends to infinity as soon as any of the independent variables goes to infinity.[6]

• 3. *Green's function*

It is possible to treat the boundary value problem for the inhomogeneous equation, $L(u) = -f$, in much the same way as the homogeneous case, $L(u) = 0$. For the inhomogeneous case it is sufficient to consider only the case of

[6] It can easily be verified that the hypotheses for the application of this theorem are satisfied.

zero boundary conditions, since different boundary conditions can be taken care of by adding a suitable solution of the homogeneous equation. To solve the linear system of equations representing the boundary value problem, $L(u) = -f$, we first choose as the function $f(x, y)$ a function which assumes the value $-1/h^2$ at the point $x = \xi$, $y = \eta$ of the mesh. If $K(x, y; \xi, \eta)$ is the solution (vanishing on the boundary) of this difference equation which depends on the parametric point (ξ, η), then the solution for an arbitrary boundary condition can be represented by the sum

$$u(x, y) = h^2 \sum_{(\xi,\eta)\,\mathrm{in}\,G_h} K(x, y; \xi, \eta) f(\xi, \eta).$$

The function $K(x, y; \xi, \eta)$ which depends on the points (x, y) and (ξ, η) is called the Green's function of the differential expressions $L(u)$. If we call the Green's function for the adjoint expression $M(v)$, $\bar{K}(x, y; \xi, \eta)$, then the equivalence

$$K(\bar{\xi}, \bar{\eta}; \xi, \eta) = \bar{K}(\xi, \eta; \bar{\xi}, \bar{\eta}) \quad \text{holds,}$$

as can be seen to follow immediately from Green's formula (3) when $u = K(x, y; \xi, \eta)$, and $v = \bar{K}(x, y; \bar{\xi}, \bar{\eta})$. For a self-adjoint difference expression the above relation gives the symmetric expression

$$K(\bar{\xi}, \bar{\eta}; \xi, \eta) = K(\xi, \eta; \bar{\xi}, \bar{\eta}).$$

• 4. *Eigenvalue problems*

Self-adjoint difference expressions, $L(u)$, give rise to eigenvalue problems of the following type: find the value of a parameter λ, the eigenvalue, such that in G_h, a solution, the eigenfunction can be found for the difference equation $L(u) + \lambda u = 0$, where u is to be zero on the boundary, Γ_h.

The eigenvalue problem is equivalent to finding the principle axes of the quadratic form $B(u, u)$. Exactly as many eigenvalues and corresponding eigenfunctions exist as there are interior mesh points of the region G_h. The system of eigenfunctions and eigenvalues (and a proof of their existence) is given by the minimum problem:

Among all functions, $\varphi(x, y)$, vanishing on the boundary, and satisfying the orthogonality relation

$$h^2 \sum_{G_h} \sum \varphi u^{(\nu)} = 0, \qquad (\nu = 1, \cdots, m-1)$$

and normalized such that

$$h^2 \sum_{G_h} \sum \varphi^2 = 1,$$

the function, $\varphi = u$, is to be found for which the sum

$$h^2 \sum_{G_h} \sum B(\varphi, \varphi)$$

assumes its minimum value. The value of this minimum is the mth eigenvalue, and the function for which it is assumed is the mth eigenfunction.[7]

Section 3.[8] *Connections with the problem of the random walk*
The theme of the following is related to a question from the theory of probability, namely the problem of the random walk in a bounded region.[9] We consider the lines of a mesh region G_h as paths along which a particle can move from one grid point to a neighboring one. In this net of streets the particle can wander aimlessly, and it can choose at random one of the four directions leading from each intersection of paths of the net—all four directions being equally probable. The walk ends as soon as a boundary point of G_h is reached because here the particle must be absorbed.

We ask:

1) What is the probability $w(P; R)$ that a random walk starting from a point P reaches a particle point R of the boundary?
2) What is the mathematical expectation $v(P; Q)$ that a random walk starting from P reaches a point Q of G_h without touching the boundary?

This probability or mathematical expectation, respectively, will be defined more precisely by the following process. Assume that at the point P there is a unit concentration of matter. Let this matter diffuse into the mesh with constant velocity, traveling say a mesh width in unit time. At each meshpoint let exactly one-fourth of the matter at the point diffuse outwards in each of the four possible directions. The matter which reaches a boundary point is to remain at that point. If the point of origin P is itself a boundary point, then the matter never leaves that point.

We define the probability $w(P; R)$ that a random walk starting from P reaches the boundary point R (without

---

[7] From the orthogonality condition on the eigenfunctions,

$$h^2 \sum_{G_h} \sum u^{(\nu)} u^{(\mu)} = 0, \qquad (\nu \neq \mu)$$

it follows that each function, $g(x, y)$, which vanishes on the boundary of the mesh can be expanded in terms of the eigenfunctions in the form

$$y = \sum_{\nu=1}^{N} c^{(\nu)} u^{(\nu)},$$

where the coefficients are determined from the equations

$$c^{(\nu)} = h^2 \sum_{G_h} \sum g u^{(\nu)}.$$

In this way in particular the following representation for the Green's functions may be derived,

$$K(x, y; \xi, \eta) = -\sum_{\nu=1}^{N} \frac{u^{(\nu)}(x, y) \cdot u^{(\nu)}(\xi, \eta)}{\lambda^{(\nu)}}.$$

[8] Section 3 is not prerequisite to Section 4.

[9] The present treatment is essentially different from the familiar treatments which can be carried through, say for example in the case of Brownian motion for molecules. The difference lies precisely in the way in which the boundary of the region enters.

previously attaining the boundary), as the amount of matter which accumulates at this boundary point over an infinite amount of time.

We define the probability $E_n(P; Q)$ that the walk starting from the point P reaches the point Q in exactly n steps without touching the boundary by the amount of matter which accumulate in n units of time provided P and Q are both interior points. If either P or Q are boundary points then E_n is set equal to zero.

The value $E_n(P; Q)$ is exactly equal to $1/4^n$ times the number of paths of n steps leading from P to Q without touching the boundary. Thus $E_n(P; Q) = E_n(Q; P)$.

We define the mathematical expectation $v(P; Q)$ that one of the paths considered above leads from P to the point Q by the infinite sum of all of these possibilities,

$$v(P; Q) = \sum_{\nu=0}^{\infty} E_\nu(P; Q) \qquad \text{(Note 10)},$$

i.e., for the interior points P and Q, the sum of all the concentrations which have passed through the point Q at different times. This will be assigned the expected value 1 for a concentration originating at Q.

If the amount arriving at the boundary point R in exactly n steps is designated as $F_n(P; R)$, then the probability $w(P; R)$ is given by the series

$$w(P; R) = \sum_{\nu=0}^{\infty} F_\nu(P; R).$$

All the terms of this series are positive and the partial sum is bounded by one (since the concentration reaching the boundary can be made up of only part of the initial concentration), and therefore the convergence of the series is assured.

Now it is easy to see that the probability $E_n(P; Q)$, that is, the concentration reaching the point Q in exactly n steps tends to zero as n increases. For if at any point Q, from which the boundary point R can be reached in m steps, we have $E_n(P; Q) > \alpha > 0$, then at least $\alpha/4^m$ of the concentration at Q will reach the point R after m steps. However, since the sum $\sum_{\nu=0}^{\infty} F_\nu(P; R)$ converges, the concentration reaching R goes to zero with increasing time, where the value of $E_n(P; Q)$ must itself vanish as time increases; that is, the probability of an infinitely long walk remaining in the interior of the region is zero.

From this it follows that the entire concentration eventually reaches the boundary; or in other words that the sum $\sum_R w(P; R)$ over all the boundary points R is equal to one.

The convergence of the infinite series for the mathematical expectation

$$v(P; Q) = \sum_{\nu=0}^{\infty} E_\nu(P; Q)$$

remains to be shown.

[10] The convergence will be shown below.

To this end we remark that the quantity $E_n(P; Q)$ satisfies the following relations

$$E_{n+1}(P; Q) = \tfrac{1}{4}\{E_n(P; Q_1) + E_n(P; Q_2) + E_n(P; Q_3) + E_n(P; Q_4)\}, \quad [n \geq 1],$$

where Q_1 through Q_4 are the four neighboring points of the interior point Q. That is, the concentration at the point Q at the $n + 1^{\text{st}}$ step consists of 1/4 times the sum of the concentrations at the four neighboring points at the nth step. If one of the neighbors of Q, for example $Q_1 = R$, is a boundary point then it follows that no concentration flows from this boundary point to Q since the expression $E_n(P; R)$ is zero in this case. Furthermore, for an interior point, $E_0(P; P) = 1$ and of course $E_1(P; Q) = 0$.

From these relationships we find for the partial sum

$$v_n(P; Q) = \sum_{\nu=0}^{n} E_\nu(P; Q)$$

the equation

$$v_{n+1}(P; Q) = \tfrac{1}{4}\{v_n(P; Q_1) + v_n(P; Q_2) + v_n(P; Q_3) + v_n(P; Q_4)\},$$

for $P \neq Q$. For the case of $P = Q$,

$$v_{n+1}(P; P) = 1 + \tfrac{1}{4}\{v_n(P; P_1) + v_n(P; P_2) + v_n(P; P_3) + v_n(P; P_4)\},$$

that is, the expectation that a point goes back into itself consists of the expectation that a nonvanishing path leads from P back again to itself—namely,

$$\tfrac{1}{4}\{v_n(P; P_1) + v_n(P; P_2) + v_n(P; P_3) + v_n(P; P_4)\},$$

together with the expectation unity that expresses the initial position of the concentration itself at this point.

The quantity $v_n(P; Q)$ thus satisfies the following difference equation[11]

$$\Delta v_n(P; Q) = \frac{4}{h^2} E_n(P; Q), \qquad \text{for} \quad P \neq Q,$$

$$\Delta v_n(P; Q) = \frac{4}{h^2} \{E_n(P; Q) - 1\}, \quad \text{for} \quad P = Q.$$

$v_n(P; Q)$ is equal to zero when Q is a boundary point.

[11] This defines the Δ-operation for the variable point Q. This equation can be interpreted as an equation of the heat conduction type. That is, if the function $v_n(P; Q)$ is considered, not as a function of the index n as in our presentation above, but rather as a function of time, t, which is proportional to n, so that $t = n\tau$ and $v_n(P; Q) = v(P; Q; t) = v(t)$, then the above equations can be written in the following form:

$$\Delta v(t) = \frac{4\tau}{h^2} \cdot \frac{v(t + \tau) - v(t)}{\tau} \qquad \text{for} \quad P \neq Q,$$

$$\Delta v(t) = \frac{4\tau}{h^2} \cdot \left(\frac{v(t + \tau) - v(t)}{\tau} - \frac{1}{\tau}\right) \quad \text{for} \quad P = Q.$$

For a similar difference equation which has a parabolic differential equation as its limiting form, see Section 6 of the second half of the paper.

The solution of this boundary value problem for arbitrary right-hand side is uniquely determined as we have explained earlier (Section 2, Part 1), and depends continuously on the right-hand side. Since the variables $E_n(P; Q)$ tend to zero, the solution $v_n(P; Q)$ converges to the solution $v(P; Q)$ of the difference equation

$$\Delta v(P; Q) = 0 \qquad \text{for} \quad P \neq Q$$

$$\Delta v(P; Q) = -\frac{4}{h^2} \quad \text{for} \quad P = Q,$$

with boundary values $v(P; R) = 0$.

Thus we see that the mathematical expectation $v(P; Q)$ exists and is none other than the Green's function for the difference equation $\Delta u = 0$, except for a factor of 4. The symmetry of the Green's function is an immediate consequence of the symmetry of the quantity $E_n(P; Q)$ which was used to define it.

The probability $w(P; R)$ satisfies, with respect to P, the relation

$$w(P; R) = \tfrac{1}{4}\{ w(P_1; R) + w(P_2; R) + w(P_3; R) + w(P_4; R)\},$$

and thus the difference equation $\Delta w = 0$. That is, if P_1, P_2, P_3, P_4, are the four neighboring points of the interior point P, then each path from P to R must pass through one of these four directions, and each of the four is equally likely. Furthermore, the probability of going from one boundary point R to another R' is $w(R; R') = 0$ unless the two points R and R' coincide, in which case $w(R; R) = 1$. Thus $w(P; R)$ is that solution of the boundary value problem $\Delta w = 0$ which assumes the value 1 at the boundary point R and the value 0 at all other points of the boundary. Therefore the solution of the boundary value problem for an arbitrary boundary value $u(R)$ has the simple form $u(P) = \sum_R w(P; R)u(R)$, where the sum is to be extended over all the boundary points.[12] If the function $u \equiv 1$ is substituted for u in this expression, then we check the relation $1 = \sum_R w(P; R)$.

The interpretation given above for Green's function as an expectation yields immediately further properties. We mention only the fact that the Green's function decreases if one goes from the region G to a subregion $\bar{G}$ lying within G; that is, the number of possible paths for steps on the mesh leading from one point P to another Q (without touching the boundary), decreases as the region decreases.

Of course, corresponding relationships hold for more than two independent variables. We note only that other elliptic difference equations admit a similar probability interpretation.

If the limit for vanishing mesh width is considered by methods given in the following section, then the Green's function on the mesh goes over to the Green's function of the potential equation except for a numerical factor; a similar relationship holds between the expression $w(P; R)/h$ and the normal derivative of the Green's function at the boundary of the region. In this way, for example, the Green's function for the potential equation could be interpreted as the specific mathematical expectation of going from one point to another,[13] without reaching the boundary.

In going over to the limit of a continuum from the mesh, the influence of the direction in the mesh prescribed for the random walk vanishes. This fact suggests that it would be of some interest to consider limiting cases of more general random walks for which the limitations on the direction of the walk are relaxed. This lies outside of the scope of this presentation, however, and we can only hope to renew the question at some other opportunity.

Section 4. The solution of the differential equation as a limiting form of the solution of the difference equation

• 1. *The boundary value problem of potential theory*

In letting the solution of the difference equation tend to the solution of the corresponding differential equation we shall relinquish the greatest possible degree of generality with regard to the boundary and boundary values in order to demonstrate more clearly the character of our method.[14] Therefore we assume that we are given a simply connected region G with a boundary formed of a finite number of arcs with continuously turning tangents. Let $f(x, y)$ be a given function which is continuous and has continuous partial derivatives of first and second order in a region containing G. If G_h is the mesh region with mesh width h belonging to the region G, then let the boundary value problem for the difference equation $\Delta u = 0$ be solved for the same boundary values which the function $f(x, y)$ assumes on the boundary; let $u_h(x, y)$ be the solution. We shall prove that as $h \to 0$ the function $u_h(x, y)$ converges to a function $u(x, y)$ which satisfies in G the partial differential equation $(\partial^2 u/\partial x^2) + (\partial^2 u/\partial y^2) = 0$ and takes the value of $f(x, y)$ at each of the points of the boundary. We shall show further that for any region lying entirely within G the difference quotients of u_h of arbitrary order tend uniformly towards the corresponding partial derivatives of $u(x, y)$.

In the convergence proof it is convenient to replace the boundary condition $u = f$ by the weaker requirement that

$$\frac{1}{r}\iint_{S_r} (u - f)^2 \, dxdy \to 0 \quad \text{as} \quad r \to 0,$$

[12] Moreover it is easy to show that the probability $w(P; R)$ of reaching the boundary is the boundary expression $\Re(K(P, Q))$, constructed from the Green's function $K(P; Q)$ in terms of Q, where $u(x, y)$ is to be identified with $w(P, Q)$, and $v(x, y)$ with $v(P, Q)$ in Green's formula (5).

[13] Here the *a priori* expectation of reaching a certain area element is understood to be equal to the area of the element.

[14] We mention however that carrying through our method for more general boundaries and boundary values in no way causes any fundamental difficulty.

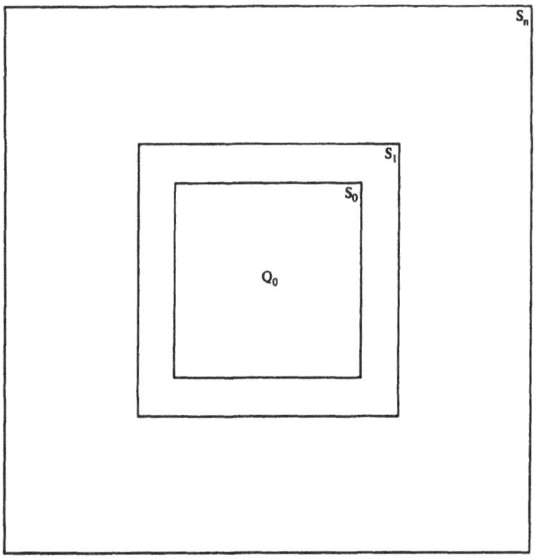

Figure 1

where S_r is that strip of G whose points are at a distance from the boundary smaller than r.[15] The convergence proof depends on the fact that for any subregion G^* lying entirely within G, the function $u_h(x, y)$ and each of its difference quotients is bounded and uniformly continuous as $h \to 0$ in the following sense: For each of these functions, say $w_h(x, y)$, there exists a $\delta(\varepsilon)$ depending only on the subregion and not on h such that

$$|w_h(P) - w_h(P_1)| < \varepsilon$$

provided the mesh points P and P_1 lie in the same subregion of G_h and are separated from each other by a distance less than $\delta(\varepsilon)$.

Once uniform continuity in this sense (equicontinuity) has been established we can in the usual way select from the functions u_h a subsequence of functions which tend uniformly in any subregion G^* to a limit function $u(x, y)$, while the difference quotients of u_h tend uniformly towards the partial derivatives of u. The limit function then possesses derivatives of arbitrarily high order in any proper subregion G^* of G and satisfies $\nabla^2 u = 0$ in this region. If we can show also that u satisfies the boundary condition we can regard it as the solution of our boundary value problem for the region G. Since this solution is uniquely determined, it is clear that not only a partial sequence of the functions u_h, but this sequence of functions itself possesses the required convergence properties.

The uniform continuity (equicontinuity) of our quantities may be established by proving the following lemmas.

1) As $h \to 0$ the sums over the mesh region $h^2 \sum_{G_h} \sum u^2$ and $h^2 \sum_{G_h} \sum (u_x^2 + u_y^2)$ remain bounded.[16]
2) If $w = w_h$ satisfies the difference equation $\Delta w = 0$ at a mesh point of G_h, and if, as $h \to 0$ the sum

$$h^2 \sum_{G_h^*} \sum w^2,$$

extended over a mesh region G_h^* associated with a subregion G^* of G remains bounded, then for any fixed subregion G^{**} lying entirely within G^* the sum

$$h^2 \sum_{G_h^{**}} \sum (w_x^2 + w_y^2)$$

over the mesh region G_h^{**} associated with G^{**} likewise remains bounded as $h \to 0$. Using 1) there follows from this, since all of the difference quotients w of the function u_h again satisfy the difference equation $\Delta w = 0$, that each of the sums $h^2 \sum_{G_h^*} \sum w^2$ is bounded.
3) From the boundedness of these sums there follows finally the boundedness and uniform continuity of all the difference quotients themselves.

• 2. *Proof of the lemmas*

The proof of 1) follows from the fact that the functional values u_h are themselves bounded. For the greatest (or least) value of the function is assumed on the boundary[17] and so is bounded by a prescribed finite value. The boundedness of the sum $h^2 \sum_{G_h} \sum (u_x^2 + u_y^2)$ is an immediate consequence of the minimum property of our mesh function formulated in Part 2 of Section 2 which gives in particular

$$h^2 \sum_{G_h} \sum (u_x^2 + u_y^2) \leq h^2 \sum_{G_h} \sum (f_x^2 + f_y^2),$$

but as $h \to 0$ the sum on the right tends to the integral

$$\iint_G \left[\left(\frac{\partial f}{\partial x}\right)^2 + \left(\frac{\partial f}{\partial y}\right)^2 dxdy \right]$$

which, by hypothesis, exists.

[15] The weaker boundary value requirement does in fact provide the unique characterization of the solution, as can be seen from the easily proved theorem: If the boundary condition above is satisfied for $f(x, y) = 0$ for a function satisfying the equation

$$\frac{\partial^2 u}{\partial x^2} + \frac{\partial^2 u}{\partial y^2} = 0$$

in the interior of G and if

$$\iint_G \left[\left(\frac{\partial u}{\partial x}\right)^2 + \left(\frac{\partial u}{\partial y}\right)^2 \right] dxdy$$

exists, then $u(x, y)$ is identically zero. (See Courant, "Über die Lösungen der Differentialgleichungen der Physik," *Math Ann.* 85, 296 ff.)

In the case of two independent variables the boundary values are actually attained, as follows from the weaker requirement; but in the case of more variables the corresponding result cannot in general be expected since there can exist exceptional points on the boundary at which the boundary value is not taken on even though a solution exists under the weaker requirement.

[16] Here and in the following we drop the index h from the grid functions.

[17] We note, however, with a view to carrying over the method to other differential equations, that we can relax these conditions. To this end we need only to bring into play the inequality (15) or to use the reasoning of the alternative (see Part 4, Section 4).

To prove 2) we consider the quadratic sum

$$h^2 \sum_{\text{int}.Q_1}\sum (w_x^2 + w_{\bar{x}}^2 + w_y^2 + w_{\bar{y}}^2),$$

where the summation extends over all the interior points of a square Q_1, (see Fig. 1). We denote the values of the function on the boundary S_1 of the square Q_1, by w_1, and those on the boundary S_0, of Q_0, by w_0. Then Green's formula gives

$$h^2 \sum_{\text{int}.Q_1}\sum (w_x^2 + w_{\bar{x}}^2 + w_y^2 + w_{\bar{y}}^2) \tag{8}$$
$$= \sum_{S_1} w^2 - \sum_{S_0} w^2 - \sum_{C_1} w^2$$

where S_1 and S_0 are respectively the boundary of Q_1 and the square boundary of the lattice points Q_0 lying within S_1, while C_1 consists of the four corner points of the boundary of Q_1.

We now consider a sequence of concentric squares $Q_0, Q_1, \cdots, Q_n$ with boundaries $S_0, S_1, \cdots, S_n$, where each boundary is separated from the next by a mesh width. Applying the formula to each of these squares and observing that we have always

$$2h^2 \sum_{Q_0}\sum (w_x^2 + w_y^2)$$
$$\leq h^2 \sum_{Q_k}\sum (w_x^2 + w_{\bar{x}}^2 + w_y^2 + w_{\bar{y}}^2), \qquad (k \geq 1)$$

we obtain

$$2h^2 \sum_{Q_0}\sum (w_x^2 + w_y^2)$$
$$\leq \sum_{S_k} w^2 - \sum_{S_{k-1}} w^2 - \sum_{C_k} w^2 \qquad (1 \leq k < n),$$

where C_k consists of the four corner points of the boundary Q_k.

We strengthen the inequality by neglecting the last non-positive term on the right and we then add the n inequalities to obtain

$$2nh^2 \sum_{Q_0}\sum (w_x^2 + w_y^2) \leq \sum_{S_n} w^2 - \sum_{S_0} w^2 \leq \sum_{S_n} w^2.$$

Summing this inequality from $n = 1$ to $n = N$ we get

$$N^2h^2 \sum_{Q_0}\sum (w_x^2 + w_y^2) \leq \sum_{Q_N}\sum w^2.$$

Diminishing the mesh width h we can make the squares Q_0 and Q_n converge towards two fixed squares lying within G and having corresponding sides separated by a distance a. In this process $Nh \to a$ and we have, independent of the mesh width

$$h^2 \sum_{Q_0}\sum (w_x^2 + w_y^2) \leq \frac{h^2}{a^2} \sum_{Q_N}\sum w^2. \tag{9}$$

For sufficiently small h this inequality holds of course not only for two squares Q_0 and Q_N but with a change in the constant, a, for any two subregions of G such that one is contained entirely within the other. Thus lemma (2) is proved.[18]

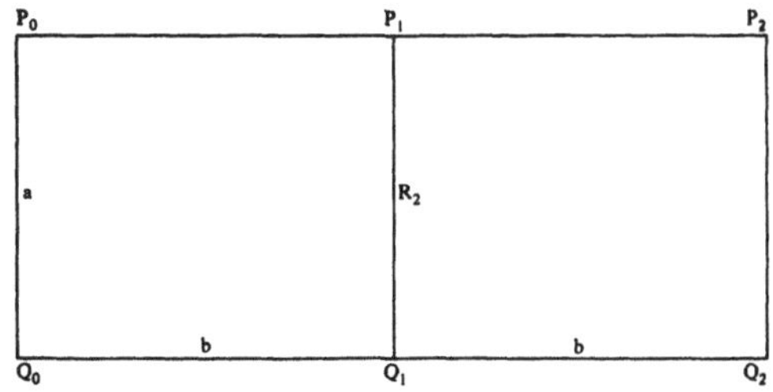

Figure 2

In order to prove the third result, that u_h and all its partial difference quotients w_h remain bounded and uniformly continuous as $h \to 0$, we consider a rectangle R (Fig. 2) with corners P_0, Q_0, P, Q and with sides P_0Q_0 and PQ of length a parallel to the x-axis.

We start with the relation

$$w(Q_0) - w(P_0) = h \sum_{PQ} w_x - h^2 \sum_{R}\sum w_{xy},$$

and the inequality

$$|w(Q_0) - w(P_0)| \leq h \sum_{PQ} |w_x| + h^2 \sum_{R}\sum |w_{xy}|, \tag{11}$$

which is a consequence of it.

We then let the side PQ of the rectangle vary between an initial line P_1Q_1, a distance b from P_0Q_0 and a final line P_2Q_2 a distance $2b$ from P_0Q_0, and we sum the corresponding $(b/h) + 1$ inequalities (11). We obtain the estimate

$$|w(P_0) - w(Q_0)| \leq \frac{1}{b + h} h^2 \sum_{R_2}\sum |w_x|$$
$$+ h^2 \sum_{R_2}\sum |w_{xy}|$$

where the summations extend over the entire rectangle, $R_2 = P_0Q_0P_2Q_2$. From Schwarz's inequality it then follows that,

$$|w(P_0) - w(Q_0)| \leq \frac{1}{b} \sqrt{2ab} \sqrt{h^2 \sum_{R_2}\sum w_x^2} \tag{12}$$
$$+ \sqrt{2ab} \sqrt{h^2 \sum_{R_2}\sum w_{xy}^2}.$$

[18] If we do not assume that $\Delta w = 0$, then in place of the inequality (9) we find:

$$h^2 \sum_{G^{**}}\sum (w_x^2 + w_y^2)$$
$$\leq c_1h^2 \sum_{G^*}\sum w^2 + c_2h^2 \sum_{G^*}\sum (\Delta w)^2 \tag{10}$$

for suitable constants c_1 and c_2 independent of h, where G^{**} lies entirely within G^*, and G^* in turn is contained in the interior of G.

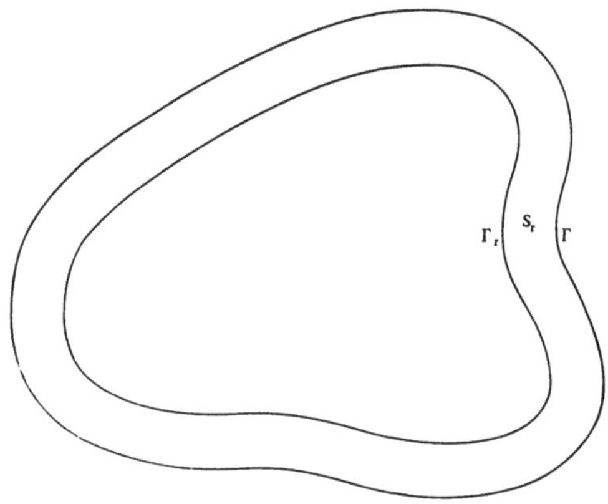

Figure 3

Since, by hypothesis, the sums which occur here multiplied by h^2 remain bounded, it follows that as $a \to 0$ the difference $|w(P_0) - w(Q_0)| \to 0$ independently of the mesh-width, since for each subregion G^* of G the quantity b can be held fixed. Consequently the uniform continuity (equicontinuity) of $w = w_h$ is proved for the x-direction. Similarly it holds for the y-direction and so also for any subregion G^* of G. The boundedness of the function w_h in G^* finally follows from its uniform continuity (equicontinuity) and the boundedness of $h^2 \sum_{G^*} \sum w_h^2$.

By this proof we establish the existence of a subsequence of functions u_h which converge towards a limit function $u(x, y)$ and which do so uniformly together with all their difference quotients, in the sense discussed above for every interior subregion of G. This limit function $u(x, y)$ has throughout G continuous partial derivatives of arbitrary order, and satisfies there the potential equation:

$$\frac{\partial^2 u}{\partial x^2} + \frac{\partial^2 u}{\partial y^2} = 0.$$

• 3. *The boundary condition*

In order to prove that the solution satisfies the boundary condition formulated above we shall first of all establish the inequality

$$h^2 \sum_{S_{r,h}} \sum v^2 \leq Ar^2h^2 \sum_{S_{r,h}} \sum (v_x^2 + v_y^2) + Brh \sum_{\Gamma_h} v^2 \qquad (13)$$

where $S_{r,h}$ is that part of the mesh region G_h which lies inside a boundary strip S_r. This boundary strip S_r consists of all points of G whose distance from the boundary is less than r; it is bounded by Γ and another curve Γ_r (Fig. 3). The constants A and B depend only on the region and not
224 on the function v nor the mesh width h.

In order to prove the above inequality, we divide the boundary, Γ, of G into a finite number of pieces for which the angle of the tangent with one of the x- or y-axes is greater than some positive value (say 30°). Let γ, for instance, be a piece of Γ which is sufficiently steep (in the above sense) relative to the x-axis (see Fig. 4). Lines parallel to the x-axis intersecting γ will cut a section γ_r from the neighboring curve Γ_r, and will define together with γ and γ_r a piece s_r of the boundary strip S_r. We use the symbol $s_{r,h}$ to denote the portion of G_h contained in s_r and denote the associated portion of the boundary Γ_h by γ_h.

We now imagine a parallel to the x-axis to be drawn through a mesh point P_h of $s_{r,h}$. Let it meet the boundary γ_h in a point $\bar{P}_h$. The portion of this line which lies in $s_{r,h}$ we call $p_{r,h}$. Its length is certainly smaller than cr, where the constant c depends only on the smallest angle of inclination of a tangent γ to the x-axis.

Between the values of v at P_h and $\bar{P}_h$ we have the relation

$$v(P_h) = r(\bar{P}_h) \pm h \sum_{P_h\bar{P}_h} v_x.$$

Squaring both sides and applying Schwarz's inequality, we obtain

$$v(P_h)^2 \leq 2v(\bar{P}_h)^2 + 2cr \cdot h \sum_{p_{r,h}} v_x^2.$$

Summing with respect to P_h in the x-direction, we get

$$h \sum_{p_r} v^2 \leq 2crv(\bar{P}_h)^2 + 2c^2r^2h \sum_{p_r} v_x^2.$$

Summing again in the y-direction we obtain the relation

$$h \sum_{s_{r,h}} \sum v^2 \leq 2cr \sum_{\Gamma_h} v(\bar{P}_h)^2 + 2c^2r^2h \sum_{s_{r,h}} \sum v_x^2. \qquad (14)$$

Writing down the inequalities associated with the other portions of Γ and adding all the inequalities together we obtain the desired inequality (13).[19]

We next set $v_h = u_h - f_h$ so that $v_h = 0$ on Γ_h.

Then since $h^2 \sum_{G_h} \sum (v_x^2 + v_y^2)$ remains bounded as $h \to 0$, we obtain from (13)

$$\frac{h^2}{r} \sum_{S_{r,h}} \sum v^2 \leq \kappa r, \qquad (16)$$

where κ is a constant which does not depend on the function v or the mesh size. Extending the sum on the left to the difference $S_{r,h} - S_{\rho,h}$ of two boundary strips, the inequality (16) still holds with the same constant κ and we can pass to the limit $h \to 0$.

[19] By similar reasoning we can also establish the inequality

$$h^2 \sum_{G_h} \sum v^2 \leq c_1 h \sum_{\Gamma_h} v^2 + c_2 h^2 \sum_{G_h} \sum (v_x^2 + v_y^2) \qquad (15)$$

in which the constants c_1 and c_2 depend only on the region G and not on the mesh division.

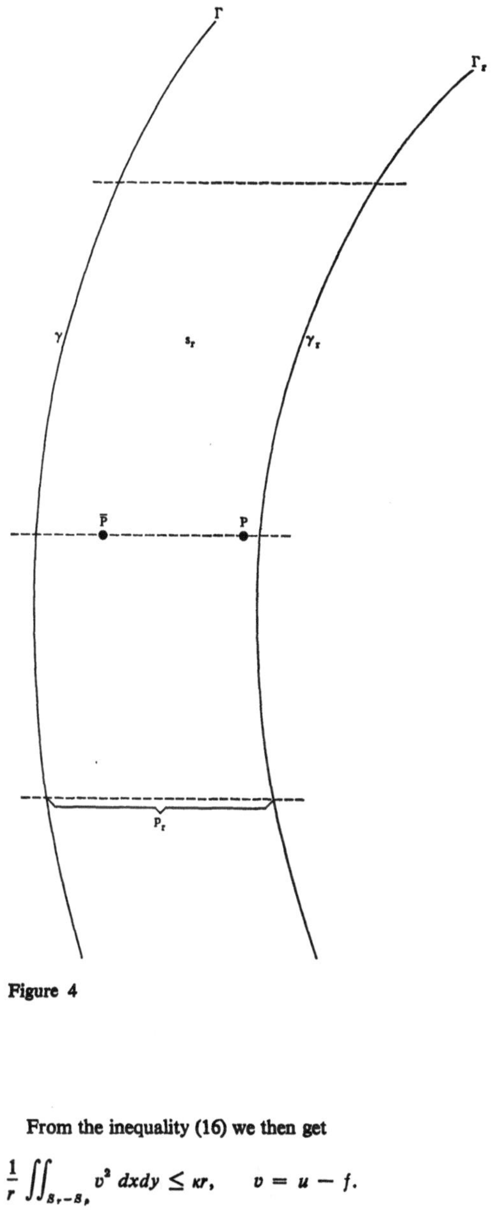

Figure 4

From the inequality (16) we then get

$$\frac{1}{r}\iint_{S_r - S_\rho} v^2\,dxdy \le \kappa r, \qquad v = u - f.$$

Now letting the narrower boundary strip S_ρ approach the boundary we obtain the inequality

$$\frac{1}{r}\iint_{S_r} v^2\,dxdy = \frac{1}{r}\iint_{S_r} (u-f)^2\,dxdy \le \kappa r$$

which states that the limit function u satisfies the prescribed boundary condition.

• 4. *Applicability of the method to other problems*

Our method is based essentially on the inequalities arising from the lemmas stated previously since the principal points of the proofs follow from the two last lemmas in Part 1 of Section 4.[20] No use is made of special fundamental solutions or special properties of the difference expression, and therefore the method can be carried over directly to the case of arbitrarily many independent variables as well as to the eigenvalue problem,

$$\frac{\partial^2 u}{\partial x^2} + \frac{\partial^2 u}{\partial y^2} + \lambda u = 0.$$

The same sort of convergence relations will obtain in this case as above.[21] Also the method applies to linear partial differential equations of other types, in particular its application to equations with variable coefficients requires only some minor modifications. The essential difference in this case lies only in proving the boundedness of $h^2 \sum \sum u_h^2$ which of course does not always hold for an arbitrary linear problem. However in case this sum is not bounded it can be shown that the general boundary value problem for the differential equation in question also possesses effectively no solutions, but that in this case there exist nonvanishing solutions of the corresponding homogeneous problem, i.e., eigenfunctions.[22]

• 5. *The boundary value problem* $\Delta\Delta u = 0$

In order to show that the method can be carried over to the case of differential equations of higher order, we will treat briefly the boundary value problem of the differential equation:

$$\frac{\partial^4 u}{\partial x^4} + 2\frac{\partial^4 u}{\partial x^2\,\partial y^2} + \frac{\partial^4 u}{\partial y^4} = 0.$$

We seek, in G, a solution of this equation for which the values of u and its first derivative are given on the boundary, being specified there by some function $f(x, y)$.

To this end we assume as previously that $f(x, y)$ together with its first and second derivatives is continuous in that region of the plane containing the region G.

We replace our differential equation problem by the new problem of solving the difference equation $\Delta u = 0$ in the mesh region G subject to the condition that at the points of the boundary strip $\Gamma_h + \Gamma_h'$ the function u takes on the values $f(x, y)$. From Section 2 we know that this boundary value problem has a unique solution. We will show that as the mesh size decreases, this solution, in each

[20] For an application of corresponding integral inequalities see K. Friedrichs, "Die Rand- und Eigenwertprobleme aus der Theorie der elastischen Platten", *Math. Ann.* 98, 222 (1926).

[21] Similarly one proves at the same time that every solution of such a differential equation problem has derivatives of every order.

[22] See Courant-Hilbert, *Methoden der Mathematischen Physik* 1, Ch. III, Section 3, where the theory of integral equations is handled with the help of the corresponding principle of the alternative. See also the Dissertation (Göttingen) of W. v. Koppenfels, which will appear soon.

interior subregion of G, converges to the solution of the differential equation, and that all of its difference quotients converge to the corresponding partial derivatives.

We note first that for the solution $u = u_h$, the sum

$$h^2 \sum_{G_h'} \sum (u_{xx}^2 + 2u_{xy}^2 + u_{yy}^2)$$

remains bounded as $h \to 0$. That is, by applying the minimum requirements on the solution (Part 2, Section 2) one finds that this sum is not larger than the corresponding sum

$$h^2 \sum_{G_h'} \sum (f_{xx}^2 + 2f_{xy}^2 + f_{yy}^2),$$

and this converges as $h \to 0$ to

$$\iint_G \left(\frac{\partial^2 f}{\partial x^2} + 2\frac{\partial^2 f}{\partial x \partial y} + \frac{\partial^2 f}{\partial y^2}\right) dxdy$$

which exists, by hypothesis.

From the boundedness of the sum

$$h^2 \sum_{G_h'} \sum (u_{xx}^2 + 2u_{xy}^2 + u_{yy}^2)$$

follows immediately the boundedness of $h^2 \sum_{G_h'} \sum (\Delta u)^2$ and also that of

$$h^2 \sum_{G_h'} \sum (u_x^2 + u_y^2) \quad \text{and} \quad h^2 \sum_{G_h} \sum u^2.$$

That is, for arbitrary w the following inequality holds (see Footnote 19),

$$h^2 \sum_{G_h} \sum w^2 \leq ch^2 \sum_{G_h} \sum (w_x^2 + w_y^2) + ch \sum_{\Gamma_h} w^2. \tag{15}$$

Then if one substitutes the first difference quotients of w for w itself in this inequality and applies the expression over the subregion of G_h for which they are defined, one finds the further inequality,

$$h^2 \sum_{G_h} \sum (w_x^2 + w_y^2) \leq ch^2 \sum_{G_h'} \sum (w_{xx}^2 + 2w_{xy}^2 + w_{yy}^2) + ch \sum_{\Gamma_h + \Gamma_h'} (w_x^2 + w_y^2),$$

where again the constant c is independent of the function and of the mesh size. We apply this inequality to $w = u_h$ and thus find the boundedness of the sum over $\Gamma_h + \Gamma_h'$ on the right-hand side; by definition these boundary sums converge to the corresponding integral containing $f(x, y)$. Thus from the boundedness of

$$h^2 \sum_{G_h'} \sum (u_{xx}^2 + 2u_{xy}^2 + u_{yy}^2)$$

follows the boundedness of $h^2 \sum_{G_h} \sum (u_x^2 + u_y^2)$ and thence that of $h^2 \sum_{G_h} \sum u^2$.

For the third step we substitute one after the other the expressions $\Delta u, \Delta u_x, \Delta u_y, \Delta u_{xx}, \cdots$, for w in the inequality

$$h^2 \sum_{G^{**}} \sum (w_x^2 + w_y^2) \leq ch^2 \sum_{G^*} \sum w^2 + ch^2 \sum_{G^*} \sum (\Delta w)^2$$

(see Part 2, Section 4) where G^* is a subregion of G containing G^{**} in its interior. The expressions introduced all satisfy the equation $\Delta w = 0$. It follows then that for each expression in turn and for all subregions G^* of G that the sums, $h^2 \sum_{G^*} \sum (w_x^2 + w_y^2)$, that is, $h^2 \sum_{G^*} \sum (\Delta u_x^2 + \Delta u_y^2)$, $h^2 \sum_{G^*} \sum (\Delta u_{xx}^2 + \Delta u_{xy}^2), \cdots$ are bounded together with the sums:

$$h^2 \sum_{G_h} \sum u^2, \qquad h^2 \sum_{G_h} \sum (u_x^2 + u_y^2),$$

$$\text{and} \quad h^2 \sum_{G_h} \sum (\Delta u)^2,$$

which are already known to be bounded.

Finally we substitute into the inequality (10), in place of w, the sequence of functions $u_{xx}, u_{xy}, u_{yy}, u_{xxx}, \cdots$, for which

$$h^2 \sum_{G_h^*} \sum (\Delta w)^2, \quad \text{i.e.} \quad h^2 \sum_{G_h^*} \sum (\Delta u_{xx})^2, \cdots$$

are bounded as shown above. We then find that for all subregions the sums

$$h^2 \sum_{G_h^*} \sum (u_{xxx}^2 + u_{xxy}^2), \quad h^2 \sum_{G_h^*} \sum (u_{xyx}^2 + u_{xyy}^2), \cdots$$

remain bounded.

From these facts we can now conclude as previously that from our sequence of mesh functions a subsequence can be chosen which in each interior subregion of G converges (together with all its difference quotients) uniformly to a limit function (or respectively its derivatives) which is continuous in the interior of G.

We have yet to show that this limit function which obviously satisfies the differential equation $\Delta\Delta u = 0$ also takes on the prescribed boundary conditions. For this purpose we say here only that, analogous to the foregoing, the expressions

$$\iint_{S_r} (u - f)^2 \, dxdy \leq cr^2,$$

$$\iint_{S_r} \left[\left(\frac{\partial u}{\partial x} - \frac{\partial f}{\partial x}\right)^2 + \left(\frac{\partial u}{\partial y} - \frac{\partial f}{\partial y}\right)^2\right] dxdy \leq cr^2$$

hold.[23] That the limit function fulfills these conditions may be seen by carrying over the treatment in Part 3, Section 4 to the function u and its first difference quotient.

[23] That the boundary values for the function and its derivatives actually are assumed is not difficult to prove. See for instance the corresponding treatment of K. Friedrichs, *loc. cit.*

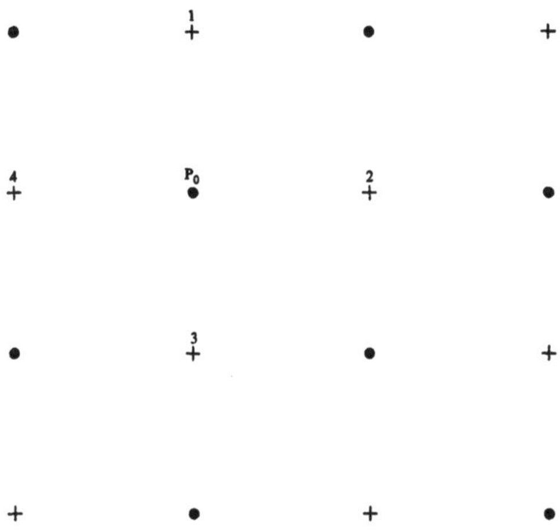

Figure 5

From the uniqueness of our boundary value problem we see furthermore that not only a selected subsequence, but also the original sequence of functions u possesses the asserted convergence properties.

II. The hyperbolic case

Section 1. The equation of the vibrating string

In the second part of this paper we shall consider the initial value problem for linear hyperbolic partial differential equations. We shall prove that under certain hypotheses the solutions of the difference equations converge to the solutions of the differential equations as the mesh size decreases.

We can discuss the situation most easily by considering the example of the approximation to the solution of the wave equation

$$\frac{\partial^2 u}{\partial t^2} - \frac{\partial^2 u}{\partial x^2} = 0 \tag{1}$$

We limit ourselves to the particular initial value problem where the value of the solution u, and its derivatives are given on the line $t = 0$.

In order to find the corresponding difference equation, we construct in the (x, t)-plane a square grid with lines parallel to the axes and with mesh width h. Following the notation of the first part of the paper we replace the differential equation (1) by the difference equation $u_{\bar{t}t} - u_{\bar{x}x} = 0$. If we select a grid point, P_0, then the corresponding difference equation relates the value of the function at this point to the values at four neighboring points. If we characterize the four neighboring values by the four indices 1, 2, 3, 4 (cf. Fig. 5), then the difference equation assumes the simple form

$$u_1 + u_3 - u_2 - u_4 = 0. \tag{2}$$

Note that the value of the function u at the point P_0 does not appear itself in the equation.

We consider the grid split up into two subgrids, indicated in Fig. 5 by dots and crosses respectively. The difference equation connects the values of the function over each of the subgrids separately, and so we shall consider only one of the two grids. As initial condition the values of the function are prescribed on the two rows of the grid, $t = 0$ and $t = h$. We can give the solution of this initial value problem explicitly; that is, we express the value of the solution at any point S in terms of the values given along the two initial rows. One can see at once that the value at a point of the row $t = 2h$ is uniquely determined by only the three values at the points close to it in the two first rows. The value at a point of the fourth row is uniquely determined by the values of the solution at three particular points in the second and third rows, and through them it is related to certain values in the first two rows. In general to a point S there will correspond a certain region of dependence in the first two rows; it may be found by drawing the lines $x + t =$ const. and $x - t =$ const., through the point S and extending them until they meet the second row at the points α and β respectively (cf. Fig. 6). The triangle $S\alpha\beta$ is called the *triangle of determination* because all the values of u in it remain unchanged provided the values on the first two rows of it are held fixed. The sides of the triangle are called *lines of determination.*

If one denotes the differences of u in the direction of the lines of determination by u' and $u^{\backprime}$, that is,

$$u'_1 = u_1 - u_4, \qquad u^{\backprime}_1 = u_1 - u_2$$

$$u'_2 = u_2 - u_3, \qquad u^{\backprime}_4 = u_4 - u_3,$$

Figure 6

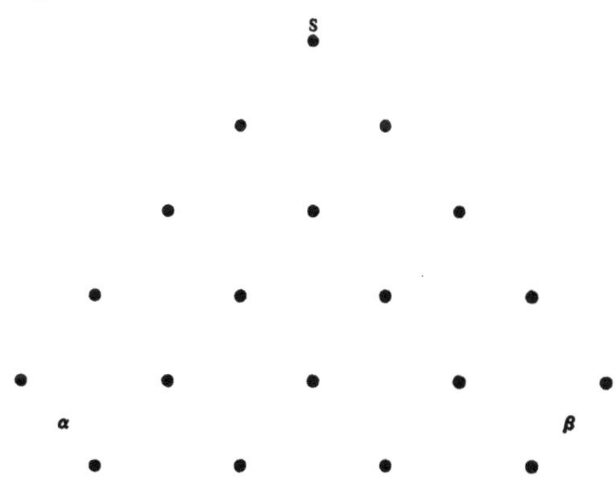

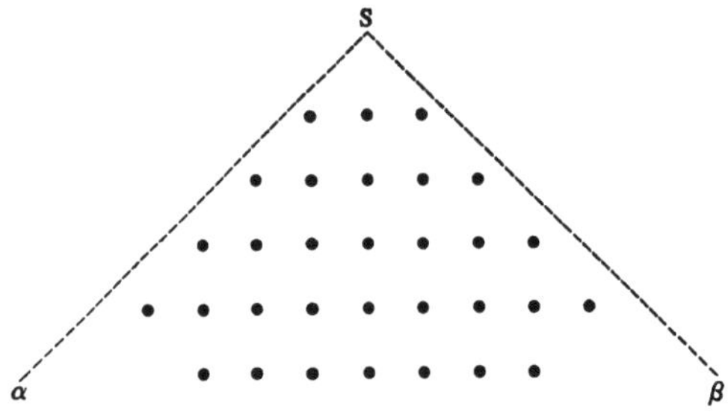

Figure 7

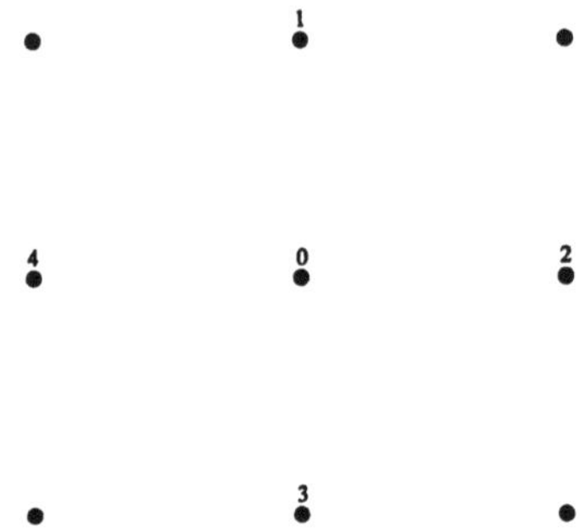

Figure 8

then the difference equation assumes the form

$$u_1' = u_2',$$

i.e., along a line of determination the differences with respect to the other direction of determination are constant, and thus are equal to one of the *given* differences between the value at two points on the first two rows. Moreover the difference $u_S - u_\alpha$ is a sum of differences u' along the determining line αS, so that using the remark above, we can obtain the final result (in obvious notation):

$$u_S = u_\alpha + \sum_{\alpha_1}^{\beta_1} u'. \tag{3}$$

We now let h go to zero, and let the prescribed values on the second and first rows converge uniformly to a twice continuously differentiable function, $f(x)$, and the difference quotients $u'/h\sqrt{2}$ there converge uniformly to a continuously differentiable function $g(x)$. Evidently the right-hand side of (3) goes over uniformly to the expression

$$f(x - t) + \frac{1}{\sqrt{2}} \int_{x-t}^{x+t} g(\xi)\, d\xi \tag{4}$$

if S converges to the point (x, t). This is the well-known expression for the solution of the wave equation (1) with initial values $u(x, 0) = f(x)$ and $\partial u(x, 0)/\partial t = f'(x) + \sqrt{2}g(x)$. Thus it is shown that as $h \rightarrow 0$ the solution of the difference equation converges to the solution of the differential equation provided the initial values converge appropriately (as above).

Section 2. On the influence of the choice of mesh. The domains of dependence of the difference and differential equations

The relationships developed in Section 1 suggest the following considerations.

In the same way that the solution of a linear hyperbolic equation at a point S depends only on a certain part of the initial line—namely the "domain of dependence" lying between the two characteristics drawn through S, the solution of the difference equation has also at the point S a certain domain of dependence defined by the lines of determination drawn through S. In Section 1 the directions for the lines of determination of the difference equation coincided with the characteristic directions for the differential equation so that the domains of dependence coincided in the limit. This result, however, was based essentially on the orientation of the mesh in the (x,t)-plane, and depended furthermore on the fact that a square mesh had been chosen. We shall now consider a more general rectangular mesh with mesh size equal to h (time interval) in the t-direction and equal to kh (space interval) in the x-direction, where k is a constant. The domain of dependence for the difference equation, $u_{tt} - u_{xx} = 0$ for this mesh will now either lie entirely within the domain of dependence of the differential equation, $\partial^2 u/\partial t^2 - \partial^2 u/\partial x^2 = 0$, or on the other hand will contain this latter region inside its own domain according as $k < 1$ or $k > 1$ respectively.

From this follows a remarkable fact: if for the case $k < 1$ one lets $h \rightarrow 0$, then the solution to the difference equation in general *cannot* converge to the solution of the differential equation. In this case a change in the initial values of the solution of the differential equation in the neighborhood of the endpoints α and β of the domain of dependence (see Fig. 7) causes, according to formula (4), a change in the solution itself at the point (x, t). For the solution of the difference equation at the point S, however, the changes at the points α and β are not relevant since these points lie outside of the domain of dependence of the difference equations. That convergence does occur for the case $k > 1$ will be proved in Section 3. See for example Fig. 9.

If we consider the differential equation

$$2\frac{\partial^2 u}{\partial t^2} - \frac{\partial^2 u}{\partial x^2} - \frac{\partial^2 u}{\partial y^2} = 0 \tag{5}$$

in two space variables, x and y, and time, t, and if we replace it by the corresponding difference equation on a rectilinear grid, then in contrast to the case of only two independent variables it is impossible to choose the mesh division so that the domain of dependence of the difference and differential equations coincide, since the domain of dependence of the difference equation is a quadrilateral while that of the differential equation is a circle. Later (cf. Section 4) we shall choose the mesh division so that the domain of determination of the difference equation contains that of the differential equation in its interior, and shall show that once again convergence occurs.

On the whole an essential result of this section will be that in the case of each linear homogeneous hyperbolic equation of second order one can choose the mesh so that the solution of the difference equation converges to the solution of the differential equation as $h \to 0$, (see for instance Sections 3, 4, 7, 8).

Section 3. Limiting values for arbitrary rectangular grids

Now we consider further the wave equation

$$\frac{\partial^2 u}{\partial t^2} - \frac{\partial^2 u}{\partial x^2} = 0, \tag{1}$$

but impose it now on a rectangular grid with time interval h and space interval kh. The corresponding difference equation is

$$L(u) = \frac{1}{h^2}(u_1 - 2u_0 + u_3) - \frac{1}{k^2h^2}(u_2 - 2u_0 + u_4) = 0, \tag{6}$$

where the indices represent a "fundamental rhombus" with midpoint P_0 and corners P_1, P_2, P_3, P_4 (see Fig. 8). According to the equation $L(u) = 0$ the value of the function u at a point S can be represented by its values on that

Figure 9

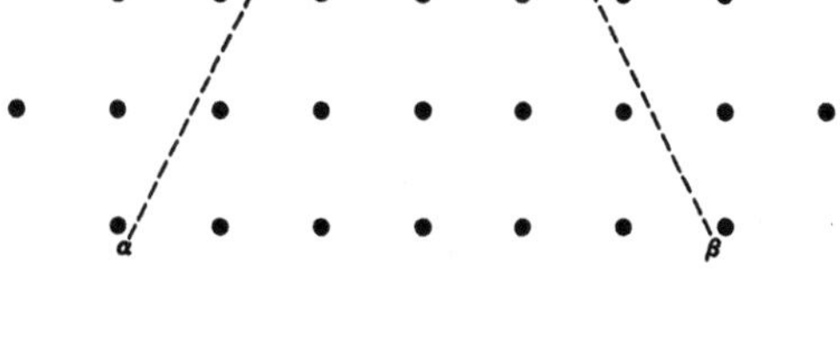

section of the initial rows $t = 0$ and $t = h$ cut out by lines of determination through S parallel to the sides of an elementary rhombus. We assume that the initial values are prescribed in such a way that as $t \to 0$ for fixed k the first difference quotients formed from them converge uniformly to given continuous functions on the line $t = 0$. The initial values can be used to form an explicit representation of the solution of the difference equation (corresponding to (3) in Section 1); however it is too complicated to yield a limiting value easily as $h \to 0$. Thus we will try another approach which will also make it possible for us to treat the general problem.[24]

We multiply the difference expression $L(u)$ by $(u_1 - u_3)$ and form the product using the following identities:

$$(u_1 - u_3)(u_1 - 2u_0 + u_3) = (u_1 - u_0)^2 - (u_0 - u_3)^2, \tag{7}$$

$$(u_1 - u_3)(u_2 - 2u_0 + u_4) = (u_1 - u_0)^2 - (u_0 - u_3)^2 - \tfrac{1}{2}[(u_1 - u_2)^2 + (u_1 - u_4)^2 - (u_2 - u_3)^2 - (u_4 - u_3)^2]. \tag{8}$$

Then we obtain

$$2(u_1 - u_3)L(u) = \frac{2}{h^2}\left(1 - \frac{1}{k^2}\right)[(u_1 - u_0)^2 - (u_0 - u_3)^2] + \frac{1}{h^2k^2}[(u_1 - u_2)^2 + (u_1 - u_4)^2 - (u_2 - u_3)^2 - (u_4 - u_3)^2]. \tag{9}$$

The product (9) is now summed over all elementary rhombuses of the domain of determination, $S\alpha\beta$. The quadratic difference terms on the right-hand side always appear with alternate signs in two neighboring rhombuses so that they cancel out for any two rhombuses belonging to the triangle $S\alpha\beta$. Only the sums of squared differences over the "boundary" of the triangle remain, and we obtain the relation:

$$\begin{aligned} h^2 \sum_{S\alpha\beta} 2\frac{u_1 - u_3}{h} L(u) &= h \sum_{S\alpha}\left[2\left(1 - \frac{1}{k^2}\right)\left(\frac{\dot{u}}{h}\right)^2 + \frac{1}{k^2}\left(\frac{u'}{h}\right)^2\right] \\ &+ h \sum_{S\beta}\left[2\left(1 - \frac{1}{k^2}\right)\left(\frac{\dot{u}}{h}\right)^2 + \frac{1}{k^2}\left(\frac{u^{\backprime}}{h}\right)^2\right] \\ &- h \sum_{\mathrm{I}\ \mathrm{II}}\left[2\left(1 - \frac{1}{k^2}\right)\left(\frac{\dot{u}}{h}\right)^2 \right. \\ &\left. + \frac{1}{k^2}\left(\frac{u'}{h}\right)^2 + \frac{1}{k^2}\left(\frac{u^{\backprime}}{h}\right)^2\right]. \end{aligned} \tag{10}$$

[24] For the following see K. Friedrichs and H. Lewy, "Über die Eindeutigkeit...etc.," *Math. Ann.* 98, 192 ff. (1928), where a similar transformation is used for integrals.

229

Here u' and $u^{\backprime}$ denote differences in the direction of determination defined in Section 1, while $\dot{u}$ designates the difference of the functional values at two neighboring points on a mesh line parallel to the t axis. The range in $\sum_{S\alpha}$ over which $(u')^2$ is taken is the outermost boundary line $S\alpha$ and its nearest parallel neighbor found by shifting the points of $S\alpha$ downward by the amount h. There is a similar range for $(u^{\backprime})^2$ in $\sum_{S\beta}$, and so all of the differences, u', $u^{\backprime}$, and $\dot{u}$ appear once and only once.

For the solution to the problem $L(u) = 0$ the right-hand side of (10) vanishes. The sum over the initial rows I and II which occurs there remains bounded as $h \to 0$ (for fixed k); specifically it goes over into an integral of the prescribed function along the initial line. Consequently the sums over $S\alpha$ and $S\beta$ in (10) also remain bounded. If now $k \geq 1$ as we must require (see previous discussion), then $1 - 1/k^2$ is non-negative, and we find that the individual sums

$$h \sum_{S\alpha} \left(\frac{u'}{h}\right)^2, \qquad h \sum_{S\beta} \left(\frac{u^{\backprime}}{h}\right)^2,$$

extended over any line of determination whatever, remain bounded.

From this we can derive the "uniform continuity" (equicontinuity) (cf. Section 4 of the first part of the paper) of the sequence of grid functions in all directions in the plane;[25] since the values of u on the initial line are bounded, there must exist a subset which converges uniformly to a limit function $u(x, t)$.

Both the first and second difference quotients of the function u also satisfy the difference equation $L(u) = 0$. Their initial values can be expressed through the equation $L(u) = 0$ in terms of the first, second and third difference quotients of u involving initial values at points on the two initial lines I and II only. We require that they tend to continuous limit functions, that is, that the given initial values $u(x, 0)$, $u_t(x, 0)$ be three times or respectively twice continuously differentiable with respect to x.

From this we can apply the convergence considerations set forth above to the first and second difference quotients of u, as well as to u itself, and we can choose a subsequence such that these difference quotients converge uniformly to certain functions, which must be the first or respectively second derivatives of the limit function $u(x, t)$. Hence the limit function satisfies the differential equation $\partial^2 u/\partial t^2 - \partial^2 u/\partial x^2 = 0$ which results as the limit of the difference equation $L(u) = 0$; it represents indeed the solution of the initial value problem. Since such a solution is uniquely determined, every subsequence of mesh functions converges, and therefore the sequence itself converges to the limit function.

[25] If S_1 and S_2 are two points at a distance δ, then one connects them by a path of two segments, S_1S and SS_2, where the former is parallel to one line of determination and the latter to the other. Then one finds the appraisal,

$$|u_{S_1} - u_S| \leq |u_{S_1} - u_S| + |u_S - u_{S_2}|$$
$$\leq \sqrt{\delta}\sqrt{h \sum_{S_1S} \left(\frac{u'}{h}\right)^2} + \sqrt{\delta}\sqrt{h \sum_{SS_2} \left(\frac{u^{\backprime}}{\delta}\right)^2}.$$

Section 4. The wave equation in three variables

We treat next the wave equation,

$$2\frac{\partial^2 u}{\partial t^2} - \frac{\partial^2 u}{\partial x^2} - \frac{\partial^2 u}{\partial y^2} = 0 \tag{11}$$

and consider its relation to the observations on the domain of dependence discussed in Section 2. The domain of dependence of the differential equation (11) is a circular cone with axis parallel to the t-direction and with apex angle α, where $\tan \alpha = 1/\sqrt{2}$. In any rectilinear grid parallel to the axes we introduce the corresponding difference equation

$$2u_{t\bar{t}} - u_{x\bar{x}} - u_{y\bar{y}} = 0. \tag{12}$$

This equation relates the functional values of u at points of an elementary tetrahedron. In fact it allows the value of the function u at a point S to be expressed uniquely in terms of the values of the function at certain points of the two initial planes $t = 0$ and $t = h$. At each point S we obtain a pyramid of determination which cuts out from the two base planes two rhombuses as domains of dependence.

If we let the mesh widths tend to zero, keeping their ratios fixed, then we can expect convergence of the sequence of mesh functions to the solution of the differential equation only provided the pyramid of determination contains the cone of determination of the differential equation in its interior. The simplest grid with this property will be one constructed in such a way that the pyramid of determination is tangent to the exterior of the cone of determination. Our differential equation is chosen so that this occurs for a grid of cubes parallel to the axes.

The difference equation (12), in the notation of Fig. 10, assumes for such a grid the form:

$$L(u) = \frac{2}{h^2}(u'_\alpha - 2u_0 + u_\alpha) - \frac{1}{h^2}(u_1 - 2u_0 + u_3)$$
$$- \frac{1}{h^2}(u_2 - 2u_0 + u_4), \tag{13}$$

in which the functional value, u_0, at the midpoint actually cancels out. The values of the solution on the two initial planes must be the values of a function having four continuous derivatives with respect to x_1y_2t.

For the convergence proof we again use the method developed in Section 3. We construct the triple sum

$$h^2 \sum\sum\sum 2\frac{u'_\alpha - u_\alpha}{h} L(u) = 0$$

for the solution to the difference equation, where the summation is to be extended over all elementary octahedrons of the pyramid of determination emanating from

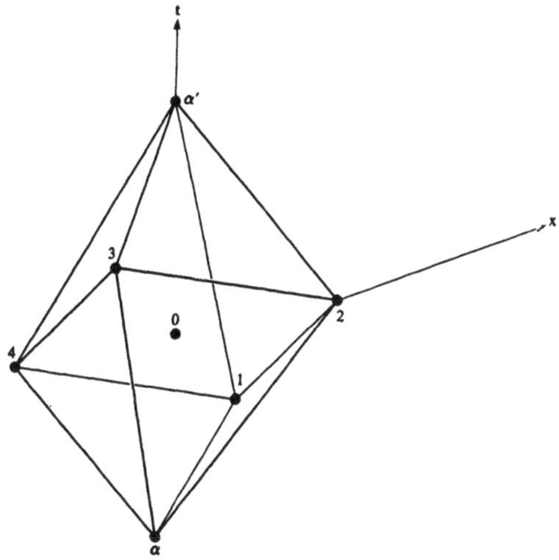

Figure 10

the point S. Then almost exactly as before we find that the values of the function u at the interior points of the pyramid of determination cancel out in the summation and that only the values on the two pyramids called F, and on the two base surfaces I and II remain.

If we denote by u' the difference of the values of the function at two points connected by a line of an elementary octahedron, then we can write the result as

$$\sum_F \sum (u')^2 - \sum_{I} \sum_{II} (u')^2 = 0, \tag{14}$$

where the sum is extended over all surfaces containing differences u'; each such difference is to appear only once.[26] The double sum over the two initial surfaces stays bounded since it goes over into an integral of the initial values. Therefore the sum over the "surface of determination" F remains bounded.

We now apply these results not to u itself, but to its first, second and third difference quotients, which themselves satisfy the difference equation (13). Their initial values can be expressed using only values on the first two initial planes by means of (13) using first through fourth difference quotients. If $w = w_h$ is one of the difference quotients of any order up to third order, then we know that the sum $h^3 \sum \sum (w'/h)^2$ extended over a surface of determination remains bounded. From this it follows, through exactly the methods used in Section 4 of the first part of the paper, that the function u and its first and second difference quotients are uniformly continuous (equicontinuous). Thus there exists a sequence of mesh widths decreasing to zero such that these quantities, which are bounded initially, converge to continuous limit functions and, in fact, converge to the solution of the differential equation and to the first and second derivatives of this solution, all exactly as we found earlier (Section 3).

[26] The grid ratio has been chosen in such a way that the differences between values of u appearing on the two neighboring surfaces in F do not occur, (as they did in the general case in one dimension treated in Section 3).

Appendix. *Supplements and generalizations*

Section 5. Example of a differential equation of first order

We have seen in Section 2 that in the case when the region of dependence of the differential equation covers only a part of the region of dependence of the difference equation, the influence of the rest of the region is not included in the limit. We can demonstrate this phenomenon explicitly by the example of the differential equation of first order, $\partial u/\partial t = 0$ if we replace it by the difference equation

$$2u_t - u_x + u_{\bar{x}} = 0. \tag{15}$$

In the notation of Fig. 5 this becomes

$$u_1 = \frac{u_2 + u_4}{2}. \tag{16}$$

As before, the difference equation connects only the points of a submesh with one another. The initial value problem consists of assigning as initial values for u at points $x = 2ih$ on the row $t = 0$ the values, f_{2i}, assumed there by a continuous function $f(x)$.

We consider the point S at a distance $2nh$ up along the t-axis. It is easy to verify that the solution u at S is represented as

$$u_S = \sum_{i=-n}^{n} \frac{1}{2^{2n}} \binom{2n}{n+i} f_{2i}. \tag{17}$$

As the mesh size decreases, that is as $n \to \infty$, the sum on the right-hand side tends simply to the value f_0. This can be demonstrated from the continuity of $f(x)$ and from the behavior of the binomial coefficients as n increases (see the following paragraph).

Section 6. The equation of heat conduction

The difference equation (16) of Section 5 can also be interpreted as the analogue of an entirely different differential equation, namely the equation of heat conduction,

$$2\frac{\partial u}{\partial t} - \frac{\partial^2 u}{\partial x^2} = 0. \tag{18}$$

In any rectangular mesh with mesh spacing l and h in the time and space directions, respectively, the corresponding difference equation becomes

$$2\left(\frac{u_1 - u_0}{l}\right) = \left(\frac{u_2 + u_4 - 2u_0}{h^2}\right). \tag{19}$$

In the limit as the mesh size goes to zero the difference equation preserves its form only if l and h^2 are decreased 231 proportionately. In particular if we set $l = h^2$, then the value u_0 drops out of the equation and the difference equation becomes

$$u_1 = \frac{u_2 + u_4}{2}. \tag{16}$$

The solution to (16) is given by formula

$$u(0, t) = \sum_{i=-n}^{n} \frac{1}{2^{2n}} \begin{bmatrix} 2n \\ n + i \end{bmatrix} f_{2i}.$$

As the mesh width decreases, a point ξ on the x-axis is always characterized by the index

$$2i = \xi/h. \tag{20}$$

The mesh width h is related to the ordinate t of a particular point by

$$2nh^2 = t. \tag{21}$$

We shall investigate what happens to formula (17) as $h \to 0$, that is $n \to \infty$. Using (21) we write (17) in the form

$$u(0, t) = \sum_{i=-n}^{n} \frac{\sqrt{2n}}{2 \cdot 2^{2n} \sqrt{t}} \begin{bmatrix} 2n \\ n + i \end{bmatrix} f_{2i} \cdot 2h. \tag{22}$$

For the coefficient of $2hf_{2i} = 2hf(\xi)$ we use the abbreviation

$$\frac{1}{2\sqrt{t}} g_{2n}(\xi) = \frac{\sqrt{2n}}{2 \cdot 2^{2n} \sqrt{t}} \begin{bmatrix} 2n \\ n + \dfrac{\xi}{\sqrt{2t}} \sqrt{n} \end{bmatrix}.$$

The limiting value of the coefficient, which is usually determined by using Stirling's formula, we will calculate here by considering the function $g_{2n}(\xi)$ as the solution of an ordinary difference equation which approaches a differential equation as $h \to 0$. As the difference equation one finds

$$\frac{1}{2h} [g_h(\xi + 2h) - g_h(\xi)] = -\frac{1}{2h} g_h(\xi) \frac{2i + 1}{n + i + 1}$$

(in which we have written $g_h(\xi)$ instead of $g_{2n}(\xi)$). Or

$$\frac{1}{2h} [g_h(\xi + 2h) - g_h(\xi)] = -g_h(\xi) \frac{\xi + h}{t + h\xi + 2h^2}.$$

$g_h(\xi)$ satisfies the normalization condition

$$\sum_{i=-n}^{n} g_h(\xi) \cdot 2h = 2\sqrt{t}.$$

This sum is over the region of dependence of the difference equation, and as $h \to 0$ this covers the entire x-axis.

It can be shown easily that $g_h(\xi)$ converges uniformly to the solution $g(x)$ of the differential equation

$$g'(x) = -g(x)x/t$$

with the auxiliary condition

$$\int_{-\infty}^{\infty} g(x)\, dx = 2\sqrt{t}.$$

From formula (22) after passing to the limit we find

$$u(0, t) = \int_{-\infty}^{\infty} \frac{1}{\sqrt{2\pi t}} e^{-\xi^2/2t} f(\xi)\, d\xi,$$

which is the known solution of the heat conduction equation.

The results of this section can be carried over directly to the case of the differential equation,

$$4\frac{\partial u}{\partial t} - \frac{\partial^2 u}{\partial x^2} - \frac{\partial^2 u}{\partial y^2} = 0$$

and so on for even more independent variables.

Section 7. The general homogeneous linear equation of second order in the plane

We consider the differential equation

$$\frac{\partial^2 u}{\partial t^2} - k^2 \frac{\partial^2 u}{\partial x^2} + \alpha \frac{\partial u}{\partial t} + \beta \frac{\partial u}{\partial x} + \gamma u = 0. \tag{23}$$

The coefficients are assumed to be twice continuously differentiable with respect to x and t, while the initial values on the line $t = 0$ are assumed three times continuously differentiable with respect to x. We replace the differential equation by the difference equation

$$L(u) = u_{t\bar{t}}(x, t) - k^2 u_{x\bar{x}}(x, t) + \alpha u_t + \beta u_x + \gamma u = 0 \tag{24}$$

in a grid with time mesh width h and space mesh width ch so that in a neighborhood of the appropriate part of the initial value line the inequality $1 - k^2/c^2 > \varepsilon > 0$ holds for the constant, c. The initial values are to be chosen as in Section 3.

For the proof of convergence we again transform the sum,

$$h^2 \sum_{S\alpha\beta} \sum 2 \frac{u_1 - u_3}{h} L(u)$$

by using identities (7) and (8). In addition to a sum (see for example (10)) over the doubled boundary of the triangle $S\alpha\beta$ (Fig. 6) one obtains a sum over the entire triangle $S\alpha\beta$ because of the variability of the coefficient k and the presence of lower order derivatives in the differential equation. By using the differentiability of k and the Schwarz inequality one can show that this latter sum is bounded from above by

$$Ch^2 \sum_{S\alpha\beta} \sum \left[\left(\frac{u^{\backprime}}{h}\right)^2 + \left(\frac{u'}{h}\right)^2 + \left(\frac{\dot{u}}{h}\right)^2 + u^2 \right]$$

where the constant C is independent of the function u,

the mesh width h, and, in a certain neighborhood of the initial line, also independent of the point S.

Again we can estimate an upper bound for $h^2 \sum_{S\alpha\beta} \sum u^2$ by[27]

$$C_1 h^2 \sum_{S\alpha\beta}\sum \left(\frac{\dot{u}}{h}\right)^2 + C_2 h \sum_{\mathrm{I\,II}} u^2,$$

where the same properties hold for C_1 and C_2 as are stated above for C.

Thus we obtain an inequality of the form

$$\begin{aligned} h \sum_{S\alpha}&\left[2\left(1-\frac{k^2}{c^2}\right)\left(\frac{\dot{u}}{h}\right)^2 + \frac{k^2}{c^2}\left(\frac{u'}{h}\right)^2\right] \\ &+ h \sum_{S\beta}\left[2\left(1-\frac{k^2}{c^2}\right)\left(\frac{\dot{u}}{h}\right)^2 + \frac{k^2}{c^2}\left(\frac{u^{\backprime}}{h}\right)^2\right] \\ &\le C_3 h^2 \sum_{S\alpha\beta}\sum\left[\left(\frac{\dot{u}}{h}\right)^2 + \left(\frac{u'}{h}\right)^2 + \left(\frac{u^{\backprime}}{h}\right)^2\right] + D, \qquad (25) \end{aligned}$$

where D, for all points S and mesh widths h, is a fixed bound for the sums over the initial line.

As vertices of our triangles we choose a sequence of points $S_0, S_1, \cdots, S_n = S$ lying on a line parallel to the t-axis. By summing the corresponding sequence of inequalities (25) after multiplying by h we obtain the following inequality

$$\begin{aligned} h^2 \sum_{SS_0\alpha}\sum&\left[2\left(1-\frac{k^2}{c^2}\right)\left(\frac{\dot{u}}{h}\right)^2 + \frac{k^2}{c^2}\left(\frac{u'}{h}\right)^2\right] \\ &+ h^2 \sum_{SS_0\beta}\sum\left[2\left(1-\frac{k^2}{c^2}\right)\left(\frac{\dot{u}}{h}\right)^2 + \frac{k^2}{c^2}\left(\frac{u^{\backprime}}{h}\right)^2\right] \\ &\le nh^3 C_3 \sum_{S\alpha\beta}\sum\left[\left(\frac{\dot{u}}{h}\right)^2 + \left(\frac{u'}{h}\right)^2 + \left(\frac{u^{\backprime}}{h}\right)^2\right] + nhD. \end{aligned} \qquad (26)$$

Now if we notice that one can express a difference u' or $u^{\backprime}$ in terms of two differences $\dot{u}$ and a difference $u^{\backprime}$ or respectively u', then we see that the left-hand side of (26) is larger than the simpler form

$$C_4 h^2 \sum_{S\alpha\beta}\sum\left[\left(\frac{\dot{u}}{h}\right)^2 + \left(\frac{u'}{h}\right)^2 + \left(\frac{u^{\backprime}}{h}\right)^2\right],$$

with a suitable constant C_4.

Then by confining the discussion to a sufficiently small neighborhood, $0 \le t \le nh = \delta$ of the initial line where δ is small enough so that

$$C_4 - nhC_3 = C_5 > 0,$$

we find from (26),

$$C_5 h^2 \sum_{S\alpha\beta}\sum\left[\left(\frac{\dot{u}}{h}\right)^2 + \left(\frac{u'}{h}\right)^2 + \left(\frac{u^{\backprime}}{h}\right)^2\right] \le \frac{C_4}{C_3} D. \qquad (27)$$

The bound given by (27) when combined with (25) gives a bound on

$$h \sum_{S\alpha}\left(\frac{u'}{h}\right)^2 + h \sum_{S\beta}\left(\frac{u^{\backprime}}{h}\right)^2,$$

from which, as in Section 3, one can prove the uniform continuity of u.

We apply the inequality (25) now, instead of to the function u itself, to its first and second difference quotients, w, which also satisfy difference equations whose second-order terms are the same as those of (24). In the rest of the terms there will appear lower order differences of u which cannot be expressed in terms w, but they will appear in the above argument in the form of quadratic double sums multiplied by h^2. This is enough to let us reach the same conclusions for the difference equation for w as we found above for u. So we can conclude from this the uniform continuity (equicontinuity) and boundedness of the function u and its first and second derivatives. Consequently a subsequence exists which converges uniformly to the solution of the initial value problem for the differential equation. Again from the uniqueness of the solution we find that the sequence of functions itself converges.

In all of this the assumption must be made that the difference quotients up to third order involving values on the two initial lines converge to continuous limit functions.[28]

Section 8. The initial value problem for an arbitrary linear hyperbolic differential equation of second order

We shall now show that the methods developed so far are adequate for solving the initial value problem for an arbitrary homogeneous linear hyperbolic differential equation of second order. It suffices to limit the discussion to the case of three variables. The development can be extended immediately to the case of more variables. It is easy to see that a transformation of variables can reduce the most general problem of this type to the following: a function $u(x, y, t)$ is to be found which satisfies the differential equation

$$\begin{aligned} u_{tt} &- (au_{xx} + 2bu_{xy} + cu_{yy}) \\ &+ \alpha u_t + \beta u_x + \gamma u_y + \delta u = 0, \end{aligned} \qquad (28)$$

and which, together with its first derivative, assumes prescribed values on the surface $t = 0$. The coefficients in Eq. (28) are functions of the variables x, y, and t and satisfy the condition

$$a > 0, \quad c > 0, \quad ac - b^2 > 0.$$

We assume that the coefficients are three times differentiable with respect to x, y, and t, and that the initial

[27] For proof one makes use of the inequality used in Footnote 25.

[28] This assumption and also the assumptions on the differentiability of the coefficients of the differential equation, and further on the restriction to a sufficiently small region of the initial line can be weakened in special cases.

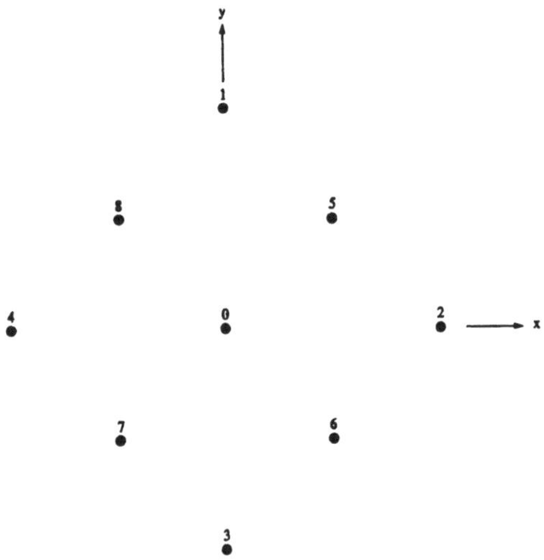

Figure 11

values u and u_t are respectively four and three times continuously differentiable with respect to x and y.

The coordinates x and y can be drawn from a given point on the initial plane in such a way that $b = 0$ there. Then of course in a certain neighborhood of this point the conditions

$$a - |b| > 0, \qquad c - |b| > 0$$

hold. We restrict our investigation to this neighborhood. We can choose a three times continuously differentiable function $d > 0$ so that

$$\left.\begin{array}{l} a - d \\ c - d \\ d - |b| \end{array}\right\} > \varepsilon > 0 \tag{29}$$

for some constant ε. Then we put the differential equation into the form

$$\begin{aligned} u_{tt} - (a - d)u_{xx} - (c - d)u_{yy} & \\ - \tfrac{1}{2}(d + b)(u_{xx} + 2u_{xy} + u_{yy}) & \\ - \tfrac{1}{2}(d - b)(u_{xx} - 2u_{xy} + u_{yy}) & \\ + \alpha u_t + \beta u_x + \gamma u_y + \delta u = 0. & \end{aligned} \tag{30}$$

We now construct in the space a grid of points, $t = lh$, $x + y = mkh$, $x - y = nkl\,(l, m, n = \cdots -1, 0, 1, 2 \cdots)$ and we replace Eq. (30) by a difference equation $L(u) = 0$ over this mesh. We do this by assigning to each point P_0 the following neighborhood: The point $P_{\alpha'}$ or the point P_α which lies at a distance h or $-h$ respectively along the t-axis from P_0; also the points $P_1, \cdots, P_8$ which lie in the same (x, y)-plane with P_0 (see Fig. 11). These points constitute an "elementary octahedron" with vertices $P_{\alpha'}$, P_α, P_1, P_2, P_3, P_4. For each grid point lying in the interior of G we replace the derivatives appearing in Eq. (30) by difference quotients over the elementary octahedron about P_0.

We replace

$$\begin{aligned} u_{tt} \quad & \text{by} \quad \frac{1}{h^2}(u_{\alpha'} - 2u_0 + u_\alpha) \\ u_{xx} \quad & \text{by} \quad \frac{1}{k^2h^2}(u_2 - 2u_0 + u_4) \\ u_{yy} \quad & \text{by} \quad \frac{1}{k^2h^2}(u_1 - 2u_0 + u_3) \\ u_{xx} + 2u_{xy} + u_{yy} \quad & \text{by} \quad \frac{4}{k^2h^2}(u_6 - 2u_0 + u_8) \\ u_{xx} - 2u_{xy} + u_{yy} \quad & \text{by} \quad \frac{4}{k^2h^2}(u_5 - 2u_0 + u_7). \end{aligned}$$

The first derivatives in (30) are replaced by the corresponding difference quotients in the elementary octahedron. The coefficients of the difference equation are given the values assumed by the coefficients of the differential equation at the point P_0.

On the initial planes $t = 0$ and $t = h$ we assume that the values of the function are assigned in such a way that as the mesh size approaches zero for fixed k, the function approaches the prescribed initial value, and the difference quotients over the two planes up through differences of fourth order converge uniformly to the prescribed derivatives.

The solution of the difference equation $L(u) = 0$ at a point is uniquely determined by the values on the two base surfaces of the pyramid of determination passing through the point.

To prove convergence we construct a sum

$$h^3 \sum \sum \sum 2\,\frac{u_{\alpha'} - u_\alpha}{h}\,L(u)$$

over all the elementary octahedrons of the pyramid of determination, and we transform it by using identities (7) and (8). This gives one space summation multiplied by h^3 and quadratic in the first difference quotients, and also over a double surface a sum which is multiplied by h^2 and contains the squares of all the difference quotients of the type $u_\alpha - u_0$, $u_\alpha - u_1, \cdots, u_\alpha - u_8$. In this expression according to (29) the coefficients are larger than some fixed positive constant in all those cases where the ratio $1/k$ between the time and space mesh sizes is taken sufficiently small.

From here on one can proceed exactly as before (Sections 7, 4) and prove that the solution of the difference equation converges to the solution of the differential equation.

(Submitted to Math. Ann. September 1, 1927)

Mathematische Annalen
99 (1928)

Das Anfangswertproblem einer beliebigen nichtlinearen hyperbolischen Differentialgleichung beliebiger Ordnung in zwei Variablen. Existenz, Eindeutigkeit und Abhängigkeitsbereich der Lösung.

Von

Kurt Friedrichs in Aachen und Hans Lewy in Göttingen.

Soll eine partielle Differentialgleichung, die man für die zeitliche Änderung irgendeines physikalischen Zustandes abgeleitet hat, wirklich das Gesetz dieser Zustandsänderung darstellen, so wird man von ihr verlangen, daß sie folgende mathematischen Eigenschaften besitzt. Ist zu irgendeiner Zeit irgendein Zustand vorgegeben, so besitzt die Differentialgleichung eine Lösung, die den Zustand auch für die folgende Zeit darstellt. Und zwar darf es nur eine Lösung geben, denn sonst würde die Differentialgleichung den Zustand nicht vollständig bestimmen. Soll ferner eine räumlich begrenzte Änderung eines vorgegebenen Zustandes sich nur mit endlicher Geschwindigkeit auswirken, so darf der Wert der Lösung in einem Punkte nicht von sämtlichen Anfangswerten abhängen, sondern nur von einem räumlich begrenzten Teil. Von einer solchen Eigenschaft kann man natürlich nur dann sprechen, wenn man auch nicht analytische Anfangswerte zuläßt.

Die linearen hyperbolischen Differentialgleichungen zweiter Ordnung erfüllen bekanntlich diese Forderungen. Für beliebige hyperbolische Differentialgleichungen zweiter Ordnung in zwei Veränderlichen wurde dies Verhalten von H. Lewy[1]) nachgewiesen. Für lineare Differentialgleichungen höherer Ordnung mit konstanten Koeffizienten hat G. Herglotz[2]) explizite Darstellungen der Lösung entwickelt, in denen die gekennzeichnete Art der Abhängigkeit von den Anfangswerten zum Ausdruck kommt. In

---

[1]) H. Lewy, Über das Anfangswertproblem einer hyperbolischen nichtlinearen Differentialgleichung ..., Math. Annalen 98.

[2]) G. Herglotz, Sächsische Akademieberichte 1926, S. 287.

unserer Arbeit beweisen wir, daß unsere Forderungen für *jede hyperbolische Differentialgleichung beliebiger Ordnung in zwei unabhängigen Veränderlichen* erfüllt sind.

Das Anfangswertproblem einer Differentialgleichung n-ter Ordnung $F = 0$ für eine Funktion $u(x, y)$ der beiden Variablen x, y formulieren wir folgendermaßen. Auf einer Kurve[3]) in der (x, y)-Ebene seien die Werte der Funktion u und ihrer Ableitungen bis zur n-ten Ordnung derartig vorgegeben, daß sie der gegebenen Differentialgleichung genügen. Längs dieser Kurve sollen für die vorgegebenen Anfangswerte $u, u_x, u_y, u_{xx}, \ldots, u_{y^n} = \frac{\partial^n u}{\partial y^n}$ eine Reihe von Streifenbedingungen erfüllt sein, d. h.

$$\begin{aligned} du &= u_x\,dx + u_y\,dy, \\ du_x &= u_{xx}\,dx + u_{xy}\,dy, \\ &\cdots\cdots\cdots \\ du_{y^{n-1}} &= u_{y^{n-1}x}\,dx + u_{y^n}\,dy. \end{aligned}$$

Um das Anfangsproblem zu lösen, leiten wir ein — für eine Lösung u von $F = 0$ und ihre Ableitungen $u_x, u_y, \ldots$ sicher erfülltes — System von Differentialgleichungen ab, in dem die Größen $x, y, u, u_x, \ldots, u_{y^n}$ als unbekannte Funktionen aufgefaßt werden und ihre ersten Ableitungen nach einer „charakteristischen Richtung" auftreten. Dieses Gleichungssystem nennen wir das System der *„charakteristischen Gleichungen"*. Wir ersetzen nun unser Anfangswertproblem für $F = 0$ durch ein Anfangswertproblem für ein System von charakteristischen Differentialgleichungen. Alle solchen Gleichungen, die sich für *eine* charakteristische Richtung ergeben, würden nicht ausreichen, um diese Funktionen völlig zu bestimmen. Da es aber, wie wir fordern, n verschiedene reelle charakterstische Richtungen gibt, so können wir die *Gesamtheit* der charakteristischen Gleichungen, die in allen Richtungen gelten, als ein System *partieller* Differentialgleichungen auffassen, welches jedoch jetzt *mehr* Differentialgleichungen als gesuchte Funktionen enthält[4]). Aus dieser Gesamtheit suchen wir ein System von ebensoviel Gleichungen wie Funktionen heraus, das bei vorgegebenen Anfangswerten gerade eine Lösung zuläßt. Wir weisen die Existenz dieser Lösung mit Hilfe eines Differenzenverfahrens nach und zeigen, daß die Lösung des ausgewählten Systems der charakteristischen Gleichungen auch die Lösung des ursprünglichen Anfangswertproblems für $F = 0$ liefert.

---

[3]) Die Tangente dieser Kurve muß stetig und darf nirgends charakteristisch sein vgl. § 2).

[4]) Vgl. H. Lewy, loc. cit.

Aus unserem Konvergenzbeweis ergibt sich unmittelbar das Resultat, daß der Wert der Lösung in einem Punkte von den Vorgaben nur auf einem Teil der Anfangskurve abhängt; d. h. daß ihr Wert sich nicht ändert, wenn wir die Vorgaben außerhalb dieses Teils der Anfangskurve ändern. Den *Abhängigkeitsbereich* der Lösung erhalten wir folgendermaßen: Wir ziehen von dem betreffenden Punkte aus die charakteristischen Linien, bis sie die Anfangskurve treffen. Das von den beiden *äußersten* Schnittpunkten begrenzte Stück gibt das Abhängigkeitsgebiet.

1. Teil.

§ 1.

Die charakteristischen Differentialgleichungen[5]).

Wir gehen aus von einer Differentialgleichung n-ter Ordnung

$$F(x, y, u, u_x, u_y, u_{xx}, \ldots) = 0, \tag{1}$$

in der sämtliche Ableitungen der Funktion $u(x, y)$ bis zur n-ten Ordnung vorkommen können. Für die höchsten Ableitungen führen wir folgende Abkürzungen ein:

$$u_{x^n} = [0],$$
$$u_{x^{n-1}y} = [1],$$
$$\cdot\ \cdot\ \cdot\ \cdot\ \cdot\ \cdot$$
$$u_{y^n} = [n].$$

Ferner verwenden wir die Abkürzungen:

$$\frac{\partial F}{\partial [0]} = F_0,$$
$$\cdot\ \cdot\ \cdot\ \cdot\ \cdot$$
$$\frac{\partial F}{\partial [n]} = F_n.$$

Verstehen wir nun unter $u(x, y)$ eine Lösung der Differentialgleichung (1), so muß für sie auch die durch totale Differentiation nach x entstehende Gleichung

$$\frac{dF}{dx} = F_0 [0]_x + F_1 [1]_x \cdots + F_n [n]_x + [F]_x = 0 \tag{2}$$

erfüllt sein. Dabei bedeutet

$$[F]_x = \frac{\partial F}{\partial x} + \frac{\partial F}{\partial u} u_x + \ldots + \frac{\partial F}{\partial u_{y^{n-1}}} u_{y^{n-1}x}.$$

---

[5]) Die im folgenden verwandte Methode läßt sich unmittelbar zur Aufstellung der Charakteristikentheorie auch für Differentialgleichungen in mehr unabhängigen Variablen und für Systeme von Differentialgleichungen benutzen.

Wir betrachten nun irgendeine Richtung in der (x, y)-Ebene und bezeichnen die Ableitungen der Ausdrücke $x, y, u, u_x, \ldots$ nach einem Parameter in dieser Richtung mit $x', y', u', u_x', \ldots$. Dann gelten die folgenden Streifenrelationen

$$\begin{aligned} [0]' &= [0]_x x' + [0]_y y', \\ [1]' &= [1]_x x' + [1]_y y', \\ &\cdots\cdots\cdots \\ [n-1]' &= [n-1]_x x' + [n-1]_y y', \end{aligned}$$

die wir wegen

$$[0]_y = [1]_x, \; [1]_y = [2]_x, \; \ldots$$

auch in der Form

$$(3) \qquad \begin{aligned} x'[0]_x + y'[1]_x - [0]' &= 0, \\ x'[1]_x + y'[2]_x - [1]' &= 0, \\ \cdots\cdots\cdots& \\ x'[n-1]_x + y'[n]_x - [n-1]' &= 0 \end{aligned}$$

schreiben können. Die n Gleichungen (3) bilden zusammen mit Gleichung (2) ein System linearer Gleichungen für die $n+1$ Größen $[0]_x, [1]_x, \ldots, [n]_x$. Im allgemeinen wird die Determinante dieser Gleichungen von null verschieden sein. *Verschwindet aber die Determinante, so verschwindet auch,* da das Gleichungssystem ja eine Lösung hat, *jede $n+1$ reihige Determinante der Matrix*

$$\left\| \begin{matrix} F_0 & F_1 & F_2 & \ldots & F_{n-1} & F_n & -[F]_x \\ x' & y' & 0 & \ldots & 0 & 0 & [0]' \\ 0 & x' & y' & \ldots & 0 & 0 & [1]' \\ \cdot & \cdot & \cdot & \cdot & \cdot & \cdot & \cdot \\ 0 & 0 & 0 & \ldots & x' & y' & [n-1]' \end{matrix} \right\| .$$

Das Verschwinden der Determinante

$$(4) \qquad \Delta_n \equiv \begin{vmatrix} F_0 & F_1 & \ldots & F_{n-1} & F_n \\ x' & y' & \ldots & 0 & 0 \\ \cdot & \cdot & \cdot & \cdot & \cdot \\ 0 & 0 & \ldots & x' & y' \end{vmatrix} = 0$$

stellt eine homogene Bedingung für x', y', d. h. eine Bedingung für die Richtung dar, nach der die durch den Strich angedeuteten Ableitungen zu bilden sind. Das Verschwinden der übrigen $(n+1)$-reihigen Determinanten, z. B. der durch Auslassung der vorletzten Spalte entstehenden

$$(5) \qquad \begin{vmatrix} F_0 & F_1 & \ldots & F_{n-1} & -[F]_x \\ x' & y' & \ldots & 0 & [0]' \\ \cdot & \cdot & \cdot & \cdot & \cdot \\ 0 & 0 & \ldots & x' & [n-1]' \end{vmatrix} = 0$$

stellt dann eine Bedingung für die Ableitungen der Größen $[0], [1], \ldots, [n-1]$ nach dieser Richtung dar. Die Gleichung (4) $\Delta_n = 0$ nennen wir die ***charakteristische Hauptgleichung*** und jede Richtung $y' : x'$ der (x, y) Ebene, die ihr genügt, eine ***charakteristische Richtung.*** Die Gleichung $\Delta_n = 0$ ist vom n-ten Grade in $\varrho = y' : x'$. Wir verlangen nun von unserer Differentialgleichung $F = 0$ und ihrer Lösung u, daß die zugehörige charakteristische Hauptgleichung $\Delta_n = 0$ ***n und nur n verschiedene reelle Wurzeln*** besitzt[6]). Diese Voraussetzung ist nichts anderes als unsere Forderung, daß die Gleichung $F = 0$ hyperbolisch ist.

Die Gleichung (5) repräsentiert demgemäß n einzelne Gleichungen, je nachdem auf welche der n charakteristischen Richtungen sich die durch den Strich angedeutete Ableitung bezieht.

Geht man anstatt von der Gleichung (2), die aus $\frac{dF}{dx} = 0$ entstanden ist, von der Gleichung $\frac{dF}{dy} = 0$ aus, und fügt die n Streifenrelationen

$$\begin{gathered} x'[0]_y + y'[1]_y - [1]' = 0, \\ \cdots\cdots\cdots\cdots \\ x'[n-1]_y + y'[n]_y - [n]' = 0 \end{gathered}$$

hinzu, so erhält man ein Gleichungssystem für die $n+1$ Unbekannten $[0]_y, [1]_y, \ldots, [n]_y$. Seine Matrix lautet:

$$\left\| \begin{matrix} F_0 & F_1 & \ldots & F_{n-1} & F_n & -[F]_y \\ x' & y & \ldots & 0 & 0 & [1]' \\ \cdot & \cdot & \cdot & \cdot & \cdot & \cdot \\ 0 & 0 & \ldots & x' & y' & [n]' \end{matrix} \right\|.$$

Die Determinante der $n+1$ Gleichungen ist wieder Δ_n. Kennzeichnet der Strich wieder die Ableitung nach einer charakteristischen Richtung, so verschwindet Δ_n, und wir schließen dann wie früher auf das gleichzeitige Bestehen z. B. der Relation

$$(6) \qquad \left| \begin{matrix} F_0 & F_1 & \ldots & F_{n-1} & -[F]_y \\ x' & y' & \ldots & 0 & [1]' \\ \cdot & \cdot & \cdot & \cdot & \cdot \\ 0 & 0 & \ldots & x' & [n]' \end{matrix} \right| = 0,$$

[6]) Wir setzen voraus, daß für keine charakteristische Richtung $x' = 0$ oder $y' = 0$ ist, was darauf hinauskommt, daß F_0 und F_n nicht gleich null sind. Da die Größen $F_0, F_1, \ldots, F_n$ nicht alle gleich null sein können, läßt sich dies für Umgebungen jedes Punktes immer dadurch erreichen, daß man an Stelle von x, y neue unabhängige Veränderliche ξ, η einführt und $F = 0$ als Differentialgleichung für u als Funktion von ξ und η auffaßt; denn dabei gehen die charakteristischen Richtungen in sich über.

die n einzelne Gleichungen repräsentiert, je nachdem auf welche der n charakteristischen Richtungen sich der Strich bezieht.

Die Determinante Δ_n können wir in die Form

$$(7) \qquad \frac{\Delta_n}{(x')^n} = \Delta_n(\varrho) = F_0 \varrho^n - F_1 \varrho^{n-1} + - \ldots + (-1)^n F_n$$

setzen, während wir an Stelle der Gleichungen (5) und (6) auch

$$(8) \qquad \Delta_{n-1}(\varrho)[n-1]' - \Delta_{n-2}(\varrho)[n-2]' + - \ldots + (-1)^{n-1}\Delta_0(\varrho)[0]' - (-1)^n [F]_x x' = 0,$$

$$(9) \qquad \Delta_{n-1}(\varrho)[n]' - \Delta_{n-2}(\varrho)[n-1]' + - \ldots + (-1)^{n-1}\Delta_0(\varrho)[1]' - (-1)^n [F]_y x' = 0$$

schreiben können. Dabei bedeuten $\Delta_0(\varrho)$, $\Delta_1(\varrho)$, ... die Determinanten

$$(10) \qquad \begin{aligned} \Delta_0(\varrho) &= F_0, \\ \Delta_1(\varrho) &= \begin{vmatrix} F_0 & F_1 \\ 1 & \varrho \end{vmatrix}, \\ \Delta_2(\varrho) &= \begin{vmatrix} F_0 & F_1 & F_2 \\ 1 & \varrho & 0 \\ 0 & 1 & \varrho \end{vmatrix}, \\ &\ldots\ldots\ldots\ldots \end{aligned}$$

Wir merken noch für später die Relationen an:

$$(10\text{a}) \qquad \Delta_{k+1}(\varrho) - \varrho\,\Delta_k(\varrho) = (-1)^{k+1} F_{k+1}.$$

§ 2.

Die Bedeutung der Bedingung, daß die Anfangskurve nicht charakteristisch ist.

Es ist nach dem Vorangehenden plausibel, welche Bedingung man für die Anfangskurve $x(t)$, $y(t)$ stellen wird, wenn man in dem in der Einleitung formulierten Anfangswertproblem für $F = 0$ *beliebige* Werte von u und seinen Ableitungen bis zur n-ten Ordnung vorgeben will, die nur der Gleichung $F = 0$ und den Streifenbedingungen längs der Anfangskurve zu genügen brauchen. Wir werden nämlich verlangen, daß die Anfangskurve nirgends charakteristisch ist, d. h. daß auf ihr der durch Gleichung (4) definierte Ausdruck Δ_n nirgends verschwindet. Würde nämlich Δ_n an einer Stelle der Anfangskurve gleich null sein, so müßten dort auch noch die Gleichungen (5) und (6) erfüllt sein, wenn es eine Lösung von $F = 0$ mit diesen Anfangswerten gibt. Das würde aber im allgemeinen eine neue Bedingung für die Anfangswerte darstellen.

§ 3.

Das Anfangswertproblem der charakteristischen Differentialgleichungen.

Eine Lösungsfläche $u(x, y)$ unserer Differentialgleichung $F = 0$ wird von n Scharen von charakteristischen Linien überdeckt. An Stelle der Linien $x =$ konst. und $y =$ konst. führen wir die beiden „äußersten" dieser Charakteristikenscharen als Koordinatenlinien $\alpha =$ konst. und $\omega =$ konst. ein. Hierbei bezieht sich der Begriff der beiden äußersten Scharen auf die Anfangslinie, und zwar in folgendem Sinne. Denken wir uns in einem Punkte P der Anfangslinie eine Richtung aus der Richtung der Anfangslinie — die selbst nicht charakteristisch ist — um zwei Rechte gedreht, so wird sie dabei einmal mit sämtlichen n charakteristischen Richtungen zusammenfallen. Die erste und die letzte, mit der sie zusammenfällt, nennen wir die beiden äußersten charakteristischen Richtungen. Lassen wir den Punkt P auf der Anfangslinie wandern, so ändert sich jede charakteristische Richtung stetig[7]), und es gehen äußerste Richtungen in äußerste Richtungen über.

Wir wählen die Parameter α und ω so, daß in einer Ebene mit den rechtwinkligen Koordinaten α, ω der Geraden $\alpha + \omega = 0$ die Anfangskurve entspricht. Wir fassen nun die Größen $x, y, u, u_x, \ldots, [n]$ als voneinander unabhängige Funktionen der Parameter α und ω auf; d. h. wir *fordern nicht*, daß $u_x = \frac{\partial u}{\partial x}, \ldots, [n] = \frac{\partial^n u}{\partial y^n}$ ist. Dann stellen die charakteristischen Gleichungen (4), (5) und (6) und die Streifengleichungen

$$u' = u_x x' + u_y y'$$
$$\cdot \quad \cdot \quad \cdot \quad \cdot \quad \cdot \quad \cdot \quad \cdot \quad \cdot \quad \cdot$$
$$u'_{y^{n-1}} = [n-1]\, x' + [n]\, y'$$

ein System von Differentialgleichungen erster Ordnung für diese Funktionen dar. Wir können nämlich den Strich einmal als Ableitung nach dem Parameter α, das andere Mal als Ableitung nach ω deuten; der Strich kann aber auch die Ableitungen nach den $n-2$ anderen charakteristischen Richtungen andeuten; diese setzen sich aus den Ableitungen nach α und ω linear zusammen.

Wir können jetzt das Anfangswertproblem für die Differentialgleichung $F = 0$ *ersetzen* durch das Anfangswertproblem des Systems der charakteristischen Differentialgleichungen erster Ordnung für die Funktionen $x, y, u, \ldots, [n]$ der Variablen α und ω; natürlich mit denselben Anfangsbedingungen auf der Geraden $\alpha + \omega = 0$. Die Anzahl der charakteristischen Gleichungen und Streifenbedingungen ist aber größer als die Anzahl der

[7]) Falls die Anfangskurve eine stetige Tangente besitzt und die Größen $F_0, F_1, \ldots, F_n$ längs der Anfangskurve stetig sind.

gesuchten Funktionen, so daß wir uns veranlaßt sehen, ein System von ebensoviel Gleichungen als gesuchten Funktionen auszuwählen. Wir können das so einrichten, daß das Gleichungssystem in den Ableitungen linear wird. Dies geschieht auf folgende Weise.

Wir denken uns die charakteristische Hauptgleichung (4) in ihre Linearfaktoren zerlegt:

$$\Delta_n = F_0(y' - \varrho_\alpha x')(y' - \varrho_\beta x') \dots (y' - \varrho_\omega x') = 0 .$$

Die Größen $\varrho_\alpha, \varrho_\beta, \dots, \varrho_\omega$ sind also bekannte Funktionen der gesuchten Funktionen $x, y, \dots, [n]$. Die Indizes sind dabei so gewählt, daß der Linearfaktor $y' - \varrho_\alpha x'$ bzw. $y' - \varrho_\omega x'$ verschwindet, wenn für y' und x' die Ableitungen nach der Richtung α bzw. ω eingesetzt werden. Wir bezeichnen mit dem oberen rechten Index α oder ω die Ableitungen nach dem Parameter α bzw. ω und wählen die beiden Gleichungen

$$\text{(11)} \qquad \boxed{\begin{aligned} y^\alpha - \varrho_\alpha x^\alpha &= 0 \\ y^\omega - \varrho_\omega x^\omega &= 0 \end{aligned}}$$

für unser zu bildendes Gleichungssystem aus.

Die $n-2$ anderen Linearfaktoren von Δ_n dienen uns dazu, Ableitungen nach den $n-2$ anderen charakteristischen Richtungen — die wir entsprechend durch Indizes $\beta, \gamma, \dots$ kennzeichnen wollen — durch die Ableitungen nach α und ω linear auszudrücken. Ist nämlich $f(\alpha, \omega)$ irgendeine Funktion von α und ω, so ist z. B. $f^\beta = f^\alpha \alpha^\beta + f^\omega \omega^\beta$. Das Verhältnis $\alpha^\beta : \omega^\beta$ berechnen wir dabei aus der Gleichung

$$y^\beta - \varrho_\beta x^\beta = (y^\alpha - \varrho_\beta x^\alpha)\alpha^\beta + (y^\omega - \varrho_\beta x^\omega)\omega^\beta = 0 .$$

Ferner wählen wir die n Gleichungen aus, die aus der Gleichung (8) entstehen, indem wir ϱ und den Strich auf die n verschiedenen charakteristischen Richtungen $\alpha, \beta, \dots, \omega$ beziehen und demgemäß schreiben

$$\text{(12)} \qquad \boxed{\begin{aligned} &\Delta_{n-1}(\varrho_\alpha)[n-1]^\alpha - \Delta_{n-2}(\varrho_\alpha)[n-2]^\alpha + \dots - (-1)^n [F]_x x^\alpha = 0 \\ &\Delta_{n-1}(\varrho_\beta)[n-1]^\beta - \Delta_{n-2}(\varrho_\beta)[n-2]^\beta + \dots - (-1)^n [F]_x x^\beta = 0 \\ &\dots\dots\dots\dots\dots\dots\dots\dots\dots\dots \\ &\Delta_{n-1}(\varrho_\omega)[n-1]^\omega - \Delta_{n-2}(\varrho_\omega)[n-2]^\omega + \dots - (-1)^n [F]_x x^\omega = 0 \end{aligned}} .$$

Weiter benutzen wir die Gleichung (9) in einem einzigen Exemplar, nämlich dem auf α bezüglichen:

$$\text{(13)} \qquad \boxed{\Delta_{n-1}(\varrho_\alpha)[n]^\alpha - \Delta_{n-2}(\varrho_\alpha)[n-1]^\alpha + \dots - (-1)^n [F]_y x^\alpha = 0} .$$

Endlich benutzen wir noch sämtliche Streifengleichungen in der Richtung α für die Größen $u, u_x, u_y, \dots, u_{x^{n-1}}, u_{x^{n-2}y}, \dots, u_{y^{n-1}}$, die — es sei noch einmal betont — als voneinander unabhängige Funktionen (vgl. S. 206) von α und ω fungieren:

$$(14)\qquad \boxed{\begin{array}{c} {}^{\alpha}U \equiv u^{\alpha} - u_x x^{\alpha} - u_y y^{\alpha} = 0 \\ \cdots\cdots\cdots\cdots \\ {}^{\alpha}U_{y^{n-1}} \equiv u^{\alpha}_{y^{n-1}} - [n-1]x^{\alpha} - [n]y^{\alpha} = 0 \end{array}}\,.$$

Hier ist $u^{\alpha} = \dfrac{\partial u}{\partial \alpha}, \dots, u^{\alpha}_{y^{n-1}} = \dfrac{\partial u_{y^{n-1}}}{\partial \alpha}$ gesetzt.

Dieses lineare partielle Differentialgleichungssystem (11), (12), (13), (14) mit ebensoviel Gleichungen wie gesuchten Funktionen hat bei dreimal beschränkt differenzierbar auf der Anfangslinie $\alpha + \omega = 0$ vorgegebenen Anfangswerten in einer gewissen Umgebung dieser Linie eine eindeutige mit stetigen Ableitungen bis zur zweiten Ordnung versehene Lösung: $x, y, u, \dots, [n]$ als Funktionen von α, ω. Unser Gleichungssystem ist nämlich ein Spezialfall eines allgemeineren solchen Systems, von dem wir in Teil II die Existenz der Lösung beweisen, unter der Voraussetzung, daß seine *Determinante* (vgl. § 4) auf der Anfangslinie *nicht verschwindet.*

Es wird sich schließlich zeigen, daß die Lösung des Anfangswertproblems der charakteristischen Differentialgleichungen die gesuchte Lösung der Gleichung $F = 0$ liefert.

§ 4.

Die Determinante verschwindet nicht.

Die Determinante unseres Gleichungssystems erklären wir folgendermaßen: Wir setzen in unsere Differentialgleichungen für die verschiedenen Ableitungen von x, nämlich $x^{\alpha}, x^{\beta}, \dots, x^{\omega}$, immer denselben Buchstaben ξ, für $y^{\alpha}, y^{\beta}, \dots, y^{\omega}$ denselben Buchstaben η, und entsprechend für die Ableitungen aller unserer gesuchten Funktionen. Wir bekommen so ein lineares homogenes Gleichungssystem für die „Unbekannten" $\xi, \eta, \dots$; seine Determinante bezeichnen wir als die Determinante unseres Differentialgleichungssystems. Sie wird von null verschieden sein, sobald wir wissen, daß dieses lineare Gleichungssystem *nur die eine* Auflösung $\xi = 0,\ \eta = 0, \dots$ besitzt.

Um dies einzusehen, beachten wir, daß aus den beiden aus (11) entstehenden Gleichungen

$$\eta - \varrho_{\alpha}\xi = 0 \quad \text{und} \quad \eta - \varrho_{\omega}\xi = 0$$

$\eta = \xi = 0$ folgt, da ϱ_{α} nicht gleich ϱ_{ω} ist. Benutzen wir dieses Resultat, so sagen die Streifenbedingungen (14) nichts anderes aus, als daß die dort auftretenden den Größen $u^a, \ldots, u^a_{y^{n-1}}$ entsprechenden Unbekannten null sind. Wir haben nur noch das Verschwinden derjenigen Unbekannten nachzuweisen, die den Ableitungen von $[0]$, $[1]$, ... entsprechen. Dazu schreiben wir uns die Determinante der Gleichungen (12) auf:

$$(15) \qquad \begin{vmatrix} \Delta_{n-1}(\varrho_\alpha) & \Delta_{n-2}(\varrho_\alpha) & \cdots & \Delta_0(\varrho_\alpha) \\ \Delta_{n-1}(\varrho_\beta) & \Delta_{n-2}(\varrho_\beta) & \cdots & \Delta_0(\varrho_\beta) \\ \cdot & \cdot & \cdot & \cdot \\ \Delta_{n-1}(\varrho_\omega) & \Delta_{n-2}(\varrho_\omega) & \cdots & \Delta_0(\varrho_\omega) \end{vmatrix}.$$

$\Delta_k(\varrho)$ ist ein Polynom k-ten Grades in ϱ, und der Koeffizient der höchsten Potenz ist F_0 (vgl. (10)). Die Determinante (15) läßt sich also auf die Form bringen:

$$F_0^n \cdot \begin{vmatrix} \varrho_\alpha^{n-1} & \varrho_\alpha^{n-2} & \cdots & \varrho_\alpha & 1 \\ \varrho_\beta^{n-1} & \varrho_\beta^{n-2} & \cdots & \varrho_\beta & 1 \\ \cdot & \cdot & \cdot & \cdot & \cdot \\ \varrho_\omega^{n-1} & \varrho_\omega^{n-2} & \cdots & \varrho_\omega & 1 \end{vmatrix}.$$

Sie ist von null verschieden, da die n Größen $\varrho_\alpha, \varrho_\beta, \ldots, \varrho_\omega$ voneinander verschieden sind und $F_0 \neq 0$ vorausgesetzt wurde. (Vgl. Anm. *) S. 204.) Hieraus folgt, daß die den Ableitungen von $[0]$, $[1]$, ... entsprechenden Unbekannten verschwinden mit Ausnahme der letzten, die der Ableitung von $[n]$ entspricht. Diese kommt in der Gleichung (13) mit dem Koeffizienten $\Delta_{n-1}(\varrho_\alpha)$ vor. Es ist aber $\varrho_\alpha \cdot \Delta_{n-1}(\varrho_\alpha) + (-1)^n F_n = \Delta_n(\varrho_\alpha) = 0$ (vgl. (10a) und (4)), und unsere letzte Unbekannte verschwindet also auch, da wir $F_n \neq 0$ vorausgesetzt haben.

Damit ist also bewiesen, daß die Determinante des ausgewählten Gleichungssystems auf der Anfangslinie von null verschieden ist.

§ 5.

Die Lösung des charakteristischen Systems liefert die Lösung der Gleichung $F = 0$.

Um nachzuweisen, daß die Lösungen des Anfangswertproblems unserer charakteristischen Differentialgleichungen auch die Lösungen des Anfangswertproblems von $F = 0$ liefern, haben wir zwei Dinge zu zeigen.

1. Die Gleichung $F = 0$ ist überall für unsere Funktionen $x, y, u, u_x, \ldots$ erfüllt.

2. Die Funktion $u_{y^k x^l}$ ist wirklich die entsprechende *Ableitung von*

u nach x und y [8]); dazu genügt es nachzuweisen, daß unsere Streifenrelationen, die in Richtung α erfüllt sind (vgl. (14)), auch nach der ω-Richtung gelten:

$$^{\omega}U \equiv u^{\omega} - u_x x^{\omega} - u_y y^{\omega} = 0$$
$$\cdots\cdots\cdots\cdots\cdots$$
$$^{\omega}U_{x^{n-1}} = 0 \quad \ldots \quad ^{\omega}U_{y^{n-1}} = 0\,.$$

Um $F = 0$ nachzuweisen, addieren wir zur Gleichung $(12)_\alpha$ die mit ϱ_α multiplizierte Gleichung (13). Beachten wir die Relationen (vgl. (10a)

$$(16) \qquad \Delta_{k+1}(\varrho_\alpha) - \varrho_\alpha \Delta_k(\varrho_\alpha) = (-1)^{k+1} F_{k+1}$$

und $\Delta_n(\varrho_\alpha) = 0$, so erhalten wir

$$\begin{aligned} 0 &= F_n[n]^\alpha + F_{n-1}[n-1]^\alpha + \ldots + F_0[0]^\alpha + [F]_x x^\alpha + [F]_y y^\alpha \\ &\equiv F_n[n]^\alpha + F_{n-1}[n-1]^\alpha + \ldots + F_0[0]^\alpha \\ &\qquad + F_x x^\alpha + F_y y^\alpha + F_u u_x x^\alpha + F_u u_y y^\alpha + \ldots\,. \end{aligned}$$

Der letzte Ausdruck ist aber, wie man unter Beachtung der Streifenrelationen (14) erkennt, die Ableitung von F nach dem Parameter α. Da aber F für die Anfangswerte verschwindet, verschwindet es demnach für unsere Lösung überhaupt.

Um die Gültigkeit der Streifenrelationen nach der ω-Richtung abzuleiten, setzen wir voraus, daß alle gesuchten Funktionen $x, y, \ldots, [n]$ stetige gemischte zweite Ableitungen nach α und ω besitzen. Unter dieser Voraussetzung gelten — wenn wir für v irgendeine unserer Funktionen einsetzen — folgende identischen Relationen [9]):

$$^{\alpha}V^{\omega} - {}^{\omega}V^{\alpha} \equiv v_x^{\alpha} x^{\omega} + v_y^{\alpha} y^{\omega} - v_x^{\omega} x^{\alpha} - v_y^{\omega} y^{\alpha},$$

$$(17) \qquad ^{\alpha}V^{\omega} - {}^{\omega}V^{\alpha} \equiv {}^{\alpha}V_x x^{\omega} + {}^{\alpha}V_y y^{\omega} - {}^{\omega}V_x x^{\alpha} - {}^{\omega}V_y y^{\alpha},$$

$$(18) \qquad ^{\alpha}V^{\omega} - {}^{\omega}V^{\alpha} \equiv \left(\frac{\partial}{\partial y} v_x - \frac{\partial}{\partial x} v_y\right)(y^{\alpha} x^{\omega} - x^{\alpha} y^{\omega}).$$

In (18) sind v_x und v_y als Funktionen von x und y gedacht (vgl. Anm. [8])).

Für die Größen $^{\omega}U$, $^{\omega}U_x$, ..., $^{\omega}U_{y^{n-1}}$, deren Verschwinden wir nachweisen wollen, können wir nun ein gewöhnliches Differentialgleichungssystem aufstellen. Zu dem Zweck betrachten wir die Gleichung $(12)_\alpha$, die wir wieder in der alten Form (5)

---

[8]) Man kann an Stelle von α und ω die Größen x, y als unabhängige Variable einführen, da $\dfrac{\partial(x,y)}{\partial(\alpha,\omega)} = x^{\alpha} x^{\omega}(\varrho_{\omega} - \varrho_{\alpha}) \neq 0$ ist.

[9]) $^{\alpha}V^{\omega}$ bedeutet $\dfrac{\partial}{\partial\omega}(^{\alpha}V)$; entsprechend später $^{\alpha}U_x^{\omega} = \dfrac{\partial}{\partial\omega}(^{\alpha}U_x), \ldots$.

$$(19)\qquad \begin{vmatrix} F_0 & F_1 & \dots & F_{n-1} & -[F]_x \\ x^a & y^a & \dots & 0 & [0]^a \\ \cdot & \cdot & \dots & \cdot & \cdot \\ 0 & 0 & \dots & x^a & [n-1]^a \end{vmatrix} = 0$$

aufgeschrieben denken. Von der letzten Spalte subtrahieren wir die nullte, erste, ..., $(n-1)$-te Spalte, nachdem wir sie bzw. mit $[0]_x$, $[1]_x$, ..., $[n-1]_x$ multipliziert haben. Außerdem subtrahieren wir die in der letzten Spalte mit $[n]_x$ multiplizierte Determinante (4):

$$(20)\qquad \begin{vmatrix} F_0 & F_1 & \dots & F_{n-1} & F_n \\ x^a & y^a & \dots & 0 & 0 \\ \cdot & \cdot & \dots & \cdot & \cdot \\ 0 & 0 & \dots & x^a & y^a \end{vmatrix}$$

Dadurch wird der Wert von (19) nicht verändert, weil die Determinante (20) verschwindet.

Wir erhalten so aus (19), indem wir $[0]^a = [0]_x x^a + [0]_y y^a, \dots$ ausschreiben,

$$(21)\qquad \begin{vmatrix} F_0 & \dots & F_{n-1} & -(F_x + F_u u_x + \dots + F_0[0]_x + \dots + F_n[n]_x) \\ x^a & \dots & 0 & ([0]_y - [1]_x)\, y^a \\ \cdot & \dots & \cdot & \cdot \\ 0 & \dots & x^a & ([n-1]_y - [n]_x)\, y^a \end{vmatrix} = 0.$$

Das oberste Glied der letzten Spalte unterscheidet sich um

$$F_u\left(u_x - \frac{\partial u}{\partial x}\right) + \dots + F_{u_{y^{n-1}}}\left(u_{y^{n-1}x} - \frac{\partial}{\partial x}\, u_{y^{n-1}}\right)$$

von dem Ausdruck $-\dfrac{dF}{dx}$, der nach der oben bewiesenen Gleichung $F = 0$ verschwindet.

Die hier auftretenden Größen $u_x - \dfrac{\partial u}{\partial x}$, ..., $u_{y^{n-1}x} - \dfrac{\partial}{\partial x} u_{y^{n-1}}$ lassen sich durch die Größen ${}^{\omega}U$, ..., ${}^{\omega}U_{y^{n-1}}$ ausdrücken. Es ist nämlich identisch (vgl. (14) und S. 210)

$${}^{a}Uy^{\omega} - {}^{\omega}Uy^{a} = \left(u_x - \frac{\partial u}{\partial x}\right)(x^{\omega} y^{a} - y^{\omega} x^{a})$$

$$\dots\dots\dots\dots\dots\dots$$

$${}^{a}U_{y^{n-1}}y^{\omega} - {}^{\omega}U_{y^{n-1}}y^{a} = \left(u_{y^{n-1}x} - \frac{\partial}{\partial x}\, u_{y^{n-1}}\right)(x^{\omega} y^{a} - y^{\omega} x^{a}),$$

und die Größen ${}^{a}U$, ..., ${}^{a}U_{x^{n-1}}$ verschwinden (vgl. (14)).

14*

Die Größen $\frac{\partial}{\partial y}[0] - \frac{\partial}{\partial x}[1], \ldots, \frac{\partial}{\partial y}[n-1] - \frac{\partial}{\partial x}[n]$ drücken wir vermöge der Gleichung (18) durch ${}^{\omega}U_{x^{n-1}}, \ldots, {}^{\omega}U_{y^{n-1}}$ aus, indem wir dort bzw. $v = u_{x^{n-1}}, \ldots, v = u_{y^{n-1}}$ setzen und ${}^{a}U_{x^{n-1}} = 0, \ldots, {}^{a}U_{y^{n-1}} = 0$ beachten.

Machen wir in der Gleichung (21) die eben angegebenen Einsetzungen, und multiplizieren wir sie noch mit $\frac{1}{y^{\alpha}}(x^{\omega} y^{\alpha} - y^{\omega} x^{\alpha})$, so erhalten wir die Gleichung

$$\begin{vmatrix} F_0 & \ldots & F_{n-1} & F_u {}^{\omega}U + F_{u_x} {}^{\omega}U_x + \ldots + F_{u_{y^{n-1}}} {}^{\omega}U_{y^{n-1}} \\ x^{\alpha} & \ldots & 0 & -{}^{\omega}U^{\alpha}_{x^{n-1}} \\ \cdot & \ldots & \cdot & \\ 0 & \ldots & x^{\alpha} & -{}^{\omega}U^{\alpha}_{y^{n-1}} \end{vmatrix} = 0,$$

die wir auch in der Form — vgl. die Ableitung der Gleichung $(12)_\alpha$ aus (5) —

$$(22)_\alpha \qquad \Delta_{n-1}(\varrho_\alpha)\, {}^{\omega}U^{\alpha}_{y^{n-1}} - \Delta_{n-2}(\varrho_\alpha)\, {}^{\omega}U^{\alpha}_{y^{n-2}x} + \ldots - (-1)^n (F_u {}^{\omega}U + \ldots) x^{\alpha} = 0$$

schreiben können. Gehen wir an Stelle von $(12)_\alpha$ von $(12)_\beta, \ldots, (12)_\omega$ aus, so erhalten wir auf dieselbe Weise die Gleichungen

$$(22)_\beta \qquad \Delta_{n-1}(\varrho_\beta)\, {}^{\omega}U^{\alpha}_{y^{n-1}} - \Delta_{n-2}(\varrho_\beta)\, {}^{\omega}U^{\alpha}_{y^{n-2}x} + \ldots - (-1)^n (F_u {}^{\omega}U + \ldots) x^{\beta} = 0$$

$$\cdots\cdots\cdots\cdots\cdots\cdots$$

$$(22)_\omega \qquad \Delta_{n-1}(\varrho_\omega)\, {}^{\omega}U^{\alpha}_{y^{n-1}} - \Delta_{n-2}(\varrho_\omega)\, {}^{\omega}U^{\alpha}_{y^{n-2}x} + \ldots - (-1)^n (F_u {}^{\omega}U + \ldots) x^{\beta} = 0.$$

Schließlich betrachten wir sämtliche Gleichungen, die wir aus (17) dadurch erhalten, daß wir für v der Reihe nach die Größen $u, u_x, \ldots, u_{y^{n-2}}$ einsetzen und die Gleichungen (14) berücksichtigen, nämlich

$$(23) \qquad \begin{aligned} {}^{\omega}U^{\alpha} &= {}^{\omega}U_x x^{\alpha} + {}^{\omega}U_y y^{\alpha} \\ &\cdots\cdots\cdots\cdots \\ {}^{\omega}U^{\alpha}_{y^{n-2}} &= {}^{\omega}U_{y^{n-2}x} x^{\alpha} + {}^{\omega}U_{y^{n-1}} y^{\alpha}. \end{aligned}$$

Die Gleichungen (22) und (23) sind in der Tat lineare homogene Differentialgleichungen erster Ordnung für die Größen ${}^{\omega}U, \ldots, {}^{\omega}U_{y^{n-1}}$, und zwar genau soviel Gleichungen wie Unbekannte. Es sind *gewöhnliche* Differentialgleichungen, da sie nur die Ableitungen nach α enthaten. Die Determinante dieses Gleichungssystems, d. h. die Determinante der Koeffizienten der *Ableitungen* der gesuchten Funktionen, ist offenbar das Produkt der Determinante der Gleichungen (22) und des Gleichungssystems (23). Die Determinante des letzteren Systems hat den Wert 1, und die Determinante der Gleichungen (22) ist gerade die Determinante (15).

Wir haben somit ein lineares homogenes gewöhnliches Differentialgleichungssystem mit nicht verschwindender Determinante für unsere Funktionen ${}^{\omega}U, \ldots, {}^{\omega}U_{y^{n-1}}$ aufgestellt. Ihre Anfangswerte sind aber null. (Denn auf der Anfangslinie sind die Streifenrelationen nach der Richtung der Anfangslinie vgl. (§ 2) und nach der von ihr verschiedenen (vgl. § 2) α-Richtung erfüllt, also auch in der ω-Richtung.) Jedes stetige Lösungssystem solcher Differentialgleichungen mit verschwindenden Anfangswerten verschwindet aber überall. Damit sind sämtliche Streifenrelationen als erfüllt erkannt, und die Größen $u_x, \ldots, u_{y^n} = [n]$ sind in der Tat die entsprechenden Ableitungen der Funktion $u(x, y)$.

Im ersten Teil dieser Arbeit haben wir somit für eine Lösung unserer Differentialgleichung $F = 0$ ein System von charakteristischen Differentialgleichungen erster Ordnung abgeleitet, denen sie genügen muß. Aus diesen charakteristischen Differentialgleichungen konnten wir ein System mit nicht verschwindender Determinante auswählen, dessen Lösung bei denselben Anfangsbedingungen zu einer Lösung der Gleichung $F = 0$ führt. Daß dieses System der ausgewählten Differentialgleichungen *wirklich eine Lösung besitzt, und nur eine einzige* in einem gewissen Bereiche, wird sich aus dem allgemeinen Existenzsatze des zweiten Teiles ergeben.

Aus ihm folgt, daß es ein sich an die Anfangskurve anschließendes Gebiet der (x, y)-Ebene gibt, innerhalb dessen die Differentialgleichung $F = 0$ hyperbolisch bleibt und wo eine und nur eine mit stetigen Ableitungen bis zur $n + 2$-ten Ordnung versehene Lösung unseres Anfangswertproblems existiert; ferner, daß der Wert in einem Punkte P nur von demjenigen Teil der Anfangskurve abhängt, der von den beiden äußersten Charakteristiken durch P aus ihr ausgeschnitten wird.

Zum Schluß stellen wir die *Voraussetzungen* zusammen, die wir machen, ohne zu behaupten, daß wir nicht mit etwas schwächeren Bedingungen hätten auskommen können. Der für die Anfangswerte als hyperbolisch vorausgesetzte Ausdruck F soll nach seinen Argumenten $x, y, u, \ldots, [n]$ viermal beschränkt differenzierbar sein. Die Anfangswerte $x, y, u, \ldots, [n]$ sollen als Funktionen eines Parameters auf der nirgends charakteristisch gerichteten Anfangskurve dreimal beschränkt differenzierbar sein und dort der Gleichung $F = 0$ und den Streifenrelationen längs der Kurve genügen.

2. Teil.

Bevor wir unseren allgemeinen Existenzsatz formulieren, behandeln wir einen typischen *Spezialfall.* Wir betrachten ein homogenes lineares Differentialgleichungssystem erster Ordnung für drei Funktionen u, v, w zweier Variablen α und ω einer rechtwinkligen α, ω-Ebene:

$$
\begin{aligned}
&a_{11} u^\alpha + a_{12} v^\alpha + a_{13} w^\alpha = 0 \\
(24)\qquad &a_{21} u^\beta + a_{22} v^\beta + a_{23} w^\beta = 0 \\
&a_{31} u^\omega + a_{32} v^\omega + a_{32} w^\omega = 0 .
\end{aligned}
$$

Die Größen a_{ik} sollen dreimal beschränkt differenzierbare Funktionen der gesuchten Funktionen u, v, w sein. $u^\alpha, v^\alpha, w^\alpha$ und $u^\omega, v^\omega, w^\omega$ bedeuten die Ableitungen von u, v, w nach α und ω, während u^β, v^β, w^β die linearen Kombinationen

$$
\begin{aligned}
u^\beta &= \sigma u^\alpha + \tau u^\omega \\
v^\beta &= \sigma v^\alpha + \tau v^\omega \\
w^\beta &= \sigma w^\alpha + \tau w^\omega
\end{aligned}
$$

sind, wo σ und τ dreimal beschränkt differenzierbare Funktionen von u, v und w sind.

Das Anfangswertproblem, das wir lösen wollen, besteht darin, daß man auf einem Stück der Geraden $\alpha + \omega = 0$ die Werte von u, v, w dreimal beschränkt differenzierbar vorgibt, so daß dort die Determinante

$$
|a_{ik}| = \begin{vmatrix} a_{11} & a_{12} & a_{13} \\ a_{21} & a_{22} & a_{23} \\ a_{31} & a_{32} & a_{33} \end{vmatrix}
$$

nicht verschwindet. Die Größen u^β, v^β, w^β können wir als die Ableitungen von u, v, w nach einer dritten Richtung in der (α, ω)-Ebene auffassen. Wir stellen die Forderung, daß für Punkte der Anfangsgeraden $\alpha + \omega = 0$ diese dritte Richtung zwischen der α- und der ω-Richtung liegt, d. h. daß die α- und ω-Richtung äußerste Richtungen sind (vgl. § 3 S. 206). Diese Forderung ist damit gleichbedeutend, daß auf der Anfangsgeraden die Größen σ *und* τ *positiv* sind.

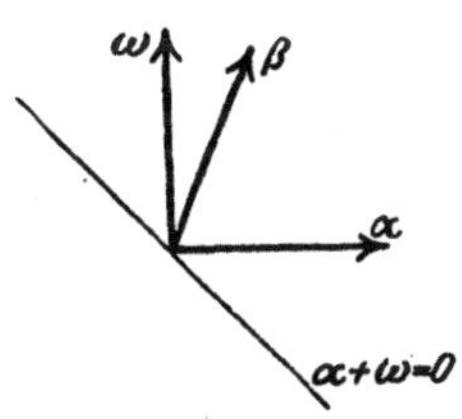

Fig. 1.

Zur Lösung des Anfangswertproblems[9a]) überdecken wir die (α, ω)-Ebene mit einem quadratischen Gitter, das aus den Geraden $\alpha = nh$,

[9a]) Während der Korrektur machte uns Herr Courant darauf aufmerksam, daß dieses Problem auch durch sukzessive Approximationen gelöst werden kann, indem man analog wie im folgenden vorgeht.

$\omega = mh$ besteht. $m, n = \ldots -1, 0, +1, \ldots$. Es sei ABB_1 ein rechtwinkliges Dreieck aus unserem Gitter (vgl. Fig. 3), dessen Seite BA in der positiven α-Richtung zeigt, B_1A in der positiven ω-Richtung, und wo die Länge dieser Seiten die Maschenweite h ist. Dann ersetzen wir unser Differentialgleichungssystem durch das folgende Differenzengleichungssystem:

$$
\begin{aligned}
&a_{11B}(u_A - u_B) + a_{12B}(v_A - v_B) + a_{13B}(w_A - w_B) = 0\\
&\quad a_{21B}[(\sigma_B + \tau_B)u_A - \sigma_B\, u_B - \tau_B\, u_{B_1}]\\
&\quad + a_{22B}[(\sigma_B + \tau_B)\, v_A - \sigma_B\, v_B - \tau_B\, v_{B_1}]\\
&\quad + a_{23B}[(\sigma_B + \tau_B)\, w_A - \sigma_B\, w_B - \tau_B\, w_{B_1}] = 0\\
&a_{31B}(u_A - u_{B_1}) + a_{32B}(v_A - v_{B_1}) + a_{33B}(w_A - w_{B_1}) = 0\,.
\end{aligned}
\tag{25}
$$

Kennt man die Werte von u, v, w in den Punkten B und B_1 und ist die Determinante

$$
|a_{ik}|_B \neq 0, \quad \text{ferner} \quad \sigma_B \geqq 0, \quad \tau_B \geqq 0, \quad \sigma_B + \tau_B \neq 0, \tag{26}
$$

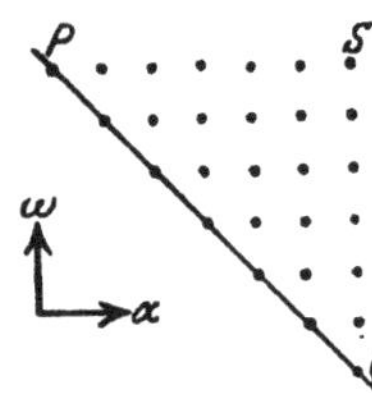

Fig. 2.

so können wir aus unseren Gleichungen (25) die Werte von u_A, v_A, w_A berechnen. Geht man also von den bekannten Werten auf einem Stück PQ der Anfangslinie aus, so kann man infolgedessen die Lösung der Differenzengleichung in allen Gitterpunkten desjenigen Dreiecks PQS berechnen, das man erhält, wenn man von den Punkten P und Q die Parallelen zur positiven α- bzw. ω-Richtung zieht, vorausgesetzt, daß überall in PQS unsere obigen Bedingungen (26) für irgendeinen Gitterpunkt B aus PQS erfüllt bleiben.

Wären die Koeffizienten a_{ik}, σ und τ konstant, so könnte man zeigen, daß im ganzen Dreieck PQS die Linearkombinationen

$$
\begin{aligned}
&a_{11}\, u + a_{12}\, v + a_{13}\, w,\\
&a_{21}\, u + a_{22}\, v + a_{23}\, w,\\
&a_{31}\, u + a_{32}\, v + a_{33}\, w
\end{aligned}
$$

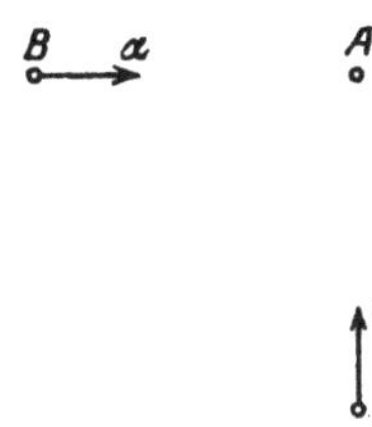

Fig. 3.

unterhalb derjenigen Schranke bleiben, der sie für ihre Anfangswerte unterworfen sind. Die erste Linearkombination ist im Punkte A eines Dreiecks ABB_1 ebenso groß wie im Punkte B, und die dritte ist im Punkte A ebenso groß wie im Punkte B_1, während die zweite Linearkombination

$$
\begin{aligned}
&a_{21}\, u_A + a_{22}\, v_A + a_{23}\, w_A\\
&= \frac{\sigma}{\sigma+\tau}(a_{21}\, u_B + a_{22}\, v_B + a_{23}\, w_B) + \frac{\tau}{\sigma+\tau}(a_{21}\, u_{B_1} + a_{22}\, v_{B_1} + a_{23}\, w_{B_1})
\end{aligned}
$$

zwischen ihren Werten in den Punkten B und B_1 liegt, da σ und τ nicht negativ sind[10]). Da die Determinante der drei Linearformen nicht verschwindet, ergibt sich, daß die drei Funktionen u, v, w selber unterhalb einer Schranke bleiben, die nur von ihren Werten auf der Anfangslinie abhängt und nicht von der Größe der zugrunde gelegten Mascheneinteilung. Durch einen entsprechenden Schluß findet man, daß auch die Werte der Differenzenquotienten von u, v, w im ganzen Dreieck PQS beschränkt bleiben, wenn sie am Anfang beschränkt sind.

Im Falle nichtkonstanter Koeffizienten beginnen wir gleich damit, die Beschränktheit der ersten Differenzenquotienten aus ihrer Beschränktheit am Anfang abzuleiten.

Auf dem betrachteten Teil der Anfangsgeraden erfüllen die Anfangswerte von u, v, w ein Kurvenstück im dreidimensionalen (u, v, w)-Raum. Wir denken uns dieses Kurvenstück innerhalb eines solchen abgeschlossenen Gebietes G im (u, v, w)-Raum gelegen, daß die Größen a_{ik}, σ, τ mit ihren Ableitungen bis zur dritten Ordnung absolut für jeden Punkt unseres Gebietes beschränkt sind, daß die Determinante $|a_{ik}|$ in G absolut oberhalb einer Schranke $\delta > 0$ bleibt und daß in G $\sigma \geqq 0$, $\tau \geqq 0$ und $\sigma + \tau \neq 0$ ist.

Wir wollen jetzt zeigen, daß, solange unsere Lösungen u, v, w innerhalb des Gebietes G liegen, ihre ersten bis dritten Differenzenquotienten unabhängig von der Intervalleinteilung beschränkt bleiben, wenn sie es auf der Anfangsgeraden sind.

Vorerst bemerken wir folgendes: Setzen wir voraus, daß die Werte von u, v, w in den Punkten $B, B_1, \ldots$ innerhalb von G liegen. Ist dann $N > 0$ eine Schranke für den Betrag der *Differenzenquotienten* in der zur Anfangsgeraden *parallelen Reihe* der Punkte $B, B_1, \ldots$, so liegen die ersten Differenzenquotienten *zwischen Punkten der Reihe* $A, A_1, \ldots$ *und* $B, B_1, \ldots$, z. B. $\frac{u_A - u_B}{h}$, $\frac{u_A - u_{B_1}}{h}$ usw., unterhalb der Schranke pN, wo p nur vom Gebiete G abhängt. Wir können nämlich die Gleichungen (25) beispielsweise so umschreiben, daß sie die Differenzen $u_A - u_B$, $v_A - v_B$, $w_A - w_B$ durch $u_B - u_{B_1}$, $v_B - v_{B_1}$, $w_B - w_{B_1}$ ausdrücken. Liegen ferner die „Linearkombinationen“

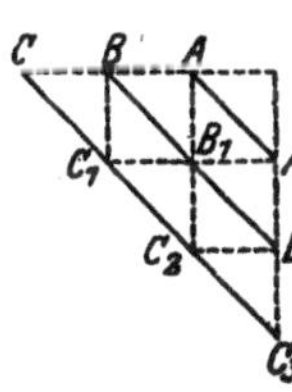

Fig. 4.

[10]) Ein solcher Schluß ließe sich nicht unmittelbar auf die Funktionen selbst anwenden, da z. B. der Wert von u im Punkte A sich außer durch den Wert von u in B und B_1 noch durch erste Differenzen von v und w zwischen den Punkten B und B_1 ausdrückt.

$$a_{11\,B}\left(\frac{u_A-u_{A_1}}{h}\right)+a_{12\,B}\left(\frac{v_A-v_{A_1}}{h}\right)+a_{13\,B}\left(\frac{w_A-w_{A_1}}{h}\right),$$

$$a_{21\,B}\left(\frac{u_A-u_{A_1}}{h}\right)+a_{22\,B}\left(\frac{v_A-v_{A_1}}{h}\right)+a_{23\,B}\left(\frac{w_A-w_{A_1}}{h}\right),$$

$$a_{31\,B}\left(\frac{u_A-u_{A_1}}{h}\right)+a_{32\,B}\left(\frac{v_A-v_{A_1}}{h}\right)+a_{33\,B}\left(\frac{w_A-w_{A_1}}{h}\right)$$

absolut unterhalb einer Schranke M, so liegen die Differenzenquotienten

$$\frac{u_A-u_{A_1}}{h},\qquad \frac{v_A-v_{A_1}}{h},\qquad \frac{w_A-w_{A_1}}{h}$$

absolut unterhalb der Schranke $q\,M$, wo q nur vom Gebiet G abhängt. Die Beschränktheit *dieser* Linearkombinationen können wir aber sukzessive nachweisen.

Wir können nämlich aus den Differenzengleichungen (25) Gleichungen herstellen, in denen unsere obigen Linearkombinationen für den Punkt A mit den entsprechenden Linearkombinationen für die Punkte B und B_1 verknüpft werden. Allerdings werden dabei noch Zusatzglieder auftreten, die jedoch von geringerer Größenordnung sind.

Zu dem Zweck ziehen wir von der Differenzengleichung $(25)_2$ im Punkte A

$$\begin{aligned}&\sigma_B\left\{a_{21\,B}\left(\frac{u_A-u_B}{h}\right)+a_{22\,B}\left(\frac{v_A-v_B}{h}\right)+a_{23\,B}\left(\frac{w_A-w_B}{h}\right)\right\}\\ &+\tau_B\left\{a_{21\,B}\left(\frac{u_A-u_{B_1}}{h}\right)+a_{22\,B}\left(\frac{v_A-v_{B_1}}{h}\right)+a_{23\,B}\left(\frac{w_A-w_{B_1}}{h}\right)\right\}\end{aligned}=0$$

dieselbe Gleichung für den Punkt A_1 ab:

$$\begin{aligned}&\sigma_{B_1}\left\{a_{21\,B_1}\left(\frac{u_{A_1}-u_{B_1}}{h}\right)+a_{22\,B_1}\left(\frac{v_{A_1}-v_{B_1}}{h}\right)+a_{23\,B_1}\left(\frac{w_{A_1}-w_{B_1}}{h}\right)\right\}\\ &+\tau_{B_1}\left\{a_{21\,B_1}\left(\frac{u_{A_1}-u_{B_2}}{h}\right)+a_{22\,B_1}\left(\frac{v_{A_1}-v_{B_2}}{h}\right)+a_{23\,B_1}\left(\frac{w_{A_1}-w_{B_2}}{h}\right)\right\}\end{aligned}=0.$$

Die Differenz formen wir nun folgendermaßen um, wobei wir nur die Glieder hinschreiben, in denen u vorkommt:

$$\text{27 a)}\quad\begin{aligned}&\sigma_B\,a_{21\,B}\,\frac{1}{h}(u_A-u_B-u_{A_1}+u_{B_1})+(\sigma_B\,a_{21\,B}-\sigma_{B_1}a_{21\,B_1})\left(\frac{u_{A_1}-u_{B_1}}{h}\right)\\ &+\tau_B\,a_{21\,B}\,\frac{1}{h}(u_A-u_{B_1}-u_{A_1}+u_{B_2})+(\tau_B\,a_{21\,B}-\tau_{B_1}a_{21\,B_1})\left(\frac{u_{A_1}-u_{B_2}}{h}\right)\end{aligned}+\dots=0,$$

oder schließlich (vgl. Fig. 4)

$$\text{27 b)}\quad\begin{aligned}&(\sigma_B+\tau_B)\,a_{21\,B}\left(\frac{u_A-u_{A_1}}{h}\right)-\sigma_B\,a_{21\,C}\left(\frac{u_B-u_{B_1}}{h}\right)-\tau_B\,a_{21\,C_1}\left(\frac{u_{B_1}-u_{B_2}}{h}\right)\\ &-h\,\sigma_B\left(\frac{a_{21\,B}-a_{21\,C}}{h}\right)\left(\frac{u_B-u_{B_1}}{h}\right)-h\,\tau_B\left(\frac{a_{21\,B}-a_{21\,C_1}}{h}\right)\left(\frac{u_{B_1}-u_{B_2}}{h}\right)\\ &+h\left(\frac{\sigma_B\,a_{21\,B}-\sigma_{B_1}a_{21\,B_1}}{h}\right)\left(\frac{u_{A_1}-u_{B_1}}{h}\right)+h\left(\frac{\tau_B\,a_{21\,B}-\tau_{B_1}a_{21\,B_1}}{h}\right)\left(\frac{u_{A_1}-u_{B_2}}{h}\right)\end{aligned}+\dots=0.$$

Analoge Umformungen erhalten wir für die Glieder mit v und w und ferner aus den Gleichungen $(25)_1$ und $(25)_3$.

Haben wir bis zur Parallelreihe der B zur Anfangsgeraden nachgewiesen, daß die Lösungen u, v, w innerhalb des Gebietes G liegen, und wissen wir ferner, daß die Linearkombinationen

$$(28)\qquad \begin{array}{ll} a_{11\,C}\left(\dfrac{u_B - u_{B_1}}{h}\right) + \cdots, & a_{21\,C_1}\left(\dfrac{u_{B_1} - u_{B_2}}{h}\right) + \cdots, \\[2ex] a_{21\,C}\left(\dfrac{u_B - u_{B_1}}{h}\right) + \cdots, & a_{31\,C_1}\left(\dfrac{u_{B_1} - u_{B_2}}{h}\right) + \cdots \end{array}$$

unterhalb der Schranke M liegen, so entnehmen wir aus den drei Gleichungen (27b), daß die Linearkombinationen

$$a_{11\,B}\left(\frac{u_A - u_{A_1}}{h}\right) + \cdots,$$

$$a_{21\,B}\left(\frac{u_A - u_{A_1}}{h}\right) + \cdots,$$

$$a_{31\,B}\left(\frac{u_A - u_{A_1}}{h}\right) + \cdots$$

unterhalb der Schranke

$$M + h\,c\,M^2$$

liegen, wo $c > 0$ nur vom Gebiet G abhängt. Wir brauchen nämlich in (27b) die Differenzenquotienten der Koeffizienten nur nach dem Mittelwertsatz durch die Differenzenquotienten der Funktionen u, v, w abzuschätzen, unsere Vorbemerkungen (S. 216) zu beachten und entsprechend wie oben (S. 216) die Positivität von σ und τ zu benutzen.

Da diese drei Linearkombinationen auf der ersten Parallelreihe beschränkt bleiben, weil die ersten Differenzenquotienten der Anfangswerte beschränkt sind, so folgt, daß diese Kombinationen innerhalb desjenigen sich an die Gerade $\alpha + \omega = 0$ anschließenden Gebietes der (α, ω)-Ebene beschränkt bleiben, in dem die Lösungen u, v, w innerhalb von G liegen[11]). Damit ergibt sich nach den Vorbemerkungen von S. 216, daß die ersten

---

[11]) Bezeichnen wir nämlich die Schranken für unsere Kombinationen auf der ersten, zweiten, dritten, ..., n-ten Parallelreihe mit $M_1, M_2, M_3, \ldots, M_n$, so finden wir für die Größen M_k die Differenzengleichung

$$\frac{M_{k+1} - M_k}{h} = c\,M_k^2,$$

deren Lösung für jede Maschenweite majorisiert wird durch die Lösung der Differentialgleichung

$$\frac{dM}{dt} = c\,M^2.$$

Differenzenquotienten selber in diesem Gebiete unabhängig von der Maschenweite beschränkt bleiben.

So ist also für das *Wachstum* der Funktionen u, v, w auf den Linien $\alpha =$ konst. und $\omega =$ konst. eine von der Intervalleinteilung unabhängige Schranke gegeben, woraus folgt, daß es ein an die Anfangsgerade anschließendes Gebiet B der (α, ω)-Ebene gibt, innerhalb dessen die Werte von u, v, w im Gebiet G bleiben, wie klein auch h gewählt ist.

Wir haben die Beschränktheit der ersten Differenzenquotienten im Gebiete B nachgewiesen, ohne die vorausgesetzte Beschränktheit des zweiten und dritten Differenzenquotienten am Anfang zu benutzen. Durch einen ähnlichen Schluß, indem man z. B. auf die Gleichung (27b) dasselbe Verfahren anwendet, durch das (27b) aus $(25)_2$ entstanden ist, kann man die Beschränktheit aller zweiten und aller dritten Differenzenquotienten im Gebiet B nachweisen, indem man diese Voraussetzungen mitbenutzt.

Aus der Beschränktheit der dritten Differenzenquotienten folgt, daß die zweiten Differenzenquotienten hinsichtlich h im ganzen Gebiet B „gleichartig stetig" sind. Es existiert also nach dem bekannten Häufungssatze für Funktionenfolgen eine Teilfolge aus der Folge der Funktionen u, v, w bei immer kleiner werdender Maschenweite, so daß die zweiten Differenzenquotienten gleichmäßig gegen stetige Grenzfunktionen konvergieren. Die Summen, durch die sich die Funktionen u, v, w und ihre ersten Differenzenquotienten aus den zweiten Differenzenquotienten darstellen, konvergieren dabei gleichmäßig gegen Integrale, so daß wir wissen, daß die stetigen Grenzfunktionen der zweiten Differenzenquotienten die entsprechenden zweiten Ableitungen der Grenzfunktionen von u, v, w sind. Diese Grenzfunktionen genügen infolgedessen unseren Differentialgleichungen (24).

Da jede Lösung unserer Differenzengleichungen (25) in einem Punkte S nur von demjenigen Teil PQ der Anfangslinie, vgl. Fig. 2, S. 215, abhängt, der sich ergibt, wenn wir vom Punkte S aus die Geraden $\alpha =$ konst. und $\omega =$ konst. ziehen, bis sie die Anfangsgerade $\alpha + \omega = 0$ schneiden, so folgt, daß auch unsere Lösungen der Differentialgleichungen im Punkte S nur von demselben Stück der Anfangsgerade abhängen.

Daß unsere Lösungen aber die einzigen mit stetigen zweiten Ableitungen versehenen sind, erkennen wir folgendermaßen: Wir differenzieren unsere Differentialgleichungen (24) nach der zur Anfangsgeraden parallelen Richtung und ersetzen die Differentialquotienten durch Differenzenquotienten. Der Fehler ε_h, den wir dabei begehen, geht gleichmäßig im ganzen Gebiet B mit h gegen null. Wir subtrahieren die so entstehenden Differenzengleichungen für die Lösungen u, v, w unserer Differentialgleichungen von den entsprechenden Differenzengleichungen (27a) für die Lösungen $\bar{u}, \bar{v}, \bar{w}$ unserer Differenzengleichungen (25). Die so entstehenden Gleichungen formen wir genau so um, wie wir (27a) in (27b) verwandelten. Wir erhalten dann eine Gleichung von ähnlichem Typus in $u-\bar{u}$, $v-\bar{v}$, $w-\bar{w}$ wie (27b) in u, v, w. Wir finden aus ihr: Ist M eine Schranke für die Linearkombinationen (28), genommen allerdings für die Differenzenquotienten der Funktionen $u-\bar{u}$, $v-\bar{v}$, $w-\bar{w}$ auf der n-ten Reihe, so liegen die Linearkombinationen auf der $n+1$-ten Reihe unterhalb der Schranke $M+phM+qh^2+rh\varepsilon_h$, wobei die Konstanten p, q, r nur vom Gebiet G und den Schranken für die ersten und zweiten Differenzenquotienten der Funktionen u, v, w und $\bar{u}, \bar{v}, \bar{w}$ abhängen. Da nun unsere Linearkombinationen offenbar am Anfang verschwinden, so sind sie im ganzen Gebiet höchstens von der Größenordnung $h+\varepsilon_h$.

Daraus läßt sich folgern, daß die Folge der Lösungen $\bar{u}, \bar{v}, \bar{w}$ unserer Differenzengleichungen gegen jede mit stetigen zweiten Ableitungen versehene Lösungen u, v, w unserer Differentialgleichungen konvergiert; es kann also nur ein solches Lösungssystem bei vorgegebenen Anfangswerten geben.

Die Übertragung unseres Existenz- und Eindeutigkeitsbeweises auf den Fall von mehr als drei Funktionen ist unmittelbar klar. Wir können also folgenden allgemeinen Satz aussprechen:

Es seien auf einem Stück der Anfangsgeraden $\alpha+\omega=0$ die Werte von n Funktionen $u_1, \ldots, u_n$ der beiden Veränderlichen α und ω dreimal beschränkt differenzierbar vorgegeben. Für diese Funktionen seien die folgenden n Differentialgleichungen gestellt:

$$\sum_{i=1}^{n} a_{1i}\, u_i^{\alpha} = 0,$$

$$\sum_{i=1}^{n} a_{2i}\, u_i^{\beta} = 0,$$

$$\cdots\cdots$$

$$\sum_{i=1}^{n} a_{ni}\, u_i^{\omega} = 0,$$

wo u_i^{α} und u_i^{ω} die Ableitungen von u_i nach α und ω bedeuten, wo

$$\begin{aligned} u_i^{\beta} &= \sigma_2 u_i^{\alpha} + \tau_2 u_i^{\omega}, \\ u_i^{\gamma} &= \sigma_3 u_i^{\alpha} + \tau_3 u_i^{\omega}, \\ &\cdots\cdots \end{aligned} \qquad \begin{pmatrix} \sigma_2 \geqq 0, & \tau_2 \geqq 0, & \sigma_2+\tau_2 \neq 0 \\ \sigma_3 \geqq 0, & \tau_3 \geqq 0, & \sigma_3+\tau_3 \neq 0 \end{pmatrix} \\ \cdots\cdots$$

ist, und wo $\sigma_2, \tau_2, \ldots$ und die Koeffizienten a_{ik} dreimal beschränkt differenzierbare Funktionen von $u_1, \ldots, u_n$ sind; und zwar innerhalb eines gewissen Bereiches G, der die Anfangswerte im Inneren enthält. Ferner sei die Determinante $|a_{ik}|$ für die Anfangswerte von 0 verschieden.

Dann gibt es ein und nur ein zweimal stetig differenzierbares Lösungssystem $u_1, \ldots, u_n$. Der Wert dieser Lösungen in einem Punkte S hängt nur von demjenigen Teil der Anfangswerte ab, der von den beiden Geraden $\alpha =$ konst. und $\omega =$ konst. durch diesen Punkt S aus der Anfangsgeraden ausgeschnitten wird. D. h. ändern wir die Vorgaben außerhalb dieses Stückes der Anfangsgeraden, so ändert sich der Wert im Punkte S und im ganzen Dreieck SPQ nicht mehr (vgl. Fig. 2, S. 215). Damit ist der im 1. Teil angekündigte Existenz- und Eindeutigkeitsbeweis nachgeholt.

(Eingegangen am 25. 2. 1927.)

Mathematische Annalen
101 (1929)

Neuer Beweis des analytischen Charakters der Lösungen elliptischer Differentialgleichungen[1]).

Von

Hans Lewy in Göttingen.

In einer Note in den Göttinger Nachrichten 1927 Heft 2 habe ich einen neuen Beweis des analytischen Charakters der Lösungen elliptischer analytischer Differentialgleichungen in einem Spezialfall ausgeführt. Hier folgt der Beweis des allgemeinen Theorems für Differentialgleichungen zweiter Ordnung in zwei unabhängigen Variablen, der wie jener darauf beruht, die vorgelegte Lösung u einer elliptischen analytischen Differentialgleichung zunächst für komplexe Werte der unabhängigen Variablen x und y zu ermitteln. Diese Fortsetzung ins Komplexe wird hier aufgefaßt als Lösung eines Anfangswertproblems eines „kanonisch-hyperbolischen Systems" von Differentialgleichungen, d. h. eines solchen, wie es z. B. auch zur Lösung des Anfangswertproblems einer hyperbolischen Differentialgleichung verwendet wird. Es wird sodann nachgewiesen, daß in einem vierdimensionalen Gebiet der Variablen x_1, x_2, y_1, y_2 (Real- und Imaginärteile von x und y), das das reelle Ebenenstück $x_2 = 0$, $y_2 = 0$, in welchem die Lösung u ursprünglich vorliegt, im Inneren enthält, die Cauchy-Riemannschen Differentialgleichungen bezüglich x_1, x_2 und bezüglich y_1, y_2 erfüllt sind. Das bedeutet aber, daß u analytisch von x und y in jenem Gebiet, also auch schon in seinem ursprünglichen Definitionsbereich abhängt.

§ 1.

Charakteristische Differentialgleichungen und Fortsetzung ins Komplexe.

Sei

$$F(x, y, u, p, q, r, s, t) = 0$$

eine analytisch von den Argumenten abhängige Differentialgleichung zweiter

---

[1]) Bezüglich der älteren Beweise sei auf den Enzyklopädiebericht von Lichtenstein 2 C 12, S. 1320 und auf die jüngst erschienenen Arbeiten von T. Rádó, Math. Zeitschr. **25**, S. 514 ff., und S. Bernstein, Math. Zeitschr. **28**, S. 330 ff., verwiesen.

Ordnung, in der

$$p = u_x, \quad q = u_y, \quad r = u_{xx}, \quad s = u_{xy}, \quad t = u_{yy}$$

wie üblich zur Abkürzung gesetzt sind. Für eine vorgelegte Lösung u sei die Differentialgleichung in einem Gebiet der x, y-Ebene vom elliptischen Typus, d. h. es sei dort

$$4\frac{\partial F}{\partial r}\frac{\partial F}{\partial t} - \left(\frac{\partial F}{\partial s}\right)^2 > 0,$$

woraus insbesondere $\frac{\partial F}{\partial r} \neq 0$, $\frac{\partial F}{\partial t} \neq 0$ folgt. Man verstehe nun unter den stetigen Funktionen $\varrho_1(x, y)$ und $\varrho_2(x, y)$ die beiden sicher verschiedenen komplexen Wurzeln der Gleichung

$$\frac{\partial F}{\partial r}\varrho^2 - \frac{\partial F}{\partial s}\varrho + \frac{\partial F}{\partial t} = 0;$$

unter den Differentiationsoperatoren $'$ bzw. $`$ die folgenden linearen Kombinationen der Symbole $\frac{\partial}{\partial x}$ und $\frac{\partial}{\partial y}$:

$$' = \frac{\partial}{\partial x} + \varrho_1\frac{\partial}{\partial y}, \qquad ` = \frac{\partial}{\partial x} + \varrho_2\frac{\partial}{\partial y}.$$

Dann folgt

$$y' - \varrho_1 x' = 0 \quad \text{und} \quad y` - \varrho_2 x` = 0.$$

Aus dem Bestehen dieser Gleichungen folgen jetzt in bekannter Weise rein algebraisch noch weitere charakterische Differentialgleichungen für die Größen x, y, u, p, q, r, s, t. Diese sowohl wie die beiden schon genannten $y' - \varrho_1 x' = 0$ und $y` - \varrho_2 x` = 0$ lassen sich nach geeigneter Auswahl in die folgende Form setzen:

$$(1) \qquad \begin{aligned} &\sum_{k=1}^{8} a_{ik}\varphi_k' = 0 \qquad && (i = 1, 2, \ldots, 6), \\ &\sum_{k=1}^{8} a_{ik}\varphi_k` = 0 \qquad && (i = 7, 8), \end{aligned}$$

in der die $\varphi_1, \varphi_2, \ldots, \varphi_8$ zur übersichtlicheren Schreibweise für $x, y, \ldots, t$ stehen, die a_{ik} vermöge unserer Voraussetzung über F analytische Funktionen der Argumente $\varphi_1, \ldots, \varphi_8$ sind und wo die Determinante $|a_{ik}|$ nicht verschwindet[2]). Das Nichtverschwinden der Determinante $|a_{ik}|$ folgt aus

[2]) Eine nähere Begründung findet man bei Hans Lewy, „Über das Anfangswertproblem einer hyperbolischen nichtlinearen partiellen Differentialgleichung zweiter Ordnung mit zwei Veränderlichen“, Math. Annalen **98**, S. 179 ff., und bei K. Friedrichs und Hans Lewy, „Lösung des Anfangswertproblems einer beliebigen nichtlinearen hyperbolischen Differentialgleichung beliebiger Ordnung in zwei Variablen“, Math. Annalen **99**, S. 200 ff.

den nach der Voraussetzung des elliptischen Charakters erfüllten Ungleichungen

$$4\frac{\partial F}{\partial r}\frac{\partial F}{\partial t}-\left(\frac{\partial F}{\partial s}\right)^2\neq 0,\qquad \frac{\partial F}{\partial r}\neq 0,\qquad \frac{\partial F}{\partial t}\neq 0.$$

Um die in der Einleitung angekündigte Fortsetzung ins Komplexe auszuführen, werden wir alle Größen φ als Funktionen von vier unabhängigen Variablen $\xi_1, \xi_2, \eta_1, \eta_2$ angeben, die wir so wählen, daß für $\xi_2=\eta_2=0$ die Ebene ξ_1, η_1 mit der x, y-Ebene zusammenfällt, d. h. so, daß

$$x(\xi_1, 0, \eta_1, 0)=\varphi_1(\xi_1, 0, \eta_1, 0)=\xi_1,$$
$$y(\xi_1, 0, \eta_1, 0)=\varphi_2(\xi_1, 0, \eta_1, 0)=\eta_1$$

ist. Dadurch sind in der Ebene $\xi_2=\eta_2=0$ die sämtlichen φ, die vorher Funktionen von x, y waren, nun als dieselben Funktionen von ξ_1 und η_1 bekannt. Um nun diese $\varphi_1, \ldots, \varphi_8$ als Funktionen der vier Argumente $\xi_1, \xi_2, \eta_1, \eta_2$ zu bestimmen, greifen wir ein ins Innere des Definitionsbereichs unserer Lösung fallendes Geradenstück $\eta_1=$ konst. unserer ξ_1, η_1-Ebene heraus. Die $\varphi_1, \ldots, \varphi_8$ sind dort also Funktionen von ξ_1. Wir bestimmen jetzt die φ in einer gewissen zweidimensionalen beiderseitigen Umgebung dieses Geradenstückes in einer ξ_1, η_2-Ebene. Wir benutzen zur Fortsetzung in diese das obige System (1) der charakteristischen Differentialgleichungen

$$\sum_k a_{ik}\varphi_k'=0 \qquad (i=1,\ldots,6);$$
$$\sum_k a_{ik}\varphi_k^{\backprime}=0 \qquad (i=7,8),$$

geben aber den Operatoren nun die Bedeutung[3])

$$(*)\qquad \begin{aligned} {}' &= \frac{\partial}{\partial\xi_1}+\frac{\partial}{\partial\eta_2}\\ {}^{\backprime} &= -\frac{\partial}{\partial\xi_1}+\frac{\partial}{\partial\eta_2}. \end{aligned}$$

Man hat hier ein Anfangswertproblem für Funktionen der beiden unabhängigen Variablen ξ_1, η_2 vor sich, dessen Lösbarkeit von mir in der zitierten Arbeit über das hyperbolische Anfangswertproblem bewiesen wurde. Allerdings werden bei dieser Fortsetzung die Funktionen φ möglicherweise komplex werden, da schon für $\xi_2=\eta_2=0$ nicht alle a_{ik} reell sind. Die Bedeutung der a_{ik} für komplexe Werte der φ anzugeben, hat keine Schwierigkeit, da die a_{ik} analytisch von den φ abhängen; wir erklären die a_{ik} für komplexe Werte der φ einfach als analytische Fortsetzungen der

---

[3]) Wesentlich an dieser Festsetzung der Bedeutung der Operatoren ist nur, daß sie Differentiationen in zwei verschiedenen Richtungen der ξ_1, η_2-Ebene bedeuten und daß keine dieser Richtungen parallel zur „Anfangsgeraden" $\eta_2=0$ ist.

40*

a_{ik} für reelle Werte der φ. Um nur den zitierten Satz[4]) unmittelbar anzuwenden, zerlege man die Funktionen φ in Real- und Imaginärteile

$$\varphi = \overline{\varphi} + i\tilde{\varphi}$$

und benutze zur Fortsetzung das aus dem obigen System durch Aufspaltung nach Real- und Imaginärteil entstehende neue System von doppelt soviel Differentialgleichungen für doppelt soviel Unbekannte. Dabei sind die Anfangswerte der $\overline{\varphi}$ selbstverständlich gleich denen der φ selbst und die der $\tilde{\varphi}$ gleich Null zu nehmen. Das neue System hat natürlich wieder eine Determinante, die für die Anfangswerte nicht verschwindet[5]).

---

[4]) H. Lewy, loc. cit. Im folgenden werden noch die folgenden Eigenschaften von Lösungen eines solchen „kanonisch hyperbolischen" Systems benutzt, die ich hier zusammenstellen möchte:

1. Man kann eine etwa aus zwei kongruenten Trapezen (vgl. Figur) bestehende beiderseitige Umgebung des Geradenstückes $\eta_2 = \text{konst.}$ angeben, in der die Funktionen φ existieren und den obigen Differentialgleichungen gehorchen; die Größe der Trapeze hängt dabei, was die Anfangswerte betrifft, nur ab

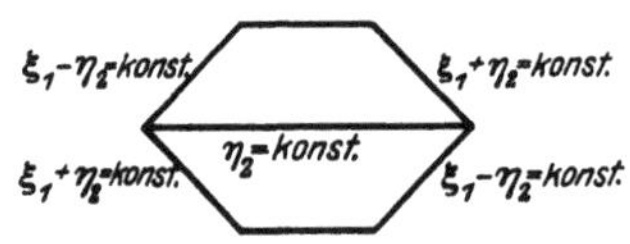

α) von oberen Schranken der $\left|\frac{\partial \overline{\varphi}_i}{\partial \eta_2}\right|$, $\left|\frac{\partial \tilde{\varphi}}{\partial \eta_2}\right|$ auf dem herausgegriffenen Geradenstück; β) von oberen und unteren Schranken für die $\overline{\varphi}_i$, $\tilde{\varphi}_i$ selbst; γ) von einer unteren Schranke für den Abstand der durch die Anfangswerte φ_i, $\tilde{\varphi}_i$ gegebenen Kurve des 16-dimensionalen $\overline{\varphi}_i$, $\tilde{\varphi}_i$-Raumes vom nächsten singulären Punkt der Funktionen $a_{ik}(\varphi_1, \varphi_2, \ldots, \varphi_8)$ und von der nächsten Nullstelle der Determinante $|a_{ik}|$. Es wird dabei vorausgesetzt daß alle betrachteten Anfangswerte der φ einem genügend kleinen Bereich des φ-Raumes angehören.

2. Sind die Anfangswerte der φ und ihre ersten Ableitungen stetige Funktionen von einem oder mehreren Parametern, so folgt aus dem soeben Gesagten, daß man die Größe der trapezförmigen Umgebung des Geradenstücks für die verschiedenen Werte des Parameters gleichmäßig nach unten abschätzen kann, und zwar durch eine positive Schranke.

3. Stetige n-malige Differenzierbarkeit der φ_i nach den Variablen der Differentialgleichungen und nach Parametern bleibt in der besagten Umgebung des Geradenstücks erhalten, wenn sie für die Anfangswerte gesichert ist.

Soweit diese Eigenschaften der Lösungen kanonisch hyperbolischer Systeme nicht schon in meiner Arbeit genannt und bewiesen werden, sind sie jedenfalls durch leichte Modifikation des dortigen Existenzbeweises zu erhalten, oder — wie ich dies ln Göttinger Vorlesungen ausgeführt habe — indem man völlig analog zu den Methoden verfährt, die zum Beweise der entsprechenden Behauptungen in der Theorie der gewöhnlichen Differentialgleichungen benutzt werden. Vgl. auch die Fußnote S. 610.

[5]) Denn sonst würde ein mit der neuen Matrix gebildetes linear homogenes Gleichungssystem eine nicht triviale Lösung besitzen, also auch ein linear homogenes Gleichungssystem mit der alten Matrix, was unmöglich ist, weil dessen Determinante nicht verschwindet.

Das Verfahren der Fortsetzung, das soeben für ein Geradenstück $\eta_1 = \text{konst.}$ der Ebene $\xi_2 = \eta_2 = 0$ als Anfangskurve beschrieben wurde, wende man nun auf *alle* parallelen Geradenstücke $\eta_1 = \text{konst.}$ dieser Ebene, soweit sie in den Definitionsbereich der Lösung u fallen, an. Man bestimmt so die Funktionen φ in einem dreidimensionalen Gebiet[6]) des $\xi_1, \xi_2, \eta_1, \eta_2$-Raumes, nämlich einem solchen, in dem $\xi_2 = 0$ gilt. Dies Gebiet enthält das Ebenenstück der Ebene $\xi_2 = \eta_2 = 0$ im Inneren.

Nunmehr geht man von den sämtlichen Geraden $\xi_1 = \text{konst.}$, $\eta_2 = \text{konst.}$ des soeben betrachteten dreidimensionalen Gebietes $\xi_2 = 0$ des $\xi_1, \xi_2, \eta_1, \eta_2$-Raumes aus und setzt die dort bekannten Funktionen $\varphi_1, \ldots, \varphi_8$ in ein umgebendes vierdimensionales Raumstück[6]) fort, indem man wieder das System

$$\sum_k a_{ik} \varphi_k' = 0 \qquad (i = 1, \ldots, 6),$$

$$\sum_k a_{ik} \varphi_k^{\backprime} = 0 \qquad (i = 7, 8)$$

benutzt, jetzt aber den Operatoren die folgende Bedeutung gibt:

$$(**) \qquad \begin{aligned} {}' &= \frac{\partial}{\partial \xi_2} - \frac{\partial}{\partial \eta_1}, \\ {}^{\backprime} &= + \frac{\partial}{\partial \xi_2} + \frac{\partial}{\partial \eta_1}. \end{aligned}$$

§ 2.

Rückkehr zu den alten Variablen.

Ein Hauptpunkt der folgenden Untersuchung ist der Nachweis, daß man anstatt $\xi_1, \xi_2, \eta_1, \eta_2$ auch x_1, x_2, y_1, y_2 als unabhängige Variabe einführen kann, wenn man sich auf eine hinreichend kleine vierdimensionale Umgebung eines Punktes x, y des ursprünglichen Definitionsbereiches von u beschränkt, was wir natürlich tun wollen. Es genügt dazu wegen der Stetigkeit[7]), das Nichtverschwinden der Determinante

$$\Delta = \frac{\partial(x_1, x_2, y_1, y_2)}{\partial(\xi_1, \xi_2, \eta_1, y_2)}$$

in dem betreffenden Punkte der „Anfangsebene" nachzuweisen. Da dort die $\xi_1, \xi_2, \eta_1, \eta_2$ bzw. mit den x_1, x_2, y_1, y_2 übereinstimmen, hat man dort die Gleichungen für acht von den in Frage kommenden Ableitungen:

$$(2) \qquad \left\{ \begin{array}{llll} \dfrac{\partial x_1}{\partial \xi_1} = 1, & \dfrac{\partial x_2}{\partial \xi_1} = 0, & \dfrac{\partial y_1}{\partial \xi_1} = 0, & \dfrac{\partial y_2}{\partial \xi_1} = 0; \\[2ex] \dfrac{\partial x_1}{\partial \eta_1} = 0, & \dfrac{\partial x_2}{\partial \eta_1} = 0, & \dfrac{\partial y_1}{\partial \eta_1} = 1, & \dfrac{\partial y_2}{\partial \eta_1} = 0. \end{array} \right.$$

---

[6]) Vgl. die Fußnote [4]).

[7]) Vgl. Anm. [4]).

Spaltet man weiterhin die Funktionen ϱ_1 und ϱ_2, vgl. S. 610, in dem betreffenden Punkte nach Real- und Imaginärteil, also

$$\varrho_1 = \sigma_1 + i\,\sigma_2, \quad \varrho_2 = \sigma_1 - i\,\sigma_2, \quad \sigma_2 \neq 0,$$

so liefern die Gleichungen $y' - \varrho_1 x' = 0$ und $y' - \varrho_2 x' = 0$ in den verschiedenen Bedeutungen (*) und (**) der Differentiationsoperatoren die folgenden vier Gleichungen (unter Benutzung von (2))

$$(3)\quad \begin{cases} \dfrac{\partial y_1}{\partial \eta_2} + i\,\dfrac{\partial y_2}{\partial \eta_2} - (\sigma_1 + i\,\sigma_2)\left(\;1 + \dfrac{\partial x_1}{\partial \eta_2} + i\,\dfrac{\partial x_2}{\partial \eta_2}\right) = 0, \\ \dfrac{\partial y_1}{\partial \eta_2} + i\,\dfrac{\partial y_2}{\partial \eta_2} - (\sigma_1 - i\,\sigma_2)\left(-1 + \dfrac{\partial x_1}{\partial \eta_2} + i\,\dfrac{\partial x_2}{\partial \eta_2}\right) = 0; \end{cases}$$

$$(4)\quad \begin{cases} \dfrac{\partial y_1}{\partial \xi_2} + i\,\dfrac{\partial y_2}{\partial \xi_2} - 1 - (\sigma_1 + i\,\sigma_2)\left(\dfrac{\partial x_1}{\partial \xi_2} + i\,\dfrac{\partial x_2}{\partial \xi_2}\right) = 0, \\ \dfrac{\partial y_1}{\partial \xi_2} + i\,\dfrac{\partial y_2}{\partial \xi_2} + 1 - (\sigma_1 - i\,\sigma_2)\left(\dfrac{\partial x_1}{\partial \xi_2} + i\,\dfrac{\partial x_2}{\partial \xi_2}\right) = 0. \end{cases}$$

Die ersten beiden liefern durch Subtraktion

$$2\,\sigma_1 + 2\,i\,\sigma_2\left(\frac{\partial x_1}{\partial \eta_2} + i\,\frac{\partial x_2}{\partial \eta_2}\right) = 0,$$

$$\frac{\partial x_1}{\partial \eta_2} + i\,\frac{\partial x_2}{\partial \eta_2} = +\,i\,\frac{\sigma_1}{\sigma_2}; \qquad \frac{\partial x_1}{\partial \eta_2} = 0; \qquad \frac{\partial x_2}{\partial \eta_2} = \frac{\sigma_1}{\sigma_2}.$$

Weiter erhält man

$$\frac{\partial y_1}{\partial \eta_2} + i\,\frac{\partial y_2}{\partial \eta_2} = \frac{(\sigma_1 + i\,\sigma_2)(\sigma_2 + i\,\sigma_1)}{\sigma_2} = \frac{i(\sigma_1^2 + \sigma_2^2)}{\sigma_2}; \qquad \frac{\partial y_1}{\partial \eta_2} = 0; \qquad \frac{\partial y_2}{\partial \eta_2} = \frac{\sigma_1^2 + \sigma_2^2}{\sigma_2}.$$

Ähnlich entsteht aus den Gleichungen (4) durch Subtraktion

$$2 + 2\,i\,\sigma_2\left(\frac{\partial x_1}{\partial \xi_2} + i\,\frac{\partial x_2}{\partial \xi_2}\right) = 0; \qquad \frac{\partial x_1}{\partial \xi_2} + i\,\frac{\partial x_2}{\partial \xi_2} = \frac{i}{\sigma_2}; \qquad \frac{\partial x_1}{\partial \xi_2} = 0, \qquad \frac{\partial x_2}{\partial \xi_2} = \frac{1}{\sigma_2},$$

woraus dann

$$\frac{\partial y_1}{\partial \xi_2} + i\,\frac{\partial y_2}{\partial \xi_2} = 1 + \frac{(\sigma_1 + i\,\sigma_2)\cdot i}{\sigma_2} = i\,\frac{\sigma_1}{\sigma_2}; \qquad \frac{\partial y_1}{\partial \xi_2} = 0, \qquad \frac{\partial y_2}{\partial \xi_2} = \frac{\sigma_1}{\sigma_2}$$

folgt. Die Matrix der Ableitungen der $x_1\,x_2\,y_1\,y_2$ nach den $\xi_1\,\xi_2\,\eta_1\,\eta_2$ sieht also in dem ins Auge gefaßten Punkte der Anfangsebene $\xi_2 = \eta_2 = 0$ so aus:

	x_1	x_2	y_1	y_2
ξ_1	1	0	0	0
ξ_2	0	$\frac{1}{\sigma_2}$	0	$\frac{\sigma_1}{\sigma_2}$
η_1	0	0	1	0
η_2	0	$\frac{\sigma_1}{\sigma_2}$	0	$\frac{\sigma_1^2+\sigma_2^2}{\sigma_2}$

Wie man sieht, ist die Determinante ungleich null, nämlich gleich 1; man kann also in einer vierdimensionalen Umgebung des betreffenden Punktes der x, y-Ebene die x_1, x_2, y_1, y_2 als neue Koordinaten einführen.

Allgemein werden die Differentiationssymbole nach $\xi_1, \xi_2, \eta_1, \eta_2$ lineare Kombinationen derer nach x_1, x_2, y_1, y_2. Schreibt man

$$(5)\quad \frac{\partial}{\partial \xi_1} + \frac{\partial}{\partial \eta_2} = A_1 \frac{\partial}{\partial x_1} + B_1 \frac{\partial}{\partial y_1} + C_1\left(\frac{\partial}{\partial x_1} + i\frac{\partial}{\partial x_2}\right) + D_1\left(\frac{\partial}{\partial y_1} + i\frac{\partial}{\partial y_2}\right),$$

so wird aus $\left(\frac{\partial}{\partial \xi_1} + \frac{\partial}{\partial \eta_2}\right)(y_1 + i y_2) - \varrho_1\left(\frac{\partial}{\partial \xi_1} + \frac{\partial}{\partial \eta_2}\right)(x_1 + i x_2) = 0$ wegen der Tatsache, daß der Operator $\frac{\partial}{\partial x_1} + i\frac{\partial}{\partial x_1}$ null ergibt, wenn er auf eine analytische Funktion von $x_1 + i x_2$ (hier $x_1 + i x_2$ selbst) ausgeübt wird, und wegen der entsprechenden Tatsache für den Operator $\frac{\partial}{\partial y_1} + i\frac{\partial}{\partial y_2}$

$$B_1 - \varrho_1 A_1 = 0,$$

und A_1 ist ungleich null, weil sonst der Operator (5) nicht reell sein kann, außer wenn C_1 und D_1 auch verschwinden; das ist aber unmöglich, weil nicht $\frac{\partial}{\partial \xi_1} = -\frac{\partial}{\partial \eta_2}$ ist.

Man findet demnach

$$(6)\quad \frac{\partial}{\partial \xi_1} + \frac{\partial}{\partial \eta_2} = A_1 \frac{\partial}{\partial x_1} + \varrho_1 A_1 \frac{\partial}{\partial y_1} + C_1\left(\frac{\partial}{\partial x_1} + i\frac{\partial}{\partial x_2}\right) + D_1\left(\frac{\partial}{\partial y_1} + i\frac{\partial}{\partial y_2}\right),$$

und ähnlich

$$(7)\quad -\frac{\partial}{\partial \xi_1} + \frac{\partial}{\partial \eta_2} = A_2 \frac{\partial}{\partial x_1} + \varrho_2 A_2 \frac{\partial}{\partial y_1} + C_2\left(\frac{\partial}{\partial x_1} + i\frac{\partial}{\partial x_2}\right) + D_2\left(\frac{\partial}{\partial y_1} + i\frac{\partial}{\partial y_2}\right),$$

$$(8)\quad \frac{\partial}{\partial \xi_2} - \frac{\partial}{\partial \eta_1} = E_1 \frac{\partial}{\partial x_1} + \varrho_1 E_1 \frac{\partial}{\partial y_1} + G_1\left(\frac{\partial}{\partial x_1} + i\frac{\partial}{\partial x_2}\right) + H_1\left(\frac{\partial}{\partial y_1} + i\frac{\partial}{\partial y_2}\right),$$

$$(9)\quad \frac{\partial}{\partial \xi_2} + \frac{\partial}{\partial \eta_1} = E_2 \frac{\partial}{\partial x_1} + \varrho_2 E_2 \frac{\partial}{\partial y_1} + G_2\left(\frac{\partial}{\partial x_1} + i\frac{\partial}{\partial x_2}\right) + H_2\left(\frac{\partial}{\partial y_1} + i\frac{\partial}{\partial y_2}\right).$$

Alle diese Größen $A_1, C_1, D_1, \ldots, H_2$ sind nichts weiter als lineare Kombinationen der Differentialquotienten der x_1, x_2, y_1, y_2 nach den $\xi_1, \xi_2, \eta_1, \eta_2$; z. B. ist offenbar

$$i C_1 = \left(\frac{\partial}{\partial \xi_1} + \frac{\partial}{\partial \eta_2}\right) x_2, \qquad i D_1 = \left(\frac{\partial}{\partial \xi_1} + \frac{\partial}{\partial \eta_2}\right) y_2,$$

$$i C_2 = \left(-\frac{\partial}{\partial \xi_1} + \frac{\partial}{\partial \eta_2}\right) x_2, \qquad i D_2 = \left(-\frac{\partial}{\partial \xi_1} + \frac{\partial}{\partial \eta_2}\right) y_2,$$

$$i G_1 = \left(\frac{\partial}{\partial \xi_2} - \frac{\partial}{\partial \eta_1}\right) x_2, \qquad i H_1 = \left(\frac{\partial}{\partial \xi_2} - \frac{\partial}{\partial \eta_1}\right) y_2,$$

$$i G_2 = \left(\frac{\partial}{\partial \xi_2} + \frac{\partial}{\partial \eta_1}\right) x_2, \qquad i H_2 = \left(\frac{\partial}{\partial \xi_2} + \frac{\partial}{\partial \eta_1}\right) y_2,$$

und es ist außerdem wichtig, nochmals hervorzuheben, daß A_1, A_2, E_1, E_2 sämtlich ungleich null sind.

Für das Folgende wesentlich sind die Werte der C_1, D_1, C_2, D_2, G_1, G_2 H_1, H_2 in der Anfangsebene $x_2 = y_2 = 0$. Man hat dort unter Benutzung unserer Differentiationstafel

$$(10)\qquad \begin{aligned} i\,C_1 &= \frac{\sigma_1}{\sigma_2}, & i\,D_1 &= \frac{\sigma_1^2+\sigma_2^2}{\sigma_2},\\ i\,C_2 &= \frac{\sigma_1}{\sigma_2}, & i\,D_2 &= \frac{\sigma_1^2+\sigma_2^2}{\sigma_2},\\ i\,G_1 &= \frac{1}{\sigma_2}, & i\,H_1 &= \frac{\sigma_1}{\sigma_2},\\ i\,G_2 &= \frac{1}{\sigma_2}, & i\,H_2 &= \frac{\sigma_1}{\sigma_2}. \end{aligned}$$

§ 3.

In der Anfangsebene gelten nun die charakteristischen Differentialgleichungen (1) in dreierlei Form, entsprechend den drei Bedeutungen des Operators $'$ bzw. $`$. Die erste auf S. 610 erwähnte Bedeutung von $'$ bzw. $`$ lautete

$$' = \frac{\partial}{\partial x_1} + \varrho_1 \frac{\partial}{\partial y_1},$$
$$` = \frac{\partial}{\partial x_1} + \varrho_2 \frac{\partial}{\partial y_2},$$

die zweite war die von (*), die dritte die von (**). Man subtrahiere nun von dem System (1) in der zweiten Bedeutung das mit A_1 bzw. A_2 multiplizierte der ersten Bedeutung und erhält so unter Berücksichtigung von (6) und (7) und von (10)

$$\sum_{k=1}^{8} a_{ik}\left[\frac{\sigma_1}{i\,\sigma_2}\left(\frac{\partial}{\partial x_1} + i\,\frac{\partial}{\partial x_2}\right) + \frac{\sigma_1^2+\sigma_2^2}{i\,\sigma_2}\left(\frac{\partial}{\partial y_1} + i\,\frac{\partial}{\partial y_2}\right)\right]\varphi_k = 0,$$

wobei der Index i von 1 bis 8 läuft. Aus dem Nichtverschwinden der Determinante $|a_{ik}|$ in der Anfangsebene folgt nunmehr

$$(11)\qquad \left[\frac{\sigma_1}{\sigma_2}\left(\frac{\partial}{\partial x_1} + i\,\frac{\partial}{\partial x_2}\right) + \frac{\sigma_1^2+\sigma_2^2}{\sigma_2}\left(\frac{\partial}{\partial y_1} + i\,\frac{\partial}{\partial y_2}\right)\right]\varphi_k = 0 \qquad (k = 1, \ldots, 8).$$

Ähnlich folgert man, wenn man das charakteristische System in seiner ersten Bedeutung mit E_1 bzw. E_2 multipliziert und von dem charakteristischen System in der dritten Bedeutung (**) abzieht und wiederum die Werte (10) der Koeffizienten G_1, G_2, H_1, H_2 in der Anfangsebene heranzieht:

$$(11)\qquad \left[\frac{1}{\sigma_2}\left(\frac{\partial}{\partial x_1} + i\,\frac{\partial}{\partial x_2}\right) + \frac{\sigma_1}{\sigma_2}\left(\frac{\partial}{\partial y_1} + i\,\frac{\partial}{\partial y_2}\right)\right]\varphi_k = 0 \qquad (k = 1, \ldots, 8).$$

Für jedes einzelne φ_k nun stellen die beiden Gleichungen (10) und (11) zwei Gleichungen dar, aus denen wegen des Nichtverschwindens von

$$\begin{vmatrix} \frac{\sigma_1}{\sigma_2} & \frac{\sigma_1^2+\sigma_2^2}{\sigma_2} \\ \frac{1}{\sigma_2} & \frac{\sigma_1}{\sigma_2} \end{vmatrix} = -1$$

$$\left(\frac{\partial}{\partial x_1} + i\frac{\partial}{\partial x_2}\right)\varphi_k = 0 \quad \text{und} \quad \left(\frac{\partial}{\partial y_1} + i\frac{\partial}{\partial y_2}\right)\varphi_k = 0$$

folgt. (Das darf nicht verwundern, denn die Bedeutung der Operatoren bzw. ' war natürlich dementsprechend festzulegen.)

Somit ist das Erfülltsein der Cauchy-Riemannschen Differentialgleichungen in der Anfangsebene $x_2 = y_2 = 0$, oder um in der Sprache der $\xi_1, \xi_2, \eta_1, \eta_2$ zu reden, in der Ebene $\xi_2 = \eta_2 = 0$ gesichert. Wir weisen zunächst ihre Gültigkeit in jenem dreidimensionalen Raumstück $\xi_2 = 0$ nach, in das wir zuerst vermöge der Bedeutung (*) der Operatoren ' bzw. ' fortgesetzt haben. Wir setzen zur Abkürzung

$$V_x = \frac{\partial}{\partial x_1} + i\frac{\partial}{\partial x_2}, \qquad V_y = \frac{\partial}{\partial y_1} + i\frac{\partial}{\partial y_2}.$$

Wir dividieren die Gleichungen des charakteristischen Systems in der besagten Bedeutung bzw. durch A_1 und A_2, die ja beide nicht verschwinden (vgl. S. 615) und wenden auf das so umgeformte System die Operatoren V_x und V_y an. Da die a_{ik} analytisch von den $\varphi_1, \ldots, \varphi_8$ abhängen, ist

$$V_x a_{ik} = \sum_l \frac{\partial a_{ik}}{\partial \varphi_l} V_x \varphi_l,$$
$$V_y a_{ik} = \sum_l \frac{\partial a_{ik}}{\partial \varphi_l} V_y \varphi_l.$$

(In diesen Gleichungen wird in der komplexen Differenzierbarkeit der a_{ik} zum einzigen wesentlichen Male die analytische Abhängigkeit der Differentialgleichung $F = 0$ von ihren Argumenten benutzt.) Weiter ist beispielsweise

$$\begin{aligned} V_x \frac{\varphi_k'}{A_1} &= V_x\left(\frac{\partial}{\partial x_1} + \varrho_1\frac{\partial}{\partial y_1} + \frac{C_1}{A_1}V_x + \frac{D_1}{A_1}V_y\right)\varphi_k \\ &= \left(\frac{\partial}{\partial x_1} + \varrho_1\frac{\partial}{\partial y_1} + \frac{C_1}{A_1}V_x + \frac{D_1}{A_1}V_y\right)V_x\varphi_k \\ &\qquad + V_x\frac{C_1}{A_1}V_x\varphi_k + V_x\frac{D_1}{A_1}V_y\varphi_k + \frac{\partial\varphi_k}{\partial y_1}\cdot\sum_{l=1}^{8}\frac{\partial\varrho_1}{\partial\varphi_l}V_x\varphi_l \\ &= \frac{1}{A_1}\cdot(V_x\varphi_k)' + V_x\frac{C_1}{A_1}V_x\varphi_k + V_x\frac{D_1}{A_1}V_y\varphi_k + \frac{\partial\varphi_k}{\partial y_1}\sum_{l=1}^{8}\frac{\partial\varrho_1}{\partial\varphi_l}V_x\varphi_l. \end{aligned}$$

Wir sehen also, daß die Anwendung des Operators ∇_x auf das durch A_1 bzw. A_2 dividierte Differentialgleichungssystem zu dem folgenden Differentialgleichungssystem führt:

$$
(12) \quad \begin{aligned}
&\sum_k a_{ik} (\nabla_x \varphi_k)' + \sum_k (b_{ik} \nabla_x \varphi_k + c_{ik} \nabla_y \varphi_k) = 0 \qquad (i = 1, \ldots, 6),\\
&\sum_k a_{ik} (\nabla_x \varphi_k)^{\backprime} + \sum_k (b_{ik} \nabla_x \varphi_k + c_{ik} \nabla_y \varphi_k) = 0 \qquad (i = 7, 8).
\end{aligned}
$$

Eine ähnliche Anwendung des Operators ∇_y bewirkt die Gleichungen

$$
(13) \quad \begin{aligned}
&\sum_k a_{ik} (\nabla_y \varphi_k)' + \sum_k (d_{ik} \nabla_x \varphi_k + e_{ik} \nabla_y \varphi_k) = 0 \qquad (i = 1, \ldots, 6),\\
&\sum_k a_{ik} (\nabla_y \varphi_k)^{\backprime} + \sum_k (d_{ik} \nabla_x \varphi_k + e_{ik} \nabla_y \varphi_k) = 0 \qquad (i = 7, 8).
\end{aligned}
$$

Dieses System von 16 Differentialgleichungen für die 16 Funktionen $\nabla_x \varphi_k$, $\nabla_y \varphi_k$ ist ein kanonisch hyperbolisches System mit nichtverschwindender Determinante. Die Determinante ist nämlich gleich $|a_{ik}|^2$. Wir haben in den Ausdrücken $\nabla_x \varphi_k$, $\nabla_y \varphi_k$ Lösungen eines Anfangswertproblems dieses Systems vor uns, nämlich eines solchen, dessen Anfangswerte sämtlich null sind. Eine Lösung dieses Problems wird

$$\nabla_x \varphi_k \equiv 0, \qquad \nabla_y \varphi_k \equiv 0,$$

wie aus der Homogenität der Differentialgleichungen folgt. Aus dem für diese Systeme gültigen Eindeutigkeitssatze [8]) folgt nunmehr, daß es nur diese Lösung gibt, also daß die Cauchy-Riemannschen Differentialgleichungen in dem betrachteten dreidimensionalen Gebiet gelten.

Ganz ebenso beweist man nunmehr die Gleichungen

$$\nabla_x \varphi_k \equiv 0, \qquad \nabla_y \varphi_k \equiv 0$$

in dem zu betrachtenden vierdimensionalen Gebiet, indem man sich erinnert, daß die Fortsetzung der φ_k in dieses erfolgte, indem man von allen Geraden $\xi_1 = \text{konst.}$, $\eta_2 = \text{konst.}$ des obigen dreidimensionalen Gebietes ausging. Auf diesen sind aber die $\nabla_x \varphi_k \equiv 0$, $\nabla \varphi_k \equiv 0$, wie eben bewiesen. Eine analoge Schlußweise führt nun zu der Behauptung.

Damit ist der Nachweis des analytischen Charakters der vorgelegten Lösung unserer Differentialgleichung beendet.

---

[8]) Ein solcher Eindeutigkeitssatz steht in meiner zitierten Abhandlung. Es läßt sich die Eindeutigkeit schon unter schwächeren Voraussetzungen als den dort angegebenen beweisen, wenn man die in der Theorie der gewöhnlichen Differentialgleichungen üblichen Methoden benutzt; vgl. hierzu auch O. Perron, „Über Existenz und Nichtexistenz von Integralen partieller Differentialgleichungen im reellen Gebiet", Math. Zeitschr. 27, S. 549 ff., sowie Haar, C. R. 187 (2. 6. 1928), und meine anfangs zitierte Abhandlung in den Göttinger Nachrichten 1927, S. 184 ff. Hier liegen die Dinge wegen der Linearität der Differentialgleichungen übrigens besonders einfach.

§ 4.

Zum Schluß noch ein Wort über die *Differenzierbarkeitsvoraussetzungen,* denen die vorgelegte Lösung $u(x, y)$ gehorchen muß. Zur Fortsetzung in den vierdimensionalen Raum ist es hinreichend, die Funktionen φ in der reellen Anfangsebene als zweimal stetig differenzierbar vorauszusetzen. Die Funktionen φ sind dann überall zweimal stetig nach den Argumenten differenzierbar und der im vorigen Paragraphen benutzte Eindeutigkeitssatz bezieht sich dann auf stetig differenzierbare Funktionen $V_x \varphi_l$, $V_y \varphi_l$ (vgl. hierzu die Anmerkung S. 618). Wir unterlassen es, zu untersuchen, wie weit unsere Voraussetzungen, die damit eine *viermalige stetige Differenzierbarkeit von u* bedeuten, gemildert werden können, ohne die Einfachheit des Gedankenganges zu beeinträchtigen.

Übrigens lassen sich die Überlegungen dieser Arbeit auf die Differentialgleichungen höherer Ordnung und auf Systeme von Differentialgleichungen erster Ordnung in zwei Variablen verallgemeinern.

(Eingegangen am 23. 10. 1928.)

RENDICONTI DELLA R. ACCADEMIA NAZIONALE DEI LINCEI
Classe di Scienze fisiche, matematiche e naturali.
Estratto dal vol. XI, serie 6ª, 1º sem., fasc. 2 - Roma, gennaio 1930-VIII.

Matematica. — *Sulla unicità della soluzione del problema di Cauchy per un'equazione ellittica del secondo ordine in due variabili.* Nota di H. LEWY, presentata[1] dal Socio T. LEVI-CIVITA.

Sia data un'equazione differenziale

$$F(x, y, u, u_x, u_y, u_{xx}, u_{xy}, u_{yy}) = 0 \tag{1}$$

e una sua soluzione $u(x, y)$, per la quale l'equazione risulti ellittica. Sia la F una funzione analitica di tutti i suoi argumenti – in un certo campo dello spazio delle $x, y, u, u_x, \cdots, u_{yy}$, il quale in particolare contiene i valori $x(s), y(s), u(s), u_x(s), \cdots, u_{yy}(s)$, che assumono queste espressioni su una certa curva C del piano (x, y), essendo s l'arco della C.

I valori assunti sulla C da $x, y, u, u_x, \cdots, u_{yy}$, – i cosidetti valori « iniziali » - determinano univocamente la soluzione $u(x, y)$ della (1)?

Ecco la classica risposta a questa domanda: Supponiamo che la $u(x, y)$ esista in un certo intorno (bilaterale) della C; che $u(x, y)$ sia funzione analitica di x, y e che la curva C sia analitica anch'essa. Allora un semplice calcolo delle derivate successive di u – per mezzo dei dati valori iniziali e della (1) – ci dà inivocamente lo sviluppo della $u(x, y)$ in una serie di potenze nell'intorno di un punto generico della C. Questo è il risultato di Cauchy-Kowalewski, concernente l'unicità della soluzione di (1).

(1) Nella seduta del 5 gennaio 1930.

Ma l'aspetto della questione si muta, quando rinunziamo all'ipotesi del carattere analitico, sia della $u(x, y)$, sia dei valori iniziali, ivi compresa la C.

Per il caso d'una equazione ellittica lineare, supponendo altresì la C analitica, il problema fu riportato dall'Holmgren (1), al risultato oramai classico del Picard, che tutte le soluzioni di tali equazioni sono analitiche.

Hadamard, nelle sue « Lectures on Cauchy's problem » (2) e altrove (3) rilevò la differenza del nostro problema dalle analoghe questioni di unicità per equazioni differenziali ordinarie e parziali del primo ordine. Fece osservare che, mentre di solito – oltre all'unicità – ha luogo e vien dimostrato con gli stessi mezzi il fatto della dipendenza continua delle soluzioni dai dati iniziali, anzi nel caso nostro non esiste in generale tale dipendenza continua, e ne diede un semplice esempio.

Per evitare la difficoltà indicata dall'Hadamard e dimostrare l'unicità nel caso generale, ho scelto un metodo diverso da quegli usuali, del quale voglio dare qui un cenno riassuntivo. Le mie ipotesi sulla C e la u sono 1. che i valori iniziali $x(s), y(s), u(s), \cdots, u_{yy}(s)$ ammettono derivate continue del secondo ordine rispetto all'arco s di C; e 2. che la soluzione $u(x, y)$ di (1) esiste almeno da una parte della C, avendo là derivate continue fino al quarto ordine.

Consideriamo nel piano x, y un sistema $\sum$ di curve dipendenti da due parametri per modo che per due generici punti del piano ne passi una ed una sola e che la C appartenga al sistema. Prendiamo una generica di queste curve. I valori ivi assunti da $x, y, u, \cdots, u_{yy}$ si possono manifestamente pensare come funzioni di un parametro; e questo a sua volta lo possiamo interpretare come distanza sopra una retta $\lambda + \mu = \text{cost.}$ in un piano cartesiano (λ, μ) (4).

Seguendo un metodo recentemente esposto (5) si potrà risolvere in un modo ed in uno solo un certo problema caratteristico associato coll'equazione $F = 0$:

$$(2) \qquad \begin{cases} \displaystyle\sum_{k=1}^{8} a_{ik} \frac{\partial}{\partial \lambda} \varphi_k = 0 & (i = 1, \cdots, 6), \\ \displaystyle\sum_{k=1} a_{ik} \frac{\partial}{\partial \mu} \varphi_k = 0 & (i = 7, 8), \end{cases}$$

ove le funzioni da determinare $\varphi_1, \varphi_2, \cdots, \varphi_8$ stanno per $x, y, u, \cdots u_{yy}$,

(1) Holmgren, *Om Cauchy's problem vid de lineära partiella differentialequationerna af 2: dra ordningen*, « Ark. f. Math. Ast. Fys. », t. 2, n. 24.

(2) Pagina 33.

(3) *Osservazione su una Nota di Haar*, « C. R. », 187 (2 giugno 1928).

(4) L'introduzione delle λ, μ serve soltanto per facilitare la descrizione della costruzione che segue.

(5) Vedi Hans Lewy, *Neuer Beweis des analytischen Charakters der Lösungen elliptischer Differentialgleichungen*, « Math. Ann. », 101, spec., pp. 610–612.

le a_{ik} sono funzioni analitiche delle $\varphi_1, \cdots, \varphi_8$, definite dalla F, e tali che il determinante $|a_{ik}|$ è differente da zero.

Osserviamo che questo è un problema iniziale iperbolico, di cui le curve caratteristiche sono le rette $\lambda =$ cost. e $\mu =$ cost. Le due caratteristiche che passano per un generico punto λ, μ, segheranno la suddetta retta iniziale $\lambda + \mu =$ cost. in due punti, ai quali corrisponderanno certi valori di x, y, diciamo

$$X_1, Y_1 \quad \text{e} \quad X_2, Y_2 .$$

Consideriamo adesso le nostre funzioni $\varphi_1, \cdots, \varphi_8$ – dopo averle sottoposte allo stesso processo di prolungamento per tutte le curve del sistema $\sum$ – e scriviamole come funzioni delle quattro variabili X_1, Y_1 e X_2, Y_2, che abbiamo testè introdotte. Se $X_1 = X_2, Y_1 = Y_2$, le $\varphi_1, \varphi_2, \cdots, \varphi_8$ si riducono alle funzioni date $x, y, u, \cdots, u_{yy}$.

Decomposte le φ_1 e φ_2 in parti reali e parti immaginarie:

$$\varphi_1 = x = x_1 + ix_2$$

$$\varphi_2 = y = y_1 + iy_2$$

è facile provare, che invece di X_1, Y_1, X_2, Y_2, si possano introdurre anche le x_1, x_2, y_1, y_2 quali variabili indipendenti; e va rilevato che, in questo passaggio, si deve profittare essenzialmente del fatto che la (1) è *ellittica*. Ora si dimostra - seguendo sempre una via simile a quella della Memoria suddetta - che le φ dipendono analiticamente da $x = x_1 + ix_2$ e $y = y_1 + iy_2$, essendo soddisfatte le equazioni differenziali di monogeneità

$$\left(\frac{\partial}{\partial x_1} + i\frac{\partial}{\partial x_2}\right)\varphi = 0 , \left(\frac{\partial}{\partial y_1} + i\frac{\partial}{\partial y_2}\right)\varphi = 0 ;$$

questo invece è sostanzialmente conseguenza della *regolarità* della F.

Ciò posto, ritorniamo alle variabili X_1, Y_1 e X_2, Y_2 e domandiamoci come si comportano le funzioni φ al variare delle X_2 e Y_2, fissando per il momento i valori di X_1 e Y_1 e viceversa. Occupiamoci da prima di $\varphi_1 = x$ e $\varphi_2 = y$. Esse costituiscono una superficie S (o rispettivamente $\bar{S}$) caratteristica della (1), passante per il punto $x = X_1, x = Y_1$ (rispettivamente $x = X_2, y = Y_2$). Ma oltre a ciò - ed in conseguenza del suaccennato modo di costruirle - le S e $\bar{S}$ saranno superficie caratteristiche della $u(x, y)$, considerata come funzione analitica delle due variabili complesse x e y (1); il che permette di compendiarne le equazioni sotto la forma

$$y = f(x),$$

dove f è funzione analitica dell'argomento complesso x. Ne segue che tutte

(1) Vedi T. Levi-Civita, *Sulle funzioni di due o più variabili complesse*, « Acc. Lincei », serie 5ª, vol. XIV, pp. 203-209.

le funzioni $\varphi_1, \varphi_2, \cdots, \varphi_8$ diventano sulla S (rispettivamente $\bar{S}$) funzioni analitiche della x.

In particolare, prendiamo come X_1, Y_1 le coordinate di un punto della curva iniziale C. Sulla S corrispondente, i valori di $\varphi_1, \cdots, \varphi_8$ si determinano mediante il suddetto problema iperbolico (2); e fin quando X_2, Y_2 restano sulla C, *i soli valori iniziali delle φ sulla* C bastano a farci conoscere

$$\varphi(X_1, Y_1; X_2, Y_2).$$

Così queste $\varphi(X_1, Y_1; X_2, Y_2)$ saranno determinate univocamente su una curva della S e, usufruendo allora della loro regolarità testè stabilita rispetto alla variabile indipendente x sulla S, le dette funzioni $\varphi(X_1, Y_1; X_2, Y_2)$ rimangono individuate univocamente in ogni punto di S almeno in un certo intorno della detta curva, il che è quanto dire per valori quali si vogliono delle X_2, Y_2 (in un certo campo).

Lo stesso vale per le funzioni $\varphi(X_1, Y_1; X_2, Y_2)$ quando X_2, Y_2 appartengono alla C e le X_1, Y_1 sono arbitrarie.

Siamo ora in grado di dimostrare l'unicità della φ per tutte le coppie identiche $X_1 = X_2 = X, Y_1 = Y_2 = Y$ le quali offrono un interesse speciale, perchè le $\varphi(X, Y; X, Y)$ sono addirittura le funzioni originali $x, y, u(x, y), \cdots, u_{yy}(x, y)$; basta ricordare la costruzione delle funzioni $\varphi(X_1, Y_1; X_2, Y_2)$. Alla schiera di curve a due parametri nel piano x, y, accennata di sopra, appartiene in particolare la curva che congiunge il punto X, Y con un punto arbitrario (1) X′, Y′ della C. Questa curva venne rappresentata su una retta $\lambda + \mu =$ cost. del piano λ, μ e furono trasferiti lì anche i valori della φ. Poi si determinarono i valori delle φ nel piano delle λ, μ secondo il problema iniziale iperbolico. Ma adesso conosciamo univocamente i valori delle φ su quelle due caratteristiche $\lambda =$ cost. e $\mu =$ cost., che passano per il punto λ, μ corrispondente al punto X′, Y′ della C, perchè lì le φ si scrivono sotto la forma $\varphi(X', Y'; X_2, Y_2)$ rispettivamente $\varphi(X_1, Y_1; X', Y')$. Oramai basta osservare che questi dati sono sufficienti a determinare la soluzione delle equazioni caratteristiche (2) nel quadrato, formato da quelle rette $\lambda =$ cost. e $\mu =$ cost., che passano per quei punti del piano λ, μ, che sono i rappresentanti dei punti X′, Y′ e X, Y. Con ciò le funzioni

$$\varphi(X, Y; X, Y)$$

rimangono determinate univocamente, soltanto mediante i valori che esse assumono quando X, Y è un punto della C, c. d. d.

(1) P. es., il punto di minima distanza.

Roma, 1930. — Dott. G. Bardi, Tip. della R. Accad. dei Lincei.

Mathematische Annalen
104 (1931)

Eindeutigkeit der Lösung des Anfangsproblems einer elliptischen Differentialgleichung zweiter Ordnung in zwei Veränderlichen.

Von

Hans Lewy in Göttingen.

Einleitung.

Gegenstand der folgenden Untersuchung ist eine Differentialgleichung zweiter Ordnung für eine Funktion u der beiden Variablen x und y:

(1) $$F(x, y, u, p, q, r, s, t) = 0,$$

in der wie üblich p, q, r, s, t für die Ableitungen erster und zweiter Ordnung stehen:

(2) $$u_x = p, \quad u_y = q; \qquad u_{xx} = r, \quad u_{xy} = s, \quad u_{yy} = t;$$

ferner ein beliebiges Kurvenstück C der x, y-Ebene, auf dessen einer Seite — in einer gewissen Umgebung — eine Lösung von (1) existiert, und zwar derart, daß

1. sie auf C gewisse Anfangswerte u, p, q, r, s, t annimmt;

2. die Funktion F für diese Anfangswerte und eine gewisse Umgebung derselben im Raume der komplexen Argumente x, y, u, p, q, r, s, t eine regulär analytische Funktion von ihnen ist;

3. die Differentialgleichung für die Anfangswerte elliptisch ausfällt, das bedeutet, daß dort die Ungleichung

(3) $$4\frac{\partial F}{\partial r}\frac{\partial F}{\partial t} - \left(\frac{\partial F}{\partial s}\right)^2 > 0 \quad \left(\text{also auch } \frac{\partial F}{\partial r} \neq 0, \ \frac{\partial F}{\partial t} \neq 0\right)$$

gilt.

Wir fragen, ob die Lösung u von (1) durch ihre Anfangswerte auf C eindeutig bestimmt ist; genauer beschränken wir uns darauf, festzustellen, ob es eine endliche Umgebung von C gibt, in der irgend zwei solche Lösungen — infolge der ihnen auferlegten Anfangsbedingungen notwendig übereinstimmen.

Bevor wir diese Frage und zwar im Sinne der Eindeutigkeit entscheiden, wird es nützlich sein, ihre Beziehung zu den bekannten Untersuchungen von Cauchy-Kowalewski klarzulegen. Gestattet die vorgelegte

Lösung in einem Punkte x_0, y_0 der Anfangskurve eine Entwicklung in eine Potenzreihe nach $x - x_0$ und $y - y_0$, so folgt aus der Differentialgleichung sofort eine rekursive Berechnung der Koeffizienten dieser Reihe, *falls die Anfangskurve C analytisch ist und die Anfangswerte der Funktionen u, p, q, r, s, t selbst analytisch von einem passend gewählten Parameter auf C, etwa der Bogenlänge, abhängen.* Um also unter dieser Voraussetzung über die Anfangsbedingungen die Eindeutigkeit der Lösung $u(x, y)$ garantieren zu können, wäre für die Anwendung der Cauchy-Kowalewskischen Gedankengänge immer noch die zusätzliche Bedingung notwendig, daß die Lösung auf der Anfangskurve C und in derjenigen Umgebung von C, in der sie auf ihre Eindeutigkeit zu untersuchen ist, regulär analytisch ist im Sinne einer Funktion der beiden Variablen x und y. Dies erfordert gewiß die Existenz von u auf beiden Seiten der Anfangskurve C; andererseits folgt aus der beiderseitigen Existenz der Lösung nach einem bekannten Satze über den analytischen Charakter der Lösungen elliptischer Differentialgleichungen, daß die Lösung u auf der Anfangskurve selbst und in einer gewissen Umgebung von ihr eine analytische Funktion von x und y ist, weil dann die Punkte der Anfangskurve *innere* Punkte des Existenzbereiches von u sind. Nun ist schließlich auch die beiderseitige Existenz der Lösung unter den obigen Voraussetzungen über die Anfangsbedingungen keine neue Einschränkung, weil ja hier eine beliebige auf der einen Seite von C bestehende Lösung auf der anderen Seite durch die von dem hier gültigen Cauchy-Kowalewskischen Existenzsatz gelieferte fortgesetzt werden kann. Man gelangt so bei analytischer Anfangskurve und analytischen Anfangswerten unter Zuhilfenahme des Satzes vom analytischen Charakter mittels der Sätze von Cauchy-Kowalewski zu der gesuchten Eindeutigkeit.

Handelt es sich speziell um eine lineare Differentialgleichung, so führen diese Überlegungen auch ohne die Voraussetzung der Analytizität von Anfangskurve und Anfangswerten zum Ziel. Man kann sich nämlich auf verschwindende Anfangswerte und homogene Gleichungen beschränken. Man kann dann eine einseitig existierende Lösung der Differentialgleichung auf der anderen Seite von C durch $u \equiv 0$ fortsetzen und berechnet nunmehr die Koeffizienten der Potenzreihenentwicklung in einem Punkte der Anfangskurve sämtlich zu null. Damit ist ein Resultat von Holmgren[1]) bewiesen, betreffend die Eindeutigkeit bei linearen elliptischen Differentialgleichungen.

In diesen Zusammenhang gehört auch ein bekannter Satz über analytische Funktionen eines komplexen Arguments und sein Beweis. Da er für die folgenden Untersuchungen der entscheidende Hilfssatz ist, möge er

[1]) Holmgren benutzt außer dem Satz vom analytischen Charakter noch gewisse Integralgleichungen beim Beweis „Om Cauchy's problem vid de lineära partiella differentialequationerna ...", Ark. f. Math. Astr. Fys. 2, S. 24.

hier genannt sein: Es gibt höchstens eine analytische Funktion, die auf einer Seite eines Kurvenstücks der komplexen Zahlenebene in dessen Umgebung existiert und bei Annäherung an die Kurve gleichmäßig gegen dort vorgegebene stetige „Anfangswerte" strebt.

Wenden wir uns nun zu den nichtlinearen Differentialgleichungen, um auch dort die Eindeutigkeit der Lösung des Anfangsproblems nachzuweisen, ohne jedoch Anfangskurve oder Anfangswerte als analytisch vorauszusetzen.

Hadamard hat auf eine Eigentümlichkeit hingewiesen, die das allgemeine Problem der Eindeutigkeit bei elliptischen Differentialgleichungen unterscheidet von dem analogen Problem etwa bei gewöhnlichen oder partiellen Differentialgleichungen erster Ordnung. Er bemerkt nämlich[2]), daß in den beiden letzten Fällen — das gilt übrigens auch für die hyperbolischen Differentialgleichungen — die Tatsache der Eindeutigkeit und auch die dafür bekannten Beweise lediglich Spezialisierungen des allgemeinen Stetigkeitsprinzips sind, das besagt, daß die Lösungen sich stetig ändern, wenn ihre Anfangswerte stetig variieren. Andererseits aber zeigt er an einem eleganten Beispiel, daß dieses Prinzip für elliptische Differentialgleichungen, allgemein zu reden, *nicht* gültig ist. Man darf daher mit Hadamard annehmen, daß auch die Übertragung jener Beweise auf unseren Fall schwerlich zum Ziele führen wird und es geraten ist, einen anderen Weg einzuschlagen. Ein solcher bietet sich in einer ähnlichen Fortsetzung der Lösung ins Komplexe[3]), wie ich sie unlängst zum Beweise des analytischen Charakters der Lösungen elliptischer Differentialgleichungen angewandt habe. Ich kann dadurch die allgemeine Eindeutigkeitsfrage auf den erwähnten Satz über Funktionen eines komplexen Arguments zurückführen.

Bei der folgenden Darstellung ließ es sich leider nicht vermeiden, verschiedene Umstände, die bereits beim Beweise des analytischen Charakters auftraten, in modifizierter Form zu wiederholen; insbesondere erscheint der analytische Charakter der zu besorgenden Fortsetzung ins Komplexe wiederum als Nebenresultat und Hauptstütze des Beweises der Eindeutigkeit; jedoch steht diesem Nachteil die Annehmlichkeit gegenüber, daß beide Eigenschaften mit derselben Methode abgeleitet werden; beide werden durch diese Fortsetzung mit dem Existenz- und Eindeutigkeitssatz im hyperbolischen Falle in enge Beziehung gesetzt, und man gewinnt so eine vergleichende Übersicht über die so verschiedenen Grundeigenschaften der Lösungen von elliptischen und hyperbolischen Differentialgleichungen, die sich mühelos als nur verschiedene Seiten ein und derselben Sache aufklären.

---

[2]) Hadamard, „Lectures on Cauchy's Problem" S. 33 und Zusatz zu einer Note von Haar, Comptes Rendus **187** (2. 6. 1928).

[3]) Vgl. Hans Lewy, Neuer Beweis des analytischen Charakters ..., Math. Ann. **101** (1929), S. 609.

22*

§ 1.

Fortsetzung ins Komplexe.

1. Einführung passender Variablen. Differenzierbarkeitsvoraussetzungen.

Da bekanntlich der elliptische Typus von F gegen Koordinatentransformationen der x, y mit nicht verschwindender Determinante invariant ist, können wir von vornherein annehmen, daß die Kurve C in einem inneren Punkt P von ihr durch den Ursprung $x = 0$, $y = 0$ geht und sich in einer Umgebung von P in die Gestalt

$$y = f(x)$$

setzen läßt. Wir verlangen, daß $f(x)$ zweimal stetig nach x, daß die vorgelegte Lösung $u(x, y)$ viermal stetig nach x und y differenzierbar sei (diese Voraussetzung soll auch noch für die Punkte der Anfangskurve gültig sein).

Führen wir folgende Variable ein:

$$\alpha = x, \quad \beta = y - f(x).$$

Die Determinante der Transformation ist $\neq 0$; es wird eine Umgebung von P auf eine Umgebung des Ursprungs des (kartesischen) α, β-Systems eindeutig bezogen. Durch eventuelles Ändern des Vorzeichens von β bewirkt man, daß diejenige Seite von C, auf der die Voraussetzung die Existenz der Lösung von (1) garantiert, auf einen Teil der Halbebene $\beta \geqq 0$ abgebildet wird. Alle im folgenden früher oder später vorzunehmenden Operationen beziehen sich dann immer auf eine *passend klein* gewählte Halbumgebung $\beta \geqq 0$ von $\alpha = 0$, $\beta = 0$, die ein endliches, den Ursprung im Innern enthaltendes Stück der Geraden als Teil ihres Randes besitzt; wir werden diese Einschränkung als selbstverständlich unterstellen und an den betreffenden Stellen nicht mehr ausdrücklich darauf hinweisen.

2. Wie die Charakteristikentheorie zeigt, gehorcht jede dreimal stetig differenzierbare Lösung $u(x, y)$ von (1) dem folgenden Differentialgleichungssystem:

$$(4)\quad \begin{cases} (4_1) \qquad y' - \varrho_1 x' = 0, \\ (4_2) \qquad y^{\backprime} - \varrho_2 x^{\backprime} = 0, \\ \varrho_1 F_r r' + F_t s' + (F_p r + F_q s + F_u p + F_x) y' = 0, \\ \varrho_1 F_r s' + F_t t' + (F_p s + F_q t + F_u q + F_y) y' = 0, \\ \varrho_2 F_r r^{\backprime} + F_t s^{\backprime} + (F_p r + F_q s + F_u p + F_x) y^{\backprime} = 0, \\ u' - p x' - q y' = 0, \\ p' - r x' - s y' = 0, \\ q' - s x' - t y' = 0. \end{cases}$$

Hierbei ist

(5) $\varrho_1 = \frac{1}{2F_r}(F_s + i\sqrt{4F_rF_t - F_s^2}), \qquad \varrho_2 = \frac{1}{2F_r}(F_s - i\sqrt{4F_rF_t - F_s^2}),$

und die Operatoren bedeuten

(6) $' = \frac{\partial}{\partial x} + \varrho_1 \frac{\partial}{\partial y}, \qquad ` = \frac{\partial}{\partial x} + \varrho_2 \frac{\partial}{\partial y}.$

Die Determinante[4]) des Systems (4) ist gleich $(\varrho_1 - \varrho_2)^2 F_r F_t^2$, also wegen (3) von null verschieden; und dies gilt jedenfalls für die Anfangskurve, also aus Stetigkeitsgründen auch in einer passenden Halbumgebung von C.

3. Wir haben jetzt gewisse Funktionen $x(\alpha_1, \beta_1; \alpha_2, \beta_2)$, $y(\alpha_1, \beta_1; \alpha_2, \beta_2)$, $u(\alpha_1, \beta_1; \alpha_2, \beta_2), \ldots$ zu definieren. Dazu nehmen wir zwei (einander und der Anfangskurve hinreichend benachbarte) Punkte α_1, β_1 und α_2, β_2 derart, daß die Verbindungsstrecke ganz dem Existenzbereich der vorgelegten Lösung, die als Funktion von α und β zu denken ist, angehört. Da die Punkte dieser Strecke sich in die Form

$$\lambda\alpha_1 + \mu\alpha_2, \; \lambda\beta_1 + \mu\beta_2 \qquad (0 \leqq \lambda \leqq 1, \; \lambda + \mu = 1)$$

setzen lassen, sind die Funktionen $x(\alpha, \beta)$, $y(\alpha, \beta)$, $u(\alpha, \beta), \ldots$ dort auch als Funktionen auf der Geraden $\lambda + \mu = 1$ einer $\lambda\mu$-Ebene auffaßbar, und sie werden gemäß den Voraussetzungen über die Anfangskurve und über die Lösung $u(x, y)$ von (1) als solche zweimal stetig nach der Bogenlänge differenzierbar sein. Wir stellen nun ein hyperbolisches, nach der Theorie eindeutig lösbares Anfangswertproblem für (komplexwertige) Funktionen der reellen Variablen λ und μ, indem wir eine Lösung des Systems (4) suchen, wobei wir jedoch den Operatoren $'$ und $`$ die Bedeutung

(7) $' = \frac{\partial}{\partial \lambda}, \qquad ` = \frac{\partial}{\partial \mu}$

beilegen und als Anfangswerte die Werte der Funktionen $x, y, u, \ldots$ auf der Geraden $\lambda + \mu = 1$ geben. Der Wert der zugehörigen Lösungen im Punkte $\lambda = 1$, $\mu = 1$ ist dann eindeutig durch die Anfangswerte gegeben, da die Charakteristiken durch ihn, d. h. die Geraden $\lambda = 1$ und $\mu = 1$ aus der Anfangsgeraden $\lambda + \mu = 1$ gerade dasjenige Stück ausschneiden, auf dem wir die Werte der Funktionen vorgeben. Wir wollen den Wert der Funktionen $x, y, u, \ldots$ im Punkte $\lambda = 1$, $\mu = 1$, insofern sie ja ihrer Konstruktion nach von den beiden Punkten α_1, β_1 und α_2, β_2 abhängen, mit $x(\alpha_1, \beta_1; \alpha_2, \beta_2)$, $y(\alpha_1, \beta_1; \alpha_2, \beta_2)$, $u(\alpha_1, \beta_1; \alpha_2, \beta_2), \ldots$ bezeichnen.

---

[4]) Vgl. für diesen Begriff Hans Lewy, „Über das Anfangswertproblem bei einer hyperbolischen nichtlinearen Differentialgleichung ...", Math. Annalen **98**, S. 186ff., oder Kurt Friedrichs und Hans Lewy, „Lösung des Anfangswertproblems einer beliebigen hyperbolischen Differentialgleichung ...", Math. Annalen **98**, S. 208. Dort auch eine bequeme Ableitung der Differentialgleichungen (4).

Offenbar handelt es sich um eine vierparametrige Schar von Anfangsproblemen; und die bezüglichen Anfangswerte sind nach den vier Parametern der Schar: $\alpha_1, \beta_1, \alpha_2, \beta_2$ zweimal stetig differenzierbar; wir schließen hieraus dasselbe für die Lösungen und finden so, daß die Funktionen $x(\alpha_1, \beta_1; \alpha_2, \beta_2)$, $y(\alpha_1, \beta_1; \alpha_2, \beta_2)$, $u(\alpha_1, \beta_1; \alpha_2, \beta_2), \ldots$ zweimal stetig differenzierbare Funktionen der Argumente sind.

4. Es ist

$$x(\alpha, \beta; \alpha, \beta) = x(\alpha, \beta), \quad y(\alpha, \beta; \alpha, \beta) = y(\alpha, \beta), \quad u(\alpha, \beta; \alpha, \beta) = u(\alpha, \beta), \ldots.$$

In der Tat stellt ja $x \equiv x(\alpha, \beta), \ldots, u \equiv u(\alpha, \beta), \ldots$ für alle λ, μ des in Betracht kommenden Dreiecks $\lambda + \mu \geqq 1$, $\lambda \leqq 1$, $\mu \leqq 1$ eine Lösung des zugehörigen Anfangsproblems dar, in dem nämlich die obigen beiden Punkte α_1, β_1 und α_2, β_2 in den einen Punkt α, β zusammenfallen; in der analytischen Formulierung des Anfangsproblems auf S. 329 wurde keine Voraussetzung über Nichtzusammenfallen der beiden Punkte gemacht. Nach dem für das System (4) gültigen Eindeutigkeitssatze ist also $x \equiv x(\alpha, \beta), \ldots, u \equiv u(\alpha, \beta)$ *die* Lösung. Daher gilt speziell

$$x(\alpha, \beta; \alpha, \beta) = x(\alpha, \beta), \ldots, u(\alpha, \beta; \alpha, \beta) = u(\alpha, \beta), \ldots.$$

5. Ich behaupte, daß die Funktionen $x(\alpha_1, \beta_1; \alpha_2, \beta_2), \ldots, t(\alpha_1, \beta_1; \alpha_2, \beta_2)$ *selbst* als Lösungen des Systems (4) auffaßbar sind, wenn man die Operatoren $'$ und $`$ geeignet als Differentialquotienten nach den Variablen $\alpha_1 \beta_1$ und $\alpha_2 \beta_2$ erklärt.

Dies einzusehen, greifen wir eins der obigen Anfangsprobleme heraus, sagen wir dasjenige, das uns die Werte $x(\alpha_1 \beta_1; \alpha_2 \beta_2), \ldots$ liefert. Wir hatten die Strecke der $\alpha\beta$-Ebene, die die Punkte $\alpha_1 \beta_1$ und $\alpha_2 \beta_2$ verbindet, affin auf die Strecke mit den Endpunkten $\lambda = 1$, $\mu = 0$ und $\lambda = 0$, $\mu = 1$ der $\lambda\mu$-Ebene abgebildet und die Funktionen $x(\alpha, \beta)$, $y(\alpha, \beta), \ldots$ auf der ersten Strecke als Funktionen der Punkte der zweiten geschrieben; die Werte $x(\alpha_1 \beta_1; \alpha_2 \beta_2), \ldots$ waren die Werte in $\lambda = \mu = 1$ der Lösungen des oben erklärten Anfangswertproblems der $\lambda\mu$-Ebene mit den soeben genannten Anfangswerten auf der Geraden $\lambda + \mu = 1$. Fassen wir aber einen beliebigen Punkt des Dreiecks $\lambda = 1, \mu = 0$; $\lambda = 0, \mu = 1$ $\lambda = \mu = 1$, der nicht auf der Geraden $\lambda + \mu = 1$ liegen möge, ins Auge, etwa den Punkt $\lambda = \Lambda$, $\mu = M$. In ihm nehmen die Lösungen des Anfangsproblems gewisse Werte an, die nach dem allgemeinen Satze für Systeme (4) in der Bedeutung (7) nur von einem Teile der Anfangswerte abhängen, nämlich von derjenigen Strecke, die auf der Geraden $\lambda + \mu = 1$ von den Geraden $\lambda = \Lambda$ und $\mu = M$ ausgeschnitten wird. Macht man jetzt eine Ähnlichkeitstransformation der $\lambda\mu$-Ebene, die die beiden Endpunkte in die Punkte $\lambda = 1$, $\mu = 0$; $\lambda = 0$, $\mu = 1$ wirft, d. h. setzt man

$$\lambda' = \frac{\lambda + M - 1}{\Lambda + M - 1}; \qquad \mu' = \frac{\mu + \Lambda - 1}{\Lambda + M - 1}$$

und schreibt die Funktionen $x(\lambda, \mu), \ldots$ als entsprechende Funktionen der λ', μ', so werden $\left(\text{wegen der Proportionalität der Operatoren } \frac{\partial}{\partial \lambda}, \frac{\partial}{\partial \mu} \text{ mit } \frac{\partial}{\partial \lambda'}, \frac{\partial}{\partial \mu'} \text{ und der Homogenität des Systems (4) in den Ableitungen}\right)$ diese Funktionen der λ', μ' gerade die Lösungen des Anfangsproblems aus der Schar, das für $\lambda' = \mu' = 1$ die Funktionen

$$x(\Lambda \alpha_1 + (1 - \Lambda)\alpha_2,\ \Lambda \beta_1 + (1 - \Lambda)\beta_2;\ M\alpha_2 + (1 - M)\alpha_1,\ M\beta_2 + (1 - M)\beta_1),$$

liefert (dies gilt nach 4. auch noch für $\Lambda + M = 1$). Die Argumente in (8) sind offenbar die Koordinaten zweier Punkte der $\alpha_1 \beta_1$ und $\alpha_2 \beta_2$ verbindenden Geraden. Bezeichnet φ den Winkel, den die positive Richtung der α-Achse der $\alpha\beta$-Ebene mit der Richtung von $\alpha_1 \beta_1$ nach $\alpha_2 \beta_2$ bildet (α_1, β_1 wieder $\neq \alpha_2, \beta_2$ vorausgesetzt), so wird also für die Funktionen (8) das Gleichungssystem (4) gelten in der Bedeutung

$$(8'') \qquad {}' = \frac{\partial}{\partial \Lambda}, \qquad {}^{\backprime} = \frac{\partial}{\partial M}$$

oder auch wegen der Homogenität allgemein mit

$$(^*) \qquad {}' = \cos\varphi \frac{\partial}{\partial A_1} + \sin\varphi \frac{\partial}{\partial B_1}, \qquad {}^{\backprime} = \cos\varphi \frac{\partial}{\partial A_2} + \sin\varphi \frac{\partial}{\partial B_2},$$

wo wir die neuen Symbole $\frac{\partial}{\partial A_1}, \frac{\partial}{\partial B_1}, \frac{\partial}{\partial A_2}, \frac{\partial}{\partial B_2}$ einführen mußten, um die Differentiation nach dem ersten, zweiten, dritten und vierten Argumente der Funktionen (8) auszudrücken. Die beiden letzten Gleichungen sind die in 5. aufgestellte Behauptung.

6. Speziell interessiert, was man für $\Lambda = 1, M = 0$ und $\Lambda = 0, M = 1$ schließen kann. Beachtet man, daß der Winkel φ bei festem $\alpha_1 \beta_1$ mit dem Punkte $\alpha_2 \beta_2$ variiert, so erhält man, für $\alpha_1 \beta_1$ wieder $\alpha \beta$ schreibend, aus der Gleichung (4_1)

$$0 = y' - \varrho_1 x' = \left(\cos\varphi \frac{\partial}{\partial A_1} + \sin\varphi \frac{\partial}{\partial B_1}\right) y - \varrho_1 \left(\cos\varphi \frac{\partial}{\partial A_1} + \sin\varphi \frac{\partial}{\partial B_1}\right) x$$

die beiden neuen

$$(9) \qquad \frac{\partial}{\partial A_1} y(\alpha, \beta; \alpha, \beta) - \varrho_1 \frac{\partial}{\partial A_1} x(\alpha, \beta; \alpha, \beta) = 0,$$

$$(10) \qquad \frac{\partial}{\partial B_1} y(\alpha, \beta; \alpha, \beta) - \varrho_1 \frac{\partial}{\partial B_1} x(\alpha, \beta; \alpha, \beta) = 0.$$

Aus (4_2) $y^{\backprime} - \varrho_2 x^{\backprime} = 0$ folgt entsprechend

$$(11) \qquad \frac{\partial}{\partial A_2} y(\alpha, \beta; \alpha, \beta) - \varrho_2 \frac{\partial}{\partial A_2} x(\alpha, \beta; \alpha, \beta) = 0,$$

$$(12) \qquad \frac{\partial}{\partial B_2} y(\alpha, \beta; \alpha, \beta) - \varrho_2 \frac{\partial}{\partial B_2} x(\alpha, \beta; \alpha, \beta) = 0.$$

Weiter gilt selbstverständlich die Differentiationsregel

$$\left(\frac{\partial}{\partial A_1}+\frac{\partial}{\partial A_2}\right)x(\alpha,\beta;\alpha,\beta)=\frac{\partial}{\partial\alpha}x(\alpha,\beta);\quad \left(\frac{\partial}{\partial B_1}+\frac{\partial}{\partial B_2}\right)x(\alpha,\beta;\alpha,\beta)=\frac{\partial}{\partial\beta}x(\alpha,\beta)$$

$$\left(\frac{\partial}{\partial A_1}+\frac{\partial}{\partial A_2}\right)y(\alpha,\beta;\alpha,\beta)=\frac{\partial}{\partial\alpha}y(\alpha,\beta);\quad \left(\frac{\partial}{\partial B_1}+\frac{\partial}{\partial B_2}\right)y(\alpha,\beta;\alpha,\beta)=\frac{\partial}{\partial\beta}x(\alpha,\beta).$$

Das sind zusammen 16 Gleichungen für die Ableitungen von $x(\alpha,\beta;\alpha,\beta)$, $y(\alpha,\beta;\alpha,\beta)$, die als komplexwertig zu betrachten sind, nach A_1, A_2, B_1, B_2. Spaltet man in Real- und Imaginärteil und setzt demgemäß

$$x = x_1 + i x_2, \qquad \bar{x} = x_1 - i x_2,$$

$$y = y_1 + i y_2, \qquad \bar{y} = y_1 - i y_2,$$

so kann man die obigen Gleichungen benutzen, um das Nichtverschwinden der Determinante (für die Kurve C der reellen xy-Ebene)

$$(13) \qquad \frac{\partial(x_1,x_2,y_1,y_2)}{\partial(A_1,B_1,A_2,B_2)} \quad \text{oder} \quad \frac{\partial(x,\bar{x},y,\bar{y})}{\partial(A_1,B_1,A_2,B_2)}$$

zu beweisen und sich so das Recht verschaffen, (in einer gewissen Umgebung von ihr) x_1, x_2, y_1, y_2 als neue Variable an Stelle von $\alpha_1\beta_1\alpha_2\beta_2$ einzuführen. Man findet nämlich eindeutig durch eine kleine Rechnung

$$\frac{\partial x(\alpha,\beta;\alpha,\beta)}{\partial A_1}=\frac{\dfrac{\partial y(\alpha,\beta)}{\partial\alpha}-\varrho_2\dfrac{\partial x(\alpha,\beta)}{\partial\alpha}}{\varrho_1-\varrho_2} \qquad \frac{\partial y(\alpha,\beta;\alpha,\beta)}{\partial A_1}=\varrho_1\frac{\partial x(\alpha,\beta;\alpha,\beta)}{\partial A_1}$$

$$\frac{\partial x(\alpha,\beta;\alpha,\beta)}{\partial B_1}=\frac{\dfrac{\partial y(\alpha,\beta)}{\partial\beta}-\varrho_2\dfrac{\partial x(\alpha,\beta)}{\partial\beta}}{\varrho_1-\varrho_2} \qquad \frac{\partial y(\alpha,\beta;\alpha,\beta)}{\partial B_1}=\varrho_1\frac{\partial x(\alpha,\beta;\alpha,\beta)}{\partial B_1}$$

$$\frac{\partial x(\alpha,\beta;\alpha,\beta)}{\partial A_2}=\frac{-\dfrac{\partial y(\alpha,\beta)}{\partial\alpha}+\varrho_1\dfrac{\partial x(\alpha,\beta)}{\partial\alpha}}{\varrho_1-\varrho_2}=\frac{\partial\bar{x}(\alpha,\beta;\alpha,\beta)}{\partial A_1} \qquad \frac{\partial y(\alpha,\beta;\alpha,\beta)}{\partial A_2}=\varrho_2\frac{\partial x(\alpha,\beta;\alpha,\beta)}{\partial A_2}$$

$$\frac{\partial x(\alpha,\beta;\alpha,\beta)}{\partial B_2}=\frac{-\dfrac{\partial y(\alpha,\beta)}{\partial\beta}+\varrho_1\dfrac{\partial x(\alpha,\beta)}{\partial\beta}}{\varrho_1-\varrho_2}=\frac{\partial\bar{x}(\alpha,\beta;\alpha,\beta)}{\partial B_1} \qquad \frac{\partial y(\alpha,\beta;\alpha,\beta)}{\partial B_2}=\varrho_2\frac{\partial x(\alpha,\beta;\alpha,\beta)}{\partial B_2}$$

und weiter unter Beachtung der für $\alpha_1=\alpha_2=\alpha$, $\beta_1=\beta_2=\beta$ sicher richtigen Gleichung $\bar{\varrho}_1=\varrho_2$

$$\frac{\partial(x,\bar{x},y,\bar{y})}{\partial(A_1,B_1,A_2,B_2)}=\begin{vmatrix} \dfrac{\partial x}{\partial A_1} & \dfrac{\partial x}{\partial B_1} & \dfrac{\partial x}{\partial A_2} & \dfrac{\partial x}{\partial B_2} \\ \dfrac{\partial\bar{x}}{\partial A_1} & \dfrac{\partial\bar{x}}{\partial B_1} & \dfrac{\partial\bar{x}}{\partial A_2} & \dfrac{\partial\bar{x}}{\partial B_2} \\ \varrho_1\dfrac{\partial x}{\partial A_1} & \varrho_1\dfrac{\partial x}{\partial B_1} & \varrho_2\dfrac{\partial x}{\partial A_2} & \varrho_2\dfrac{\partial x}{\partial B_2} \\ \varrho_2\dfrac{\partial\bar{x}}{\partial A_1} & \varrho_2\dfrac{\partial\bar{x}}{\partial B_1} & \varrho_1\dfrac{\partial\bar{x}}{\partial A_2} & \varrho_1\dfrac{\partial\bar{x}}{\partial B_2} \end{vmatrix}=-\frac{\partial(x,\bar{x})}{\partial(A_1,B_1)}\cdot\frac{\partial(x,\bar{x})}{\partial(A_2,B_2)}(\varrho_1-\varrho_2)^2,$$

und wenn zur Abkürzung

$$x_\alpha = \frac{\partial x(\alpha,\beta)}{\partial \alpha}, \quad y_\alpha = \frac{\partial y(\alpha,\beta)}{\partial \alpha}, \ldots$$

stehen,

$$(14) \qquad \frac{\partial(x,\bar{x})}{\partial(A_1,B_1)} = \frac{-1}{\varrho_1-\varrho_2}(x_\alpha y_\beta - x_\beta y_\alpha),$$

$$\frac{\partial(x,\bar{x})}{\partial(A_2,B_2)} = \frac{\partial(\bar{x},x)}{\partial(A_1,B_1)} = \frac{1}{\varrho_1-\varrho_2}(x_\alpha y_\beta - x_\beta y_\alpha).$$

Man stellt übrigens leicht auch fest

$$(15) \qquad \frac{\partial(x_2,y_2)}{\partial(A_1,B_1)} = -\frac{1}{4}(x_\alpha y_\beta - x_\beta y_\alpha), \quad \frac{\partial(x_2,y_2)}{\partial(A_2,B_2)} = -\frac{1}{4}(x_\alpha y_\beta - x_\beta y_\alpha).$$

Damit ist also für die Punkte der Anfangskurve C, d. h. für die Gerade $\beta_1=\beta_2=0$ der Ebene $\alpha_1=\alpha_2$, $\beta_1=\beta_2$ das Nichtverschwinden der Determinante (13) (als Folge des elliptischen Charakters von (1), der in den Beziehungen $\varrho_1=\bar{\varrho}_2 \neq \varrho_2$ benutzt wird) nachgewiesen und ist dann wegen der Stetigkeit der Ableitungen nach $\alpha_1\,\beta_1\,\alpha_2\,\beta_2$ auch in hinreichender Nähe dieser Punkte gesichert.

Weiter folgt aus den vorletzten beiden der obigen Gleichungen, daß man auf den Flächen $\alpha_1=$ konst., $\beta_1=$ konst. und $\alpha_2=$ konst., $\beta_2=$ konst. x_1 und x_2 als unabhängige Veränderliche einführen kann.

§ 2.

Analytischer Charakter der Funktionen $u(x,y)$, $p(x,y)$, ... [5]).

Eine Ableitung $\frac{\partial}{\partial A_1}, \frac{\partial}{\partial A_2}, \frac{\partial}{\partial B_1}, \frac{\partial}{\partial B_2}$ kann nunmehr als lineare Kombination der Ableitungen $\frac{\partial}{\partial x_1}, \frac{\partial}{\partial x_2}, \frac{\partial}{\partial y_1}, \frac{\partial}{\partial y_2}$ geschrieben werden. Das charakteristische Gleichungssystem gilt für $\alpha_1=\alpha_2$, $\beta_1=\beta_2$ (vgl. S. 331) mit den folgenden Bedeutungen der Operatoren

$$' = \frac{\partial}{\partial A_1}, \quad ' = \frac{\partial}{\partial B_1}; \qquad ` = \frac{\partial}{\partial A_2}, \quad ` = \frac{\partial}{\partial B_2}$$

und auch noch mit

$$' = \frac{\partial}{\partial x_1} + \varrho_1 \frac{\partial}{\partial y_1}, \qquad ` = \frac{\partial}{\partial x_1} + \varrho_2 \frac{\partial}{\partial y_1}$$

(vgl. (6); man beachte $x_2=0$, $y_2=0$).

Es wird nun für $\alpha_1=\alpha_2$, $\beta_1=\beta_2$

$$\frac{\partial}{\partial A_1} = K_1\frac{\partial}{\partial x_1} + K_2\frac{\partial}{\partial y_1} + K_3\left(\frac{\partial}{\partial x_1} + i\frac{\partial}{\partial x_2}\right) + K_4\left(\frac{\partial}{\partial y_1} + i\frac{\partial}{\partial y_2}\right),$$

[5]) Vgl. hierzu Hans Lewy, „Neuer Beweis des analytischen Charakters", Math. Annalen **101**, S. 615 ff. Die obigen Überlegungen wiederholen im wesentlichen das dort Gesagte für unsern Fall, sind daher kurz gefaßt.

und wegen Gleichung (9) folgt $\varrho_1 K_1 = K_2$, wobei $K_1 \neq 0$ ist, weil sonst der reelle Operator $\frac{\partial}{\partial A_1}$ notwendig komplex würde; ähnlich ist

$$\frac{\partial}{\partial B_1} = L_1 \frac{\partial}{\partial x_1} + L_2 \frac{\partial}{\partial y_1} + L_3 \left(\frac{\partial}{\partial x_1} + i \frac{\partial}{\partial x_2}\right) + L_4 \left(\frac{\partial}{\partial y_1} + i \frac{\partial}{\partial y_2}\right)$$

mit $\varrho_1 L_1 = L_2$, $L_1 \neq 0$.

Analog kommt

$$\frac{\partial}{\partial A_2} = M_1 \frac{\partial}{\partial x_1} + \varrho_1 M_1 \frac{\partial}{\partial y_1} + M_3 \left(\frac{\partial}{\partial x_1} + i \frac{\partial}{\partial x_2}\right) + M_4 \left(\frac{\partial}{\partial y_1} + i \frac{\partial}{\partial y_2}\right), \quad M_1 \neq 0,$$

$$\frac{\partial}{\partial B_2} = N_1 \frac{\partial}{\partial x_1} + \varrho_1 N_1 \frac{\partial}{\partial y_1} + N_3 \left(\frac{\partial}{\partial x_1} + i \frac{\partial}{\partial x_2}\right) + N_4 \left(\frac{\partial}{\partial y_1} + i \frac{\partial}{\partial y_2}\right), \quad N_1 \neq 0.$$

Es folgt nun hieraus

$$\begin{array}{cc|cc} K_3 = \frac{1}{i} \frac{\partial x_2}{\partial A_1}, & K_4 = \frac{1}{i} \frac{\partial y_2}{\partial A_1} & M_3 = \frac{1}{i} \frac{\partial x_2}{\partial A_2}, & M_4 = \frac{1}{i} \frac{\partial y_2}{\partial A_2} \\ L_3 = \frac{1}{i} \frac{\partial x_2}{\partial B_1}, & L_4 = \frac{1}{i} \frac{\partial y_2}{\partial B_1} & N_3 = \frac{1}{i} \frac{\partial x_2}{\partial B_2}, & N_4 = \frac{1}{i} \frac{\partial y_2}{\partial B_2} \end{array}$$

und somit nach (15) für reelle x, y (oder $\alpha_1 = \alpha_2$, $\beta_1 = \beta_2$) und daher für eine gewisse Umgebung infolge der Stetigkeit

$$K_3 L_4 - K_4 L_3 \neq 0, \quad M_3 N_4 - M_4 N_3 \neq 0.$$

Wir schreiben jetzt die charakteristischen Gleichungen in der Form

$$\sum_{k=1}^{8} a_{ik} \chi'_k = 0 \qquad (i = 1, \ldots, 6),$$

$$\sum_{k=1}^{8} a_{ik} \chi^{`}_k = 0 \qquad (i = 7, 8),$$

dabei vertreten die χ_k die Funktionen x, y, u, p, q, r, s, t, und die a_{ik} sind die entsprechenden Koeffizienten der Gleichungen (4), deren Reihenfolge der einfacheren Schreibweise wegen etwas abgeändert erscheint; es ist, wie schon seinerzeit bemerkt wurde, $|a_{ik}| \neq 0$. Deuten wir wieder den Operator $'$ als $\frac{\partial}{\partial A_1}$ und $`$ als $\frac{\partial}{\partial A_2}$, und subtrahieren wir von dem so entstehenden System (4) beziehungsweise dieselben (mit K_1 resp. M_1 multiplizierten) Gleichungen in der Bedeutung (6) der Operatoren, so kommt das für $\alpha_1 = \alpha_2$, $\beta_1 = \beta_2$ gültige Resultat

$$\sum_k a_{ik} \left[K_3 \left(\frac{\partial}{\partial x_1} + i \frac{\partial}{\partial x_2}\right) + K_4 \left(\frac{\partial}{\partial y_1} + i \frac{\partial}{\partial y_2}\right)\right] \chi_k = 0 \quad (i = 1, \ldots, 6),$$

$$\sum_k a_{ik} \left[M_3 \left(\frac{\partial}{\partial x_1} + i \frac{\partial}{\partial x_2}\right) + M_4 \left(\frac{\partial}{\partial y_1} + i \frac{\partial}{\partial y_2}\right)\right] \chi_k = 0 \quad (i = 7, 8),$$

ähnlich erhält man auch

$$\sum_k a_{ik}\left[L_3\left(\frac{\partial}{\partial x_1}+i\frac{\partial}{\partial x_2}\right)+L_4\left(\frac{\partial}{\partial y_1}+i\frac{\partial}{\partial y_2}\right)\right]\chi_k=0 \quad (i=1,\ldots,6),$$

$$\sum_k a_{ik}\left[N_3\left(\frac{\partial}{\partial x_1}+i\frac{\partial}{\partial x_2}\right)+N_4\left(\frac{\partial}{\partial y_1}+i\frac{\partial}{\partial y_2}\right)\right]\chi_k=0 \quad (i=7,8).$$

Dies sind lineare Gleichungen für $\left(\frac{\partial}{\partial x_1}+i\frac{\partial}{\partial x_2}\right)\chi_k$ und $\left(\frac{\partial}{\partial y_1}+i\frac{\partial}{\partial y_2}\right)\chi_k$, sie besitzen wegen $K_3L_4-K_4L_3\neq 0$ und $M_3N_4-M_4N_3\neq 0$ die einzige Lösung

$$\nabla_x\chi_k\equiv\left(\frac{\partial}{\partial x_1}+i\frac{\partial}{\partial x_2}\right)\chi_k=0,\qquad \nabla_y\chi_k\equiv\left(\frac{\partial}{\partial y_1}+i\frac{\partial}{\partial y_2}\right)\chi_k=0.$$

D. h. die Cauchy-Riemannschen Differentialgleichungen gelten jedenfalls für $\alpha_1=\alpha_2$, $\beta_1=\beta_2$.

Es folgt nunmehr wie in der zitierten Arbeit durch Differentiation des Systems der charakteristischen Differentialgleichungen, daß die Größen $\nabla_x\chi_k=\left(\frac{\partial}{\partial x_1}+i\frac{\partial}{\partial x_2}\right)\chi_k$ und $\nabla_y\chi_k=\left(\frac{\partial}{\partial y_1}+i\frac{\partial}{\partial y_2}\right)\chi_k$ ein linear homogenes hyperbolisches Anfangswertproblem mit nicht verschwindender Determinante lösen, das als *eine* Lösung sicher $\nabla_x\chi_k\equiv 0$, $\nabla_y\chi_k\equiv 0$ zuläßt. Die Anwendung des Eindeutigkeitssatzes für die hyperbolischen Anfangswertprobleme liefert, daß notwendig somit $\nabla_x\chi_k\equiv\nabla_y\chi_k\equiv 0$ gilt; und zwar in dem ganzen vierdimensionalen Bereich, in dem man sich durch den angegebenen Fortsetzungsprozeß die Lösungen χ_k beschafft hat. Diese sind daher dort analytische Funktionen der beiden komplexen Veränderlichen x und y.

§ 3.

Beweis der Eindeutigkeit der Lösung von $F=0$.

Während die Ausführungen des vorigen Paragraphen, die dem Nachweis des analytischen Charakters der besorgten Fortsetzung der Lösung ins Komplexe dienten, von der speziellen Konstruktion dieser Fortsetzung keinen Gebrauch machten, kommen wir nun zu einer neuen Betrachtung, die, die Kenntnis der Lösung auf der Kurve C wesentlich benutzend, unsere Funktionen im vierdimensionalen Bereich auf eine andere Weise allmählich aufbaut und so die Eindeutigkeit der vorgelegten Lösung $u(x, y)$ von $F=0$ ergeben wird. Dabei wird sich auch die Bedeutung einiger früher eingeführter Bezeichnungen, soweit die vorangehenden Überlegungen dies nicht schon taten, motivieren.

Ein erster Schritt ist der Nachweis, daß die Flächen $\alpha_1=$ konst., $\beta_1=$ konst. charakteristische Flächen im Sinne der regulären Funktionen zweier komplexen Veränderlichen sind, wie er in einer Arbeit von Levi-Civita[6]) erklärt wurde, ebenso die Flächen $\alpha_2 = \text{konst.}$, $\beta_2 = \text{konst.}$ Genauer: die Flächen $\alpha_1 = \text{konst.}$, $\beta_1 = \text{konst.}$ bzw. $\alpha_2 = \text{konst.}$, $\beta_2 = \text{konst.}$ lassen sich in die Gestalt

$$y = f(x)$$

setzen, wo f eine analytische Funktion des komplexen Arguments x ist.

Nehmen wir eine Fläche $\alpha_1 = \text{konst.}$, $\beta_1 = \text{konst.}$ Führen wir den Winkel φ wie auf S. 331 und außerdem $R = \sqrt{(\alpha_2 - \alpha_1)^2 + (\beta_2 - \beta_1)^2}$ ein, so sind mit

$$\alpha_2 = \alpha_1 + R\cos\varphi\,, \qquad \beta_2 = \beta_1 + R\sin\varphi$$

natürlich auch $x(\alpha_1, \beta_1; \alpha_2, \beta_2), \ldots, t(\alpha_1, \beta_1; \alpha_2, \beta_2)$ zweimal stetig differenzierbare Funktionen von R und φ, die für $R = 0$ offenbar konstant in φ sind. (4_2) nimmt nun nach S. 331 (*) die Gestalt

$$y' - \varrho_2 x' = \frac{\partial y}{\partial R} - \varrho_2 \frac{\partial x}{\partial R} = 0$$

an. Für $R = 0$ ist der Ausdruck

$$A \equiv \frac{\partial y}{\partial \varphi} - \varrho_2 \frac{\partial x}{\partial \varphi}$$

null; wir zeigen, daß A einer linear homogenen Differentialgleichung nach R genügt, womit auch für $R \neq 0$

$$A \equiv \frac{\partial y}{\partial \varphi} - \varrho_2 \frac{\partial x}{\partial \varphi} \equiv 0$$

nachgewiesen sein wird.

Es ist nämlich

$$\frac{\partial \varrho_2}{\partial \varphi} = \frac{\partial x}{\partial \varphi}\left(\frac{\partial \varrho_2}{\partial x} + \frac{\partial \varrho_2}{\partial u}\frac{\partial u}{\partial x} + \ldots\right) + \frac{\partial y}{\partial \varphi}\left(\frac{\partial \varrho_2}{\partial y} + \frac{\partial \varrho_2}{\partial u}\frac{\partial u}{\partial y} + \ldots\right),$$

$$\begin{aligned}\frac{\partial \varrho_2}{\partial R} &= \frac{\partial x}{\partial R}\left(\frac{\partial \varrho_2}{\partial x} + \frac{\partial \varrho_2}{\partial u}\frac{\partial u}{\partial x} + \ldots\right) + \frac{\partial y}{\partial R}\left(\frac{\partial \varrho_2}{\partial y} + \frac{\partial \varrho_2}{\partial u}\frac{\partial u}{\partial y} + \ldots\right)\\ &= \frac{\partial x}{\partial R}\left(\frac{\partial \varrho_2}{\partial x} + \frac{\partial \varrho_2}{\partial u}\frac{\partial u}{\partial x} + \ldots\right) + \varrho_2\frac{\partial x}{\partial R}\left(\frac{\partial \varrho_2}{\partial y} + \frac{\partial \varrho_2}{\partial u}\frac{\partial u}{\partial y} + \ldots\right);\end{aligned}$$

also

$$\begin{aligned}\frac{\partial A}{\partial R} &= \frac{\partial}{\partial R}\left(\frac{\partial y}{\partial \varphi} - \varrho_2\frac{\partial x}{\partial \varphi}\right) = \frac{\partial}{\partial \varphi}\left(\frac{\partial y}{\partial R} - \varrho_2\frac{\partial x}{\partial R}\right) + \frac{\partial \varrho_2}{\partial \varphi}\frac{\partial x}{\partial R} - \frac{\partial \varrho_2}{\partial R}\frac{\partial x}{\partial \varphi}\\ &= \frac{\partial x}{\partial R}\left(\frac{\partial \varrho_2}{\partial y} + \frac{\partial \varrho_2}{\partial u}\frac{\partial u}{\partial y} + \ldots\right)\left(\frac{\partial y}{\partial \varphi} - \varrho_2\frac{\partial x}{\partial \varphi}\right) = \text{Faktor} \times A,\end{aligned}$$

q. e. d.

---

[6]) Levi-Civita, Sulle funzioni di due o più variabili complesse. Rend. Acc. Linc. **14**, pp. 492—99.

Man bekommt somit für $\alpha_1, \beta_1 \neq \alpha_2, \beta_2$

$$\frac{\partial y}{\partial \alpha_2} - \varrho_2 \frac{\partial x}{\partial \alpha_2} = \frac{\partial y}{\partial \beta_2} - \varrho_2 \frac{\partial x}{\partial \beta_2} = 0,$$

eine Relation, deren Gültigkeit für $\alpha_1 = \alpha_2$, $\beta_1 = \beta_2$ früher (S. 331, (11) (12)) bewiesen wurde. Daraus folgt, wenn man an Stelle von α_2, β_2 die unabhängigen Variablen x_1, x_2 einführt, was ja nach § 1 (Ende) möglich ist,

$$\frac{d}{d x_1} y - \varrho_2 \frac{d}{d x_1} x = 0,$$

$$\frac{d}{d x_2} y - \varrho_2 \frac{d}{d x_2} x = 0,$$

also, wegen $\left(\frac{d}{d x_1} + i \frac{d}{d x_2}\right) x = 0$,

$$\left(\frac{d}{d x_1} + i \frac{d}{d x_2}\right) y = 0,$$

und dies ist gerade die Behauptung.

Somit schreiben sich auch u, p, q, r, s, t auf der Fläche $\alpha_1 =$ konst., $\beta_1 =$ konst. als analytische Funktionen des (unabhängigen) Argumentes $x_1 + i x_2$. Denn man hat

$$\left(\frac{d}{d x_1} + i \frac{d}{d x_2}\right) u = \left(\frac{\partial}{\partial x_1} + i \frac{\partial}{\partial x_2}\right) u + \frac{\partial u}{\partial y}\left(\frac{d}{d x_1} + i \frac{d}{d x_2}\right) y = 0$$

nach § 2 und dem soeben Bewiesenen.

Ich wähle nun für α_1, β_1 zunächst einen Punkt der Anfangskurve $\beta = 0$. Von den Funktionen $x(\alpha_1, 0; \alpha_2, \beta_2)$, $y(\alpha_1, 0; \alpha_2, \beta_2)$, $u(\alpha_1, 0; \alpha_2, \beta_2), \ldots$ kennen wir eindeutig die Werte für $\beta_2 = 0$, da wir die Werte der $x(\alpha, \beta)$, $y(\alpha, \beta)$, $u(\alpha, \beta), \ldots$ für $\beta = 0$ kennen. Um dies einzusehen, genügt es, sich an die Definition der Funktionen der vier Argumente α_1, β_1, α_2, β_2 zu erinnern. Die den verschiedenen Werten von α_2 entsprechenden Funktionswerte

$$x_1 + i x_2 = x(\alpha_1, 0; \alpha_2, 0)$$

bilden in der Fläche $\alpha_1 =$ konst., $\beta_1 =$ konst. $= 0$, in der x_1 und x_2 als unabhängige Variable genommen sein mögen, eine Kurve K (das Bild der Geraden $\beta_2 = 0$ einer α_2, β_2-Ebene bei der Parameterdarstellung durch α_2, β_2). Welches nun auch die Funktionen $x, y, u, p, q, r, s, t\,(\alpha, \beta)$ sonst sein mögen; da sie für $\beta = 0$ übereinstimmen, so stimmen sie[7]) auf der Geraden $\alpha_1 =$ konst., $\beta_1 = 0$, $\beta_2 = 0$ überein, weil sie dort Lösungen desselben Anfangswertproblems sind. Weiter sind sie, als Funktionen der komplexen Variablen $x = x_1 + i x_2$ in der Fläche $\alpha_1 =$ konst., $\beta_1 = 0$ betrachtet, in einer gewissen Halbumgebung der Kurve K regulär, sind

[7]) bzw. ihre oben konstruierten Fortsetzungen.

auf K noch stetig und nehmen daher bei Annäherung an K gleichmäßig gewisse (soeben als eindeutig erkannte) Randwerte an. Sie werden also auf der ganzen Fläche $\alpha_1 =$ konst., $\beta_1 = 0$ übereinstimmen, vermöge des in der Einleitung zitierten Satzes über reguläre Funktionen eines komplexen Arguments. Das gibt die Eindeutigkeit der Funktionen

$$x(\alpha_1, 0; \alpha_2, \beta_2), \quad y(\alpha_1, 0; \alpha_2, \beta_2), \quad u(\alpha_1, 0; \alpha_2, \beta_2), \ldots$$

Ähnlich wird dasselbe für

$$x(\alpha_1, \beta_1; \alpha_2, 0), \quad y(\alpha_1, \beta_1; \alpha_2, 0), \quad u(\alpha_1, \beta_1; \alpha_2, 0), \ldots$$

bewiesen.

Nunmehr ist es leicht, die Eindeutigkeit in dem Umfange zu beweisen, wie sie in der Einleitung behauptet wurde. Um nämlich zu zeigen, daß die Funktionen $x(\alpha, \beta; \alpha, \beta)$, $y(\alpha, \beta; \alpha, \beta)$, $u(\alpha, \beta; \alpha, \beta), \ldots$ für hinreichend kleine Werte α, β eindeutig bestimmt sind, verbinde man in der α, β-Ebene den Punkt α, β gradlinig mit einem Punkte der Anfangskurve, etwa mit $\alpha_1, \beta_1 = 0$. Die Werte

$$x(0, 0; \mu\alpha, \mu\beta), \quad y(0, 0; \mu\alpha, \mu\beta), \ldots \qquad (1 \geqq \mu \geqq 0)$$

und

$$x(\lambda\alpha, \lambda\beta; 0, 0), \quad y(\lambda\alpha, \lambda\beta; 0, 0), \ldots \qquad (1 \geqq \lambda \geqq 0)$$

sind bereits eindeutig festgelegt, wie soeben gezeigt. Allgemein gehorchen die Funktionen von λ und μ

$$x(\lambda\alpha, \lambda\beta; \mu\alpha, \mu\beta), \ldots$$

dem Differentialgleichungssystem (4) mit der Festsetzung

$$' = \frac{\partial}{\partial\lambda} \qquad ` = \frac{\partial}{\partial\mu}$$

(vgl. S. 331, (8'')). Man hat hier demnach ein Anfangswertproblem vorliegen, in dem die gesuchten Funktionen auf den beiden Charakteristiken $\mu = 0$ und $\lambda = 0$, $1 \geqq \lambda \geqq 0$ bzw. $1 \geqq \mu \geqq 0$ bekannt sind. Ihre Werte sind dann nach dem allgemeinen Satze über solche hyperbolische Differentialgleichungen in dem ganzen Quadrate $0 \leqq \mu \leqq 1$, $0 \leqq \lambda \leqq 1$ eindeutig bestimmt; für den Punkt $\lambda = 1$, $\mu = 1$ bekommt man speziell die Eindeutigkeit von

$$x(\alpha, \beta; \alpha, \beta), \quad y(\alpha, \beta; \alpha, \beta), \ldots$$

Dies ist genau das Ziel der Untersuchung; denn wenn man wieder auf der Fläche $\alpha_1 = \alpha_2$, $\beta_1 = \beta_2$ die reellen Veränderlichen x und y einführt, so heißt das, die Funktionen

$$u(x, y), \quad p(x, y), \ldots$$

sind durch ihre Anfangswerte auf der Kurve C eindeutig festgelegt.

Bemerkung: Wünscht man eine genauere Festsetzung der in dem Beweise auftretenden *Umgebungen,* so überzeugt man sich leicht von folgendem: Da es zu jedem inneren Punkt P der Kurve C eine gewisse Halbumgebung gibt, in der zwei Lösungen der Gleichung (1), die auf ihre Übereinstimmung zu untersuchen sind, existieren, so gibt es auch in der α, β-Ebene einen von $\beta = 0$ begrenzten Halbkreis mit dem Bilde jenes Punktes P als Mittelpunkt, so daß dort beide Lösungen existieren; wählt man einen genügend kleinen konzentrischen Halbkreis, so sind dort alle Prozesse der Fortsetzung für beide Lösungen erlaubt; desgleichen alle weiteren Schlüsse betreffend die Eindeutigkeit der Lösungen der vorkommenden hyperbolischen Anfangswertprobleme; die diesbezüglichen Verhältnisse sind ähnlich denen bei gewöhnlichen Differentialgleichungen.

(Eingegangen am 20. 5. 1930.)

Nachrichten der Gesellschaft der Wissenschaften zu Göttingen
1932

Über fortsetzbare Anfangsbedingungen bei hyperbolischen Differentialgleichungen in drei Veränderlichen.

Von

Kurt Friedrichs in Braunschweig und **Hans Lewy** in Göttingen.

Vorgelegt von R. Courant in der Sitzung am 26. Februar 1932.

Diese Note handelt von den Anfangsbedingungen, welche der Lösung einer hyperbolischen Differentialgleichung für eine Funktion von drei Variablen auferlegt werden. Wir beschränken uns der größeren Einfachheit zuliebe auf die Behandlung der Differentialgleichung

$$(1) \qquad \frac{\partial^2 u}{\partial t^2} - \frac{\partial^2 u}{\partial x^2} - \frac{\partial^2 u}{\partial y^2} = 0$$

für eine Funktion $u(t, x, y)$, die auf der Ebene $t = 0$ die Werte $u(0, x, y) = \mathfrak{u}(x, y)$, deren Ableitung nach t dort die Werte $\frac{\partial u}{\partial t}(0, x, y) = \mathfrak{u}_t(x, y)$ annimmt; $\mathfrak{u}, \mathfrak{u}_t$ sind die vorgegebenen „Anfangswerte"; wir setzen noch $\mathfrak{u}_x = \frac{\partial}{\partial x}\mathfrak{u}(x, y)$, $\mathfrak{u}_y = \frac{\partial}{\partial y}\mathfrak{u}(x, y)$.

Die Lösung dieses Problems liefert bekanntlich die „Volterrasche Formel"

$$(2) \qquad u(t,x,y) = \frac{1}{2\pi t}\iint \frac{1}{\sqrt{\Gamma}}\{\mathfrak{u} + t\,\mathfrak{u}_t + (X-x)\,\mathfrak{u}_x + (Y-y)\,\mathfrak{u}_y\}\,dX\,dY.$$

Dabei ist $\Gamma = t^2 - (X-x)^2 - (Y-y)^2$; die Argumente von $\mathfrak{u}$ und $\mathfrak{u}_t, \mathfrak{u}_x, \mathfrak{u}_y$ sind X, Y; die Integration nach diesen Variablen X, Y erstreckt sich über den Kreis $\Gamma \geqq 0$. Dieser Kreis wird vom „charakteristischen" Kegel durch t, x, y aus der Ebene $t = 0$ ausgeschnitten; er stellt das „Abhängigkeitsgebiet" des Punktes t, x, y dar. Mit einem von t, x, y unabhängigen Integrationsgebiet schreibt sich die Formel

$$u(t, x, y)$$

$$(3) \quad = \frac{1}{2\pi} \int_0^{2\pi} \int_0^1 \frac{1}{\sqrt{1-\varrho^2}} [\mathfrak{u} + t\mathfrak{u}_t + t\varrho \cos\vartheta\, \mathfrak{u}_x + t\varrho \sin\vartheta\, \mathfrak{u}_y]\, \varrho\, d\varrho\, d\vartheta,$$

wo die Argumente von $\mathfrak{u}, \mathfrak{u}_t, \mathfrak{u}_x, \mathfrak{u}_y$ sind: $x + t\varrho \cos\vartheta$ und $y + t\varrho \sin\vartheta$.

In diese Darstellung der Lösung u gehen nun auch die ersten Ableitungen der Anfangswerte von u ein. In den durch Differention nach t, x oder y aus (3) entstehenden Darstellungen für die Ableitungen n-ter Ordnung treten die Ableitungen $(n+1)$-ter Ordnung der Anfangswerte auf.

Dieses — für hyperbolische Differentialgleichungen in drei und mehr Variabeln typische — Verhalten der VOLTERRAschen Lösung ist ein Anzeichen für folgende Tatsache: Ist von den Anfangswerten $\mathfrak{u}$ bezw. $\mathfrak{u}_t$ bekannt, daß sie dreimal bezw. zweimal stetig nach x, y differenzierbar sind, so gibt es wohl eine Lösung der Gleichung (1), welche auch außerhalb der Ebene $t = 0$ zweimal stetig differenzierbar ist. Aber dritte Ableitungen brauchen dann nicht überall zu existieren, wie HADAMARD bewiesen hat[1]). Anderseits ist das Interesse, die Lösung auch für nichtanalytische Anfangsdaten zu untersuchen, wohlmotiviert, wie vor allem von HADAMARD (l. c. p. 33) hervorgehoben wurde. Denn gerade die physikalisch wichtigen Konsequenzen der Differentialgleichung (1), insbesondere das Phänomen der Wellenausbreitung, können nicht durch analytische Funktionen $u(t, x, y)$ beschrieben werden.

Das geschilderte Verhalten der Ableitungen führt nun zu Unzuträglichkeiten bei der Konstruktion der Lösung nach dem „HUYGENS'schen Prinzip im weiteren Sinne". Dieses Prinzip verlangt, daß die Lösung auf der Ebene $t = t_2 > 0$, anstatt durch unmittelbare Konstruktion von $t = 0$ aus, sich auch dadurch konstruieren läßt, daß von der Ebene $t = 0$ aus zunächst die Lösung auf $t = t_1$ $(0 < t_1 < t_2)$ und dann von der neuen Anfangsebene $t = t_1$ aus die Lösung auf $t = t_2$ konstruiert wird. Soll nun die Lösung auf $t = t_2$ stetige zweite Ableitungen besitzen, so muß man, um die VOLTERRA'sche Formel anzuwenden, auf $t = t_1$ etwa stetige dritte Ableitungen voraussetzen können und dementsprechend auf $t = 0$ vierte Ableitungen; bei der unmittelbaren Konstruktion von $t = 0$ aus braucht man aber diese Voraussetzung nicht zu machen. Diese Unzuträglichkeit steigert sich noch im Falle der Gleichungen mit nicht konstanten Koeffizienten und im Falle nicht

1) J. HADAMARD, Lectures on CAUCHY's Problem p. 131—132.

2) J. HADAMARD, l. c. p. 53.

linearer Gleichungen, wo die Lösung zunächst überhaupt nur in einer genügend kleinen Umgebung einer Anfangsfläche gebildet werden kann. Man wird dann die Lösung in ihrem ganzen Existenzgebiet dadurch konstruieren wollen, daß man ihre Existenz zunächst in einer Umgebung der Anfangsfläche gewährleistet und schrittweise von einer Anfangsfläche zu einer nahegelegenen neuen übergeht. Aber je nach der Anzahl der Schritte muß man immer höhere Differenzierbarkeiten am Anfang verlangen, was der Natur der Sache widerspricht.

Der eigentliche durch die Differentialgleichung bestimmte Fortsetzungsprozeß wird aber offenbar erst dann treffend beschrieben sein, wenn es gelingt, solche Bedingungen für die Anfangswerte zu stellen, die sich beim Übergang zu einer neuen Anfangsfläche erhalten.

Solche „fortsetzbare" Bedingungen, in die nur Ableitungen beschränkter Ordnung eingehen, sollen nun im Folgenden angegeben werden. Und zwar handelt es sich übrigens um Bedingungen, welche diesen Charakter für jede — auch für nicht lineare — hyperbolische Differentialgleichungen besitzen, wie in einer demnächst erscheinenden Arbeit gezeigt wird[3]). Diese Bedingungen besagen zwar mehr als zweimalige stetige Differenzierbarkeit, enthalten aber nicht dreimalige Differenzierbarkeit der Anfangswerte. Es sind nicht Bedingungen im Kleinen, sondern ihre Formulierung bezieht sich auf ein festes Gebiet. Man kommt automatisch auf diese Bedingungen von einer „Energieintegralumformung" her, die wir schon in früheren Arbeiten zum Nachweis der Eindeutigkeit und Existenz bei linearen hyperbolischen Gleichungen herangezogen haben.

Bedingungen.

Es sei **E** ein konvexer abgeschlossener Bereich auf einer Ebene mit den rechtwinkligen Koordinaten ξ, η. Durch Operatoren Δ; Δ^2; ... werden die Differentiationen $\frac{\partial}{\partial \xi}, \frac{\partial}{\partial \eta}; \frac{\partial^2}{\partial \xi^2}, \frac{\partial^2}{\partial \eta \partial \xi}, \frac{\partial^2}{\partial \eta^2}; \ldots$ gekennzeichnet. Wir formulieren eine Eigenschaft A einer stetigen Funktion $w(\xi, \eta)$ auf **E**.

---

3) K. Friedrichs, Lösung des Anfangsproblems und Eindeutigkeit bei einer beliebigen hyperbolischen Differentialgleichung zweiter Ordnung in mehr als zwei Variablen, Math. Ann.

$\underline{A}|$ Die stetige Funktion $w(\xi, \eta)$ läßt sich gleichmäßig in $\mathbf{E}$ approximieren durch zweimal stetige differenzierbare Funktionen $w_\nu(\xi, \eta)$, so daß für geeignete $\mathbf{M}, \mathbf{N} > 0$

$$\iint\limits_{\mathbf{E}} [\Delta w_\nu]^2 \, d\xi \, d\eta \leqq \mathbf{M} \tag{4}$$

und

$$\iint\limits_{\mathbf{E}} [\Delta^2 w_\nu]^2 \, d\xi \, d\eta \leqq \mathbf{N} \tag{5}$$

gilt für jede der durch Δ, Δ^2 bezeichneten Ableitungen; dabei hängen $\mathbf{M}$ und $\mathbf{N}$ nicht von ν ab[4]).

Eine zweite Eigenschaft formulieren wir so:

$\underline{A_n}|$ Die in $\mathbf{E}$ erklärte Funktion $v(\xi, \eta)$ besitzt stetige Ableitungen n-ter Ordnung; sie läßt sich zugleich mit ihren Ableitungen bis zur n-ten Ordnung gleichmäßig in $\mathbf{E}$ approximieren durch Funktionen v_ν, welche stetige Ableitungen bis zur $n+2$-ten Ordnung besitzen. Es ist für die n-ten Ableitungen von v_ν die Eigenschaft A mit dieser Approximation erfüllt, d. h. also, es gilt

$$\iint\limits_{\mathbf{E}} [\Delta^{n+1} v_\nu]^2 \, d\xi \, d\eta \leqq \mathbf{M} \tag{4$_n$}$$

$$\iint\limits_{\mathbf{E}} [\Delta^{n+2} v_\nu]^2 \, d\xi \, d\eta \leqq \mathbf{N}. \tag{5$_n$}$$

Δ^{n+1}, Δ^{n+2} kennzeichnen sämtliche Ableitungen der $n+1$-ten und $n+2$-ten Ordnung.

Wir bemerken den folgenden elementaren

1. Hilfssatz. Besitzt v die Eigenschaft A_n, so läßt sich die Approximation auch so wählen, daß die v_ν Polonyme sind.

Man braucht nämlich nur jedes der gegebenen v_ν nach WEIERSTRASS genügend stark durch ein Polynom v_ν^* anzunähern derart, daß zugleich die Ableitungen bis zur $n+2$-ten Ordnung gleichmäßig mit approximiert werden, was bekanntlich möglich ist. Wir verwenden dann den folgenden trivialen

---

4) Es ist nicht schwer, zu erkennen, daß man anstelle der obigen Bedingungen auch andere angeben kann, insbesondere solche, die sich mit Hilfe von Begriffen der LESBESGUE'schen Theorie ausdrücken. — Übrigens ist (4) eine Folge von (5); das lehrt die Ungleichung

$$\iint\limits_{\mathbf{E}} \sum (\Delta w)^2 \, d\xi \, d\eta \leqq \text{const.} \iint\limits_{\mathbf{E}} \{\sum (\Delta^2 w)^2 + w^2\} \, d\xi \, d\eta,$$

wobei die Konstante von w nicht abhängt. Vgl. etwa K. FRIEDRICHS, „Die Rand- und Eigenwertprobleme der elastischen Platten", Math. Ann. **98** (1927), S. 247, (**64**). Wir werden dies aber nicht benutzen.

2. Existenzsatz. Sind $\mathfrak{u}(x, y)$ und $\mathfrak{u}_t(x, y)$ Polynome, so gibt es ein Polynom $u(t, x, y)$, das der Gleichung (1) genügt und für das $u(0, x, y) = \mathfrak{u}(x, y)$ und $u_t(0, x, y) = \mathfrak{u}_t(x, y)$ gilt.

Zum Beweise braucht man nur die Potenzentwicklung von $u(t, x, y)$ aufzusuchen.

Es sei das Gebiet, das wir nunmehr betrachten, ein gerader Kreiskegelstumpf K, dessen Grundfläche E der Kreis $x^2 + y^2 \leqq R^2$ der Ebene $t = 0$ ist und dessen Spitze $(t = R > 0,\ x = y = 0)$ durch eine Ebene $t = \text{const.} > 0$ abgeschnitten ist. Das Ziel dieser Note ist der folgende

3. Satz. *Es seien auf E Anfangswerte $\mathfrak{u}(x, y)$, $\mathfrak{u}_t(x, y)$ vorgegeben, welche auf E die Eigenschaft A_2 bezw. A_1 besitzen. Dann gibt es in K eine zugehörige zweimal stetig differenzierbare Lösung $u(t, x, y)$ der Gleichung* (1), *für welche auf jeder raumartigen Ebene in K u die Eigenschaft A_2, $\dfrac{\partial u}{\partial n}$ die Eigenschaft A_1 besitzt*[5]).

Dabei bezeichnen wir eine Ebene als raumartig, wenn sie gegen die x, y-Achse um weniger als $\dfrac{\pi}{4}$ geneigt ist; $\dfrac{\partial u}{\partial n}$ ist die Ableitung in Normalrichtung.

Beweis. Die Funktionen $\mathfrak{u}$ und $\mathfrak{u}_t$ sind, da sie die Eigenschaft A_2 bezw. A_1 besitzen, durch Funktionen $\mathfrak{u}_\nu$, $\mathfrak{u}_{t\nu}$ so zu approximieren, daß für passende $M, N > 0$

$$(6)\qquad \iint\limits_E \sum_D [D^3 \mathfrak{u}_\nu]^2\, dx\, dy \leqq M$$

und

$$(7)\qquad \iint\limits_E \sum_D [D^4 \mathfrak{u}_\nu]^2\, dx\, dy \leqq N$$

gilt. Dabei bedeute $D^p \mathfrak{u}_\nu$ sämtliche p-ten Ableitungen von $\mathfrak{u}_\nu$ und $(p-1)$-ten Ableitungen von $\mathfrak{u}_{t\nu}$ nach x, y. Nach dem Hilfssatz kann angenommen werden, daß $\mathfrak{u}_\nu$, $\mathfrak{u}_{t\nu}$ Polynome sind. Dann aber ist Satz 2 anwendbar; er garantiert die Existenz von beliebig oft differenzierbaren Lösungen $u_\nu(t, x, y)$ in K.

Es sei nun $D^3 u_\nu$ irgend eine der dritten Ableitungen von u_ν nach t, x und y. Die Funktion $D^3 u_\nu$ genügt auch der Differentialgleichung (1); die Werte von $D^3 u_\nu$, $\dfrac{\partial}{\partial t} D^3 u_\nu$ auf $t = 0$ sind vermöge von (1) durch die Ableitungen von $\mathfrak{u}_\nu$, $\mathfrak{u}_{t\nu}$ nach x und y linear

5) Der obige Satz bleibt richtig, wenn man überall A_2 durch A_1, A_1 durch A ersetzt und von der Lösung nur stetige erste Ableitungen und Erfüllung einer — irgendwie — integrierten Form der Differentialgleichung verlangt.

auszudrücken. Es seien allgemein die Anfangswerte von $D^3 u_\nu$ mit $D^3 \mathrm{u}_\nu$, die Anfangswerte der ersten Ableitungen $D^4 u_\nu$ von $D^3 u_\nu$ mit $D^4 \mathrm{u}_\nu$ bezeichnet. Auch für die so neu eingeführten $D^3 \mathrm{u}_\nu$, $D^4 \mathrm{u}_\nu$ gelten dann natürlich Relationen der Form (6) und (7), eventuell mit größeren M und N, die wir dann auch in (6) und (7) selber zugrunde legen wollen.

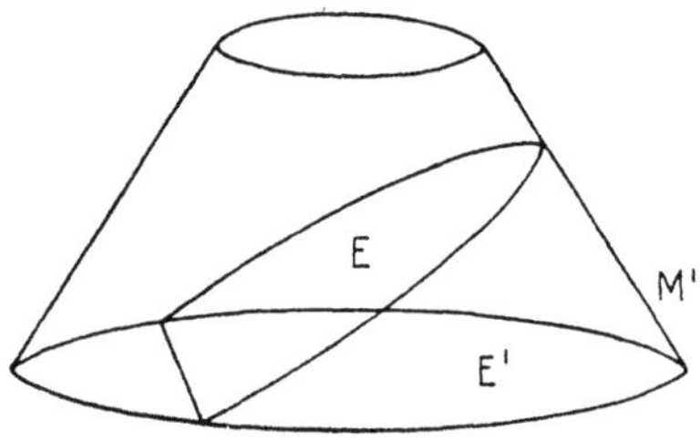

Fig. 1.

Wir wenden nun auf eine später zu spezialisierende Lösung $w(t, x, y)$ von (1) eine Energieintegralrelation an. Zu dem Zweck schneiden wir durch eine raumartige Ebene aus K ein Gebiet heraus, begrenzt von dem Anteil **E** der Ebene in K, einem Teile M' des Mantels von K und einem Teile E' des Grundkreises E. Über dieses Gebiet integrieren wir die Gleichung

$$2\frac{\partial w}{\partial t}\left(\frac{\partial^2 w}{\partial t^2} - \frac{\partial^2 w}{\partial x^2} - \frac{\partial^2 w}{\partial y^2}\right) = 0$$

und erhalten in bekannter Weise[6])

$$(8)\quad \iint\limits_{\mathbf{E}+M'} \frac{1}{\dfrac{\partial t}{\partial n}} \left[\begin{array}{c} \left(\dfrac{\partial w}{\partial t}\dfrac{\partial x}{\partial n} - \dfrac{\partial w}{\partial x}\dfrac{\partial t}{\partial n}\right)^2 + \left(\dfrac{\partial w}{\partial t}\dfrac{\partial y}{\partial n} - \dfrac{\partial w}{\partial y}\dfrac{\partial t}{\partial n}\right)^2 \\ + \left(\dfrac{\partial w}{\partial t}\right)^2\left(\left(\dfrac{\partial t}{\partial n}\right)^2 - \left(\dfrac{\partial x}{\partial n}\right)^2 - \left(\dfrac{\partial y}{\partial n}\right)^2\right) \end{array}\right] do$$

$$= \iint\limits_{E'} \left[\left(\frac{\partial w}{\partial t}\right)^2 + \left(\frac{\partial w}{\partial x}\right)^2 + \left(\frac{\partial w}{\partial y}\right)^2\right] dx\, dy.$$

Dabei ist do das Flächenelement auf M' und **E**: $\frac{\partial t}{\partial n}, \frac{\partial x}{\partial n}, \frac{\partial y}{\partial n}$ sind die Komponenten der äußeren Normalen.

Der Ausdruck

$$\left(\frac{\partial t}{\partial n}\right)^2 - \left(\frac{\partial x}{\partial n}\right)^2 - \left(\frac{\partial y}{\partial n}\right)^2$$

verschwindet auf M'; denn die Seitenlinie des Kegelstumpfes K ist um $\frac{\pi}{4}$ gegen die (x, y)-Ebene geneigt; auf **E** ist der Ausdruck aber positiv, da die Neigung von **E** geringer als $\frac{\pi}{4}$ ist. Der

6) K. FRIEDRICHS und H. LEWY, „Über die Eindeutigkeit und das Abhängigkeitsgebiet der Lösungen linearer hyperbolischer Differentialgleichungen", Math. Ann. **98** (1927) S. 195/197.

Integrand des ersten Integrals ist also nicht **kleiner** als

$$\frac{1}{C_\delta}\left[\left(\frac{\partial w}{\partial t}\right)^2+\left(\frac{\partial w}{\partial x}\right)^2+\left(\frac{\partial w}{\partial y}\right)^2\right]$$

mit derselben positiven Konstante C_δ für alle $\mathsf{E} = \mathsf{E}_\delta$, deren Neigung gegen die (x, y)-Ebene höchstens $\frac{\pi}{4} - \delta\ (\delta > 0)$ ist.

Wir verringern nun die linke Seite von (8), wenn wir den Anteil von M' weglassen und erhalten

$$\iint\limits_{\mathsf{E}_\delta}\left[\left(\frac{\partial w}{\partial t}\right)^2+\left(\frac{\partial w}{\partial x}\right)^2+\left(\frac{\partial w}{\partial y}\right)^2\right]do \leqq C_\delta \iint\limits_{E}\left[\left(\frac{\partial w}{\partial t}\right)^2+\left(\frac{\partial w}{\partial x}\right)^2+\left(\frac{\partial w}{\partial y}\right)^2\right]dx\,dy;$$

wenn wir einmal $w = D^2 u_\nu$, das andere Mal $w = D^3 u_\nu$ einsetzen [7]) und über alle Typen D^3 bezw. D^4 summieren, entstehen Ungleichungen der Form

(10) $$\iint\limits_{\mathsf{E}_\delta} \sum [D^3 u_\nu]^2\, do \leqq C_\delta \iint\limits_{E} \sum [D^3 \mathrm{u}_\nu]^2\, dx\, dy.$$

(11) $$\iint\limits_{\mathsf{E}_\delta} \sum [D^4 u_\nu]^2\, do \leqq C_\delta \iint\limits_{E} \sum [D^4 \mathrm{u}_\nu]^2\, dx\, dy.$$

Wenn wir jetzt wüßten, daß man aus den u_ν eine Teilfolge auswählen kann, die samt ersten und zweiten Ableitungen gegen eine Grenzfunktion u mit deren ersten und stetigen zweiten Ableitungen gleichmäßig konvergiert, so wäre damit unser Satz bewiesen; denn u wäre dann die eindeutig bestimmte Lösung der Differentialgleichung (1) mit den Anfangswerten u und u_t auf E, und die Formeln (10) und (11) erweisen die Gültigkeit der Eigenschaft A_2 für u und A_1 für $\frac{\partial u}{\partial n}$ auf E_δ aus ihrer Gültigkeit auf E.

Für die Möglichkeit, eine solche Teilfolge auszuwählen, berufen wir uns auf den bekannten Satz von Ascoli. Wir zeigen, daß in unserem Kegelstumpf K die zweiten Ableitungen $D^2 u_\nu$ gleichgradig stetig sind. Da ihre Werte am Anfang beschränkt sind, folgt dann nach diesem Satze, daß jede unendliche Folge der u_ν zu jedem Typus von D^2 eine Teilfolge u_μ enthält, so daß $D^2 u_\mu$ gleichmäßig in K konvergiert. Durch wiederholte Auswahl von Teilfolgen gewinnt man eine Folge u_μ, so daß alle $D^2 u_\mu$ gleichmäßig in K gegen stetige Grenzfunktionen konvergieren. Da die ersten Ableitungen $D u_\mu$ sich durch Integrale über die $D^2 u_\mu$

7) Setzt man für w eine mit ihren Ableitungen auf $t = 0$ verschwindende Lösung, so gibt (9) den soeben zitierten Beweis für die Eindeutigkeit der Lösung des Anfangsproblems.

aufbauen und da die Anfangswerte der $D u_\mu$ auf E gleichmäßig konvergieren, so streben auch die $D u_\mu$ gegen stetige Funktionen. Ähnlich folgert man gleichmäßige Konvergenz der u_μ in K gegen eine stetige Funktion u. Natürlich sind dann die Grenzfunktionen von $D^2 u_\mu$ und $D u_\mu$ die entsprechenden Ableitungen $D^2 u$ und $D u$ von u.

Die gleichgradige Stetigkeit der $D^2 u_\nu$ folgt jetzt aus den Ungleichungen (10) und (11).

Zunächst eine Hilfsüberlegung. Es heiße ein konvexer abgeschlossener Bereich einer (ξ, η)-Ebene $B = B_{r\bar{r}}$, wenn er im Inneren einen Kreis C_r vom Radius r enthält und selbst in einem Kreis $C_{\bar{r}}$ vom Radius $\bar{r}$ enthalten ist. Es sei $w(\xi, \eta)$ eine zweimal stetig differenzierbare Funktion in B. P, P_ε seien zwei Punkte vom Abstande ε[8]) in B. Dann gilt

$$(12) \qquad |w(P) - w(P_\varepsilon)| \leqq \sqrt{\varepsilon c \iint_B \{\sum [\Delta w]^2 + \sum [\Delta^2 w]^2\} \, d\xi \, d\eta},$$

wobei die Ko..stante c nur von r und $\bar{r}$ abhängt.

Zum Beweise legen wir in das Innere von C_r eine zu $P P_\varepsilon$ parallele Strecke $P' P'_\varepsilon$ auch von der Länge ε, die jedoch von $P P_\varepsilon$ einen senkrechten Abstand größer als $\frac{r}{2}$ besitzt. Das ist offensichtlich immer möglich. Der Sinus des Winkels $P' P P_\varepsilon$ besitzt eine nur von r und $\bar{r}$ abhängige positive untere Schranke. Der Abstand $PP' = l$ ist in gleicher Weise nach unten und nach oben beschränkt. Wir führen anstelle von ξ, η schiefwinklige Koordinaten σ, τ wie in der Figur ein. Dann ist

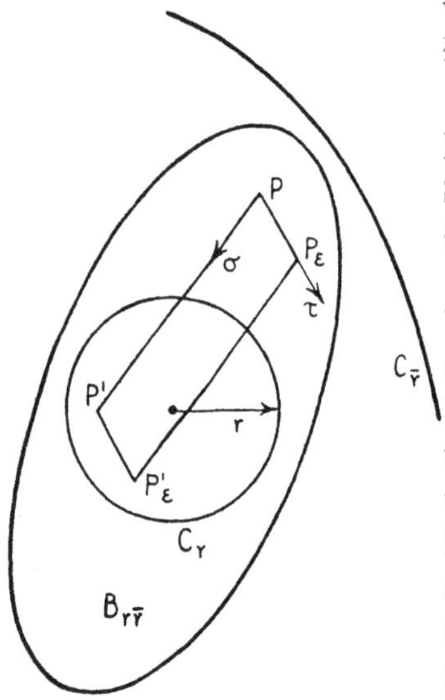

Fig. 2.

$$\begin{aligned} &|w(0, \varepsilon) - w(0, 0)| \\ &\leqq |w(\sigma, \varepsilon) - w(\sigma, 0)| + |w(\sigma, \varepsilon) - w(\sigma, 0) - w(0, \varepsilon) + w(0, 0)| \\ &\leqq \int_0^\varepsilon \left| \frac{\partial w}{\partial \tau}(\sigma, \tau) \right| d\tau + \int_0^\varepsilon \int_0^l \left| \frac{\partial^2 w}{\partial \sigma \, \partial \tau} \right| d\sigma \, d\tau. \end{aligned}$$

8) Etwa $0 < \varepsilon < \frac{r}{4}$.

Durch Mittelwertbildung in σ folgt

$$|w(0,\varepsilon)-w(0,0)| \leqq \frac{1}{l}\int_0^\varepsilon\int_0^l \left|\frac{\partial w}{\partial \tau}\right| d\sigma\, d\tau + \int_0^\varepsilon\int_0^l \left|\frac{\partial^2 w}{\partial \sigma\, \partial \tau}\right| d\sigma\, d\tau$$

und

$$|w(0,\varepsilon)-w(0,0| \leqq \sqrt{2\frac{\varepsilon}{l}\int_0^\varepsilon\int_0^l \left(\frac{\partial w}{\partial \tau}\right)^2 d\sigma\, d\tau + 2\varepsilon l\int_0^\varepsilon\int_0^l \left(\frac{\partial^2 w}{\partial \tau\, \partial \sigma}\right)^2 d\sigma\, d\tau}.$$

Wir kehren nun wieder zu den rechtwinkligen Koordinaten ξ, η zurück und schätzen $\left(\frac{\partial w}{\partial \tau}\right)^2$ und $\left(\frac{\partial^2 w}{\partial \sigma\, \partial \tau}\right)^2$ mittels $\sum (\Delta w)^2$ und $\sum (\Delta^2 w)^2$ ab. Dann erhält man (12) mit einer Konstanten c, die in der Tat nur von r und $\bar{r}$ abhängt.

Für B nehmen wir erstens einen Kreis, der durch eine Ebene $t = \text{const.}$ aus K ausgeschnitten wird und setzen für w eine der zweiten Ableitungen $D^2 u_\nu$. Die Relation (12) besagt dann wegen (10) und (11) die gleichgradige Stetigkeit der $D^2 u_\nu$ auf diesem Kreise.

Um die gleichgradige Stetigkeit in der t-Richtung zu zeigen, betrachten wir zwei Punkte $P(x, y, t)$ und $Q(x, y, t+\varepsilon)$, $\varepsilon > 0$, aus K. Man kann nun Größen $r > 0$ und $\bar{r} > r$ angeben, so daß sich durch jeden Punkt von K, also auch durch Q eine um $\frac{\pi}{8}$ gegen die x, y-Ebene geneigte Ebene legen läßt, die aus K einen Bereich $\mathsf{E}_\delta = B_{r,\bar{r}}$ ausschneidet. Es folgt also zunächst ebenso die gleichgradige Stetigkeit auf $B_{r,\bar{r}}$. Man legt nun durch P die Ebene $t = \text{const.}$ Auf ihrem Schnitt mit $B_{r,\bar{r}}$ befindet sich dann ein Punkt S, der von P und von Q höchstens $\frac{\varepsilon}{\sin\frac{\pi}{8}}\pi$ entfernt ist. Wegen

$$|w(P)-w(Q)| \leqq |w(P)-w(S| + |w(S)-w(Q)|$$

folgt demnach schließlich auch gleichgradige Stetigkeit der $D^2 u_\nu$ in der t-Richtung.

Reprinted from Duke Mathematical Journal
Vol. 1, No. 1, March, 1935

ON CONVERGENCE IN LENGTH

By C. R. Adams and Hans Lewy

1. **Introduction.** Adams and Clarkson[1] have recently considered a sequence of functions $f_n(x)$ $(n = 1, 2, 3, \cdots)$ defined on an interval[2] (a, b) and subject to the following conditions: (i) $f_n(x)$ tends to a limit function $f_0(x)$ of bounded variation; (ii) the total variation $T_a^b(f_n)$ of $f_n(x)$ on (a, b) tends to the total variation $T_a^b(f_0)$ of $f_0(x)$ on (a, b). The notation $f_n(x) - v \rightarrow f_0(x)$ was employed to describe this situation, the symbol $- v \rightarrow$ being read "converges in variation." Under these conditions each curve $y = f_n(x)$ for n sufficiently large has a finite length in the sense of Peano;[3] and the pair of conditions, (i) and (ii') $L_a^b(f_n) \rightarrow L_a^b(f_0)$, define what may be called "convergence in length", the notation $f_n(x) - l \rightarrow f_0(x)$ being used for brevity.[4]

The reader of AC may very naturally raise a question as to the relation between convergence in variation and convergence in length; to compare these notions is one of the objects of the present note.

Definition of notation. It will be convenient usually to designate by S, with or without a subscript, a set of points $a = x_0, x_1, \cdots, x_p = b$, with

$$x_0 < x_1 < \cdots < x_p.$$

In general B will stand for a broken line inscribed in a curve $y = f(x)$ and consisting of p segments, the ith segment $(i = 1, 2, \cdots, p)$ having end-points at $(x_{i-1}, f(x_{i-1}))$, $(x_i, f(x_i))$ and length b_i. The function whose graph is B will frequently be referred to as $B(x)$. Broken lines inscribed in two distinct curves and determined by the same set S will be called "corresponding".[5]

Received February 8, 1935; presented to the American Mathematical Society, February 23, 1935.

[1] C. R. Adams and J. A. Clarkson, *On convergence in variation*, Bulletin of the American Mathematical Society, vol. 40 (1934), pp. 413–417; this paper will be referred to as AC.

[2] The *closed interval* is always to be understood.

[3] According to Peano, a curve (continuous or not) is said to have finite length if the lengths of all inscribed broken lines B are bounded; by definition their least upper bound is the length of the curve.

[4] S. Bochner has imposed the condition of convergence in variation ("abgeschlossene Konvergenz") to permit passage to the limit in a Stieltjes integral: *Monotone Funktionen, Stieltjessche Integrale und harmonische Analyse*, Mathematische Annalen, vol. 108 (1933), pp. 378–410. This condition and that of convergence in length are sometimes employed in the calculus of variations; see Tonelli, *Fondamenti di Calcolo delle Variazioni*, vol. 1, Bologna, 1922.

[5] Incidentally we shall nowhere employ the fact that an absolutely continuous function is the integral of its own derivative. On the other hand, some of the results herein contained can be used as a basis for the deduction of theorems concerning summable functions; for instance, the possibility of approximating in the mean a summable function by step-functions may be concluded from Theorem 4, and the fact that an absolutely continuous function is the integral of a summable function follows easily.

2. **Preliminary remarks.** Since L and T for a function f are defined respectively as the least upper bound of L and of T for broken lines inscribed in f, we have at once the fundamental *semi-continuity relations:* that $f_n(x) \to f_0(x)$ on (a, b) implies

$$(1) \qquad \underline{\lim}_{n\to\infty} L_a^b(f_n) \geqq L_a^b(f_0), \qquad \underline{\lim}_{n\to\infty} T_a^b(f_n) \geqq T_a^b(f_0).$$

Hence the relation $f_n(x) - l \to f_0(x)$ $[f_n(x) - v \to f_0(x)]$ on (a, b) implies that relation for every subinterval. Moreover it may be observed that the relations $f_n(x) \to f_0(x)$ on a set of points everywhere dense in (a, b) and $L_a^b(f_n) \to L_a^b(f_0)$ $[T_a^b(f_n) \to T_a^b(f_0)]$, *when* $f_0(x)$ *is continuous on* (a, b), imply $f_n(x) - l \to f_0(x)$ $[f_n(x) - v \to f_0(x)]$ on (a, b).

If $f(x)$ is of bounded variation on (a, b), we have

$$(2) \qquad b - a + T_a^b(f) \geqq L_a^b(f) \geqq \{(b - a)^2 + [T_a^b(f)]^2\}^{1/2},$$

both of the inequalities flowing at once from the definitions of L and T. From (2) it follows (i) that *each of the relations* $f_n(x) - l \to c$, $f_n(x) - v \to c$ $(c =$ *constant) on* (a, b) *implies the other;* and (ii) that $L_a^x(f)$ is continuous on the left [right] at each point x where f has this property. If $f(x)$ is absolutely continuous on (a, b), both $T_a^x(f)$ and $L_a^x(f)$ are likewise.

Whether f (of bounded variation on (a, b)) is continuous or not, there always exists a sequence of inscribed broken lines B_n $(n = 1, 2, 3, \cdots)$ such that we have both $B_n - l \to f$ and $B_n - v \to f$. For the set S_n determining B_n can be so chosen that S_{n+1} includes S_n for each n, and $S = \sum_{n=1}^{\infty} S_n$ is everywhere dense in (a, b) and contains all the points of discontinuity of f. Then $B_n(x)$ converges to $f(x)$ on (a, b): at each point ξ' in S the convergence is evident, while at a point ξ not in S we have

$$\overline{\lim}_{n\to\infty} | f(\xi) - B_n(\xi) | \leqq T_{\xi'}^{\xi}(f) + \overline{\lim}_{n\to\infty} | f(\xi') - B_n(\xi') | + \overline{\lim}_{n\to\infty} T_{\xi'}^{\xi}(B_n)$$
$$\leqq 2\, T_{\xi'}^{\xi''}(f)$$

for $\xi' < \xi < \xi''$ and ξ', ξ'' in S; since f is continuous at ξ, the right-hand member tends to zero with $\xi'' - \xi'$. From (1) and the definitions of L and T follow the relations $L_a^b(B_n) \to L_a^b(f)$ and $T_a^b(B_n) \to T_a^b(f)$.

3. We first prove

THEOREM 1. *The relation* $f_n(x) - l \to f_0(x)$ *on* (a, b) *implies* $f_n(x) - v \to f_0(x)$; *the converse is not true even when* $f_n(x)$ $(n = 0, 1, 2, \cdots)$ *is assumed absolutely continuous.*

Any ϵ (> 0) being given, let B be a broken line inscribed in f_0 and satisfying the inequality $L_a^b(B) > L_a^b(f_0) - \epsilon$. For convenience we set

$$L_{x_{i-1}}^{x_i}(f_0) - b_i = \epsilon_i, \qquad x_i - x_{i-1} = d_i\,;$$

then $\Sigma \epsilon_i$ is clearly $< \epsilon$ and there exists an m such that we have

$$L_{x_{i-1}}^{x_i}(f_n) \leqq L_{x_{i-1}}^{x_i}(f_0) + \epsilon_i \qquad (i = 1, 2, \cdots, p; n > m).$$

From (2) we thus obtain for each i

$$\begin{aligned} T_{x_{i-1}}^{x_i}(f_n) &\leqq \left\{\left[L_{x_{i-1}}^{x_i}(f_n)\right]^2 - d_i^2\right\}^{1/2} \\ &\leqq [(b_i + 2\epsilon_i)^2 - d_i^2]^{1/2} \\ &\leqq (b_i^2 - d_i^2)^{1/2} + 2(b_i\epsilon_i)^{1/2} + 2\epsilon_i, \end{aligned}$$

and hence, by aid of Schwarz's inequality,

$$\begin{aligned} T_a^b(f_n) &\leqq T_a^b(f_0) + 2\left(\sum_{i=1}^p b_i \sum_{i=1}^p \epsilon_i\right)^{1/2} + 2\sum_{i=1}^p \epsilon_i \\ &\leqq T_a^b(f_0) + 2\,[\epsilon L_a^b(f_0)]^{1/2} + 2\epsilon. \end{aligned}$$

.In view of (1) the direct statement in our theorem is now proved. An example illustrating the failure of the converse, even when all the functions involved are absolutely continuous, is provided by the sequence $f_n(x)$ defined in §5 of AC.

COROLLARY.[6] $L_a^b\,[f(x)] = L_a^b\,[T_a^x(f)]$.

From (2) it follows that neither or both of these quantities are finite. In the second case the relation

$$|f(x_i) - f(x_{i-1})| \leqq T_a^{x_i}(f) - T_a^{x_{i-1}}(f)$$

shows that, of two corresponding broken lines inscribed in $f(x)$ and $T_a^x(f)$, the former has length no greater than the latter. Hence $L_a^b[f(x)]$ is $\leqq L_a^b[T_a^x(f)]$. On the other hand, for any broken line B we have $L_a^b(B) = L_a^b[T_a^x(B)]$; hence, choosing a sequence of broken lines B_n ($n = 1, 2, 3, \cdots$) with $B_n - l \rightarrow f$ we obtain

$$L_a^b(f) = \lim_{n\to\infty} L_a^b(B_n) = \lim_{n\to\infty} L_a^b\,[T_a^x(B_n)] \geqq L_a^b\,[T_a^x(f)],$$

since by Theorem 1 $T_a^x(B_n)$ tends to $T_a^x(f)$.

By virtue of Theorem 1 and the results obtained in §3 of AC, and by reasoning analogous to that employed in §3 of AC, we may readily establish the following theorems.

THEOREM 2. *The relations* $f_n(x) - l \rightarrow f_0(x)$ *on* (a, b) *and* $f_0(x') = f_0(x' - 0)$ $[f_0(x') = f_0(x' + 0)]$ *imply that* x' *is a point of uniform convergence on the left [right] for* $f_n(x)$, $T_a^x(f_n)$, *and* $L_a^x(f_n)$.

COROLLARY. *The relation* $f_n(x) - l \rightarrow f_0(x)$ *on* (a, b), *with* $f_0(x)$ *continuous, implies that the convergence of* $f_n(x)$ *to* $f_0(x)$, *of* $T_a^x(f_n)$ *to* $T_a^x(f_0)$, *and of* $L_a^x(f_n)$ *to* $L_a^x(f_0)$ *is uniform over* (a, b).

THEOREM 3. *The relations* $f_n(x) - l \rightarrow f_0(x)$ *on* (a, b) *and* $f_0(x') \neq f_0(x' - 0)$

[6] Dr. J. A. Clarkson suggested this relation and gave a proof.

$[f_0(x') \neq f_0(x' + 0)]$ imply that x' is a point of uniform convergence on the left [right] for $f_n(x)$, $T_a^x(f_n)$, and $L_a^x(f_n)$ or for none of the three, according as $f_n(x' - 0)$ $[f_n(x' + 0)]$ tends to $f_0(x' - 0)$ $[f_0(x' + 0)]$ or not.

COROLLARY. *If we have $f_n(x) - l \to f_0(x)$ on (a, b), a necessary and sufficient condition that $T_a^x(f_n) \to T_a^x(f_0)$ uniformly on (a, b), or that $L_a^x(f_n) \to L_a^x(f_0)$ uniformly on (a, b), is that $f_n(x) \to f_0(x)$ uniformly on (a, b).*

4. **Further inequalities.** If $f(x)$ and $g(x)$ are of bounded variation on (a, b), we have

$$T_a^b(f) + T_a^b(g) \geqq T_a^b(f + g); \tag{3}$$

$$L_a^b(f) + T_a^b(g) \geqq L_a^b(f + g); \tag{4}$$

$$M_2 T_a^b(f) + M_1 T_a^b(g) \geqq T_a^b(f \cdot g), \tag{5}$$

where $M_1 = \text{l.u.b.} \mid f \mid$ and $M_2 = \text{l.u.b.} \mid g \mid$;

$$[L_a^b(f)]^2 - (b - a)^2 (1 + m^2) \geqq [T_a^b(f - mx)]^2/(1 + m^2), \tag{6}$$

where $m = [f(b) - f(a)]/(b - a)$.

Relation (3) is well known and obvious; (5) is an immediate consequence of the definition of T. Relation (4) follows at once from the fact that the ith segment of a broken line inscribed in $f + g$ has length no greater than the corresponding segment inscribed in f plus the quantity $\mid g(x_i) - g(x_{i-1}) \mid$.

Proof of (6). We first establish the inequality for the case of a broken line B, denoting the derivative $B'(x)$ by $\varphi(x)$ so that $B(x) = \int_a^x \varphi\, dx$. Then (6) reduces to[7]

$$\left[\int_a^b (1 + \varphi^2)^{1/2}\, dx\right]^2 \geqq (b - a)^2(1 + m^2) + \left[\int_a^b \mid \varphi - m \mid dx\right]^2 \Big/ (1 + m^2), \tag{7}$$

where $m = \int_a^b \varphi\, dx/(b - a)$. Let us designate by $\int_+ \left[\int_-\right]$ an integral over the set of points x for which $\varphi - m$ is $\geqq 0$ $[< 0]$. By Minkowski's inequality we have

$$\int_+ (1 + \varphi^2)^{1/2}\, dx \geqq \left[\left(\int_+ dx\right)^2 + \left(\int_+ \varphi\, dx\right)^2\right]^{1/2},$$

together with the same relation for $\int_-$. Setting

$$\alpha = \int_+ dx, \qquad \beta = \int_+ \varphi\, dx, \qquad \gamma = \int_- dx, \qquad \delta = \int_- \varphi\, dx,$$

[7] It is of interest to note that, as soon as (6) is proved, (7) holds for any summable function φ.

we obtain $m = (\beta + \delta)/(\alpha + \gamma)$ and

$$\int_a^b |\varphi - m|\,dx = \int_+ (\varphi - m)\,dx - \int_- (\varphi - m)dx$$
$$= \beta - m\alpha - \delta + m\gamma$$
$$= 2(\beta\gamma - \alpha\delta)/(\alpha + \gamma)\,,$$

and (7) will be true if

$$(8)\quad |\alpha + i\beta| + |\gamma + i\delta| \geqq \left[(\alpha + \gamma)^2 + (\beta + \delta)^2 + \frac{4(\beta\gamma - \alpha\delta)^2}{(\alpha + \gamma)^2 + (\beta + \delta)^2}\right]^{1/2}$$

holds. But under a suitable orthogonal transformation of a ξ, η-plane, the two points $\xi = \alpha$, $\eta = \beta$ and $\xi = \gamma$, $\eta = \delta$ respectively go into (α', β') and (γ', δ') with $\beta' + \delta' = 0$, while (8) is invariant and reduces to the triangular inequality

$$|\alpha' + i\beta'| + |\gamma' - i\beta'| \geqq |\alpha' + \gamma' + 2i\beta'|.$$

Relation (7) having now been proved, (6) follows at once by approximating f by a sequence of inscribed broken lines B_n $(n = 1, 2, 3, \cdots)$ with $B_n - l \rightarrow f$ and taking account of (1).

5. **Addition and multiplication of sequences.** In this section we examine the question of invariance under addition and multiplication of the property of convergence in length. That *convergence in variation is not invariant under these operations, even when all the functions involved are assumed absolutely continuous,* has been shown in AC. That *convergence in length is not invariant under addition, even when the limit functions are assumed continuous and all the approximating functions absolutely continuous,* may be seen from the following example.

Let $f_0(x)$ be defined on (0, 1) as the Cantor function, continuous but not absolutely continuous,[8] and let $f_0(x) = 1$ for $1 \leqq x \leqq 2$. We first observe that, if B is any broken line inscribed in f_0 on (0, 1), we have

$$T_0^1(f_0 - B) = \int_0^1 |B'|\,dx + T_0^1(f_0) = 2;$$

this shows that $f_n - l \rightarrow f_0$ on (0, 1), with f_0 continuous and f_n $(n = 1, 2, 3, \cdots)$ absolutely continuous, does not imply $T_0^1(f_n - f_0) \rightarrow 0$. In view of a theorem of Plessner[9] there exists a $\beta > 0$ and a sequence of positive numbers α_n $(n = 1, 2, 3, \cdots)$ such that $\alpha_n \rightarrow 0$ as $n \rightarrow \infty$ and $T_0^1[f_0(x) - f_0(x + \alpha_n)]$ is $> \beta$ for each n. Let a sequence of sets S_n $(n = 1, 2, 3, \cdots)$ on (0, 1) be chosen as

[8] Any other function of this character will do equally well.

[9] Plessner, *Eine Kennzeichnung der totalstetigen Funktionen*, Crelle's Journal, vol. 160 (1929), pp. 26–32; see also Dunford, *On a theorem of Plessner*, to appear in the Bulletin of the American Mathematical Society, vol. 41 (1935).

follows: for each n let the set S_n' determine a sum approximating T_0^1 $[f_0(x) - f_0(x + \alpha_n)]$ within $\beta/2$; let sequence S_n be selected so that S_n includes $S_{n-1} + S_n'$ for each n and $S = \sum_{n=1}^{\infty} S_n$ is everywhere dense in (0, 1). Designate by B_n, B_n^*, and C_n respectively the broken lines inscribed in $f_0(x)$, $f_0(x + \alpha_n)$, and $f_0(x) - f_0(x + \alpha_n)$ and determined by S_n. We then have $B_n(x) - B_n^*(x) \equiv C_n(x)$, $B_n^*(x) - f_0(x) \equiv B_n(x) - C_n(x) - f_0(x)$, $B_n(x) \to f_0(x)$, and $| C_n(x) | \leqq \max | f_0(x) - f_0(x + \alpha_n) | \to 0$, whence $B_n^*(x) \to f_0(x)$. Moreover we have by (1)

$$L_0^1 [f_0(x)] \leqq \varliminf_{n\to\infty} L_0^1(B_n^*) \leqq \varlimsup_{n\to\infty} L_0^1(B_n^*) \leqq \varlimsup_{n\to\infty} L_0^1 [f_0(x + \alpha_n)] = L_0^1 [f_0(x)].$$

Hence we have $B_n^* - l \to f_0$. We also have $B_n - l \to f_0$. On the other hand $T_0^1(B_n - B_n^*) = T_0^1(C_n)$ is $> \beta/2$ for each n, which implies (see Theorem 1) that $B_n(x) - B_n^*(x)$ does not converge in length to the function zero.

If, however, the limit functions are *absolutely* continuous, the property of convergence in length is invariant under addition and multiplication independently of the nature of the approximating functions. To establish this fact we first prove[10]

THEOREM 4. *The relation* $f_n(x) - l \to f_0(x)$ *on* (a, b), *when* $f_0(x)$ *is absolutely continuous, implies* $T_a^b(f_n - f_0) \to 0$.

That the hypothesis of absolute continuity for f_0 cannot be dispensed with is shown by the example just described above. For the proof of the theorem we employ two lemmas concerning an absolutely continuous function f_0, as follows.

a) If f_0 is absolutely continuous, S an arbitrary set subdividing (a, b), B_0 the broken line inscribed in f_0 and determined by S, and $\mathfrak{J}_h$ $(h = 1, 2, \cdots, q)$ those subintervals in which the slope m_h of B_0 is in absolute value $\geqq M$, there exists a function $\alpha(M)$ independent of S such that

$$\sum_h T_{\mathfrak{J}_h}(f_0) \leqq \alpha(M), \qquad \text{and} \qquad \alpha(M) \to 0 \qquad \text{as} \qquad M \to \infty.$$

To prove this assertion, we observe first $\Sigma \mathfrak{J}_h M \leqq \Sigma \mathfrak{J}_h | m_h | \leqq T_a^b(f_0)$, whence $\Sigma \mathfrak{J}_h \leqq T_a^b(f_0)/M$. Then, on account of the absolute continuity of f_0, $T_a^x(f_0)$ is absolutely continuous and for any set of non-overlapping intervals the sum of whose lengths is $\leqq T_a^b(f_0)/M$, and in particular for our set of intervals $\mathfrak{J}_h$, we have

$$\Sigma T_{\mathfrak{J}_h}(f_0) \leqq \alpha(M),$$

with $\alpha(M) \to 0$ as $M \to \infty$.

b) The hypotheses of a), together with $f_n(x) - v \to f_0(x)$ on (a, b), imply

$$\varlimsup_{n\to\infty} \sum_h T_{\mathfrak{J}_h}(f_n - f_0) \leqq \varlimsup_{n\to\infty} \sum_h T_{\mathfrak{J}_h}(f_n) + \sum_h T_{\mathfrak{J}_h}(f_0) \leqq 2\alpha(M).$$

[10] A proof has been given by Tonelli under the additional assumption that all the functions $f_n(x)$ $(n = 1, 2, 3, \cdots)$ are absolutely continuous; loc. cit., pp. 186–187.

The conclusion is immediate, since only a finite number of intervals $\mathfrak{J}_h$ is involved and in each we have $T(f_n) \to T(f_0)$.

Turning now to the proof of Theorem 4, let S be any set on (a, b) determining a broken line B_n inscribed in f_n $(n = 0, 1, 2, \cdots)$; for each subinterval defined by S we clearly have

$$\begin{aligned}\overline{\lim_{n\to\infty}}\, T_{x\,i-1}^{x\,i}(f_n - f_0) &\leqq \overline{\lim_{n\to\infty}}\, T_{x\,i-1}^{x\,i}(f_n - B_n) + \overline{\lim_{n\to\infty}}\, T_{x\,i-1}^{x\,i}(B_n - B_0) \\ &\quad + T_{x\,i-1}^{x\,i}(f_0 - B_0) \\ &\leqq \overline{\lim_{n\to\infty}}\, T_{x\,i-1}^{x\,i}(f_n - B_n) + T_{x\,i-1}^{x\,i}(f_0 - B_0)\,.\end{aligned}$$

For those intervals $\mathfrak{J}_h$ for which the slope m_h of B_0 is numerically $\geqq M$, we have by Theorem 1 and lemma b)

$$\overline{\lim_{n\to\infty}}\, \Sigma\, T_{\mathfrak{J}_h}(f_n - f_0) \leqq 2\alpha(M), \qquad \alpha(M) \to 0 \text{ as } M \to \infty\,. \tag{9}$$

Denoting by Σ' the summation over only those intervals of S for which the slope m_h of B_0 is numerically $< M$, we obtain by (6) and Schwarz's inequality

$$\begin{aligned}\Sigma' T_{x\,i-1}^{x\,i}(f_0 - B_0) &\leqq (1 + M^2)^{1/2}\, \Sigma' \left\{\left[L_{x\,i-1}^{x\,i}(f_0)\right]^2 - \left[L_{x\,i-1}^{x\,i}(B_0)\right]^2\right\}^{1/2} \\ &\leqq (1 + M^2)^{1/2}\, \Sigma' \left[L_{x\,i-1}^{x\,i}(f_0) + L_{x\,i-1}^{x\,i}(B_0)\right]^{1/2} \left[L_{x\,i-1}^{x\,i}(f_0) - L_{x\,i-1}^{x\,i}(B_0)\right]^{1/2} \\ &\leqq (1 + M^2)^{1/2} K(B_0),\end{aligned} \tag{10}$$

where

$$K(B_0) = [L_a^b(f_0) + L_a^b(B_0)]^{1/2}\, [L_a^b(f_0) - L_a^b(B_0)]^{1/2}\,.$$

Similarly, on summing over these *same* intervals, we find in view of the convergence of B_n to B_0

$$\overline{\lim_{n\to\infty}}\, \Sigma' T_{x\,i-1}^{x\,i}(f_n - B_n) \leqq (1 + M^2)^{1/2}\, \overline{\lim_{n\to\infty}}\, [L_a^b(f_n) + L_a^b(B_n)]^{1/2} [L_a^b(f_n) - L_a^b(B_n)]^{1/2}.$$

Hence from $L_a^b(f_n) \to L_a^b(f_0)$ and $L_a^b(B_n) \to L_a^b(B_0)$ we infer

$$\overline{\lim_{n\to\infty}}\, \Sigma' T_{x\,i-1}^{x\,i}(f_n - B_n) \leqq (1 + M^2)^{1/2} K(B_0)\,. \tag{11}$$

From (9), (10), and (11) we obtain

$$\overline{\lim_{n\to\infty}}\, T_a^b(f_n - f_0) \leqq 2(1 + M^2)^{1/2} K(B_0) + 2\,\alpha(M).$$

This relation is independent of S; $\alpha(M)$ can be made arbitrarily small by taking M sufficiently large and, M having been fixed, the first term on the right can be made arbitrarily small by a suitable choice of S. Consequently we have $\lim_{n\to\infty} T_a^b(f_n - f_0) = 0$, and the proof of Theorem 4 is complete.

THEOREM 5. *The relations* $f_n(x) \to f_0(x)$ *on* (a, b) *and* $T_a^b(f_n - f_0) \to 0$, *with* $f_0(x)$ *of bounded variation, imply* $f_n(x) - l \to f_0(x)$.

For by (1) and (4) we have

$$L_a^b(f_0) \leqq \varliminf_{n\to\infty} L_a^b(f_n) \leqq \varlimsup_{n\to\infty} L_a^b(f_n) \leqq L_a^b(f_0) + \varlimsup_{n\to\infty} T_a^b(f_n - f_0) = L_a^b(f_0).$$

THEOREM 6. *The relations* $f_n(x) - l \to f_0(x)$ *and* $g_n(x) - l \to g_0(x)$ *on* (a, b), *with* $f_0(x)$ *and* $g_0(x)$ *absolutely continuous, imply both*

$$[f_n(x) + g_n(x)] - l \to [f_0(x) + g_0(x)] \text{ and } f_n(x)g_n(x) - l \to f_0(x)g_0(x).$$

From (3), (5), and Theorem 4 we infer

$$T_a^b(f_n + g_n - f_0 - g_0) \leqq T_a^b(f_n - f_0) + T_a^b(g_n - g_0) \to 0,$$

$$T_a^b(f_ng_n - f_0g_0) \leqq M \cdot T_a^b(f_n - f_0) + N \cdot T_a^b(g_n - g_0) \to 0,$$

M being a uniform upper bound for $| g_n |$ and N an upper bound for $| f_0 |$. The conclusions then follow from Theorem 5.

BROWN UNIVERSITY.

Transactions of the American Mathematical Society
Vol. 37, January–June, 1935

A PRIORI LIMITATIONS FOR SOLUTIONS OF MONGE-AMPÈRE EQUATIONS*

BY
HANS LEWY

Introduction. As has been well known since the fundamental papers of S. Bernstein† on the generalization of Dirichlet's principle, one of the main points in proofs of existence for elliptic equations, and the most difficult one, is to obtain suitable estimates for solutions and their derivatives *under the assumption* of their existence, the so-called a priori limitations. Since the purpose of this paper is merely to establish such limitations in the case of analytic elliptic Monge-Ampère equations, we do not intend to repeat here once more the classical reasoning of S. Bernstein.

The method used in this paper will be found different from those generally adopted for similar purposes in the case of quasi-linear equations. It seems to the author that the adaptation of a Monge-Ampère equation to a scheme permitting the application of the sharp a priori limitations for linear equations‡ is efficient as long as there exist estimates for the third derivatives, but is no longer available, if, as in the present paper, we try to *establish* estimates for the third derivatives assuming the knowledge of bounds for the absolute values of the solution and its derivatives up to the second order.

Our method consists in an analytic continuation of the solution into a complex domain similar to that which we introduced in a former paper§ in order to prove the uniqueness of Cauchy's problem. In the case of a Monge-Ampère equation we can, however, use a simpler system of characteristic equations than in the most general non-linear case of analytic elliptic equations. After this has been done a study of the behavior of the analytic continuation is undertaken in order to prove that it has a certain property of "Schlichtheit." This is the principal feature which makes the application of Cauchy's integral formula possible. This Schlichtheit is proved by some simple topological considerations.

* Presented to the Society, September 5, 1934; received by the editors October 1, 1934.

† S. Bernstein, *Sur la généralisation du problème de Dirichlet*, Mathematische Annalen, vol. 62 (1906), p. 253, and vol. 69 (1910), p. 82.

‡ See for example Julius Schauder, *Über lineare elliptische Differentialgleichungen zweiter Ordnung*, Mathematische Zeitschrift, vol. 38 (1934), p. 257.

§ Hans Lewy, *Eindeutigkeit der Lösung des Anfangsproblems einer elliptischen Differentialgleichung*, Mathematische Annalen, vol. 104 (1931), p. 325.

In a subsequent paper we intend to give an application of our present results to an important existence problem of differential geometry in the large.

1. We start with a discussion of the analytic continuation of solutions mentioned in the Introduction. Let A, B, C, D, E be analytic functions of the five complex arguments u, v, x, p, q in a complex neighborhood N of a real point $T_0(u_0, v_0, x_0, p_0, q_0)$, $p_0^2+q_0^2 \leqq 1$, which assume real values for real values of the arguments. Suppose moreover that at the point T_0 the expression

$$\Delta \equiv 4(AC + DE) - B^2 = 2\alpha^2, \quad \alpha > 0,$$

so that

$$4(AC + DE) \geqq 2\alpha^2.$$

Now let $M>0$ be a given number. Then there exists a sufficiently small subneighborhood N_ϵ of T_0 of the form

$$| u - u_0 | \leqq \epsilon, \quad | v - v_0 | \leqq \epsilon, \quad | x - x_0 - p_0(u - u_0) - q_0(v - v_0) | \leqq 4\epsilon^2 M,$$
$$| p - p_0 | \leqq 2\epsilon M, \quad | q - q_0 | \leqq 2\epsilon M, \quad \epsilon > 0,$$

and a constant $K>0$, such that at every point T of N_ϵ,

(1) $$| \Delta | > \alpha^2, \qquad | 4(AC + DE) | > \alpha^2,$$

and that all the functions A, B, C, D, E are analytic in N_ϵ and remain, together with their first partial derivatives, less than K in absolute value. We omit the obvious proof and state that the numbers α, ϵ, K are functionals of A, B, C, D, E and functions of the number M. In N_ϵ the two roots of the quadratic equation in λ

(2) $$\lambda^2 - \lambda B + (AC + DE) = 0$$

are distinct and given by

$$2\lambda_1 = B + i\Delta^{1/2} \text{ and } 2\lambda_2 = B - i\Delta^{1/2}.$$

We now take for x an analytic and real-valued function of two real variables u, v, defined in the square

S_ϵ : $$| u - u_0 | \leqq \epsilon, \qquad | v - v_0 | \leqq \epsilon,$$

and denote, as usual, by p, q, r, s, t its first and second partial derivatives. We suppose

$$x(u_0, v_0) = x_0, \qquad p(u_0, v_0) = p_0, \qquad q(u_0, v_0) = q_0,$$

and assume that in the square S_ϵ

(1.1) $$| r |, \ | s |, \ | t | \leqq M$$

and that throughout S_ϵ the Monge-Ampère equation holds:

$$(3) \qquad Ar + Bs + Ct + D(rt - s^2) = E.$$

Condition (1.1) implies that, for every point of S_ϵ, (u, v, x, p, q) lies in N_ϵ. We take two points of S_ϵ, P and Q, and map the line segment PQ affinely onto the line segment from the point $(-1, 1)$ to the point $(1, -1)$ of the line $\gamma+\delta=0$ of the γ, δ-plane. Thus to every point π of the latter segment there belongs a pair of functions $u(\pi)$, $v(\pi)$, which serve to define three more functions of π,

$$x(\pi) = x(u(\pi), v(\pi)), \qquad p(\pi) = p(u(\pi), v(\pi)), \qquad q(\pi) = q(u(\pi), v(\pi)).$$

With that segment as initial curve and $u(\pi)$, $v(\pi)$, $x(\pi)$, $p(\pi)$, $q(\pi)$ as initial values we set up the following hyperbolic system (of characteristic equations of (3)):

$$(4.1) \qquad x_\gamma - pu_\gamma - qv_\gamma = 0,$$

$$(4.2) \qquad \lambda_1 u_\gamma - Av_\gamma - Dq_\gamma = 0,$$

$$(4.3) \qquad \lambda_2 u_\delta - Av_\delta - Dq_\delta = 0,$$

$$(4.4) \qquad - Ev_\gamma + Cq_\gamma + \lambda_1 p_\gamma = 0,$$

$$(4.5) \qquad - Ev_\delta + Cq_\delta + \lambda_2 p_\delta = 0.$$

Concerning this problem we prove the following

THEOREM. *There exists a positive number $\epsilon_1<\epsilon$, which depends only on ϵ, α, K, M, such that*
(i) *a solution $u(\gamma, \delta)$, $v(\gamma, \delta)$, $\cdots$, $q(\gamma, \delta)$ of* (4.1–4.5) *exists throughout the square $|\gamma| \leqq 1$, $|\delta| \leqq 1$;*
(ii) *for every point of this square $(u, \cdots, q)$ belongs to N_ϵ;*
(iii) *the solution has continuous derivatives with respect to γ and δ and depends analytically on the coordinates of the points P and Q, used in the determination of the initial values;*
(iv) *in the square $|\gamma| \leqq 1$, $|\delta| \leqq 1$ the first derivatives of the functions u, v, x, p, q with respect to γ and δ remain bounded by $2\tau z$, $z=\overline{PQ}$, and moreover*

$$(5) \qquad |u_\gamma(\gamma, \delta) - u_\gamma(\gamma, -\gamma)| \leqq \beta z,$$

provided P and Q lie both in the square

$$S_{\epsilon_1}: \qquad |u - u_0| \leqq \epsilon_1, \qquad |v - v_0| \leqq \epsilon_1;$$

the numbers τ, β are functions of α, K, M, ϵ, to be specified later.

We first derive certain bounds for our functions $x(u, v)$, $p(u, v)$, $q(u, v)$, $r(u, v)$, $s(u, v)$, $t(u, v)$ for real u, v in S_ϵ. In view of (3),

$$\Delta = 4(A + Dt)(C + Dr) - (B - 2Ds)^2.$$

Hence, by (1),

$$4(A + Dt)(C + Dr) > \alpha^2.$$

The factor $A + Dt$, being continuous, never vanishes and may, without loss of generality, be assumed positive. We then have

$$A + Dt > \frac{\alpha^2}{4K(1 + M)}, \qquad C + Dr > \frac{\alpha^2}{4K(1 + M)}, \tag{6}$$

$$\left(\frac{\frac{B}{2} - Ds}{A + Dt}\right)^2 = \frac{-\Delta}{4(A + Dt)^2} + \frac{C + Dr}{A + Dt} < \frac{C + Dr}{A + Dt} \leqq \frac{4K^2(1 + M)^2}{\alpha^2}, \tag{6'}$$

$$|\Delta| \leqq 9K^2, \qquad |\Delta|^{1/2} \leqq 3K, \tag{7}$$

the last inequality being valid in the whole N_ϵ. We now define a non-negative angle $\phi_1 < \pi/2$ by the equation

$$\tan \phi_1 = \frac{4K(1 + M)}{\alpha}. \tag{8}$$

If $|\phi| \leqq \phi_1$, we have $\cos \phi \geqq \cos \phi_1 > 0$. If, on the other hand, $\pi - \phi_1 \geqq \phi \geqq \phi_1$, we deduce the following inequalities:

$$\tan \phi_1 \leqq \tan \phi, \quad \left|\cos \phi \cdot \left(Ds - \frac{B}{2}\right)\right| \leqq \frac{1}{2} \sin \phi \cdot (A + Dt),$$

$$\cos \phi \cdot \left(Ds - \frac{B}{2}\right) + \sin \phi \cdot (A + Dt) \geqq \frac{1}{2} \sin \phi \cdot (A + Dt) \tag{9}$$

$$\geqq \frac{1}{2} \sin \phi_1 \cdot (A + Dt) \geqq \frac{\alpha^2 \sin \phi_1}{8K(1 + M)}.$$

In order to solve the above stated hyperbolic problem, we reduce it to a system of five equations of second order. Thus we shall need to know the initial values of the first derivatives of the solution whose existence we are going to establish, along the initial line $\gamma + \delta = 0$. The equation of the line in the u, v plane joining P and Q has the form

$$v \cos \phi - u \sin \phi = \text{const.},$$

where ϕ is the angle of the direction from P to Q with the positive u axis. On the other hand let us denote by (′) the operation of differentiation with respect to the arc length of the line $\gamma + \delta = 0$. We evidently have for any function $f(\gamma, \delta)$, having continuous first derivatives,

$$f' = \frac{1}{2^{1/2}}\left(\frac{\partial}{\partial\gamma} - \frac{\partial}{\partial\delta}\right)f.$$

Thus we find

$$2^{1/2}u' = u_\gamma - u_\delta = \frac{z}{2}\cos\phi, \qquad 2^{1/2}v' = v_\gamma - v_\delta = \frac{z}{2}\sin\phi,$$
$$2^{1/2}p' = \frac{z}{2}(r\cos\phi + s\sin\phi), \quad 2^{1/2}q' = \frac{z}{2}(s\cos\phi + t\sin\phi),$$
$$2^{1/2}x' = p(u_\gamma - u_\delta) + q(v_\gamma - v_\delta).$$

This together with the equations (4.1–4.5) yields the following values for the initial values of u_γ, u_δ, v_γ, v_δ, account being taken of the non-vanishing of the determinant

$$(10) \qquad \begin{vmatrix} 1 & -p & -q & 0 & 0 \\ 0 & \lambda_1 & -A & -D & 0 \\ 0 & \lambda_2 & -A & -D & 0 \\ 0 & 0 & -E & C & \lambda_1 \\ 0 & 0 & -E & C & \lambda_2 \end{vmatrix} = (\lambda_1 - \lambda_2)^2(AC + DE)$$
$$= -\Delta(AC + DE);$$

$$(11) \qquad \begin{aligned} u_\gamma &= \frac{z}{2(\lambda_1 - \lambda_2)}[-(\lambda_2 - Ds)\cos\phi + (A + Dt)\sin\phi], \\ u_\delta &= \frac{z}{2(\lambda_1 - \lambda_2)}[-(\lambda_1 - Ds)\cos\phi + (A + Dt)\sin\phi], \\ v_\gamma &= \frac{z}{2(\lambda_1 - \lambda_2)}[-(Dr + C)\cos\phi + (\lambda_1 - Ds)\sin\phi], \\ v_\delta &= \frac{z}{2(\lambda_1 - \lambda_2)}[-(Dr + C)\cos\phi + (\lambda_2 - Ds)\sin\phi], \end{aligned}$$

$$x_\gamma = pu_\gamma + qv_\gamma, \qquad x_\delta = pu_\delta + qv_\delta,$$
$$p_\gamma = ru_\gamma + sv_\gamma, \qquad p_\delta = ru_\delta + sv_\delta,$$
$$q_\gamma = su_\gamma + tv_\gamma, \qquad q_\delta = su_\gamma + tv_\delta.$$

This leads to an important set of inequalities for the real and imaginary parts of u_γ on the initial line $\gamma + \delta = 0$. If $|\phi| \leqq \phi_1$, we have

$$(12_1) \qquad \Re(u_\gamma) = \frac{z}{4}\cos\phi \geqq \frac{z}{4}\cos\phi_1.$$

Similarly for $\pi-\phi_1\leqq\phi\leqq\pi+\phi_1$,

$$\Re(u_\gamma) = \frac{z}{4}\cos\phi \leqq -\frac{z}{4}\cos\phi_1. \tag{12$_2$}$$

If, on the other hand, $\phi_1\leqq\phi\leqq\pi-\phi_1$, we find, on account of (9) and (7),

$$-\Im(u_\gamma) \geqq \frac{\alpha^2\sin\phi_1}{96K^2(1+M)}\cdot z, \tag{12$_3$}$$

and, for $\pi+\phi_1\leqq\phi\leqq 2\pi-\phi_1$,

$$-\Im(u_\gamma) \leqq \frac{-\alpha^2\sin\phi_1}{96K^2(1+M)}\cdot z. \tag{12$_4$}$$

This may be expressed in the following form: the entire range of ϕ splits up into four sectors such that within the first sector $\Re(u_\gamma)$ is bounded below by $2\beta z$, within the third sector $\Re(u_\gamma)$ is bounded above by $-2\beta z$, within the second sector $-\Im(u_\gamma)$ is bounded below by $2\beta z$ and within the fourth sector $-\Im(u_\gamma)$ is bounded above by $-2\beta z$. It is sufficient to choose for $2\beta>0$ the smaller one of the two numbers $\frac{1}{4}\cos\phi_1$ and $\alpha^2\sin\phi_1/(96K^2(1+M))$ which both depend only on α, K, M. Hence β depends only on α, K and M.

We notice furthermore that our formulas (11) show the existence of a positive number τ, depending only on α, K, M, ϵ, such that, on $\gamma+\delta=0$,

$$|u_\gamma| < \tau z, \quad |u_\delta| < \tau z, \quad \cdots, \quad |q_\delta| < \tau z. \tag{13}$$

We now indicate briefly a method† of solving the system (4.1–4.5). Differentiate (4.1), (4.2), (4.4) with respect to δ and (4.3) and (4.5) with respect to γ. We obtain five equations containing second derivatives of the unknown functions only of the type $\partial^2/\partial\gamma\partial\delta$. We solve with respect to these, which is possible since $\Delta\neq 0$. The equations so obtained have the form

$$\Delta u_{\gamma\delta} = \text{quadratic form in } u_\gamma, u_\delta, \cdots, q_\gamma, q_\delta; \cdots, \tag{14}$$

with coefficients which are polynomials in A, B, C, D, E and their partial derivatives with respect to u, v, x, p, q, in other words with coefficients limited in absolute value by a suitable polynomial $g(K)>0$ as long as u, v, x, p, q remain within N_ϵ. Under the same condition $|\Delta|>\alpha^2$. We now apply successive approximations to system (14) with initial values as determined above by two points P and Q of S_{ϵ_1} where ϵ_1 satisfies the following inequalities:

$$\begin{gathered} \epsilon_1<\epsilon, \qquad 800g(K)\tau\cdot 2^{3/2}\epsilon_1<\alpha^2, \qquad \epsilon_1(1+8\cdot 2^{1/2}\tau)<\epsilon, \\ \epsilon_1(2M+8\cdot 2^{1/2}\tau)<2M\epsilon, \qquad \epsilon_1(4M\epsilon+24\cdot 2^{1/2}\tau)<4M\epsilon^2, \\ 800g(K)\tau^2\cdot 2^{3/2}\epsilon_1<\alpha^2\beta. \end{gathered} \tag{15}$$

† Cf. Hadamard, *Leçons sur le Problème de Cauchy*, Paris, Hermann, 1932, pp. 488–501.

Postponing the definition of the first approximation, let us suppose that any one of the successive approximations exists in $|\gamma| \leqq 1$, $|\delta| \leqq 1$, and that (u, v, x, p, q) is in N_ϵ, while the partial derivatives with respect to γ or δ are in absolute value $\leqq 2\tau z$. Then for the derivatives of the next approximation $u^*, \cdots, q^*$ which certainly exists in $|\gamma| \leqq 1$, $|\delta| \leqq 1$, we find from (14)

$$| u^*_{\gamma\delta} | \leqq \frac{g(K)}{\alpha^2} \cdot 400\tau^2 z^2.$$

On integrating this and taking into account (13) and the relation $\overline{PQ} = z \leqq 2^{3/2}\epsilon_1$, we have

$$\begin{aligned} &| u_\gamma^* | \leqq \tau z + \frac{800g(K)}{\alpha^2}\tau^2 z^2 \leqq \tau z\left(1 + \frac{800g(K)}{\alpha^2}\tau z\right) \leqq 2\tau z, \\ &| u_\delta^* | \leqq 2\tau z, \cdots, | q_\delta^* | \leqq 2\tau z \qquad \text{for } |\gamma| \leqq 1, \ |\delta| \leqq 1. \end{aligned} \tag{16}$$

On the other hand by (1.1) we have for the initial values of the u, v, x, p, q

$$\begin{aligned} &| u - u_0 | \leqq \epsilon_1, \ | v - v_0 | \leqq \epsilon_1, \ | p - p_0 | \leqq 2M\epsilon_1, \ | q - q_0 | \leqq 2M\epsilon_1, \\ &| x - x_0 - p_0(u - u_0) - q_0(v - v_0) | \leqq 4M\epsilon_1^2. \end{aligned}$$

Hence on integrating (16) and in view of (15) and $|p_0| \leqq 1$, $|q_0| \leqq 1$, we have for $|\gamma| \leqq 1$, $|\delta| \leqq 1$,

$$\begin{aligned} &| p^* - p_0 | \leqq 2M\epsilon_1 + 8\tau \cdot 2^{1/2}\epsilon_1 \leqq 2M\epsilon, \\ &| q^* - q_0 | \leqq 2M\epsilon, \\ &| u^* - u_0 | \leqq \epsilon_1 + 8\tau \cdot 2^{1/2}\epsilon_1 \leqq \epsilon, \qquad | v^* - v_0 | \leqq \epsilon, \\ &| x^* - x_0 - p_0(u^* - u_0) - q_0(v^* - v_0) | \leqq 4M\epsilon_1^2 + 24 \cdot 2^{1/2}\tau\epsilon_1 \leqq 4M\epsilon^2. \end{aligned} \tag{17}$$

This shows that all the functions u^*, v^*, x^*, p^*, q^* remain within N_ϵ, provided we define a suitable first approximation. We do this by setting as a first approximation a solution of the system

$$u_{\gamma\delta} = 0, \ v_{\gamma\delta} = 0, \cdots, \ q_{\gamma\delta} = 0,$$

with the given initial data.

Since for this solution the values of $u_\gamma, \cdots, q_\delta$ at any point γ, δ coincide with some initial value, the above inequalities (16), (17) hold for the first approximation and thus for all of them.

It may now be shown that the successive approximations converge uniformly together with their first derivatives to limit functions and we get a solution of the system (4) assuming the given initial values. This solution is the only one having continuous first derivatives and can be differentiated with respect to γ and δ as well as with respect to any parameter as many times as the initial data themselves.†

Obviously the solution satisfies the same inequalities (16) and (17) that are satisfied by all the approximations.

Moreover, by (14),

$$\left| u_\gamma(\gamma, \delta) - u_\gamma(\gamma, -\gamma) \right| = \left| \int_\delta^{-\gamma} \frac{\partial^2 u}{\partial\gamma\partial\delta} d\delta \right| \leqq 800 \frac{g(K)}{\alpha^2} \tau^2 z^2 \leqq \beta z.$$

This is the desired inequality (5). Thus all the statements of our theorem on page 419 have been proved except the analytical dependence on the coordinates of P and Q.

This last property results from the fact that our initial data are analytic functions of these four coordinates. Let us denote the coordinates of P by a_1 and b_1, those of Q by a_2 and b_2. If we give a_1, b_1, a_2, b_2 complex values, but *very near* to those real values for which we made the above construction, we get new initial data which differ as little as we please from the former ones; we may assume that the inequalities (13) remain valid for the new initial data. If now we construct the solution of the modified initial problem, we get again a solution for the whole square $|\gamma| \leqq 1$, $|\delta| \leqq 1$ and this solution stays as near as we please to the former solution on account of the continuous dependence on the initial data. We thus get a solution depending on four complex parameters and the two real variables γ, δ. Denote by $\eta = \eta_1 + i\eta_2$ any one of the complex parameters. The initial values, being analytic functions of η, satisfy the Cauchy-Riemann equations which evidently can be written in the following form using the abbreviation $\nabla = \partial/\partial\eta_1 + i\partial/\partial\eta_2$:

$$\nabla u = 0, \quad \nabla v = 0, \cdots, \quad \nabla q = 0.$$

On the other hand, on operating with ∇ on the system (4.1–4.5) we get

$$\frac{\partial}{\partial\gamma}(\nabla x) - p\frac{\partial}{\partial\gamma}(\nabla u) - q\frac{\partial}{\partial\gamma}(\nabla v) = u_\gamma \nabla p + v_\gamma \nabla q,$$

$$\cdots\cdots\cdots\cdots\cdots\cdots$$

$$-E\frac{\partial}{\partial\delta}(\nabla v) + C\frac{\partial}{\partial\delta}(\nabla q) + \lambda_2\frac{\partial}{\partial\delta}(\nabla p) = v_\gamma \frac{\partial E}{\partial u}\nabla u + \cdots,$$

$$\cdots\cdots\cdots\cdots\cdots\cdots$$

i.e., a hyperbolic system, linear and homogeneous, with a non-vanishing "determinant" (10) for the unknowns $\nabla x, \cdots, \nabla q$. Since a solution of this system is $\nabla x = \cdots = \nabla q = 0$ and the solution is uniquely determined by initial values, we conclude

† For the proof of all these statements, see Hadamard, loc. cit.

$$\nabla x = \nabla u = \cdots = \nabla q = 0.$$

This implies the desired analytic dependence on a_1, b_1, a_2, b_2 in the square $|\gamma| \leqq 1$, $|\delta| \leqq 1$, expressed by means of the Cauchy-Riemann equations.

2. We return to real values of a_1, b_1, a_2, b_2. The manifold of solutions seems to depend on these four variables and the two variables γ, δ. It is possible, however, to reduce the dependence to that of a_1, b_1, a_2, b_2 only. If we substitute for γ a variable $\gamma' = k\gamma + l$ and for δ a variable $\delta' = k'\delta + l'$ with $kk' \neq 0$, a solution of the system (4) becomes a solution of the same system in γ', δ' because of the homogeneity of (4) with respect to the derivatives $\partial/\partial\gamma$, $\partial/\partial\delta$. Now let P', Q' be two points of, and in the same order as, PQ. The range of the initial values of the problem (4) determined by P' and Q' may be transformed by the above substitution into a part of the range of the initial values of the PQ problem. The same must then be true for the range of the corresponding solutions of the $P'Q'$ problem and of the PQ problem. If therefore we agree to consider only the point $\gamma = 1$, $\delta = 1$ of the PQ problem and to write the solution at this point as a function of a_1, b_1, a_2, b_2 only, the whole square will consist of points at which the solution is the same function with different values of the argument a_1, b_1, a_2, b_2. In particular, the line $\delta = 1$ corresponds to fixed (a_1, b_1) while (a_2, b_2) varies along a line through (a_1, b_1). This shows that we have on $\delta = 1$ the following rule: The differentiation $\partial/\partial\gamma$ of our solution, considered as function of γ, δ and of the four parameters a_1, b_1, a_2, b_2, reduces to the operation

$$\frac{z}{2}\left(\cos\phi \frac{\partial}{\partial a_2} + \sin\phi \frac{\partial}{\partial b_2}\right)$$

if we write our solutions as functions of a_1, b_1, a_2, b_2 only. Notice that the solution is defined for $|a_1 - u_0| \leqq \epsilon_1$, $|b_1 - v_0| \leqq \epsilon_1$, $|a_2 - u_0| \leqq \epsilon_1$, $|b_2 - v_0| \leqq \epsilon_1$ and that the originally given functions in the real u, v-plane now appear as functions of the four arguments for coincident P and Q, i.e., for $a_1 = a_2$, $b_1 = b_2$.

3. We now set up another hyperbolic system for two functions $u(\gamma, \delta)$ and $v(\gamma, \delta)$ to be determined for $|\gamma| \leqq 1$, $|\delta| \leqq 1$. The initial values are given on $\gamma + \delta = 0$ in the same way as on page 419, but the equations are

$$(18.1) \qquad (\lambda_1 - Ds)u_\gamma - (A + Dt)v_\gamma = 0,$$

$$(18.2) \qquad (\lambda_2 - Ds)u_\delta - (A + Dt)v_\delta = 0.$$

Here the quantities x, p, q, r, s, t are to be considered as analytic functions of u and v, determined by the originally presented solution of the Monge-Ampère equation. On writing

$$(19)\qquad \rho = \frac{\lambda_1 - Ds}{A + Dt}, \qquad \bar{\rho} = \frac{\lambda_2 - Ds}{A + Dt},$$

this becomes

$$(18.3)\qquad v_\gamma - \rho u_\gamma = 0,$$

$$(18.4)\qquad v_\delta - \bar{\rho} u_\delta = 0.$$

We conclude

$$\lambda_1 u_\gamma - A v_\gamma - D q_\gamma = 0, \ \lambda_2 u_\delta - A v_\delta - D q_\delta = 0, \ - E v_\gamma + C q_\gamma + \lambda_1 p_\gamma = 0,$$

since, in view of (2) and (3),

$$(\lambda_1 r + Cs)(A + Dt) + (\lambda_1 s + Ct - E)(\lambda_1 - Ds) = 0.$$

Similarly we obtain the relation

$$- E v_\delta + C q_\delta + \lambda_2 p_\delta = 0.$$

We also mention the identity

$$x_\gamma - p u_\gamma - q v_\gamma = 0.$$

Thus the introduction of the new initial problem leads to a set of functions $u(\gamma, \delta), \cdots, q(\gamma, \delta)$ which turns out to be a solution of (4.1–4.5) with the same initial conditions. In view of the uniqueness theorem the two solutions must coincide wherever they exist simultaneously, which is true for P sufficiently near Q.

Equations (18.3) and (18.4) admit of a simple interpretation. For instance, if γ varies alone, the differentials of u and v are connected by the *ordinary* differential equation

$$dv - \rho du = 0.$$

Returning to the variables a_1, b_1, a_2, b_2, we may say that if (a_1, b_1) is fixed while (a_2, b_2) varies along any line through (a_1, b_1), then (u, v) moves on a certain two-dimensional surface M_1 of the four-dimensional complex u, v-space, called characteristic "megaline."* For (a_2, b_2) sufficiently near (a_1, b_1) the equation of M_1 has the form

$$v = \text{analytic function of } u.$$

Thus, for Q sufficiently near P, we may write the corresponding Cauchy-Riemann equations in the form

$$(20)\qquad v_{a_2} u_{b_2} - v_{b_2} u_{a_2} = 0.$$

* Hadamard, loc. cit., p. 512.

But in view of the above stated analytical dependence on (a_1, b_1, a_2, b_2) this relation holds on the whole range of the values of (a_1, b_1, a_2, b_2). Similarly we find

$$v_{a_1}u_{b_1} - v_{b_1}u_{a_1} = 0. \tag{21}$$

From (20), (21), (18.3), (18.4) we derive a fact of fundamental importance for our discussion, namely that the *quotient* $u_{a_2}:u_{b_2}$ *is independent of* a_1, b_1.

Indeed, since on M_1 we have $0 = dv - \rho du$, the relation

$$v_{a_2} - \rho u_{a_2} = 0 \tag{22.1}$$

is satisfied on M_1 and holds for any M_1, or, in other words, for any a_1, b_1. Similarly

$$v_{b_2} - \rho u_{b_2} = 0, \tag{22.2}$$

$$v_{a_1} - \bar{\rho} u_{a_1} = 0. \tag{23.1}$$

Now differentiate (23.1) with respect to a_2 and b_2, (22.1) with respect to a_1 and (22.2) with respect to a_1. This gives

$$v_{a_1a_2} - \bar{\rho} u_{a_1a_2} - \bar{\rho}_u u_{a_1}u_{a_2} - \bar{\rho}_v u_{a_1}v_{a_2} = 0,$$

$$v_{a_1b_2} - \bar{\rho} u_{a_1b_2} - \bar{\rho}_u u_{a_1}u_{b_2} - \bar{\rho}_v u_{a_1}v_{b_2} = 0,$$

$$v_{a_1a_2} - \rho u_{a_1a_2} - \rho_u u_{a_2}u_{a_1} - \rho_v u_{a_2}v_{a_1} = 0,$$

$$v_{a_1b_2} - \rho u_{a_1b_2} - \rho_u u_{b_2}u_{a_1} - \rho_v u_{b_2}v_{a_1} = 0.$$

By (20)

$$v_{a_1a_2}u_{b_2} - v_{a_1b_2}u_{a_2} - \bar{\rho}(u_{a_1a_2}u_{b_2} - u_{a_1b_2}u_{a_2}) = 0,$$

$$v_{a_1a_2}u_{b_2} - v_{a_1b_2}u_{a_2} - \rho(u_{a_1a_2}u_{b_2} - u_{a_1b_2}u_{a_2}) = 0.$$

$\rho \neq \bar{\rho}$ gives

$$u_{a_1a_2}u_{b_2} - u_{a_1b_2}u_{a_2} = 0,$$

which shows that $u_{a_2}:u_{b_2}$ does not depend on a_1.

In an analogous manner we find that $u_{a_2}:u_{b_2}$ does not depend on b_1. Hence, in order to compute the value of $u_{a_2}:u_{b_2}$ for any (a_1, b_1, a_2, b_2) we may compute it for $a_1 = a_2$, $b_1 = b_2$, i.e., for (a_2, b_2, a_2, b_2). In view of the known values (11) for the initial values of u_γ and our remark on page 425 we get

$$\begin{aligned} \frac{2}{z} u_\gamma &= \cos\phi \cdot u_{a_2} + \sin\phi \cdot u_{b_2} \\ &= \frac{1}{\lambda_1 - \lambda_2}[-(\lambda_2 - Ds)\cos\phi + (A + Dt)\sin\phi]. \end{aligned} \tag{24}$$

Taking $\phi=0$ and $\phi=\pi/2$ gives finally

$$u_{a_2}:u_{b_2} = -\frac{\lambda_2 - Ds}{A + Dt}. \tag{25}$$

We notice that this expression has a positive imaginary part. Since x, p, q are analytic functions of u and v and thus of u alone on M_1, we have on M_1 the Cauchy-Riemann equations

$$\begin{aligned} x_{a_2}u_{b_2} - x_{b_2}u_{a_2} &= 0, \\ p_{a_2}u_{b_2} - p_{b_2}u_{a_2} &= 0, \\ q_{a_2}u_{b_2} - q_{b_2}u_{a_2} &= 0, \end{aligned} \tag{23}$$

and these relations again hold for all a_1, b_1, a_2, b_2 in question.

4. We now proceed to establish a fact which forms the salient point of the present investigation.

THEOREM. *The function* $u(u_0, v_0, a_2, b_2)$ *is "schlicht" for* $|u-u_0|<2\beta\epsilon_1$, *or, in other words, the equation*

$$u(u_0, v_0, a_2, b_2) = U$$

has one and only one solution (a_2, b_2) *in* $|a_2-u_0|\leqq\epsilon_1$, $|b_2-v_0|\leqq\epsilon_1$ *provided* $|U-u_0|<2\beta\epsilon_1$.

Let us write (5) in the form

$$u_\gamma(\gamma, \delta) = u_\gamma(\gamma, -\gamma) + h, \quad |h| \leqq \beta z.$$

The value $u_\gamma(\gamma, -\gamma)$ satisfies the inequalities (12_1–12_4). Hence, for $|\gamma|\leqq 1$, $|\delta|\leqq 1$,

$$\begin{aligned} \Re(u_\gamma) &\geqq \beta z \quad &&\text{for} \quad |\phi| \leqq \phi_1, \\ \Re(u_\gamma) &\leqq -\beta z &&\text{for } \pi - \phi_1 \leqq \phi \leqq \pi + \phi_1, \\ -\Im(u_\gamma) &\geqq \beta z &&\text{for} \quad \phi_1 \leqq \phi \leqq \pi - \phi_1, \\ -\Im(u_\gamma) &\leqq -\beta z &&\text{for } \pi + \phi_1 \leqq \phi \leqq 2\pi - \phi_1. \end{aligned} \tag{26}$$

An integration from $\gamma=-1$ to $\gamma=+1$ along $\delta=1$ yields

$$\begin{aligned} \Re[u(1, 1) - u_0] &\geqq 2\beta z \quad &&\text{for} \quad |\phi| \leqq \phi_1, \\ \Re[u(1, 1) - u_0] &\leqq -2\beta z &&\text{for } \pi - \phi_1 \leqq \phi \leqq \pi + \phi_1, \\ -\Im[u(1, 1) - u_0] &\geqq 2\beta z &&\text{for} \quad \phi_1 \leqq \phi \leqq \pi - \phi_1, \\ -\Im[u(1, 1) - u_0] &\leqq -2\beta z &&\text{for } \pi + \phi_1 \leqq \phi \leqq 2\pi - \phi_1. \end{aligned} \tag{27}$$

Let us take for P the point (u_0, v_0) and let (a_2, b_2) vary in the square R: $|a_2-u_0|\leqq\epsilon_1$, $|b_2-v_0|\leqq\epsilon_1$. Imagine the complex conjugate quantity $\bar{u}(u_0, v_0,$

$a_2, b_2) - u_0$ as a vector* attached to the variable point (a_2, b_2) of R. Formulas (27) show that $\bar{u}(u_0, v_0, a_2, b_2) - u_0$ forms an angle less than π with the vector

$$a_2 - u_0 + i(b_2 - v_0)$$

which is $\neq 0$, provided $(a_2-u_0)^2+(b_2-v_0)^2 \neq 0$. If we follow $\bar{u}$ along any closed Jordan curve contained in R and such that (u_0, v_0) lies in its interior, the rotation of $\bar{u}-u_0$ after one circuit must equal that of $a_2-u_0+i(b_2-v_0)$, i.e., 2π. On the other hand the vector $\bar{u}-u_0$ depends analytically on a_2, b_2 and has for every subdomain of R only a finite number of singularities. (The same will be true for any vector $\bar{u}(u_0, v_0, a_2, b_2) - \bar{U}$, $\bar{U}$ being a constant.) We conclude that the point $a_2 = u_0$, $b_2 = v_0$ is a singularity of the field $\bar{u}(u_0, v_0, a_2, b_2) - u_0$ with index $+1$.

By (27), the expression $\bar{u} - \bar{U}$ does not vanish along the contour of R if $|U-u_0| < 2\beta\epsilon_1$, since $z = \overline{PQ} \geqq \epsilon_1$ as long as Q varies along the contour of R and P is the center (u_0, v_0) of R. For reasons of continuity, the rotation (along the contour of R) of $\bar{u} - \bar{U}$ equals that of $\bar{u} - u_0$, so that we may say that the vector field $\bar{u} - \bar{U}$ has at least one singularity within R for $|U-u_0| < 2\beta\epsilon_1$, and the sum of the indices of all singularities in R equals $+1$. On the other hand it follows from the lemma below that any such singularity has an essentially positive index, hence there can be at most one singularity, which is the desired result.

LEMMA. *Let, without loss of generality, the origin of an a, b-plane be a zero of a complex vector* $\mathfrak{u}(a, b)$, *analytic near the origin and such that the quotient* $\mathfrak{u}_a : \mathfrak{u}_b = \kappa$ *has a negative imaginary part at the origin*: $\mathfrak{J}(\kappa_0) < 0$. *Then the singularity of the origin has a positive index.*

We prove this lemma by developing $\mathfrak{u}$ in a Taylor series in $b+\kappa_0 a$ and $b-\kappa_0 a$. Let $k>0$ be the smallest degree for which there are non-vanishing terms so that the series begins as follows:

$$\mathfrak{u} = c_1(b - \kappa_0 a)^k + c_2(b - \kappa_0 a)^{k-1}(b + \kappa_0 a) + \cdots + c_{k+1}(b + \kappa_0 a)^k + \cdots.$$

Form the expression $\mathfrak{u}_a - \kappa\mathfrak{u}_b$ and develop it into a power series in a and b; it must vanish identically. On the other hand the terms of lowest degree in $\mathfrak{u}_a - \kappa\mathfrak{u}_b$ have the form

$$-2\kappa_0\left[kc_1(b - \kappa_0 a)^{k-1} + \cdots + c_k(b + \kappa_0 a)^{k-1}\right]$$

so that we conclude $c_1 = c_2 = \cdots = c_k = 0$. Thus the Taylor series of $\mathfrak{u}$ begins with the term $c_{k+1}(b+\kappa_0 a)^k$, $c_{k+1} \neq 0$, and the singularity of $\mathfrak{u}$ at $a = b = 0$ is that of $c_{k+1}(b+\kappa_0 a)^k$ and has the index $k>0$.

* For the notions concerning vector fields used in this paragraph, see for instance W. Fenchel, *Elementare Beweise und Anwendungen einiger Fixpunktsätze*, Matematisk Tidsskrift, (B), 1932, p. 66.

5. The theorem of the preceding section enables us to write the megaline M_1 in the form

$$v = \text{analytic function of } u$$

for u varying within the circle $\Gamma: |u-u_0| < 2\beta\epsilon_1$. Within R none of the partial derivatives of $u(u_0, v_0, a_2, b_2)$ with respect to a_2 or b_2 vanishes. For we conclude from (26) and (24) that there is at least a linear form in u_{a_2} and u_{b_2} which differs from zero. On the other hand the quotient $u_{a_2}:u_{b_2}$ has a positive imaginary part and cannot vanish, which proves that $u_{a_2} \neq 0$, $u_{b_2} \neq 0$.

Hence, the relations (20) and (23) yield the analytic dependence of v, x, p, q on u on the megaline M_1, provided u lies in Γ.

Integration of (13) gives, for $|a_2-u_0| \leqq \epsilon_1$, $|b_2-v_0| \leqq \epsilon_1$,

$$|v - v_0|, \cdots, |q - q_0| \leqq 2^{3/2}\tau\epsilon_1.$$

Let us denote by d/du the differential quotient on M_1 (u in Γ). Cauchy's integral formula yields for $u = u_0$

$$\rho = \frac{dv}{du} = \frac{1}{2\pi i}\oint_\Gamma \frac{(v - v_0)du}{(u - u_0)^2}, \quad \frac{d^n v}{du^n} = \frac{n!}{2\pi i}\oint_\Gamma \frac{(v - v_0)du}{(u - u_0)^{n+1}}, \cdots,$$

$$\frac{d^n q}{du^n} = \frac{n!}{2\pi i}\oint_\Gamma \frac{(q - q_0)du}{(u - u_0)^{n+1}};$$

$$\left|\frac{d^n v}{du^n}\right|, \cdots, \left|\frac{d^n q}{du^n}\right| \leqq \frac{n!}{2\pi}\cdot\frac{2\pi 2^{3/2}\tau\epsilon_1}{(2\beta\epsilon_1)^n} = \frac{n! 2^{3/2}\tau\epsilon_1}{(2\beta\epsilon_1)^n}.$$

The differentiation d/du at the point $u = u_0$ of M_1 is connected with the partial derivatives $\partial/\partial u$ and $\partial/\partial v$ at the point u_0, v_0 by the relation

$$\frac{d}{du} = \frac{\partial}{\partial u} + \rho\frac{\partial}{\partial v}.$$

Thus the above inequalities for the successive derivatives at the point u_0 yield inequalities for the partial derivatives of the given functions $x(u, v)$, $p(u, v)$, $q(u, v)$. And anyone of these inequalities yields two estimates if we take account of the reality of these functions for real u, v.

6. As mentioned in the introduction, the most important of the a priori limitations are those for the third derivatives of x. We easily can establish them at the point u_0, v_0.

We have at u_0, v_0

$$\frac{d^2p}{du^2} = \left(\frac{\partial}{\partial u} + \rho\frac{\partial}{\partial v}\right)^2 p = \left(\frac{\partial}{\partial u} + \rho\frac{\partial}{\partial v}\right)(r + \rho s)$$

$$= r_u + 2\rho s_u + \rho^2 t_u + \frac{d^2v}{du^2}\, s,$$

$$\frac{d^2q}{du^2} = \left(\frac{\partial}{\partial u} + \rho\frac{\partial}{\partial v}\right)^2 q = \qquad s_u + 2\rho t_u + \rho^2 t_v + \left(\overline{\frac{d^2v}{du^2}}\right) t,$$

and the conjugate equations

$$\left(\overline{\frac{d^2p}{du^2}}\right) = \left(\frac{\partial}{\partial u} + \bar\rho\frac{\partial}{\partial v}\right)^2 p = r_u + 2\bar\rho s_u + \bar\rho^2 t_u + \left(\overline{\frac{d^2v}{du^2}}\right) s,$$

$$\left(\overline{\frac{d^2q}{du^2}}\right) = \left(\frac{\partial}{\partial u} + \bar\rho\frac{\partial}{\partial v}\right)^2 q = \qquad s_u + 2\bar\rho t_u + \bar\rho^2 t_v + \left(\overline{\frac{d^2v}{du^2}}\right) t.$$

The determinant of this system is

$$\begin{vmatrix} 1 & 2\rho & \rho^2 & 0 \\ 1 & 2\bar\rho & \bar\rho^2 & 0 \\ 0 & 1 & 2\rho & \rho^2 \\ 0 & 1 & 2\bar\rho & \bar\rho^2 \end{vmatrix} = (\rho - \bar\rho)^4 = \frac{\Delta^2}{(A + Dt)^4}$$

and is in absolute value $>\alpha^4/[K^4(1+M)^4]$ (see (1)). Moreover $|r|$, $|s|$, $|t|$ $\leqq M$ and

$$|\rho| = \left|\frac{dv}{du}\right| \leqq 2^{1/2}\tau/\beta \text{ and } \left|\frac{d^2v}{du^2}\right|, \cdots, \left|\frac{d^2q}{du^2}\right| \leqq 2^{1/2}\tau/(\beta^2\epsilon_1).$$

Hence by solving the above linear system with respect to r_u, s_u, t_u, t_v we easily obtain upper bounds for $|r_u|$, $|s_u|$, $|t_u|$, $|t_v|$, depending on α, ϵ, K, M only. We shall not carry through the computations.

7. The estimates for the higher derivatives are not much more complicated, but for the applications it is important to know that they can be chosen in such a way that they insure a lower a priori bound for the radii of convergence of the Taylor series for x at a point u_0, v_0.

We observe that there are two megalines M_1 and M_2 through the point (u_0, v_0), conjugate to each other. The values of u, v, x, p, q at their points will be referred to by $u(u_0, v_0, a_2, b_2) \equiv u(u_0, v_0; Q), \cdots$ and $u(a_1, b_1, u_0, v_0) \equiv u(P; u_0, v_0)$ respectively. Consider the megaline passing through (u_0, v_0, Q) which consists of points $(a_1, b_1; Q)$ with variable (a_1, b_1) and fixed Q, and the megaline through $(P; u_0, v_0)$ consisting of points $(P; a_2, b_2)$ with variable

(a_2, b_2) and fixed P. The two megalines evidently intersect at (P, Q). Let ξ denote the variable u on M_1 and η the same variable on M_2. If both vary within the circle Γ of "Schlichtheit," i.e., for $|\xi-u_0|<2\beta\epsilon_1$ and $|\eta-u_0|<2\beta\epsilon_1$, they are in unique correspondence with Q and P respectively. We therefore may write $u(P, Q), \cdots$ as functions of ξ and η. We say that they are analytic functions of these complex arguments.

We have evidently

$$\xi = \xi(a_2, b_2) = \xi_1 + i\xi_2$$

and the independence of $u_{a_2}:u_{b_2}$ on a_1, b_1 shows that $u_{a_2}:u_{b_2}$ or, what is essentially the same, $u_{\xi_1}:u_{\xi_2}$ are independent on η. Hence we may compute its value by setting $\eta=u_0$. But on M_1 we have Cauchy-Riemann equations $u_{\xi_1}:u_{\xi_2}=-i$. Similarly we find the Cauchy-Riemann equations $u_{\eta_1}:u_{\eta_2}=-i$. Thus $u(\xi, \eta)$ depends analytically on ξ and η for $|\xi-u_0|<2\beta\epsilon_1$, $|\eta-u_0|<2\beta\epsilon_1$. The equations (20), (21), (23) show that for any function ω of the set u, v, x, p, q we have

$$\frac{\partial(\omega, u)}{\partial(a_2, b_2)} = 0 \quad\text{and}\quad \frac{\partial(\omega, u)}{\partial(a_1, b_1)} = 0.$$

This yields the analytic dependence of all of them on ξ and η for $|\xi-u_0|<2\beta\epsilon_1$, $|\eta-u_0|<2\beta\epsilon_1$. Moreover u, v, x, p, q admit bounds for their absolute values, depending only on ϵ, α, K, M.

On the other hand we find for $\xi=u_0$, $\eta=u_0$ the following *lower* bounds:

$$\frac{\partial u}{\partial \xi} = 1, \quad \frac{\partial u}{\partial \eta} = 1, \quad \left|\frac{\partial v}{\partial \xi}\right| \geqq \frac{\alpha}{2K(1+M)}, \quad \left|\frac{\partial v}{\partial \eta}\right| \geqq \frac{\alpha}{2K(1+M)},$$

$$\left|\frac{\partial(u, v)}{\partial(\xi, \eta)}\right| \geqq \frac{\alpha}{K(1+M)}.$$

We therefore are able to introduce u, v as independent variables* instead of ξ and η within a domain $|u-u_0|<\epsilon_2$, $|v-v_0|<\epsilon_2$, with ϵ_2 depending merely on α, ϵ, K, M. In this domain we may develop x in a power series in $u-u_0$ and $v-v_0$; it will converge there and its coefficients will have absolute bounds in terms of α, ϵ, K, M. Even the majorant series formed with these bounds instead of the derivatives of x at u_0, v_0 will converge for $|u-u_0|<\epsilon_2$, $|v-v_0|<\epsilon_2$. We state the final result in the following form:

THEOREM. *The derivatives of an analytic solution of the analytic Monge-Ampère equation* (3) *existing for* $|u-u_0|<\epsilon$, $|v-v_0|<\epsilon$ *can be developed into a power series in* $u-u_0$ *and* $v-v_0$ *whose associated radii of convergence and*

* This step is justified in §8.

whose coefficients can be bounded, the former from below, the latter from above, by certain numbers which depend only
(i) *on the bound M for the moduli of the second derivatives,*
(ii) *on the value $2\alpha^2$ of $\Delta > 0$ at $u = u_0$, $v = v_0$,*
(iii) *on the bound ϵ such that in a neighborhood N_ϵ of the 10-dimensional space of complex x, p, q, u, v around u_0, v_0, x_0, p_0, q_0 the coefficients A, B, C, D, E remain regular, and $|\Delta| > \alpha^2$,*
(iv) *on the bound K for these coefficients and their first partial derivatives in N_ϵ, provided we have $p_0^2 + q_0^2 \leqq 1$. The power series in u and v, formed with the bounds of the coefficients, has the same bounds for its associated radii of convergence.*

8. We have made use of the following lemma concerning the inversion of a system of analytic functions in the neighborhood of the origin.

LEMMA. *Let*

$$u = ax + by + \sum_{i+k \geqq 2} a_{ik} x^i y^k,$$

$$v = cx + dy + \sum_{i+k \geqq 2} c_{ik} x^i y^k.$$

Suppose $|ad - bc| > A > 0$, $0 < B_1 < |a|, |b|, |c|, |d| < B_2$. Suppose furthermore that $\sum a_{ik} x^i y^k$ and $\sum c_{ik} x^i y^k$ converge for $|x| < \rho$, $|y| < \rho$ and that there $|u|, |v| < C$. Then there exists a neighborhood of the origin $|u| + |v| < h$, and a fortiori a neighborhood $|u| < h/2$, $|v| < h/2$, for which we can solve with respect to x and y, with h depending only on A, B_1, B_2, C, ρ.

Though the proof of the possibility of inversion in a sufficiently small neighborhood is well known, we were unable to find the above estimate for the neighborhood mentioned in the literature. We therefore sketch a simple method of establishing this estimate.

We have on account of the assumptions

$$|a_{ik}| < \frac{C}{\rho^{i+k}} \quad \text{and} \quad |c_{ik}| < \frac{C}{\rho^{i+k}}.$$

Hence

$$u' = \frac{du - bv}{ad - bc} = x + \sum_{i+k \geqq 2} a_{ik}' x^i y^k,$$

$$v' = \frac{-cu + av}{ad - bc} = y + \sum_{i+k \geqq 2} c_{ik}' x^i y^k,$$

and $|a_{ik}'|, |c_{ik}'| < C'/\rho^{i+k}$ with $C' = C'(A, B_2, C)$. For $|x| + |y| \leqq \rho/2$, we conclude

$$| u' - x | + | v' - y | < \sum_{i+k \geqq 2} \frac{2C'}{\rho^{i+k}} | x |^i | y |^k \leqq (| x | + | y |)^2 M,$$

with $M = M(C', \rho) > 0$. Upon introducing the further restriction

$$| x | + | y | \leqq \min \left\{ \frac{1}{2M}, \frac{\rho}{2} \right\}$$

we get on the boundary of this domain for $| U' | + | V' | < \min \{1/(4M), \rho/4\}$,

$$| u' - x | + | v' - y | < \frac{| x | + | y |}{2},$$

$$| u' - U' - x | + | v' - V' - y | < | x | + | y |.$$

These relations show that the vector of the four-dimensional space with the components $u' - U'$, $v' - V'$ has in $|x| + |y| \leqq \min \{1/(2M), \rho/2\}$ the same sum of indices of its singularities as the vector with components x, y. This sum is $+1$. On the other hand the vector $u' - U'$, $v' - V'$, being analytic, admits only singularities with positive index. Thus there is precisely one solution, x, y, of the equations $u' = U'$, $v' = V'$ for $| U' | + | V' | < \min \{1/(4M), \rho/2\}$. This can be expressed in terms of u, v which proves our lemma.

Brown University,
Providence, R. I.

Reprinted from the Proceedings of the NATIONAL ACADEMY OF SCIENCES,
Vol. 22, No. 6, pp. 377–381. June, 1936.

GENERALIZED INTEGRALS AND DIFFERENTIAL EQUATIONS

BY HANS LEWY

DEPARTMENT OF MATHEMATICS, UNIVERSITY OF CALIFORNIA

Communicated April 21, 1936

The following theorems have been proved by the author:

1. *Generalization of the Integral.*—Theorem: In the interval $X_0 \leq x \leq X_1$ consider a function $f(x)$, approximable by continuous functions and satisfying the inequality $|f(x) - f(X_0)| < F$, and n continuous functions $g_1(x), g_2(x), \ldots, g_n(x)$ of the bounded total variations $G_1, G_2, \ldots, G_n$. Suppose that two functions a and b of $f, x, g_1, g_2, \ldots, g_n$ in the domain

$$|f - f(X_0)| \leq F, X_0 \leq x \leqq X_1, |g_1 - g_1(X_0)| \leq G_1, \ldots, |g_n - g_n(X_0)| \leq G_n$$

admit of continuous partial derivatives with respect to all arguments. Then, for every sequence of continuously differentiable functions $f_k(x)$, $k = 1,2\ldots$ satisfying the inequality $|f_k(x) - f(X_0)| < F$ and tending to $f(x)$ as k tends to ∞, the limit

$$L = \lim_{k \to \infty} \int_{X_0}^{x} a\,(f_k(x), x, g_1(x), \ldots, g_n(x))\, db(f_k(x), x, g_1(x), \ldots, (g_n(x))$$

exists.

We write

$$L = \int_{X^0}^{x} a(f(x), x, g_1(x), \ldots, g_n(x))\, db(f(x), x, g_1(x), \ldots, g_n(x)).$$

Remark. This integral is *not* identical to the Stieltjes integral. For the function $f(x) = 0$, $0 \leqq x < 1$, $f(1) = 1$, we have for instance

$$\int_0^1 f(x)\, df(x) = \frac{f^2(1) - f^2(0)}{2} = \frac{1}{2}\cdot$$

The Lesbesque-Stieltjes integral would be 1.

2. *Ordinary Differential Equations.*—Theorem 2:

Assumptions: In the interval $0 \leqslant x \leqq X$, the continuously differentiable functions $f_k(x)$, $k = 1,2,\ldots$ converge to a function $f(x)$. $f_k(0) = 0$, $k = 1,2, \ldots, \infty$. The continuously differentiable functions $g_{ik}(x) i = 1, \ldots, n;\ k = 1,2, \ldots, \infty$ tend to the continuous functions $g_i(x)$ as $k \longrightarrow \infty$. The total variations of $g_{ik}(x) - g_i(x)$ tend to zero and the total variations of $g_{ik}(x)$ are uniformly bounded by numbers $G_i > 0$, $i = 1, \ldots, n$. For some $\epsilon > 0$ the quantities $a_0, a_1, \ldots, a_n$ are functions of z, y_i, u in the domain

$$|z| \leq F,\ |y_i| \leq G_i,\ |u| < \epsilon + 2FM_0 + \sum_{l=1}^{n} (e^{2FN} - e^{FN} + M_l\, e^{FN})G_l$$

and are absolutely bounded by $M_1, M_2, \ldots, M_n$. The functions $a_0, \ldots, a_n$ are all continuously differentiable and the first derivatives of a_0 are bounded by $N > 0$ and satisfy a Lipschitz condition of exponent 1.

Assertion: The solution $u_k(x)$ of $u_k(0) = 0$ and

$$du_k(x) = a_0(f_k(x), g_{ik}(x), \ldots, g_{nk}(x), u_k(x))df_k(x) + \sum_{l=1}^{n} d_l(f_k(x), g_{ik}(x), u_k(x))dg_{lk}(x)$$

exists for $0 \leq x \leq X$ and all k, and tends to a limit function $u(x)$ as $k \longrightarrow \infty$. We have

$$|u(x)| \leq FM_0 + \sum_{l=1}^{n} (e^{2FN} - e^{FN} + M_l\, e^{FN})G_l.$$

We call $u(x)$ the solution of

$$du(x) = a_0(f(x), g_{ik}(x), u(x))\, df(x) + \sum_{l=1}^{n} a_l(f(x), g_i(x), u(x))dg_l(x) \quad (1)$$

and the initial condition $u(0) = 0$.

A similar theorem may be obtained for systems of equations of the form (1), with unknown functions $u(x), v(x), \ldots,$ *one* function $f(x)$ of Baire's first class and a certain number of functions of bounded and continuous variation $g_1(x), g_2(x), \ldots$ intervening in the equations

3. *Initial Problem for Hyperbolic Systems.*—Theorem 3: Consider the following system of partial differential equations for n functions $\varphi_1, \varphi_2, \ldots, \varphi_n$ of two independent variables α and β:

$$\sum_{k=1}^{n} a_{ik} \frac{\partial \varphi_k}{\partial \alpha} = 0,\ i = 1, \ldots, m < n;\ \sum_{k=1}^{n} a_{ik} \frac{\partial \varphi_k}{\partial \beta} = 0,\ i = m + 1, \ldots, n;$$

for $n \leqslant 2$.

The initial line is a segment σ of the line $\alpha = \beta$ and on σ the values of $\varphi_1, \ldots, \varphi_n$ are given as functions of the arc length τ. Assume that (i) for all values of τ in σ the points of a Euclidean $\varphi_1, \ldots, \varphi_n$ space with coördinates $\varphi_1(\tau), \varphi_2(\tau), \ldots, \varphi_n(\tau)$ lie within a sphere S_ρ of radius ρ; (ii) for all points of a concentric sphere $S_{2\rho}$ the determinant of the functions $a_{ik}(\varphi_1, \ldots, \varphi_n)$ satisfies the condition:

$$|a_{ik}| \neq 0;$$

(iii) throughout $S_{2\rho}$ the a_{ik} have continuous partial derivatives up to the fourth order with respect to $\varphi_1, \ldots, \varphi_n$; (iv) there exists a transformation with constant coefficients γ_{ik}, of determinant $\neq 0$, $\psi_i = \sum_{k=1}^{n} \gamma_{ik}\varphi_k$, $i = 1, 2, \ldots, n$, such that on the initial line the function ψ_1 is continuous, while $\psi_2, \ldots, \psi_n$ are continuous and of bounded variation.

Then there exists for every interior point of σ an α,β-neighborhood where a continuous solution of the initial problem exists. Under "solution" we understand the (uniquely determined) limit of solutions for approximating continuously differentiable initial values $\phi_{ij}(\tau)$, $i = 1, \ldots, n$; $j \longrightarrow \infty$ subject only to the following conditions: For every j the point with components $\phi_{ij}(\tau), \phi_{2j}(\tau), \ldots, \phi_{nj}(\tau)$ for τ in σ belongs to S_ρ; moreover

$$\psi_{1j}(\tau) = \sum_{k=1}^{n} \gamma_{1k}\, \phi_{kj}(\tau) \longrightarrow \psi_1(\tau) \text{ as } j \longrightarrow \infty$$

and the functions

$$\psi_{ij}(\tau) = \sum_{k=1}^{n} \gamma_{ik}\, \phi_{kj}(\tau),\ i = 2, \ldots, n,$$

and their total variations tend to $\psi_i(\tau)$ and its total variation as $j \longrightarrow \infty$

4. *Elliptic Equations of Monge Ampère Type.*—Theorem 4: Consider a sequence of Monge-Ampère equations

$$A_n r + B_n s + C_n t + rt - s^2 = E_n \tag{2}$$

in which, as usual, p, q, r, s, t stand for the first and second derivatives of a function $x(u, v)$ and where A_n, B_n, C_n, E_n are analytic functions of the arguments u, v, x, p, q as long as these belong to a fixed bounded domain D of a complex u, v, x, p, q, space, and where $A_n, \ldots, E_n$ are real for real u, v, x, p, q of D and such that there

$$\Delta_n = 4(A_n C_n + E_n) - B_n^2 > 0 \tag{3}$$

while for the whole of D

$$|\Delta_n| > 0.$$

Suppose that in D

$$A_n, B_n, C_n, E_n \Rightarrow A, B, C, E(u, v, x, p, q) \text{ as } n \longrightarrow \infty,$$

and

$$|\Delta| = |4(AC + E) - B^2| > 0.$$

Suppose furthermore that for every n and a circle K: $(u - u_0)^2 + (v - v_0)^2 < \epsilon$ of the real u,v-plane we know a real analytic solution $x_n(u,v)$ of (2) which together with its first derivatives $p_n(u,v)$ and $q_n(u,v)$ tends uniformly to a function $x(u,v)$ and its derivatives $p(u,v)$ and $q(u,v)$, such that for every n and u,v in K $(u, v, x_n(u,v), p_n(u,v), q_n(u,v))$ lies in D and the same is true for $(u, v, x(u,v), p(u,v), q(u,v))$.

Then, for (u,v) in K, there exist two parameters ζ and η such that u, v, $x(u,v)$, $p(u,v)$, $q(u,v)$ are analytic functions of ζ and η in $\zeta^2 + \eta^2 < 1$. The map $\zeta, \eta \longrightarrow u, v$ is one to one; x has second derivatives r, s, t and satisfies throughout K

$$Ar + Bs + Ct + rt - s^2 = E, \tag{4}$$

except possibly at points satisfying the analytic equation

$$\frac{\partial(u,v)}{\partial(\zeta,\eta)} = 0.$$

The proof is an application of the Theorem 3 of the foregoing section to the analytic continuation of the functions $x(u,v)$ into a complex domain, similar to the one the author used in former papers. This application is rendered possible by the remark, that from (3) one can deduce[1] $r_n + C_n \neq$ throughout K; hence there exists a constant k such that for all n the function $x_n(u,v) + ku^2$ is monotone on every line $v =$ const., so that, for a suitable subsequence of n, the functions u, v, $x_n(u,v) + ku^2$, $p_n(u,v) + 2ku$ tend *together with their variations* to u, v, $x(u,v) + ku^2$, $p(u,v) + 2ku$, while the functions $q_n(u,v)$ tend uniformly to $q(u,v)$. This last remark may be utilized in order to omit in the Theorem the assumption of the convergence of $p_n(u,v)$ and $q_n(u,v)$ to $p(u,v)$ and $q(u,v)$ **provided** $p(u,v)$ **and** $q(u,v)$ **are continuous,** as it can be shown that the uniform convergence of a subsequence of $p_n(u,v)$ and $q_n(u,v)$ follows for every concentric circle in K. On the other hand, after a suitable contact transformation which, as is well known, preserves the Monge-Ampère type and the elliptic character (3), even the assumption of the continuity of $p(u,v)$ and $q(u,v)$ may be obtained as consequences of the convergence of the $x_n(u,v)$. Thus we have the following

Theorem 5: Suppose that, under the assumptions of Theorem 4 as to A_n, B_n, C_n, E_n, A, B, C, E, there exists a circle K of the real u, v-plane where for every n there exists a real and analytic solution of (2) with derivatives $p_n(u,v)$ and $q_n(u,v)$ such that for u,v in K the point $(u, v, x_n(u,v), p_n(u,v), q_n(u,v))$ lies in D. Then there exists a subsequence of $x_n(u,v)$ which in concentric circle K' tends to a function $x(u,v)$ and two parameters ζ, η such that u, v, x are analytic functions of ζ, η, the map of $\zeta^2 + \eta^2 < 1$ on K' is one to one, and $x(u,v)$ satisfies (4) except possibly at points where $\dfrac{\partial(u,v)}{\partial(\zeta,\eta)} = 0$.

Theorem 5 is of great importance for the *treatment* of equation (4) *in the large*, as it is necessitated by several problems of the differential geometry in the large, especially in connection with a Lemma[2] that under certain conditions permits to rule out the possibility $\dfrac{\partial(u,v)}{\partial(\zeta,\eta)} = 0$. We intend to take up those applications at another place.

[1] Cf. F. Rellich, "Zur ersten Randwertaufgabe bei Monge-Ampèreschen Differentialgleichungen vom elliptischen Typus," *Math. Ann.*, **107**, 505–513 (1932). Hans Lewy, "A Priori Limitations for Solutions of Monge-Ampère Equations," *Trans. Am. Math. Soc.*, **37**, 417–434 (1935).

[2] Hans Lewy, "On the Non-Vanishing of the Jacobian of Certain Mappings," *Bull. Am. Math. Soc.* (in press at present).

Bulletin of the American Mathematical Society
Vol. 42, October, 1936

ON THE NON-VANISHING OF THE JACOBIAN IN CERTAIN ONE-TO-ONE MAPPINGS

BY HANS LEWY

THEOREM 1. *If $u(x, y)$ and $v(x, y)$ are harmonic, $u(0, 0) = v(0, 0) = 0$, and if there exists a neighborhood N_1 of the origin of the xy plane and a neighborhood N_2 of the origin of the uv plane such that $u(x, y)$ and $v(x, y)$ establish a mapping of N_1 onto N_2 which is one-to-one both ways, then the Jacobian $\partial(u, v)/\partial(x, y)$ does not vanish at the origin.*

PROOF. As the statement of Theorem 1 remains invariant under homogeneous linear transformations of the uv plane, we may assume, in the developments in polar coordinates for u and v, that

$$u = \sum_{i}^{\infty} [a_n r^n \cos n\theta + b_n r^n \sin n\theta], \qquad (a_i^2 + b_i^2 \neq 0),$$

$$v = \sum_{k}^{\infty} [A_n r^n \cos n\theta + B_n r^n \sin n\theta], \qquad (A_k^2 + B_k^2 \neq 0),$$

that the positive index i does not exceed k, and that for $i = k$ we have $a_k B_k - A_k b_k \neq 0$. Considering the case $i = k$ first, we may, because of the invariance mentioned, assume

$$a_k = B_k = 1, \qquad b_k = A_k = 0.$$

For small values of r, the auxiliary mapping,

$$\bar{u} = r^k \cos k\theta, \qquad \bar{v} = r^k \sin k\theta,$$

can be continuously joined with the given one for $0 \leqq t \leqq 1$ by

$$u_t = r^k \cos k\theta + t\sum_{k+1}^{\infty} (a_n r^n \cos n\theta + b_n r^n \sin n\theta),$$

$$v_t = r^k \sin k\theta + t\sum_{k+1}^{\infty} (A_n r^n \cos n\theta + B_n r^n \sin n\theta).$$

The vector (u_t, v_t) thereby never differs by more than a vector of length $r^k/2$ from the vector $(\bar{u}, \bar{v})$ whose length is r^k. Thus the index of the origin* in both fields (u, v) and $(\bar{u}, \bar{v})$ is the same, and as it is ± 1 for one-to-one mappings we conclude $i=k=1$ and our theorem follows.

In the remaining case $i<k$ we consider the Jacobian

$$J = \frac{\partial(u, v)}{\partial(x, y)} = \frac{1}{r}\,\frac{\partial(u, v)}{\partial(r, \theta)}.$$

We find by elementary computation that the development of J starts with the terms of lowest degree in r

$$kir^{i+k-1}\left[(A_k a_i + B_k b_i)\sin(i-k)\theta + (a_i B_k - b_i A_k)\cos(i-k)\theta\right]$$

or, with a suitable angle θ_0, with the term

$$kir^{i+k-1}(A_k^2 + B_k^2)^{1/2}(a_i^2 + b_i^2)^{1/2}\cos\left[(i-k)\theta - \theta_0\right].$$

Hence, for sufficiently small values of r, J assumes both positive and negative values, while in a one-to-one map the Jacobian, where it does not vanish, is of the same sign as the index of the map which is either $+1$ everywhere or -1 everywhere. Hence $i<k$ is impossible and our proof is completed.

THEOREM 2. *Let $u(x, y)$ and $v(x, y)$ be analytic functions of x and y in a neighborhood N_1 of the origin of the xy plane which they map onto a neighborhood N_2 of the origin of the uv plane in a one-to-one correspondence. Suppose, moreover, that u and v are solutions of the following equations:*

$$(1)\quad \begin{aligned}\frac{\partial^2 u}{\partial x^2} + \frac{\partial^2 u}{\partial y^2} &+ a\left[\left(\frac{\partial u}{\partial x}\right)^2 + \left(\frac{\partial u}{\partial y}\right)^2\right] + b\left(\frac{\partial u}{\partial x}\frac{\partial v}{\partial x} + \frac{\partial u}{\partial y}\frac{\partial v}{\partial y}\right)\\ &+ c\left[\left(\frac{\partial v}{\partial x}\right)^2 + \left(\frac{\partial v}{\partial y}\right)^2\right] + d\left(\frac{\partial u}{\partial x}\frac{\partial v}{\partial y} - \frac{\partial u}{\partial y}\frac{\partial v}{\partial x}\right) = 0,\end{aligned}$$

$$(2)\quad \begin{aligned}\frac{\partial^2 v}{\partial x^2} + \frac{\partial^2 v}{\partial y^2} &+ A\left[\left(\frac{\partial u}{\partial x}\right)^2 + \left(\frac{\partial u}{\partial y}\right)^2\right] + B\left(\frac{\partial u}{\partial x}\frac{\partial v}{\partial x} + \frac{\partial u}{\partial y}\frac{\partial v}{\partial y}\right)\\ &+ C\left[\left(\frac{\partial v}{\partial x}\right)^2 + \left(\frac{\partial v}{\partial y}\right)^2\right] + D\left(\frac{\partial u}{\partial x}\frac{\partial v}{\partial y} - \frac{\partial u}{\partial y}\frac{\partial v}{\partial x}\right) = 0,\end{aligned}$$

* For the topological notions used, see W. Fenchel, *Elementare Beweise und Anwendungen einiger Fixpunktsätze*, Matematisk Tidsskrift, (B), 1932, p. 66.

in which $a(u, v)$, $b(u, v)$, $\cdots$, $D(u, v)$ are analytic functions of u and v defined for (u, v) in N_2. Then the Jacobian $\partial(u, v)/\partial(x, y)$ does not vanish at the origin.

PROOF. Suppose, without loss of generality, that the power series for u and v in x and y start with non-vanishing terms of ith and kth degree, respectively, and that furthermore $i \leqq k$, $i > 0$. Then the terms of ith degree in $u(x, y)$ form a harmonic function, because this is trivial for $i = 1$, and for $i > 1$ the only terms in the development of the left hand of (1) of degree $i-2$ are furnished by those of $\partial^2 u/\partial x^2 + \partial^2 u/\partial y^2$. Introduce $V(x, y) = v(x, y) - F(u)$, where F is analytic in u and $F(0) = 0$. Instead of (1) and (2) we find

$$\begin{aligned}(3)\quad &\frac{\partial^2 u}{\partial x^2} + \frac{\partial^2 u}{\partial y^2} + [a(u, V + F(u)) + F'(u)b(u, V + F(u))\\ &\quad + F'^2(u)c(u, V + F(u))]\left[\left(\frac{\partial u}{\partial x}\right)^2 + \left(\frac{\partial u}{\partial y}\right)^2\right] + \cdots = 0,\end{aligned}$$

$$\begin{aligned}&\frac{\partial^2 V}{\partial x^2} + \frac{\partial^2 V}{\partial y^2} + F'(u)\left(\frac{\partial^2 u}{\partial x^2} + \frac{\partial^2 u}{\partial y^2}\right)\\ (4)\quad &+ [A(u, V + F(u)) + F'(u)B(u, V + F) + F'^2(u)C(u, V + F)\\ &\quad + F''(u)]\left[\left(\frac{\partial u}{\partial x}\right)^2 + \left(\frac{\partial u}{\partial y}\right)^2\right] + \cdots = 0,\end{aligned}$$

and, by substitution of (3) in (4),

$$\begin{aligned}(5)\quad &\frac{\partial^2 V}{\partial x^2} + \frac{\partial^2 V}{\partial y^2} + [A(u, V + F) + F'B + F'^2C + F'' - F'a\\ &\quad - F'^2b - F'^3c]\left[\left(\frac{\partial u}{\partial x}\right)^2 + \left(\frac{\partial u}{\partial y}\right)^2\right] + \cdots = 0,\end{aligned}$$

where the omitted terms are linear in

$$\frac{\partial u}{\partial x}\frac{\partial V}{\partial x} + \frac{\partial u}{\partial y}\frac{\partial V}{\partial y}, \quad \left(\frac{\partial V}{\partial x}\right)^2 + \left(\frac{\partial V}{\partial y}\right), \quad \text{and} \quad \frac{\partial(u, V)}{\partial(x, y)}.$$

Now notice that V vanishes for $x = y = 0$ and that in its development the lowest degree K of non-vanishing terms still is $\geqq i$.

Let us distinguish two cases, $k > i$ and $k = i$. For $k > i$, we con-

struct a function $F(u)$, solution of the ordinary differential equation of second order

$$(6)\quad \begin{aligned} &A(u, F(u)) + F'(u)B(u, F(u)) \\ &\qquad + F'^2C - F'a - F'^2b - F'^3c + F'' = 0, \end{aligned}$$

vanishing with its first derivative for $u=0$. We have $K>i$, as $F(u)$, considered as function of x and y starts with terms of degree $\geqq 2i$. Equation (5) shows now a coefficient of $(\partial u/\partial x)^2+(\partial u/\partial y)^2$ whose development in u and V has *all terms divisible* by V. Thus the x, y series for the left hand of (5) obtains all terms of degree $K-2$ from the expression $\partial^2 V/\partial x^2+\partial^2 V/\partial y^2$, whence we conclude that in V the terms of lowest order K form a harmonic function. As in the proof of Theorem 1, we conclude that the Jacobian $\partial(u, V)/\partial(x, y)$, which equals $\partial(u, v)/\partial(x, y)$, would assume both positive and negative values in every neighborhood of the origin, which leads to a contradiction. Thus $i<k$ is impossible.

In the case $i=k$ we again choose $F(u)$ as solution of (6), vanishing for $u=0$, but we try to determine $F'(u)$ in such a way that k results $>i$. If this were possible, we could apply the same argument as at the end of the last paragraph, and find the same contradiction. Hence, should $i=k=1$, we would have the desired inequality

$$\frac{\partial(u, V)}{\partial(x, y)} = \frac{\partial(u, v)}{\partial(x, y)} \neq 0$$

at the origin. If, however, $i=k>1$, we may determine the linear term of $F(u)$ such that for polar coordinates in the xy plane and suitable determination of $\theta=0$ we have

$$\begin{aligned} u &= \text{const. } r^k \cos k\theta + r^{k+1}(\cdots), \\ V &= \text{const. } r^k \sin k\theta + r^{k+1}(\cdots). \end{aligned}$$

This leads, by the same reasoning as in the proof of Theorem 1, to the conclusion that the index of the vector field (u, V) at the origin is $k\neq 1$. But together with the functions $u(x, y)$, $v(x, y)$, the functions $u(x, y)$, $V(x, y)=v(x, y)-F(u)$ also establish a one-to-one mapping of the xy plane, because the correspondence between the uv plane and the uV plane evidently is one-to-one. Hence the index ought to be ± 1, which completes our proof.

THE UNIVERSITY OF CALIFORNIA

Transactions of the American Mathematical Society
Vol. 41, No. 3, May, 1937

A PRIORI LIMITATIONS FOR SOLUTIONS OF MONGE-AMPÈRE EQUATIONS. II*

BY
HANS LEWY

In this paper we are concerned with the convergence of solutions of elliptic and analytic Monge-Ampère equations. Theorem 1 gives the principal result of this paper. The example on p. 372 indicates the possibility of certain types of singularities which for linear elliptic equations cannot occur. Theorems 2 and 3 give sufficient conditions for the analyticity of the limit function. These conditions allow applications to certain problems of the differential geometry in the large. Our method consists in introducing a regularizing contact transformation which transforms convex functions into functions with bounded second derivatives and thus makes possible the reduction of Theorem 1 to the principal result of the first part of this paper.

1. **Regularizing contact transformation.** Consider the contact transformation T of an (x, y, z)-space into a (ξ, η, ζ)-space generated by the relation

$$0 = z + \zeta - x\xi - y\eta + \frac{x^2 + y^2}{2}.$$

It leads to the following transformation of a surface $z=z(x, y)$ with continuous first derivatives z_x and z_y

$$\xi = x + z_x(x, y) \qquad \eta = y + z_y(x, y) \tag{1}$$

$$\zeta = -z + x\xi + y\eta - (x^2 + y^2)/2. \tag{2}$$

Suppose $z(x, y)$ convex. Then (1) transforms an open rectangle R' of the (x, y)-plane into a domain of the (ξ, η)-plane in a one-to-one way.

For let (ξ_1, η_1) and (ξ_2, η_2) be the images of two distinct points (x_1, y_1) and (x_2, y_2), ϑ the angle between the vector joining (x_1, y_1) to (x_2, y_2) and the positive x-axis so that the distance between these two points is

$$(x_2 - x_1) \cos \vartheta + (y_2 - y_1) \sin \vartheta > 0.$$

As the derivative of z in the direction of the above vector is $z_x \cos\vartheta + z_y \sin \vartheta$, we conclude from the convexity of $z(x, y)$

* Presented to the Society, November 28, 1936; received by the editors September 1, 1936. Part I of this paper was published in these Transactions, vol. 37 (1935), pp. 417–434.

$$z_x(x_1, y_1) \cos \vartheta + z_y(x_1, y_1) \sin \vartheta \leqq z_x(x_2, y_2) \cos \vartheta + z_y(x_2, y_2) \sin \vartheta.$$

Hence

$$0 < (x_2 - x_1) \cos \vartheta + (y_2 - y_1) \sin \vartheta \leqq (\xi_2 - \xi_1) \cos \vartheta + (\eta_2 - \eta_1) \sin \vartheta$$

which implies

$$(x_2 - x_1)^2 + (y_2 - y_1)^2 \leqq (\xi_2 - \xi_1)^2 + (\eta_2 - \eta_1)^2. \tag{3}$$

Using (3) for differentiation of (2) we obtain

$$x = \frac{\partial \zeta}{\partial \xi}, \qquad y = \frac{\partial \zeta}{\partial \eta} \tag{4}$$

which together with (3) shows that ζ has continuous derivatives in ξ and η.

If, moreover, z has continuous second derivatives with respect to (x, y), then ζ has continuous second derivatives in ξ and η all of which are numerically $\leqq 1$. For, by (1), the partial derivatives

$$\xi_x = 1 + z_{xx}, \quad \eta_x = \xi_y = z_{xy}, \quad \eta_y = 1 + z_{yy},$$

are continuous in (x, y). Hence the derivatives x_ξ, x_η, y_ξ, y_η of the inverse functions are continuous in (ξ, η) and we have†

$$\begin{aligned} \frac{\partial^2 \zeta}{\partial \xi^2} &= x_\xi = \frac{1 + z_{yy}}{(1 + z_{xx})(1 + z_{yy}) - z_{xy}^2}, \\ \frac{\partial^2 \zeta}{\partial \xi \partial \eta} &= \frac{- z_{xy}}{(1 + z_{xx})(1 + z_{yy}) - z_{xy}^2}, \cdots, \end{aligned} \tag{5}$$

$$\left|\frac{\partial^2 \zeta}{\partial \xi^2}\right|, \left|\frac{\partial^2 \zeta}{\partial \xi \partial \eta}\right|, \left|\frac{\partial^2 \zeta}{\partial \eta^2}\right| \leqq 1, \qquad \frac{\partial^2 \zeta}{\partial \xi^2} \frac{\partial^2 \zeta}{\partial \eta^2} - \left(\frac{\partial^2 \zeta}{\partial \xi \partial \eta}\right)^2 > 0. \tag{6}$$

Now consider a bounded sequence of continuously differentiable convex functions $z_n(x, y)$ defined in R'. Designate by T_n the transformation of a concentric open rectangle R, contained in R', into the (ξ, η)-plane as induced by T for the function $z_n(x, y)$, by T_n^{-1} the inverse of T_n. We proceed to show that there exists a subsequence of n such that

(i) $z_n(x, y)$ converge uniformly in R;

(ii) there exists an open set D in the (ξ, η)-plane contained in all $T_m(R)$ for m sufficiently large;

(iii) the inverse T_m^{-1} converges uniformly in D to a transformation T_∞^{-1};

(iv) $T_\infty^{-1}(D)$ contains a neighborhood of the center of R, and all those

† The denominator in (5) is >0 since the convexity of $z(x, y)$ implies $z_{xx} \geqq 0$, $z_{yy} \geqq 0$, $z_{xx}z_{yy} - z_{xy}^2 \geqq 0$.

points (ξ, η) which by T_∞^{-1} are mapped into the center of R have no limit point on the boundary of D.

The functions $z_n(x, y)$ are uniformly bounded and convex in R', hence equally continuous, so that (i) is satisfied and we have for $m>n$ in R

$$\left| z_m(x, y) - z_n(x, y) \right| < \epsilon_n$$

with $\epsilon_n \to 0$ as $n \to \infty$. If (x_0, y_0) is an arbitrary point of R, a circle γ_n about (x_0, y_0) with radius $\rho_n = 2(\epsilon_n)^{1/2}$ lies for sufficiently large n within R.

Denote by (ξ_n, η_n) the image of (x_0, y_0) in T_n, by ϑ the angle of the vector from (x_0, y_0) to a variable point (x_n', y_n') of the circumference of γ_n and the positive x-axis and by (ξ_m', η_m') the image of (x_n', y_n') in T_m. We have

$$\begin{aligned} z_{nx}(x_0, y_0)\cos\vartheta + z_{ny}(x_0, y_0)\sin\vartheta &\leqq \frac{z_n(x_n', y_n') - z_n(x_0, y_0)}{\rho_n} \\ &\leqq \frac{z_m(x_n', y_n') - z_m(x_0, y_0)}{\rho_n} + \frac{2\epsilon_n}{\rho_n} \\ &\leqq z_{mx}(x_n', y_n')\cos\vartheta + z_{my}(x_n', y_n')\sin\vartheta + \frac{2\epsilon_n}{\rho_n}. \end{aligned}$$

Hence

$$\xi_n \cos\vartheta + \eta_n \sin\vartheta < \xi_m' \cos\vartheta + \eta_m' \sin\vartheta.$$

Thus as ϑ varies from 0 to 2π the projection of the vector $(\xi_m' - \xi_n, \eta_m' - \eta_n)$ on the vector $(\cos\vartheta, \sin\vartheta)$ is positive or the angle between the two vectors is $<\pi/2$, thereby implying that the vector $(\xi_m' - \xi_n, \eta_m' - \eta_n)$ turns by 2π as the point (ξ_m', η_m') describes the image in T_m of the circumference of γ_n. Hence, for all $m>n$, (ξ_n, η_n) lies within $T_m(\gamma_n)$. Denoting by R_n the set of all those points of R whose distance from the boundary of R is $>\rho_n$, we may say that $T_m(R)$ contains $T_n(R_n)$ for $m>n$. Thus (ii) is proved with $D = T_n(R_n)$.

As the transformations T_m^{-1} are bounded and by (3) equicontinuous for all m we may pick out a suitable subsequence satisfying (iii).

Setting $(\xi_m, \eta_m) = T_m(x_0, y_0)$ and $(\xi_n', \eta_n') = T_n(x_n', y_n')$ for (x_n', y_n') on γ_n, we obtain in a way similar to the one used above

$$\xi_m \cos\vartheta + \eta_m \sin\vartheta < \xi_n' \cos\vartheta + \eta_n' \sin\vartheta$$

from which we conclude that for $m>n$ the point (ξ_m, η_m) lies in $T_n(\gamma_n)$. Now take n large enough so that there exists a point (x_0, y_0) of R_n at distance $>\rho_n$ from the boundary of R_n. As the corresponding points (ξ_m, η_m) lie all in $T_n(\gamma_n)$ there exists a converging subsequence tending to a point (ξ_0, η_0). In view of (3), the distance between $T_m^{-1}(\xi_m, \eta_m)$ and $T_m^{-1}(\xi_0, \eta_0)$ tends to zero if the distance between (ξ_m, η_m) and (ξ_0, η_0) tends to zero. Hence for our subsequence $T_m^{-1}(\xi_0, \eta_0)\to(x_0, y_0)$, or (x_0, y_0) is the image in T_∞^{-1} of a point (ξ_0, η_0) of $T_n(R_n)=D$. Furthermore, as T_m^{-1} maps $T_n(R_n)$ into a domain containing all points (x_0, y_0) of R_n at distance $>\rho_n$ from the boundary of R_n and as the mapping is one-to-one it follows that the boundary of $T_n(R_n)$ is mapped by T_m^{-1} into a set of points whose distance from the boundary of R_n is $\leqq\rho_n$. In view of the uniform convergence we conclude that T_∞^{-1} maps the boundary of $T_n(R_n)$ into a set of points distinct from the center of R. At the same time $T_\infty^{-1}T_n(R_n)$ contains a neighborhood of the center of R, namely, the set of points at distance $>\rho_n$ from the boundary of R_n, which proves (iv).

From (2) we derive by passing to the limit for $T_\infty^{-1}(\xi, \eta)=(x, y)$

$$\lim_{n\to\infty} \zeta_n(\xi, \eta) = \lim_{n\to\infty} (-z_n(x, y) + x\xi + y\eta - (x^2 + y^2)/2);$$

by (4) the derivatives of $\zeta(\xi, \eta)$ converge uniformly, hence evidently

$$\frac{\partial\zeta_n}{\partial\xi}\to\frac{\partial \lim \zeta_n}{\partial\xi}, \quad \frac{\partial\zeta_n}{\partial\eta}\to\frac{\partial \lim \zeta_n}{\partial\eta}.$$

If the limit function ζ is analytic in ξ and η, the map T_∞^{-1} is "schlicht" in the neighborhood of a point Q such that $T_\infty^{-1}(Q)$ is a point c near the center of R.

To prove this, let us assume the existence of another point Q' with $T_\infty^{-1}(Q')=c$. Call, for n in the above subsequence, σ_n the segment joining $T_n^{-1}(Q)$ to $T_n^{-1}(Q')$, and k_ρ a circle of small, arbitrary radius $\rho>0$ about Q and lying in D. The intersections of k_ρ with $T_n(\sigma_n)$ have at least one limit point L and we conclude from the uniform convergence of T_n^{-1} to T_∞^{-1} that $T_\infty^{-1}(L)=c$. Thus it is shown that if there are two points in D, Q and Q', such that $T_\infty^{-1}(Q)=T_\infty^{-1}(Q')=c$, then there exists an infinity of such points. Since the functions $x(\xi, \eta)$ and $y(\xi, \eta)$ determined by the transformation T_∞^{-1} are analytic, there exists a real curve C such that the two equations hold

$$(*) \qquad [x(\xi, \eta), y(\xi, \eta)] = c.$$

All points (ξ, η) such that (*) is satisfied lie within a closed set interior to D because T_∞^{-1} on the boundary of D differs from c as formerly proved. Taking local Puiseux developments it is easy to show that the above analytic curve has at each point a finite number of tangents and that no point of it is a "dead end," i.e., endpoint of one arc and of no other. Furthermore (*) implies that there are at most a finite number of singular points of C in D. Let us choose an arbitrary regular point on C and introduce an orientation on C in the neighborhood of this point. Continue the orientation on the same branch of the curve until we arrive at a singular point. Among the several arcs ending at the singular point we pick one distinct from the one already oriented and continue the orientation on this arc. Proceeding this way we must finally encounter a point twice, as the length of our path is bounded on account of the finiteness of the number of singular points. Hence there exists a closed circuit δ consisting only of arcs of C, in particular one bounding a simply connected domain D'. Now $T_n^{-1}(\delta)$ is the boundary of a simply connected domain $T_n^{-1}(D')$ and $T_n^{-1}(\delta)\to c$. Hence $T_n^{-1}(D')\to c$, $T_\infty^{-1}(D')=c$ which is impossible since T_∞^{-1} is analytic and does not map identically into c.

2. We first prove the following theorem.

THEOREM 1. *Let A, B, C, E be analytic functions of five complex arguments u, v, x, p, q in a bounded and closed neighborhood N of a real point u_0, v_0, x_0, p_0, q_0 and suppose A, B, C, E depending on a parameter μ and converging uniformly in N as $\mu\to\mu^*$. Assume A, B, C, E real for real u, v, x, p, q,*

$$\Delta \equiv 4(AC + E) - B^2 > 0,$$

and $|\Delta|^{-1}$ uniformly bounded in N. Suppose that there exists a rectangle R with u_0, v_0 as center and such that there exists for $\mu\neq\mu^*$ a real and analytic function $x(u, v)$ which is a solution of the Monge-Ampère equation†

$$Ar + Bs + Ct + (rt - s^2) = E; \tag{7}$$

suppose that for this solution (u, v, x, p, q) remains in N, when (u, v) ranges over R. Then there exists a real neighborhood of (u_0, v_0) such that the solutions $x(u, v)$ corresponding to a subsequence of values of μ converge uniformly to a limit function $x^*(u, v)$ as $\mu\to\mu^*$. There exists an analytic relation between u, v, x^*, and $x^*(u, v)$ has continuous first derivatives. Moreover, $x^*(u, v)$ is analytic in u and v and satisfies the limit relation (7) for $\mu\to\mu^*$ at each point where a certain not identically vanishing analytic function $G(u, v)$ does not vanish.

Since

$$0 < \Delta = 4(A + t)(C + r) - (B - 2s)^2,$$

we have $A+t$ and $C+r$ for each solution always $\neq 0$ and of the same sign. Hence in the sequence, there are infinitely many solutions for which $A+t$ is of the same sign, and we may even assume $A+t$ positive since in the opposite case we may replace A, B, C, E, and $x(u, v)$ by $-A$, $-B$, $-C$, $+E$, and $-x(u, v)$ without changing hypothesis nor conclusion of the theorem.

Let, for all μ of the subsequence,

$$|A|,\ |C|,\ \left|\frac{B}{2}\right| \leqq a.$$

† Here p, q, r, s, t denote, as usual, the partial derivatives of $x(u, v)$ of the first and second orders, respectively.

We find

$$
\begin{aligned}
0 &< (r+C)(t+A) - \left(\frac{B}{2} - s\right)^2 \\
&\leqq (r+a)(t+a) - s^2 + \frac{B^2}{4} - B\left(\frac{B}{2} - s\right) \\
&\leqq (r+a)(t+a) - s^2 + a^2 + 2a\,\frac{r+a+t+a}{2} \\
&= (r+2a)(t+2a) - s^2.
\end{aligned}
$$

Hence $x(u, v)+a(u^2+v^2)$ is convex in R, and satisfies the transformed Monge-Ampère equation with the same discriminant Δ. It is therefore legitimate to *assume henceforth, that for all μ of a subsequence $x(u, v)$ is convex*. Let us then apply the contact transformation T of the (u, v, x)-space into a (ξ, η, ζ)-space. The Monge-Ampère equation thereby is transformed into another,

$$
\tilde{A}\tilde{r} + \tilde{B}\tilde{s} + \tilde{C}\tilde{t} + \tilde{D}(\tilde{r}\tilde{t} - \tilde{s}^2) = \tilde{E}, \tag{8}
$$

with

$$
\tilde{A} = C - 1,\ \tilde{C} = A - 1,\ \tilde{B} = -B,\ \tilde{D} = 1 - E - A - C,\ \tilde{E} = -1. \tag{9}
$$

The discriminant of the new equation equals that of the former:

$$
4(\tilde{A}\tilde{C} + \tilde{D}\tilde{E}) - \tilde{B}^2 = 4(AC + E) - B^2 = \Delta.
$$

The solution $\zeta(\xi, \eta)$ exists for all μ of a suitable subsequence in a common domain D of the (ξ, η)-plane and converges with its first derivatives to a limit function ζ^* and its derivatives. Let ξ_0, η_0 be an arbitrary, but fixed point in D, π_0 and κ_0 the derivatives of $\zeta^*(\xi, \eta)$ at (ξ_0, η_0). Replace $\zeta(\xi, \eta)$ by $\zeta(\xi, \eta)-\pi_0(\xi-\xi_0)-\kappa_0(\eta-\eta_0)=\bar{\zeta}(\xi, \eta)$. Then

$$
\tilde{A}\frac{\partial^2\bar{\zeta}}{\partial\xi^2} + \tilde{B}\frac{\partial^2\bar{\zeta}}{\partial\xi\partial\eta} + \tilde{C}\frac{\partial^2\bar{\zeta}}{\partial\eta^2} + \tilde{D}\left(\frac{\partial^2\bar{\zeta}}{\partial\xi^2}\frac{\partial^2\bar{\zeta}}{\partial\eta^2} - \left(\frac{\partial^2\bar{\zeta}}{\partial\xi\partial\eta}\right)^2\right) = \tilde{E} \tag{10}
$$

and

$$
\begin{aligned}
u &= \frac{\partial\bar{\zeta}}{\partial\xi} + \pi_0, \qquad v = \frac{\partial\bar{\zeta}}{\partial\eta} + \kappa_0, \\
x &= -\bar{\zeta} - \pi_0(\xi - \xi_0) - \kappa_0(\eta - \eta_0) + \xi\left(\frac{\partial\bar{\zeta}}{\partial\xi} + \pi_0\right) + \eta\left(\frac{\partial\bar{\zeta}}{\partial\eta} + \kappa_0\right) \\
&\quad - \frac{1}{2}\left[\left(\frac{\partial\bar{\zeta}}{\partial\xi} + \pi_0\right)^2 + \left(\frac{\partial\bar{\zeta}}{\partial\eta} + \kappa_0\right)^2\right],
\end{aligned}
$$

$$p = \xi - \frac{\partial\bar{\zeta}}{\partial\xi} - \pi_0, \qquad q = \eta - \frac{\partial\bar{\zeta}}{\partial\eta} - \kappa_0.$$

For all μ of the subsequence we may assume

$$\left(\frac{\partial\bar{\zeta}(\xi_0, \eta_0)}{\partial\xi}\right)^2 + \left(\frac{\partial\bar{\zeta}(\xi_0, \eta_0)}{\partial\eta}\right)^2 \leqq 1,$$

and our formulas show that the coefficients $\tilde{A}$, $\tilde{B}$, $\tilde{C}$, $\tilde{D}$, $\tilde{E}$ considered as analytic functions of five independent variables $\xi, \eta, \bar{\zeta}, \partial\bar{\zeta}/\partial\xi, \partial\bar{\zeta}/\partial\eta$ (instead of u, v, x, p, q) exist in a complex neighborhood of ξ_0, η_0, $\bar{\zeta} = \zeta^*(\xi_0, \eta_0)$, $\partial\bar{\zeta}/\partial\xi = 0$, $\partial\bar{\zeta}/\partial\eta = 0$ and tend there uniformly to limit functions as $\mu \to \mu^*$.

By (6) we see that for all μ the second derivatives of our solution $\bar{\zeta}(\xi, \eta)$ are absolutely bounded by 1. Hence there exists a neighborhood of the point (ξ_0, η_0) of the (ξ, η)-plane in which, by the theorem on p. 432 of Part I of our paper, all $\bar{\zeta}(\xi, \eta)$ can be developed into power series in $\xi - \xi_0$ and $\eta - \eta_0$ such that there is a common majorant series for all μ which converges. Hence the limit $\zeta^*(\xi, \eta)$ of $\bar{\zeta}(\xi, \eta)$ is analytic in a neighborhood of ξ_0, η_0 and the same holds in the neighborhood of *any* interior point of D. As we have seen in §1 the transformation

$$u = \frac{\partial\zeta^*}{\partial\xi}, \qquad v = \frac{\partial\zeta^*}{\partial\eta}$$

maps the domain D into a set containing a neighborhood of the point (u_0, v_0) and there exists $\lim_{\mu\to\mu^*} x(u, v) = x^*(u, v) = -\zeta^*(\xi, \eta) + u\xi + \eta v - (u^2 + v^2)/2$. Thus it is shown that for a certain neighborhood of (u_0, v_0) the quantities u, v, x^* can be written as analytic functions of two variables ξ, η. Wherever the determinant $\partial(u, v)/\partial(\xi, \eta)$ differs from zero we may introduce u and v instead of ξ and η, and x^* becomes analytic in u and v. The relation $\partial(u, v)/\partial(\xi, \eta) = 0$ cannot be identically satisfied as the mapping $(\xi, \eta) \to (u, v)$ contains a neighborhood of (u_0, v_0). For all points with $\partial(u, v)/\partial(\xi, \eta) \neq 0$ we may differentiate with respect to u, v and find $\partial x^*(u, v)/\partial u = \xi - \partial\zeta^*/\partial\xi$, $\partial x^*/\partial v = \eta - \partial\zeta^*/\partial\eta$. Hence $\partial x^*/\partial u$ and $\partial x^*/\partial v$ are analytic functions of ξ, η and therefore continuous, even at points where $\partial(u, v)/\partial(\xi, \eta) = 0$ (since derivative at limit point equals limit of derivative). The proof of the theorem is now easily completed in view of the fact that the inverse transform of (8) is (7).

The following example shows that there are cases where there exists an analytic relation between u, v, x in a neighborhood of the origin, the function x can be considered as dependent on u and v and having continuous first derivatives there, and $x(u, v)$ is a solution of an analytic and elliptic Monge-

Ampère equation everywhere except on the u-axis where it fails to have second derivatives at all. The function is

$$x(u, v) = \frac{(3u)^{4/3}}{4} + \frac{v^2}{2},$$

the equation is

$$(-1 + p^2)r + rt - s^2 = 1.$$

3. There are wide classes of Monge-Ampère equations important in differential geometry such that the function $\partial(u, v)/\partial(\xi, \eta)$ cannot vanish at any interior point (u_0, v_0). In this case the limit function $x^*(u, v)$ is necessarily analytic.

Let us introduce two new variables α, β and solve the following initial problem for our variables ξ, η to be considered as functions of α and β:

$$\left(\frac{\partial\eta}{\partial\alpha} + i\frac{\partial\eta}{\partial\beta}\right)(\tilde{B} - 2\tilde{D}\tilde{s} + i\Delta^{1/2}) = \left(\frac{\partial\xi}{\partial\alpha} + i\frac{\partial\xi}{\partial\beta}\right)\cdot 2(\tilde{C} + \tilde{D}\tilde{r}), \tag{11}$$

$$\xi(\alpha, 0) = \alpha, \qquad \eta(\alpha, 0) = 0.$$

By the theorem of Cauchy-Kowalewski we have a solution $\xi(\alpha, \beta)$, $\eta(\alpha, \beta)$, analytic in the neighborhood of the origin. We have at the origin

$$\frac{\partial(\xi, \eta)}{\partial(\alpha, \beta)} = \frac{\partial\eta}{\partial\beta} = -2i(\tilde{C} + \tilde{D}\tilde{r})\frac{\left(1 + i\frac{\partial\xi}{\partial\beta}\right)}{(\tilde{B} - 2\tilde{D}\tilde{s} + i\Delta^{1/2})}$$

and hence may assert that there exists a neighborhood of the origin for which $\partial(\xi, \eta)/\partial(\alpha, \beta) \neq 0$ and the mapping $(\alpha, \beta) \to (\xi, \eta)$ is one-to-one. (11) may be written, with $=\partial/\partial\alpha + i\partial/\partial\beta$

$$(\tilde{B} + i\Delta^{1/2})\nabla\eta - 2\tilde{C}\nabla\xi - 2\tilde{D}\nabla\tilde{\zeta}_\xi = 0. \tag{12}$$

With the aid of $\tilde{\zeta}_\xi = u$, $\xi = u + p^*$, $\eta = v + q^*$ we obtain, in view of (9), for $\partial(u, v)/\partial(\xi, \eta) \neq 0$,

$$\begin{aligned}[(-B + i\Delta^{1/2})(1 + t^*) - 2(A - 1)s^*]\nabla v + [(-B + i\Delta^{1/2})s^* \\ - 2(A - 1)(r^* + 1) - 2(1 - E - A - C)]\nabla u = 0,\end{aligned}$$

and this is seen to be equivalent to

$$2(A + t^*)\nabla v + (-B + 2s^* - i\Delta^{1/2})\nabla u = 0 \tag{13'}$$

or

$$(-B - i\Delta^{1/2})\nabla u + 2A\nabla v + 2\nabla q^* = 0. \tag{13}$$

From (13′) we have

$$(14) \qquad (-B + i\Delta^{1/2})\nabla v + 2C\nabla u + 2\nabla p^* = 0,$$

in view of

$$\Delta = 4(A + t^*)(C + r^*) - (B - 2s^*)^2.$$

Now (ξ, η) is analytic in (α, β), and u, v, p^*, q^*, x^* are analytic in (ξ, η), hence also in (α, β) and the relations (13), (14) hold independently of $\partial(u, v)/\partial(\xi, \eta) \neq 0$. We have moreover

$$(15) \qquad \nabla x^* - p^*\nabla u - q^*\nabla v = 0.$$

(13), (14), (15) form the characteristic equations of (7) for the limit function $x^*(u, v)$. The mapping $(\alpha, \beta) \to (u, v)$ is one-to-one as both $(\alpha, \beta) \to (\xi, \eta)$ and $(\xi, \eta) \to (u, v)$ are one-to-one (see p. 368). We have

THEOREM 2. *If, in addition to the hypotheses of Theorem* 1, *we have* $\partial(u, v)/\partial(\alpha, \beta) \neq 0$, *then* $x^*(u, v)$ *is analytic in* (u, v).

The proof follows immediately from Theorem 1 and the relation

$$\frac{\partial(u, v)}{\partial(\alpha, \beta)} = \frac{\partial(u, v)}{\partial(\xi, \eta)} \frac{\partial(\xi, \eta)}{\partial(\alpha, \beta)}$$

which implies $\partial(u, v)/\partial(\alpha, \beta) = 0$ if $\partial(u, v)/\partial(\xi, \eta) = 0$.

We mention the following case which frequently presents itself in differential geometry: Let us operate with $\partial/\partial\alpha - i\partial/\partial\beta$ on (13) and (14) and take imaginary parts. On eliminating the derivatives of x^*, p^*, q^* with respect to (α, β) with the aid of (13), (14), (15), we obtain

$$\frac{\partial^2 u}{\partial \alpha^2} + \frac{\partial^2 u}{\partial \beta^2} = h_1\left[\left(\frac{\partial u}{\partial \alpha}\right)^2 + \left(\frac{\partial u}{\partial \beta}\right)^2\right] + h_2\left[\left(\frac{\partial v}{\partial \alpha}\right)^2 + \left(\frac{\partial v}{\partial \beta}\right)^2\right]$$
$$+ h_3\left(\frac{\partial u}{\partial \alpha}\frac{\partial v}{\partial \alpha} + \frac{\partial u}{\partial \beta}\frac{\partial v}{\partial \beta}\right) + h_4\frac{\partial(u, v)}{\partial(\alpha, \beta)},$$
$$\frac{\partial^2 v}{\partial \alpha^2} + \frac{\partial^2 v}{\partial \beta^2} = h_1'\left[\left(\frac{\partial u}{\partial \alpha}\right)^2 + \left(\frac{\partial u}{\partial \beta}\right)^2\right] + \cdots + h_4'\frac{\partial(u, v)}{\partial(\alpha, \beta)}.$$

Now suppose that the coefficients $h_1, h_2, h_3, h_4, h_1', h_2', h_3', h_4'$ *depend only on* u *and* v. As they are analytic in (u, v), a result of ours† shows that then necessarily $\partial(u, v)/\partial(\alpha, \beta) \neq 0$.

† *On the non-vanishing of the Jacobian in certain one-to-one mappings*, Bulletin of the American Mathematical Society, vol. 42 (1936), pp. 689–692.

Consider the special case where the coefficients of the equation (7) depend only on (u, v). Then the elimination of x^*, p^*, q^* is obviously possible. Furthermore it becomes possible to widen the conditions of Theorem 1, and to state

THEOREM 2′. *Let A, B, C, E be analytic functions of the complex variables u, v for (u, v) in $N[|u-u_0| \leqq \epsilon, |v-v_0| \leqq \epsilon]$ and suppose A, B, C, E depending on μ and converging uniformly as $\mu \to \mu^*$. Assume A, B, C, E real for real u, v,*

$$\Delta \equiv 4(AC + E) - B^2 > 0,$$

and $|\Delta|^{-1}$ uniformly bounded for (u, v) in N. Suppose that there exists a real and analytic function $x(u, v)$ which is a solution of (7) as (u, v) ranges over the real rectangle $R[|u-u_0| < \epsilon, |v-v_0| < \epsilon]$, and that $x(u, v)$ is uniformly bounded as μ varies. Then there exists a subsequence of values μ such that the corresponding solutions $x(u, v)$ converge uniformly in every closed subregion of R to an analytic limit function $x^*(u, v)$, which is a solution of the limit equation (7).

Proceeding precisely as in the proof of Theorem 1, we find a suitable constant α such that $x(u, v) + a(u^2 + v^2)$ is convex for some subsequence of values μ, and, in view of the assumptions, bounded as (u, v) ranges over R. It follows that in every closed concentric rectangle contained in R the first derivatives of $x(u, v)$ are uniformly bounded as μ varies. Thus the assumptions of Theorem 2′ imply those of Theorem 1 and, on the other hand, also those of Theorem 2.

UNIVERSITY OF CALIFORNIA,
BERKELEY, CALIF.

Reprinted from the Proceedings of the NATIONAL ACADEMY OF SCIENCES,
Vol. 24, No. 2, pp. 104–106. February, 1938.

ON THE EXISTENCE OF A CLOSED CONVEX SURFACE REALIZING A GIVEN RIEMANNIAN METRIC

BY HANS LEWY

DEPARTMENT OF MATHEMATICS, UNIVERSITY OF CALIFORNIA AT BERKELEY

Communicated January 14, 1938

Let σ_0 be the surface of the unit sphere and $(ds_0)^2$ the first fundamental form of Differential Geometry on σ_0. Denote by $(ds_1)^2$ another given positive definite form defined at each point of σ_0 which in local parameters (u, v) reduces to

$$(ds_1)^2 = E(du)^2 + 2Fdudv + G(dv)^2.$$

Suppose that the Gaussian curvature K_1 of the metric $(ds_1)^2$ is positive. Weyl[1] has attacked the following problem: Does there exist a closed convex surface σ_1 which may be mapped in a one-to-one way on σ_0 so that the first fundamental form of σ_1 is precisely $(ds_1)^2$? (Embedment problem in the large.)

Weyl shows first that, under certain conditions affecting the differentiability of E, F, G, one can indicate a family of metrics $(ds_\lambda)^2$, $0 \leqq \lambda \leqq 1$,

$$(ds_\lambda)^2 = E_\lambda d(u)^2 + 2F_\lambda dudv + G_\lambda(dv)^2$$

in the parameters (u, v) on σ_0, such that

i) E_λ, F_λ, G_λ depend analytically on λ;
ii) for all λ the Gaussian curvature K_λ of $(ds_\lambda)^2$ is positive everywhere on σ_0;
iii) for $\lambda = 0$ and $\lambda = 1$, $(ds_\lambda)^2$ reduces to $(ds_0)^2$, respectively $(ds_1)^2$.

Weyl then proves: If there exists a surface σ_{λ_0} which solves the embedment problem of $(ds_{\lambda_0})^2$ for a certain value λ_0 of λ, then there also exists a sufficiently small neighborhood of λ_0, say $|\lambda - \lambda_0| < \epsilon(\lambda_0)$, such that the embedment can be obtained for all λ of this neighborhood.

These facts alone don't permit to conclude the existence of a σ_λ for $\lambda = 1$, as the described continuity method does not exclude the possibility (π) that there may be a value λ^1 which cannot be reached from $\lambda = 0$ in finitely many steps. Weyl discusses this point in the second part of his paper, suggesting a line of attack and establishing certain estimates which he believes to be the clue to the solution. However these ideas have never been carried through.

I have been able by a different method[2] to eliminate (π) and thus to complete the solution of the embedment problem in the large. The following lines give the essential steps of the proof for the case of analytic E, F, G; in this case E_λ, F_λ, G_λ are analytic in u, v, λ.

A result of Darboux[3] shows that the distance $\rho_\lambda(u, v)$ from a fixed point P of the space to the point (u, v) of σ_λ satisfies a Monge-Ampère equation whose coefficients depend on E_λ, F_λ, G_λ and their derivatives as well as on ρ_λ, $\frac{\partial \rho_\lambda}{\partial u}$ and $\frac{\partial \rho_\lambda}{\partial v}$. The equation is elliptic because $K_\lambda > 0$. Take P as the center of the largest sphere inscribed to σ_λ. We want to prove: Let $0 \leq \lambda_1 < \lambda_2 < \ldots$ be a sequence tending to λ^1, such that $(ds_{\lambda_1})^2$, $(ds_{\lambda_2})^2$, ... can be embedded. Then $(ds_{\lambda^1})^2$ can be embedded. This fact is established as soon as we know that in the sequence ρ_{λ_1}, ρ_{λ_2}, ... of analytic functions of (u, v) there exists a subsequence converging to an analytic limit function $\rho_{\lambda^1}(u, v)$ which solves Darboux's equation for $\lambda = \lambda'$. This fact follows immediately from Theorem 2 of my paper[4] if the conditions of the theorem can be verified. They are:

i) $|\rho_{\lambda_1}|$, $|\rho_{\lambda_2}|$, ... are uniformly bounded. This is a consequence of Bonnet's theorem which states that the diameter of a closed convex surface is bounded from above by a number depending only on a (positive) lower bound of the Gaussian curvature.

ii) $\left|\frac{\partial\rho_{\lambda_1}}{\partial u}\right|$, $\left|\frac{\partial\rho_{\lambda_1}}{\partial v}\right|$, $\left|\frac{\partial\rho_{\lambda_2}}{\partial u}\right|$, ... are uniformly bounded. This follows from $(d\rho_\lambda)^2 \leqq (ds_\lambda)^2$.

iii) The coefficient of $\frac{\partial^2\rho_\lambda}{\partial u^2}\frac{\partial^2\rho_\lambda}{\partial v^2} - \left(\frac{\partial^2\rho_\lambda}{\partial u\partial v}\right)^2$ in Darboux's equation can be assumed to be $= 1$ while the other coefficients are bounded for complex values of u, v, ρ_λ, $\frac{\partial\rho_\lambda}{\partial u}$, $\frac{\partial\rho_\lambda}{\partial v}$ in a bounded neighborhood of their values under consideration. The contrary assumption is ruled out by a simple geometrical consideration showing that the radii of the maximum spheres inscribed in σ_{λ_1}, σ_{λ_2}, ... are bounded away from zero.

iiii) Elimination of ρ_λ. The independent variables (u, v) considered as functions of variables (α, β) which are characteristic with respect to the Monge-Ampère equation, satisfy two equations of the form

$$\frac{\partial^2 u}{\partial\alpha^2} + \frac{\partial^2 u}{\partial\beta^2} = h_{1\lambda}\left[\left(\frac{\partial u}{\partial\alpha}\right)^2 + \left(\frac{\partial u}{\partial\beta}\right)^2\right] + h_{2\lambda}\left(\frac{\partial u}{\partial\alpha}\frac{\partial v}{\partial\alpha} + \frac{\partial u}{\partial\beta}\frac{\partial v}{\partial\beta}\right) + h_{3\lambda}\left[\left(\frac{\partial v}{\partial\alpha}\right)^2 + \left(\frac{\partial v}{\partial\beta}\right)^2\right] + h_{4\lambda}\frac{\partial(u, v)}{\partial(\alpha, \beta)},$$

$$\frac{\partial^2 v}{\partial\alpha^2} + \frac{\partial^2 v}{\partial\beta^2} = k_{1\lambda}\left[\left(\frac{\partial u}{\partial\alpha}\right)^2 + \left(\frac{\partial u}{\partial\beta}\right)^2\right] + \ldots + k_{4\lambda}\frac{\partial(u, v)}{\partial(\alpha,\beta)}, \quad (1)$$

where $h_{1\lambda}, h_{2\lambda}, \ldots, k_{1\lambda}, k_{2\lambda}, \ldots$ depend on (u, v), but not explicitly on ρ_λ nor $\frac{\partial\rho_\lambda}{\partial u}$ nor $\frac{\partial\rho_\lambda}{\partial v}$. In our case $\alpha + i\beta$ and $\alpha - i\beta$ are parameters on the asymptotic lines and our equations (1) are identical with equations already found by Darboux.[5]

[1] Hermann Weyl, "Über die Bestimmung einer geschlossen konvexen Fläche durch ihr Linienelement." *Vierteljahrsschrift der naturf. Ges. Zürich*, **61**, 40–72 (1915).

[2] I have been in possession of this proof for several years, cf. *Bull. Am. Math. Soc.*, **42**, 824 (1936). The preparatory investigations of Monge-Ampère equations were undertaken mainly for this purpose and published in *Trans. Am. Math. Soc.*, **37**, 417–434 (1935) and **41**, 365–374 (1937).

[3] G. Darboux, *Théorie des surfaces*, vol. 3, p. 260 (Paris, 1894).

[4] Loc. cit., **41**, 373 (1937); also *Bull. Am. Math. Soc.*, **42**, 689–692 (1936).

[5] Loc. cit., p. 290.

ON DIFFERENTIAL GEOMETRY IN THE LARGE, I
(MINKOWSKI'S PROBLEM)

BY

HANS LEWY

Reprinted from the

TRANSACTIONS OF THE AMERICAN MATHEMATICAL SOCIETY

Vol. 43, No. 2, pp. 258-270

March, 1938

ON DIFFERENTIAL GEOMETRY IN THE LARGE, I (MINKOWSKI'S PROBLEM)*

BY

HANS LEWY

Introduction. Hermann Minkowski,† in a fundamental paper on convex bodies, proposed the following Problem (M): to determine a convex, three-dimensional body B whose surface admits of a given Gaussian curvature $K(n)>0$, assigned as a continuous function of the direction of the interior normal n to the surface. After having stated three obviously necessary conditions for the function $K(n)$, Minkowski proceeds to solve an analogous Problem (M′) for polyhedra. He then considers a passage to the limit among the solutions of problems (M′) approximating (M), and establishes their convergence to a convex body B_0. This construction leaves open the question as to whether B_0 is a solution of (M). Minkowski remarks that B, if it exists, is, to within a translation, the uniquely determined solution of a certain third Problem (M″) of the calculus of variations and that B_0, too, is a solution of (M″).

Now if we assume (H): the surface of B_0 is differentiable to a sufficiently high order, then B_0 solves (M). However, Minkowski does not discuss (H), but proves instead that the mixed volume $V(B_0, B_0, C)$ of B_0 with an arbitrary convex body C may be computed as though B_0 were a solution of (M) and that, furthermore, a convex body is, except for a translation, uniquely determined by its mixed volume with the totality of convex bodies. While later authors have modified Minkowski's methods, there has been no improvement of his results as far as the hypothesis (H) is concerned.

Thus Minkowski's results are open to the same criticism that could be raised against the early solutions of Plateau's problem: namely, that instead of solving the proposed problem, a more general problem is treated whose solution coincides with that of the former only if the latter solution satisfies certain highly restrictive conditions, and no indication is presented that these conditions are actually satisfied.

The present paper contains a solution of (M) for the case of analytic $K(n)$. It does not involve the Brunn-Minkowski inequalities, nor, indeed, the idea of mixed volume. It uses instead the author's results on elliptic and

* Presented to the Society, November 28, 1936; received by the editors March 1, 1937.

† Minkowski, *Werke*, pp. 231–276, entitled *Volumen und Oberfläche*. See especially §10. Also Bonnesen-Fenchel, *Theorie der konvexen Körper*, Ergebnisse, vol. 3, no. 1 (1934), which contains a bibliography of related papers.

analytic equations of the Monge-Ampère type and establishes the existence of an analytic B by a continuity method.

In order to make this treatment of (M) complete, a new proof of the uniqueness of B is included. This proof is based on a modification of a beautiful idea of Cohn-Vossen* who, in a similar problem, reduced the uniqueness problem to the determination of a certain topological index of a vector field. Our modification is such as to allow an application of this idea to a wider class of related uniqueness problems in the large.

1. Let $u(x, y)$ be a homogeneous polynomial solution of Laplace's equation of degree $n>2$. A suitable rotation of the (x, y)-system transforms $u(x, y)$ into a constant multiple of $\Re[(x+iy)^n]$, and the asymptotic directions of the surface $u=u(x, y)$, determined by

$$u_{xx}dx^2 + 2u_{xy}dxdy + u_{yy}dy^2 = 0, \tag{1}$$

undergo the same rotation as the coordinates. Assuming this rotation effected, we obtain for the asymptotic directions

$$\Re[(x + iy)^{n-2}(d(x + iy))^2] = 0,$$

or, in polar coordinates r, θ,

$$\Re[r^{n-2}e^{i(n-2)\theta}(dx + idy)^2] = 0.$$

Thus the vector $dx+idy$ of the asymptotic direction forms the angle $-(n-2)\theta/2$ or $-(n-2)\theta/2+\pi/2$ with the x-axis; the two asymptotic directions are perpendicular, and either direction turns through an angle $-(n-2)\pi$ as we follow it along a Jordan curve containing the origin in its interior. In other words: the asymptotic directions form two distinct fields of directions with the origin as a singular point of index $-(n-2)/2$. We notice furthermore that the discriminant of (1) is a constant multiple of r^{2n-4} and vanishes only for $r=0$.

2. We prove now the following lemma:

LEMMA. *Let $F(x, y, u, p, q, r, s, t)$ be analytic in the neighborhood of $(x_0, y_0, u_0, p_0, q_0, r_0, s_0, t)$ and $4(\partial F/\partial r)(\partial F/\partial t)-(\partial F/\partial s)^2>0$. Let $u(x, y)$ and its first derivatives p, q and second derivatives r, s, t be a solution of $F=0$, analytic in a neighborhood of (x_0, y_0) and such that $u(x_0, y_0)=u_0$, $p(x_0, y_0)=p_0, \cdots, t(x_0, y_0)=t_0$. Assume $u'(x, y)$ and its derivatives p', q', r', s', t' to be a second analytic solution of $F=0$, coinciding together with its derivatives p', q', r', s', t', with u and its derivatives at the point (x_0, y_0). Then the difference $U=u-u'$ represents a surface $U(x, y)$ whose Gaussian curvature is negative in a sufficiently small*

* Cohn-Vossen, *Zwei Sätze über die Starrheit der Eiflächen*, Göttinger Nachrichten, 1927, pp. 125–134.

neighborhood of (x_0, y_0), *with the exception of* (x_0, y_0) *itself, and the index of either of its distinct asymptotic directions is negative at* (x_0, y_0), *unless* $u(x, y)$ *is identically equal to* $u'(x, y)$.

In terms of the second derivatives R, S, T of $U(x, y)$ the asympototic directions of $U = U(x, y)$ are given by

$$(2) \qquad Q \equiv R dx^2 + 2S dx dy + T dy^2 = 0;$$

and (2) admits of real solutions dx, dy if $RT - S^2 < 0$. To show that $RT - S^2 < 0$ express $F(x, y, u(x, y), p(x, y), \cdots) - F(x, y, u'(x, y), p'(x, y), \cdots)$ as a power series in $(x - x_0, y - y_0)$ and observe that the terms of lowest degree are given by

$$E \equiv a\bar{R} + b\bar{S} + c\bar{T},$$

where $\bar{R}$, $\bar{S}$, $\bar{T}$ are the second derivatives of the non-vanishing terms $\bar{U}$ of lowest degree $n > 2$ in the development of U, and a, b, c are the values of $\partial F/\partial r$, $\partial F/\partial s$, $\partial F/\partial t$ at (x_0, y_0), for which by hypothesis $4ab - c^2 > 0$. Since $F(\cdots, u(x, y), \cdots) - F(\cdots, u'(x, y), \cdots) = 0$, we have $E = 0$, and this is possible only for $\bar{R}\bar{T} - \bar{S}^2 < 0$ or $\bar{R} = \bar{S} = \bar{T} = 0$. Now $\bar{R}$, $\bar{S}$, $\bar{T}$ vanish simultaneously only at (x_0, y_0). For there exists a suitable linear transformation of the (x, y)-plane with determinant 1 which leaves (x_0, y_0) invariant and transforms $E = 0$ into Laplace's equation. $\bar{U}$ is thereby transformed into a harmonic homogeneous polynomial of degree $n > 2$ in $(x - x_0, y - y_0)$, and for such polynomials we proved in §1 that the discriminant of the second derivatives vanishes only at (0, 0). Hence the development of $RT - S^2$ starts with the negative term $\bar{R}\bar{T} - \bar{S}^2$ and we conclude that $RT - S^2$ is negative for sufficiently small $|x - x_0|$, $|y - y_0|$ and vanishes only at $x = x_0$, $y = y_0$.

Since the linear transformation does not change the index of a field of directions, we conclude from §1 that the index of the field

$$(3) \qquad \bar{Q} \equiv \bar{R} dx^2 + 2\bar{S} dx dy + \bar{T} dy^2 = 0$$

is negative. Since, for sufficiently small values of $|x - x_0|$, $|y - y_0|$, the directions of the field (2) differ arbitrarily little from those of (3), the index of (3) is negative at (x_0, y_0). This completes the proof of the Lemma.

3. We may now prove the following:

THEOREM 1. *Two closed convex analytic surfaces* S *and* S' *are congruent if they possess the same positive Gaussian curvature* K *at points for which their inner normals are parallel and similarly directed.*

By parallel normals we map S and S' on the unit sphere σ. An arbitrary equator divides σ and, since the map is one-to-one, S and S', into two regions in each of which the sphere as well as S and S' assume the form $Z=Z(X, Y)$ in suitably chosen rectangular coordinates. Upon introducing

$$x = \frac{\partial Z}{\partial X}, \qquad y = \frac{\partial Z}{\partial Y} \tag{4}$$

as independent variables instead of (X, Y) and setting

$$H(x, y) = -Z(x, y) + xX(x, y) + yY(x, y), \tag{5}$$

we obtain

$$X = H_x, \qquad Y = H_y, \qquad Z = -H + xX + yY. \tag{6}$$

How S and S' satisfy the condition that their curvatures for corresponding parallel normals, i.e., for the same value of (x, y), are the same positive function $K(x, y)$, whence S and S' are solutions of

$$\left(1 + \left(\frac{\partial Z}{\partial X}\right)^2 + \left(\frac{\partial Z}{\partial Y}\right)^2\right)^2 K(x, y) = \frac{\partial^2 Z}{\partial X^2}\frac{\partial^2 Z}{\partial Y^2} - \left(\frac{\partial^2 Z}{\partial X \partial Y}\right)^2$$

or

$$(1 + x^2 + y^2)^2 K(x, y) = \frac{\partial(x, y)}{\partial(X, Y)},$$

or, finally, of

$$H_{xx}H_{yy} - H_{xy}^2 = K^{-1}(x, y)(1 + x^2 + y^2)^{-2}. \tag{7}$$

It is readily shown that the second fundamental form of the surface (6) is

$$(H_{xx}dx^2 + 2H_{xy}dxdy + H_{yy}dy^2)(1 + x^2 + y^2)^{-1/2}. \tag{8}$$

Suppose that the formulas (4) and (5) lead to a function $H(x, y)$ if applied to S, and to $H'(x, y)$ if applied to S'. Then the statement of Theorem 1 is equivalent to the equation

$$H'(x, y) = H(x, y) + l(x, y),$$

where $l(x, y)$ is a linear function of (x, y).

Consider, with Cohn-Vossen, the congruence points of S and S', i.e., those points for which their normals and their second fundamental forms coincide. If all of their points were congruence points, we should have

$$H_{xx} = H'_{xx}, \qquad H_{xy} = H'_{xy}, \qquad H_{yy} = H'_{yy},$$

and the theorem is proved. In the alternative case we shall show the existence of at least one congruence point. The two second differential forms of S and S' may both be assumed to be positive definite. The equation

$$(H_{xx} - H'_{xx})dx^2 + 2(H_{xy} - H'_{xy})dxdy + (H_{yy} - H'_{yy})dy^2 = 0 \tag{9}$$

obtained by setting the two forms equal to one another, determines two distinct directions tangential to σ at each point of σ which is not a congruence point. For the ellipses of the (dx, dy)-plane

$$H_{xx}dx^2 + 2H_{xy}dxdy + H_{yy}dy^2 = 1$$

and

$$H'_{xx}dx^2 + 2H'_{xy}dxdy + H'_{yy}dy^2 = 1$$

have, by (7), the same area $\pi(1+x^2+y^2)K^{1/2}(x, y)$. As they are concentric but not identical, they intersect in four distinct points; the ratios of their coordinates are the two distinct solutions $dx: dy$ of (9).

It is impossible to construct a field of tangential directions on the sphere without singularities, and the sum of the indices of these singularities equals 2 if there are only finitely many singularities. Now choose, at an arbitrary point of σ, one of the two directions (9) and extend, by continuity, this choice over the whole of σ. If there were no congruence points, we should obtain a field of tangential directions on σ without singularities. Hence there is at least one congruence point (x_0, y_0).

Subtracting if necessary a linear function from $H'(x, y)$ we may assume that, at (x_0, y_0), $H(x, y)$ and $H'(x, y)$ coincide with their derivatives up to the second order, without affecting the truth of (7) nor that of (9). But now our lemma implies that unless H and H' are identical, the congruence point (x_0, y_0) is isolated and has a negative index. Summing over all indices of all singularities we still obtain a negative number in contradiction to the general fact mentioned above. Hence S and S' are congruent.

THEOREM 1′. *Consider a sequence of closed convex analytic surfaces $S(\tau)$ of positive curvature such that their corresponding functions $H(x, y, \tau)$ depend analytically on (x, y, τ) for small values of τ. Suppose that all surfaces $S(\tau)$ have a common point of contact corresponding to the same point (x, y) of σ. Assume that for each point of $S(0)$ the derivative $(\partial K/\partial\tau)_{\tau=0}$ vanishes. Then we have also $(\partial H/\partial\tau)_{\tau=0}=0$.*

Abbreviate the operator $(\partial/\partial\tau)_{\tau=0}$ by the use of the symbol δ and consider the field of tangential directions dx, dy on σ, determined by

$$(9') \qquad \delta H_{xx}dx^2 + 2\delta H_{xy}dxdy + \delta H_{yy}dy^2 = 0.$$

Unless $\delta H_{xx}=\delta H_{xy}=\delta H_{yy}=0$, there are, at each point (x, y), two real and distinct directions satisfying (9′). For we deduce from (7) by differentiation

$$(7') \qquad H_{yy}\delta H_{xx} + H_{xx}\delta H_{yy} - 2H_{xy}\delta H_{xy} = 0,$$

which yields $\delta H_{xx}\delta H_{yy}-(\delta H_{xy})^2<0$ in view of $H_{xx}H_{yy}-H_{xy}^2>0$.

As in the preceding proof we conclude first the existence of a point P on σ for which $\delta H_{xx}=\delta H_{yy}=\delta H_{xy}=0$. Assume this not to be true identically.

Since an addition of a linear function of (x, y) to δH does not affect the truth of (7′) and (9′), we may assume that the field of directions determined by (9′) in the neighborhood of P corresponds to a function δH that vanishes at P together with its derivatives up to the second order. This function δH may be considered as the difference between the following two solutions of (7′): δH itself and the identically vanishing solution. From the lemma we see that P is an isolated singularity of the field defined by (9′) and that the index of P is negative. The proof of Theorem 1′ can now easily be completed by recalling the conclusion at the end of the proof of Theorem 1.

4. We state now the following:

DEFINITION. *Let $d\omega$ be the surface element of the unit sphere σ and let ξ, η, ζ be the cosines of the angle of its normal with the axes of the (X, Y, Z)-system. A function F of a point of σ is called admissible if it depends analytically on the point of σ and if*

$$\int F\xi d\omega = \int F\eta d\omega = \int F\zeta d\omega = 0. \tag{10}$$

THEOREM 2. *For every admissible positive function K on σ there exists a closed convex analytic surface S whose curvature, considered as function of the interior normal vector (of length 1), equals K.*

First we shall demonstrate the following statement:

II. *Assume that for small values of a parameter τ an admissible positive function K depends analytically on the point of σ and τ, and that for $\tau=0$ there exists a surface $S(0)$ with $K(0)$ as corresponding curvature function. Then there exists an analytic closed convex surface $S(\tau)$ with $K(\tau)$ as corresponding curvature provided $|\tau|$ is sufficiently small.*

Let M be an arbitrary but analytic function on the unit sphere σ. Continue its definition from the sphere into the three-dimensional space containing it by assuming M to be zero at the center and linear on every ray issuing from the center. M thereby becomes a homogeneous function of degree 1 in any system of rectangular coordinates (ξ, η, ζ) with the origin as center and M is analytic everywhere except at the origin. We have from the homogeneity

$$M = M_\xi \xi + M_\eta \eta + M_\zeta \zeta. \tag{11}$$

Introduce

$$x = -\xi/\zeta, \qquad y = -\eta/\zeta, \qquad \zeta H(x, y) = -M. \tag{12}$$

We obtain

$$(13) \qquad H_x = M_\xi, \qquad H_y = M_\eta.$$

On σ the coordinates (x, y) may be used for either hemisphere $\zeta<0$ or $\zeta>0$. From (13) we conclude that H_x, H_y are analytic on the whole of σ if we define them on the equator $\zeta=0$ by continuity.

We find

$$(14) \qquad \pm(\xi, \eta, \zeta) = \left(\frac{x, y, -1}{(1+x^2+y^2)^{1/2}}\right), \qquad d\omega = \frac{dxdy}{(1+x^2+y^2)^{3/2}}.$$

Whenever M is such that $\partial(H_x, H_y)/\partial(x, y) \neq 0$, the reasoning of p. 260 shows that

$$(15) \qquad \frac{\partial(H_x, H_y)}{\partial(x, y)} \cdot (1+x^2+y^2)^2 = \phi(H, H)$$

is the reciprocal Gaussian curvature of the surface

$$(16) \qquad X = M_\xi, \qquad Y = M_\eta, \qquad Z = M_\zeta,$$

whence we conclude that ϕ remains invariant under the change which x, y, and H undergo as ξ, η, ζ are subjected to an orthogonal transformation that is sufficiently near the identity. This fact is obviously independent of the condition $\partial(H_x, H_y)/\partial(x, y) \neq 0$.

The integral over an arbitrary region

$$\mathfrak{J} = \iint \mp (H_{xx}H_{yy} - H_{xy}^2)dxdy = \iint \phi(H, H)\zeta d\omega$$

can be transformed into the integral over its boundary

$$\frac{1}{2}\int (H_x dH_y + H_y dH_x).$$

As the expressions H_x, H_y are analytic on the unit sphere we can extend the integration in $\mathfrak{J}$ over the whole of σ and obtain $\mathfrak{J}=0$ since the boundary integrals over the equator, generated by the integration over the upper and lower hemispheres, annul each other.

Using the invariance of $\phi(H, H)$, we obtain similarly

$$(17) \qquad \iint \phi(H, H)\xi d\omega = \iint \phi(H, H)\eta d\omega = \iint \phi(H, H)\zeta d\omega = 0.$$

We also find for the associated bilinear form

$$\phi(H_1, H_2) = (1 + x^2 + y^2)^2(H_{1xx}H_{2yy} + H_{1yy}H_{2xx} - 2H_{1xy}H_{2xy})$$

that

$$\iint \phi(H_1, H_2)\xi d\omega = \iint \phi(H_1, H_2)\eta d\omega = \iint \phi(H_1, H_2)\zeta d\omega = 0. \tag{18}$$

In other words, $\phi(H_1, H_2)$ is admissible if M_1 and M_2 are analytic on the sphere.

After these preliminary remarks we return to the given function $K^{-1}(\tau)$ and develop it into a power series in τ,

$$K^{-1}(\tau) = \kappa(\tau) = \kappa_0 + \tau\kappa_1 + \tau^2\kappa_2 + \cdots. \tag{19}$$

We try to determine a function $M(\xi, \eta, \zeta; \tau)$, homogeneous of first degree in ξ, η, ζ,

$$M(\tau) = M_0 + M_1\tau + M_2\tau^2 + \cdots, \tag{20}$$

depending analytically on ξ, η, ζ, τ (excepting the point $\xi=\eta=\zeta=0$) and being in the following relation to $K^{-1}(\tau)$: If we introduce by (12) the quantities $x, y, H(x, y; \tau)$, then (7) holds for all sufficiently small values of $|\tau|$. Let us develop the equation (7)

$$\phi(H, H) = \kappa$$

into a power series in τ and set the coefficients of both members equal. Denoting again by the operator δ the differentiation with respect to τ at $\tau=0$, we find the equations:

$$\delta^n M = \delta^n H(x, y)/(1 + x^2 + y^2)^{1/2}, \quad M_0 = H(x, y)/(1 + x^2 + y^2)^{1/2}, \tag{21}$$

$$H_{xx}\delta H_{yy} + H_{yy}\delta H_{xx} - 2H_{xy}\delta H_{xy} = \kappa_1(x, y)(1 + x^2 + y^2)^{-2},$$

$$\begin{aligned} H_{xx}\delta^2 H_{yy} + H_{yy}\delta^2 H_{xx} - 2H_{xy}\delta^2 H_{xy} = {} & 2!\kappa_2(x, y)(1 + x^2 + y^2)^{-2} \\ & - 2(\delta H_{xx}\delta H_{yy} - \delta H_{xy}\delta H_{xy}), \end{aligned} \tag{22}$$

. . . .

Since by hypothesis $K^{-1}(\tau)$ is admissible for all τ in question, the same is true for the coefficients:

$$\iint \kappa_\nu \xi d\omega = \iint \kappa_\nu \eta d\omega = \iint \kappa_\nu \zeta d\omega = 0, \qquad (\nu = 0, 1, 2, \cdots).$$

Suppose that we have found $\delta^n M(\xi, \eta, \zeta)$ for $n<m$, in accordance with (22), and analytic on the sphere. Then the preliminary remarks show that the right-hand member in the mth equation yields zero if we multiply it in turn by

$$(1 + x^2 + y^2)^2 \xi d\omega, \qquad (1 + x^2 + y^2)^2 \eta d\omega, \qquad (1 + x^2 + y^2)^2 \zeta d\omega$$

and integrate over the whole sphere. We therefore are entitled to apply a theorem of Hilbert* in which he states the now well known alternative for elliptic differential equations on the sphere for the special case of the equation with admissible right-hand member r,

$$(1 + x^2 + y^2)^2 [H_{xx}(\delta H)_{yy} + H_{yy}(\delta H)_{xx} - 2H_{xy}(\delta H)_{xy}] = r,$$
$$\delta H \cdot (1 + x^2 + y^2)^{-1/2} = \delta M,$$

and establishes the existence of a solution δM, which is differentiable infinitely many times as the coefficients of the differential equation are. Applying Hilbert's result to the mth equation (22) we find the function $\delta^m M$. Since this equation is elliptic because $S(0)$ is convex and accordingly $H_{xx}H_{yy} - H_{xy}^2 > 0$, we conclude from the analyticity of the equation the analyticity of $\delta^m M$ on the sphere.

It remains to be seen that for sufficiently small $|\tau|$ the series

$$M = M_0 + \sum_1^\infty \frac{\delta^n M}{n!} \tau^n$$

converges. This can be true only if we eliminate the arbitrariness that affects the determination of $\delta^m M$, since we may add to $\delta^m M$ an arbitrary linear combination of ξ, η, ζ with constant coefficients and still retain a solution of the differential equation for $\delta^m M$. Denote by $L[v]$ the linear elliptic differential expression in v on the sphere which in coordinates (x, y) reduces to

$$L(v) \equiv (1 + x^2 + y^2)^2 (H_{xx}u_{yy} + H_{yy}u_{xx} - 2u_{xy}H_{xy}), \quad u = v(1 + x^2 + y^2)^{1/2}.$$

Since $L[v] = 0$ has precisely the three linearly independent solutions $v = \xi, \eta, \zeta$, $L[v]$ admits of a Green's function of the second kind $G(A; B)$, where A and B are two points of the sphere. With the aid of one such $G(A; B)$ we solve the equations (22), written in the abbreviated form

$$L[\delta^\nu M] = f^\nu, \qquad (\nu = 1, 2, \cdots),$$

by setting

$$\delta^\nu M = \iint G(A; B) f^\nu(B) d\omega_B, \tag{23}$$

and thereby disposing of the arbitrariness in the determination of $\delta^\nu M$.

We shall say "a function f on the sphere satisfies a Hölder condition of ex-

* D. Hilbert, *Grundzüge einer Allgemeinen Theorie der Linearen Integralgleichungen*, Leipzig and Berlin, 1912, p. 250.

ponent α and coefficient C" if for arbitrary distinct points Q_1 and Q_2 of spherical distance $\overline{Q_1Q_2}>0$ we have

$$|f(Q_1) - f(Q_2)| \leqq C\overline{Q_1Q_2}^{\alpha},$$

where α is a positive constant less than 1. Similarly "the derivatives of f satisfy a Hölder condition of exponent α and coefficient C" if $0<\alpha<1$ and for arbitrary $Q_1 \neq Q_2$

$$|f_s(Q_1) - f_s(Q_2)| \leqq C\overline{Q_1Q_2}^{\alpha}, \qquad |f_n(Q_1) - f_n(Q_2)| \leqq C\overline{Q_1Q_2}^{\alpha},$$

where f_s is the derivative in the direction of the great circle joining Q_1 to Q_2 and f_n is the derivative in the direction normal to this circle. Similarly for higher derivatives.

With these notations the familiar estimates of the theory of linear elliptic differential equations are applied to the integral $v(A)=\iint G(A;B)f(B)d\omega_B$ and lead to the following:

LEMMA. *If f is bounded by C and satisfies a Hölder condition of exponent α and coefficient C, then $v(A)=\iint G(A;B)f(B)d\omega_B$ and its first and second derivatives are bounded by hC and satisfy a Hölder condition of exponent α and coefficient Ch where h does not depend on f.*

From the convergence of the series (19), with the use of the Heine-Borel theorem, we derive the existence of a number $\bar{\rho}>0$ such that

$$|\kappa_\nu| < \frac{1}{\bar{\rho}^\nu}, \qquad |\kappa_\nu'| < \frac{1}{\bar{\rho}^\nu}, \qquad (\nu = 1, 2, \cdots),$$

where κ_ν' stands for the directional derivative of κ_ν at an arbitrary point in an arbitrary direction. Hence there also exists a number $\rho>0$ such that $|\kappa_\nu|<1/\rho^\nu$ and κ_ν satisfies a Hölder condition of a certain exponent α and coefficient $1/\rho^\nu$.

The first equation (22) shows that δM and its first and second derivatives are bounded by h/ρ and satisfy a Hölder condition of exponent α and coefficient h/ρ.

Consider the equation

$$z^2(\tau) = \frac{1}{64h^2} - \sum_1^\infty \frac{\tau^n}{\rho^n}. \tag{24}$$

It admits of two roots for $z(\tau)$ and in particular the root

$$z(\tau) = -\frac{1}{8h} + \sum_1^\infty \frac{c_n\tau^n}{n!\rho^n}.$$

We have the following system of recurrent formulas from (24):

$$2z(0)z'(0) = -\frac{1}{\rho} \qquad \text{or} \qquad c_1 = \frac{4h}{\rho},$$

$$2z(0)z''(0) = -\frac{2!}{\rho^2} - 2z'^2(0) \qquad \text{or} \qquad c_2 = \frac{4h\cdot 2!}{\rho^2} + 8hc_1^2,$$

$$2z(0)z'''(0) = -\frac{3!}{\rho^3} - 6z'(0)z''(0) \qquad \text{or} \qquad c_3 = \frac{4h\cdot 3!}{\rho^3} + 24hc_1c_2,$$

. . . .

On the other hand consider the successive bounds and Hölder coefficients C_ν for $\delta^\nu M$ and its first and second derivatives as obtained from (22). We have evidently the same law by which to form the successive inequalities*

$$C_1 \leqq \frac{4h}{\rho},$$

$$C_2 \leqq 4h\cdot\frac{2!}{\rho^2} + 8hC_1^2,$$

$$C_3 \leqq 4h\cdot\frac{3!}{\rho^3} + 24hC_1C_2,$$

. . . .

As all terms involved are positive we conclude $C_\nu \leqq c_\nu$, $(\nu = 1, 2, \cdots)$. Hence the series

$$M(\tau) = M(0) + \sum_1^\infty \frac{\delta^n M}{n!}\tau^n$$

converges together with its first and second derivatives with respect to ξ, η, ζ uniformly for sufficiently small $|\tau|$ since $z(\tau)$ is a majorant series. In order to complete the proof of statement II we have to show that $M(\tau)$ depends analytically on ξ, η, ζ for all sufficiently small $|\tau|$ and $(\xi, \eta, \zeta) \neq (0, 0, 0)$.

Denoting by $M_n(\tau)$ the nth partial sum of $M(\tau)$, by $\kappa_n(\tau)$ that of $\kappa(\tau)$, and by $H_n(x, y; \tau)$ the function $M_n(\tau)(1+x^2+y^2)^{1/2}$, we have for arbitrary $\epsilon > 0$ and uniformly for all points of the sphere

$$|\Delta_n| \equiv |(H_{nxx}H_{nyy} - H_{nxy}^2)(1 + x^2 + y^2)^2 - \kappa_n(\tau)| < \epsilon,$$

provided n is large enough. For Δ_n is a polynomial whose term of lowest degree in τ is of degree n; if we replace everywhere in Δ_n the δM, $\delta^2 M$, $\cdots$ and

* Observe the invariance of $\phi(\delta H, \delta H)$.

their first and second derivatives with respect to (x, y) by their upper bounds C_ν and form the sum of the absolute values of all terms thus obtained, we obtain less than the nth remainder in the similarly formed majorant of

$$(H_{xx}(\tau)H_{yy}(\tau) - H^2_{xy}(\tau))(1 + x^2 + y^2)^2, \text{ where } H(\tau) = M(\tau)\cdot(1+ x^2 + y^2)^{1/2};$$

and this majorant converges since it is a polynomial of convergent power series in τ.

Thus $\lim_{n\to\infty}H_n(x, y; \tau) = H(x, y; \tau)$ is a solution of the analytic elliptic equation

$$(H_{xx}(\tau)H_{yy}(\tau) - H^2_{xy}(\tau))(1 + x^2 + y^2) = K^{-1}(\tau)$$

and its second derivatives satisfy a Hölder condition; hence, by a well known theorem,* $H(x, y; \tau)$ is analytic in x and y for every closed bounded region of the (x, y)-plane. This result may, of course, be formulated invariantly by stating that $M(\tau)$ depends analytically on the point of the sphere for small values of $|\tau|$.

It will be observed that the proof of statement II which is thereby completed follows precisely the routine way of solving a functional equation in the neighborhood of a value τ_0 of a parameter τ entering the equation, if it can be solved for $\tau = \tau_0$. Our theorem on Monge-Ampère equations† is, however, the essential tool which makes it possible to derive our present Theorem 2 from the statement II.

Returning to the hypotheses of Theorem 2 we form, with the given positive admissible distribution of reciprocal curvature on σ, the family of positive admissible distributions

$$K^{-1}(\tau) = 1 - \tau + \tau K^{-1},$$

which for $\tau = 0$ reduces to the reciprocal curvature of σ itself and for $\tau = 1$ to that of the surface to be determined. Let τ' be the greatest value of τ, $0 \leqq \tau' \leqq 1$, such that for every positive ϵ there exists an analytic surface $S(\tau)$ of curvature $K(\tau)$ with $\tau' - \epsilon < \tau < \tau'$. We shall show that $\tau' = 1$. First of all, by II, $\tau' > 0$. Since for all values of τ in $0 \leqq \tau \leqq 1$ the curvature $K(\tau)$ is bounded from below by a fixed positive number, a theorem of Bonnet‡ shows that all existing $S(\tau)$ have a diameter which is bounded from above. Now take an arbitrary normal of σ and introduce coordinates (ξ, η, ζ) such that its inter-

* E. Hopf, *Über den funktionalen, insbesondere den analytischen Charakter der Lösungen elliptischer Differentialgleichungen zweiter Ordnung*, Mathematische Zeitschrift, vol. 34 (1931), pp. 194–233.

† Hans Lewy, *A priori limitations for solutions of elliptic Monge-Ampère equations*, II, these Transactions, vol. 41 (1937), pp. 365–374, especially Theorem 2′ on p. 374.

‡ Cf. Blaschke, *Differentialgeometrie*, vol. 1, Berlin, 1924, p. 150.

section with σ becomes (0, 0, 1). Our introduction of the (x, y)-system will give this point the coordinates (0, 0). Let $H(x, y; \tau)\ (1+x^2+y^2)^{-1/2}$ be the distance of the tangent plane of $S(\tau)$ from a fixed point for which we take the center of gravity of $S(\tau)$. Then we have $|H(x, y, \tau)| < 2\beta$ where β is the upper bound of the diameter of $S(\tau)$ and (x, y) is restricted to the circle $x^2+y^2<1$. Apply our theorem on Monge-Ampère equations to a sequence of solutions $H(x, y, \tau)$ of (7) for which the parameter τ tends to τ'. We obtain a subsequence converging to an analytic solution $H(x, y, \tau')$ of (7) with $\tau=\tau'$. Since the origin of the (x, y)-system corresponds to an arbitrary normal of σ, the Heine-Borel Lemma shows the existence of a closed analytic surface $S(\tau')$ of curvature $K(\tau')$. Now $S(\tau')$ may be made the starting point for the construction of $S(\tau)$ for infinitely many values of τ, greater than and close to τ', with the aid of II. Thus the assumption that τ' be less than 1 and at the same time the greatest value in every neighborhood of which there are smaller values of τ admitting a surface $S(\tau)$, has led to a contradiction. Hence $\tau'=1$ and $S(\tau')=S(1)$ exists.

University of California
Berkeley Calif.

Transactions of the American Mathematical Society
Vol. 43, No. 3, May, 1938

GENERALIZED INTEGRALS AND DIFFERENTIAL EQUATIONS*

BY
HANS LEWY

Introduction. The idea of the following considerations can best be explained in the simplest case of an integral

$$\int a(x, f(x))db(x, f(x))$$

in which a and b are continuously differentiable functions of two variables, $f(x)$ a continuously differentiable function. The routine estimate of $\int adb$ gives bounds depending either on $\int |df(x)|$ or on the maximum of $|df/dx|$ in the interval of integration. It is, however, possible, as shown in Theorem 1, to give a bound that is entirely independent of the derivative of $f(x)$, and, consequently, to define, by a limiting process, $\int adb$, even in the case where $f(x)$ not only has no derivative, but is no longer continuous, provided $f(x)$ belongs to Baire's first class. The same observation holds for a great number of functionals of $f(x)$ whose construction depends on the derivative of $f(x)$, but for which bounds can be found nevertheless without reference to df/dx. In this paper we are concerned mainly with ordinary differential equations (Theorems 2–3′) and systems of hyperbolic equations in two independent variables (Theorem 6 and corollaries) whose treatment is based on a detailed study of the double integral (33).

1. **Simple integrals.** The theory of the Lebesgue-Stieltjes integral contains the following statement: If in a closed interval J the function $\beta_0(x)$ is monotone and the sequence of continuous functions $\alpha_\mu(x)$ is uniformly bounded and tends to a limit function $\alpha_0(x)$, then the Stieltjes integral $\int \alpha_\mu(x)d\beta_0(x)$ tends to the Lebesgue-Stieltjes integral $\int \alpha_0(x)d\beta_0(x)$ as $\mu \rightarrow 0$. If, furthermore, a sequence of functions $\beta_\mu(x)$ of bounded variation tends to $\beta_0(x)$ as $\mu \rightarrow \infty$ so that the total variation of the difference $\beta_\mu(x) - \beta_0(x)$ tends to zero, then

$$\limsup \left| \int \alpha_\mu(x)d\beta_\mu(x) - \int \alpha_0(x)d\beta_0(x) \right|$$
$$\leqq \limsup \left\{ \max |\alpha_\mu(x)| \cdot \int |d(\beta_\mu - \beta_0)| \right\} \rightarrow 0,$$

* Presented to the Society, November 28, 1936; received by the editors May 20, 1936, and May 24, 1937.

which relation leads to the following lemma:

LEMMA 1. *If in the interval* $X_0 \leqq x \leqq X_1$ *the continuous functions* $\alpha_\mu(x)$, $\mu = 1, 2, \cdots$, *are uniformly bounded and tend to* $\alpha_0(x)$ *as* $\mu \to \infty$, *and if the functions* $\beta_\mu(x)$ *of bounded variation tend to a function* $\beta_0(x)$ *of bounded variation while the total variation of* $\beta_\mu(x) - \beta_0(x)$ *tends to zero as* $\mu \to \infty$, *then*

$$\lim_{\mu\to\infty} \int_{X_0}^{x} \alpha_\mu(x) d\beta_\mu(x) \textit{ exists and } = \int_{X_0}^{x} \alpha_0(x) d\beta_0(x),$$

where the integral on the right is the Lebesgue-Stieltjes integral.

We proceed to introduce a new notion of integral which is essentially different from the Lebesgue-Stieltjes integral and is based upon the following theorem:

THEOREM 1. *In the interval* J $(X_0 \leqq x \leqq X_1)$ *there are given a function* $f(x)$ *satisfying the inequality* $|f(x) - f(X_0)| < F$, *a sequence* $\{f_\mu(x)\}$ *of continuously differentiable functions with* $|f_\mu(x) - f(X_0)| < F$ *such that* $f_\mu(x) \to f(x)$ *as* $\mu \to \infty$, *and* n *continuous functions* $g_1(x), \cdots, g_n(x)$ *with bounded total variations* $T[g_1], \cdots, T[g_n]$. *Denote by* $f^0, g_1^0, \cdots, g_n^0$ *the values of* $f(X_0), g_1(X_0), \cdots, g_n(X_0)$ *and by* $\gamma_1, \cdots, \gamma_n$ *upper bounds of* $|g_1(x) - g_1^0|, \cdots, |g_n(x) - g_n^0|$ *in* J. *Let* $\{g_{1\mu}(x)\}, \cdots, \{g_{n\mu}(x)\}$ *be* n *sequences of continuous functions of bounded variation, tending to* $g_1(x), \cdots, g_n(x)$, *respectively, as* $\mu \to \infty$ *while the total variations of the differences* $T[g_{1\mu} - g_1], \cdots, T[g_{n\mu} - g_n]$ *tend to zero and* $|g_{i\mu}(x) - g_i^0| < \gamma_i$ *for all* x *in* J, $i = 1, 2, \cdots, n$ *and* $\mu = 1, 2, \cdots$. *Suppose that in the* $(n+2)$*-dimensional domain*

$$D: \qquad \begin{aligned} &|z - f^0| \leqq F, \ X_0 \leqq x \leqq X_1, \\ &|y_1 - g_1^0| \leqq \gamma_1, \cdots, |y_n - g_n^0| \leqq \gamma_n \end{aligned}$$

the functions $a(z, x, y_1, \cdots, y_n)$ *and* $b(z, x, y_1, \cdots, y_n)$ *are continuously differentiable. Then the Stieltjes integral*

$$(1) \quad S_\mu(x) = \int_{X_0}^{x} a(f_\mu(x), x, g_{1\mu}(x), \cdots, g_{n\mu}(x)) db(f_\mu(x), x, g_{1\mu}(x), \cdots, g_{n\mu}(x))$$

tends to a limit $L(x)$ *as* $\mu \to \infty$.

Remarks. It is clear that any function $f(x)$ with $|f(x) - f^0| < F$ which is a limit of continuous functions may be considered as a limit of continuously differentiable functions $f_\mu(x)$ with $|f_\mu(x) - f^0| < F$. Moreover, the limit $L(x)$ is independent of the approximating sequences $f_\mu(x)$ and $g_{i\mu}(x)$. For the statement of Theorem 1 implies the existence of a limit no matter which sequences are used and hence any two sequences may be considered as subsequences of a third one containing both. Consequently we may state the following as a definition:

DEFINITION.

$$L(x) = \int_{X_0}^{x} a(f(x), x, g_1(x), \cdots, g_n(x))db(f(x), x, g_1(x), \cdots, g_n(x)).$$

In order to prove Theorem 1, we first assume that $b(z, x, y_1, \cdots, y_n)$ has continuous derivatives of second order of the mixed type. We determine a function $A(z, x, y_1, \cdots, y_n)$ in D as the solution of the differential equation

$$\frac{\partial A}{\partial z} = a\frac{\partial b}{\partial z}$$

with the initial condition $A(f^0, x, y_1, \cdots, y_n) = 0$. We obtain

$$\begin{aligned} &A(z, x, y_1, \cdots, y_n) \\ &\qquad = \int_{f^0}^{z} a(z', x, y_1, \cdots, y_n)b_z(z', x, y_1, \cdots, y_n)dz', \end{aligned}$$

where we may differentiate with respect to $x, y_1, \cdots, y_n$ under the integral sign. Thus we find

$$\begin{aligned} adb &= ab_z dz + ab_x dx + \sum_{i=1}^{n} ab_{y_i} dy_i \\ &= dA + (ab_x - A_x)dx + \sum (ab_{y_i} - A_{y_i})dy_i, \end{aligned}$$

$$\begin{aligned} A_x(z, x, y_1, \cdots, y_n) &= \int_{f^0}^{z} (ab_z)_x dz' \\ &= \int_{f^0}^{z} [(ab_x)_z + (a_x b_z - a_z b_x)]dz' \\ &= a(z, x, y_1, \cdots)b_x(z, x, y_1, \cdots) \\ &\quad - a(f^0, x, y_1, \cdots)b_x(f^0, x, y_1, \cdots) + \int_{f^0}^{z} (a_x b_z - a_z b_x)dz' \end{aligned}$$

and

$$\begin{aligned} A_{y_i}(z, x, y_1, \cdots, y_n) &= a(z, x, y_1, \cdots)b_{y_i}(z, x, y_1, \cdots) \\ &\quad - a(f^0, x, y_1, \cdots)b_{y_i}(f^0, x, y_1, \cdots) \\ &\quad + \int_{f^0}^{z} (a_{y_i} b_z - a_z b_{y_i})dz'. \end{aligned}$$

Hence

$$S_\mu(x) \equiv \int_{X_0}^{x} a(f_\mu(x'), x', g_{i\mu}(x'))db(f_\mu(x'), x', g_{i\mu}(x'))$$

$$= A(f_\mu(x), x, g_{i\mu}(x)) + \int_{X_0}^{x} a(f^0, x', g_{i\mu}(x'))b_x(f^0, x', g_{i\mu}(x'))dx$$

$$(2) \qquad + \sum_{i=1}^{n} \int_{X_0}^{x} a(f^0, x', g_{1\mu}(x'), \cdots)b_{y_i}(f^0, x', g_{1\mu}(x'), \cdots)dg_{i\mu}(x')$$

$$+ \int_{X_0}^{x} dx' \int_{f^0}^{f_\mu(x')} [a_z(z', x', g_{1\mu}(x'), \cdots)b_x(z', x', g_{1\mu}(x'), \cdots) - a_x b_z(\cdots)]dz'$$

$$+ \sum_i \int_{X_0}^{x} dg_{i\mu}(x') \int_{f^0}^{f_\mu(x')} [a_z(z', x', g_{1\mu}(x'), \cdots)b_{y_i} - a_{y_i}b_z]dz'.$$

This formula, derived under the assumption that b has continuous second derivatives of mixed type, still holds under the conditions of Theorem 1. For any b which is continuously differentiable in D may be uniformly approximated by a polynomial such that its first derivatives uniformly approximate those of b. Introducing the approximations instead of b into (2) and passing to the limit we obtain again the formula (2) as both sides of (2) involve only first derivatives of b and the passage to the limit under the integral signs is legitimate in view of the uniform convergence of the derivatives of the polynomials to those of b.

In the right-hand member of (2) we can effect the passage $\mu \to \infty$ by simply cancelling all reference to μ. This may be seen as follows. An integral $\int_{f^0}^{z} a_z(z', x, y_1, \cdots, y_n)b_{y_i}dz'$, for instance, is continuous in $z, x, y_1, \cdots, y_n$. Thus $\int_{f^0}^{f_\mu(x)} a(z', x, g_{1\mu}(x), \cdots)b_{y_i}dz'$ is a continuous function of x, bounded as $\mu \to \infty$, and converging as $\mu \to \infty$. Now the convergence of

$$\int_{X_0}^{x} dg_{i\mu}(x') \int_{f^0}^{f_\mu(x)} a_z(z', x', g_{1\mu}(x'), \cdots)b_{y_i}dz'$$

follows from Lemma 1.

Thus Theorem 1 is proved.

From (2) we have the following estimate:

$$|L(x)| \leqq MN\left\{ |f(x) - f^0| + \sum_{i=1}^{n} T_{X_0}^{x}[g_i] + |x - X_0| \right\}$$

$$+ 2N^2F\left(\sum_i T_{X_0}^{x}[g_i] + |x - X_0| \right),$$

where M is an upper bound for $|a|$ and $|b|$, N an upper bound for the moduli of the first derivatives of a and b in D.

Remark. We have, for instance, for every admissible $f(x)$

$$L(x) \equiv \int_0^x f(x')df(x') = \tfrac{1}{2}(f^2(x) - f^2(0)),$$

which leads to $L(1) = \frac{1}{2}$ for $f(x) = 0$ if $0 \leqq x < 1$, $f(1) = 1$. The Lebesgue-Stieltjes integral, however, would be 1.

2. **Ordinary differential equations.** We may now prove the following theorem:

THEOREM 2. *Let the functions $f(x)$, $g_1(x), \cdots, g_n(x)$ be continuously differentiable in $0 \leqq x \leqq X$ and $f(0) = f^0$, $g_1(0) = \cdots = g_n(0) = 0$. Denote by F an upper bound of $|f(x)|$ and by $G_1, G_2, \cdots, G_n$ the total variations of $g_1(x)$, $g_2(x), \cdots, g_n(x)$ in $[0, X]$. Suppose, for $\epsilon > 0$, that in the domain*

$$D_{3,\epsilon}: \quad |z| \leqq F, \ |y_i| \leqq G_i, \ |u| < \epsilon + 3FM_0 + \sum_{i=1}^{n} (e^{2FN} - e^{FN} + M_i e^{FN}) G_i$$

the functions $a_0, a_1, \cdots, a_n$ are continuous functions of z, y_i, u satisfying the inequalities $|a_k| \leqq M_k$, $k = 0, 1, \cdots, n$, and that a_0 has continuous derivatives of first order, bounded in absolute value by N. Then the solution $u(x)$ of the equation

$$\text{(E)} \qquad \begin{aligned} du(x) &= a_0(f(x), g_1(x), \cdots, g_n(x), u(x))df(x) \\ &\quad + \sum_{i=1}^{n} a_i(f(x), g_1(x), \cdots, g_n(x), u(x))dg_i(x) \end{aligned}$$

with $u(0) = 0$ can be extended over the whole interval $0 \leqq x \leqq X$. It satisfies, moreover, the inequality

$$|u(x)| \leqq 2FM_0 + \sum_{i=1}^{n} (e^{2FN} - e^{FN} + M_i e^{FN}) G_i.$$

Let us solve the auxiliary partial equation for $A(z, y_1, \cdots, y_n, u)$

$$\text{(3)} \qquad \frac{\partial A}{\partial z} + \frac{\partial A}{\partial u} \cdot a_0 = a_0, \qquad 1 - \frac{\partial A}{\partial u} \neq 0$$

under the initial condition

$$\text{(4)} \qquad A(z, y_i, u) = 0 \text{ for } z = 0.$$

The characteristic equations of (3) are

$$\text{(5)} \qquad dz : du : dA = 1 : a_0 : a_0.$$

Consider the family C of curves satisfying the differential equation $du/dz=a_0$ and passing through any point P of the domain

$$D_{2,\epsilon}:\quad |z|\leqq F,\ |y_i|\leqq G_i,\ |u|<\epsilon+2FM_0+\sum_{i=1}^{n}(e^{2FN}-e^{FN}+M_ie^{FN})G_i.$$

Since a_0 has continuous derivatives of first order throughout $D_{2,\epsilon}$ and is bounded by M_0, there exists one and only one curve through an arbitrary point P, and the corresponding value $\bar{u}$ of u for $z=0$ lies within the range

$$|\bar{u}|<\epsilon+3FM_0+\sum_{i=1}^{n}(e^{2FN}-e^{FN}+M_ie^{FN})G_i.$$

Conversely, an arbitrary curve of C is uniquely determined by the quantities $\bar{u}, y_1, \cdots, y_n$, and a point of the curve is determined by giving in addition the corresponding value of z. On writing $u=u(z, y_i, \bar{u})$, we find bounds for the derivatives of u with respect to the arguments $\bar{u}, y_1, \cdots, y_n, z$. We have, in fact,

$$\frac{d}{dz}\frac{\partial u}{\partial \bar{u}}=\frac{\partial a_0}{\partial u}\frac{\partial u}{\partial \bar{u}}\quad\text{with}\quad \frac{\partial u}{\partial \bar{u}}=1\quad\text{for}\quad z=0,$$

hence

$$e^{-N|z|}\leqq\frac{\partial u}{\partial \bar{u}}\leqq e^{N|z|}.$$

Similarly

$$\frac{d}{dz}\frac{\partial u}{\partial y_i}=\frac{\partial a_0}{\partial y_i}+\frac{\partial a_0}{\partial u}\frac{\partial u}{\partial y_i}\quad\text{with}\quad \frac{\partial u}{\partial y_i}=0\quad\text{for}\quad z=0,$$

hence

$$\left|\frac{\partial u}{\partial y_i}\right|\leqq e^{N|z|}-1.$$

On introducing the new variables $z, y_1, \cdots, y_n, u$ instead of $z, y_1, \cdots, y_n, \bar{u}$ we find throughout $D_{2,\epsilon}$

$$\frac{\partial(z, y_i, u)}{\partial(z, y_i, \bar{u})}=\frac{\partial u}{\partial \bar{u}},\quad \frac{\partial(z, y_i, \bar{u})}{\partial(z, y_i, u)}=\frac{\partial \bar{u}}{\partial u}=\left(\frac{\partial u}{\partial \bar{u}}\right)^{-1},\quad e^{-|z|N}\leqq\frac{\partial \bar{u}}{\partial u}\leqq e^{|z|N},$$

$$\frac{\partial \bar{u}}{\partial y_i}=-\frac{\partial u}{\partial y_i}\Big/\frac{\partial u}{\partial \bar{u}},\quad \left|\frac{\partial \bar{u}}{\partial y_i}\right|\leqq e^{|z|N}(e^{|z|N}-1). \tag{6}$$

Since $\bar{u}$ is constant along any curve of C, we have

$$\frac{\partial \bar{u}}{\partial z} + a_0 \frac{\partial \bar{u}}{\partial u} = 0, \qquad \left| \frac{\partial \bar{u}}{\partial z} \right| \leqq M_0 e^{N|z|}.$$

Now put, throughout $D_{2,\epsilon}$,

$$A(z, y_i, u) = u - \bar{u}. \tag{7}$$

Evidently A satisfies (3) and (4). Furthermore we have

$$\begin{aligned} &| A | \leqq | z | M_0, & &\left| \frac{\partial A}{\partial y_i} \right| \leqq e^{|z|N}(e^{|z|N} - 1), \\ &\left| \frac{\partial A}{\partial z} \right| \leqq M_0 e^{N|z|}, & &e^{-|z|N} \leqq \left| 1 - \frac{\partial A}{\partial u} \right| \leqq e^{|z|N}. \end{aligned} \tag{8}$$

Returning to the ordinary differential equation (E), we remark that the conditions of our theorem allow us to write (E) in the form

$$\frac{du}{dx} = \phi(x, u)$$

with $\phi(x, u)$ continuous in the rectangle R determined by

$$0 \leqq x \leqq X, \quad | u | \leqq \frac{\epsilon}{2} + 2FM_0 + H, \quad H = \sum_{i=1}^{n} (e^{2FN} - e^{FN} + M_i e^{FN}) G_i.$$

The fundamental existence theorem for differential equations shows that a solution through $[x=0,\ u(0)=0]$ always may be continued until it reaches the boundary of R. Hence a solution which cannot be continued across a certain point x_1 with $0 \leqq x_1 < X$ may be assumed to exist for $0 \leqq x \leqq x_1$ and to satisfy the condition

$$| u(x_1) | = \frac{\epsilon}{2} + 2FM_0 + H.$$

Thus our theorem is proved as soon as we show the following property of $u(x)$. If, in the interval $0 \leqq x \leqq x_1$, the solution $u(x)$ of (E) satisfies the inequality

$$| u(x) | \leqq \frac{\epsilon}{2} + 2FM_0 + H,$$

it satisfies the stronger inequality

$$| u(x) | \leqq 2FM_0 + H.$$

Indeed, by (E) we have, since $z=f(x)$, $y_i=g_i(x)$, $u=u(x)$ stay in $D_{2,\epsilon}$,

$$(9)\quad \begin{aligned} d(u - A) &= (a_0 - A_z)df(x) + \sum_{i=1}^{n} (a_i - A_{y_i})dg_i(x) - A_u du \\ &= (a_0 - A_z - A_u a_0)df(x) + \sum_{i=1}^{n} (a_i(1 - A_u) - A_{y_i})dg_i(x), \end{aligned}$$

with $z=f(x)$, $y_i=g_i(x)$ and $u(0)=0$. Hence we conclude from (3) and (8)

$$(10)\qquad | u(x) - A(f(x), g_i(x), u(x)) | \leqq FM_0 + H,$$

and, by (8),

$$(11)\qquad | u(x) | \leqq 2FM_0 + H.$$

Remarks. The function $u(z, y_i, \bar{u})$ is continuous. This may be expressed by the statement: If the quantities z, y_i, $u-A(z, y_i, u)$ tend to limiting values which belong to the domain $|z| \leqq F$, $|y_i| \leqq G_i$, $|u-A(z, y_i, u)| \leqq H+FM_0$, then u itself tends to a limiting value which in absolute value does not exceed $2FM_0+H$.

Any function satisfying a Lipschitz condition of exponent 1 in z, y_i, u satisfies also a Lipschitz condition of exponent 1 in the variables z, y_i, $u-A$. This follows from (8), for we have

$$(12)\qquad | u_1 - A(z, y_i, u_1) - u_2 + A(z, y_i, u_2) | \geqq | u_2 - u_1 | e^{-FN}.$$

THEOREM 3. *Assume that in the interval* $0 \leqq x \leqq X$

(i) *the functions* $f_\mu(x)$, $\mu=1, 2, \cdots$, *are continuously differentiable,* $|f_\mu(x)| < F$, *and* $f_\mu(x)$ *converges to a function* $f(x)$ *as* $\mu \to \infty$;

(ii) *the functions* $g_{1\mu}(x), \cdots, g_{n\mu}(x)$, $\mu=1, 2, \cdots$, *are continuously differentiable,* $g_{i\mu}(x) \to g_i(x)$ *as* $\mu \to \infty$, $i=1, 2, \cdots, n$, *where* $g_i(x)$ *is continuous, and the total variations of the differences* $T[g_{i\mu}-g_i]$ *tend to zero as* $\mu \to \infty$; *furthermore* $g_{i\mu}(0)=0$ *and* $T[g_{i\mu}] \leqq G_i$ *for all* i *and* μ;

(iii) *the functions* $a_0, a_1, \cdots, a_n$ *are defined in*

$$D_{3,\epsilon}:\quad \begin{aligned} &| z | \leqq F, \quad | y_i | \leqq G_i, \quad | u | < \epsilon + 3FM_0 + H, \\ &H = \sum_{i=1}^{n} (e^{2FN} - e^{FN} + M_i e^{FN})G_i, \qquad (\epsilon > 0), \end{aligned}$$

and we have in $D_{3,\epsilon}$

$$(13)\qquad | a_0 | \leqq M_0, \ | a_1 | \leqq M_1, \cdots, | a_n | \leqq M_n.$$

Furthermore, $a_0, a_1, \cdots, a_n$ *have continuous derivatives of first order in* $D_{3,\epsilon}$, *and those of* a_0 *are bounded in absolute value by* N *and satisfy a Lipschitz condition of exponent* 1.

Then the solution of

$$(\mathrm{E}_\mu) \qquad du_\mu(x) = a_0(f_\mu(x), g_{i\mu}(x), u_\mu(x))df_\mu(x) + \sum_{i=1}^{n} a_i dg_{i\mu}(x), \quad u_\mu(0) \equiv 0,$$

exists for $0 \leqq x \leqq X$, *satisfies*

$$(14) \qquad |u_\mu(x)| \leqq 2FM_0 + H,$$

and tends to a limit function $u(x)$ *as* $\mu \to \infty$. $u(x)$ *is said to be the solution of* (E) *for the initial condition* $u(0) = 0$.

From Theorem 2 we conclude the existence of $u_\mu(x)$ and the inequality (14). The classical statement about uniqueness of the solution of the initial problem may, incidentally, be used to establish the uniqueness of $u_\mu(x)$. On putting

$$B_\mu(x) = u_\mu(x) - A(f_\mu(x), g_{i\mu}(x), u_\mu(x)),$$

we conclude from (9) and (8)

$$(15) \qquad |B_\mu(x_1) - B_\mu(x_2)| \leqq \sum_{i=1}^{n} (M_i e^{FN} + e^{2FN} - e^{FN}) \int_{x_1}^{x_2} |dg_{i\mu}(x)|.$$

Now a theorem by Adams and Clarkson† shows that the total variation between any two points x_1 and x_2, of $g_{i\mu}(x)$ tends uniformly to that of $g_i(x)$ on account of the continuity of $g_i(x)$, and of the assumption (iii) that $T[g_{i\mu} - g_i] \to 0$, $g_{i\mu} \to g_i$. Thus (15) establishes equicontinuity for $B_\mu(x)$, while (14) gives boundedness. Hence, by Ascoli's theorem, we may select a subsequence $B_{\mu'}(x)$ tending uniformly to a function $B^*(x)$. From the remark on page 444 we conclude that also the corresponding subsequence of $u_\mu(x)$, say $u_{\mu'}(x)$, converges to a function $u^*(x)$. $B^*(x)$ satisfies the following integral equation

$$(16) \qquad B^*(x) = \int_0^x \sum_{i=1}^{n} (a_i(1 - A_u) - A_{y_i})dg_i(x) - A(f^0, 0, 0, \cdots)$$

in which the expressions in a_i and A are to be considered as functions of z, y_i, u with $z = f(x)$, $y_i = g_i(x)$, $u - A = B^*(x)$. This follows from Lemma 1 and (9).

Any two subsequences of $B^*(x)$ converge to the same limit. In the opposite case we would have two functions $B^*(x)$ and $B^{**}(x)$, both satisfying (16). In view of (iii) the coefficients $a_0, a_1, \cdots, a_n$ admit of continuous derivatives with respect to z, y_i, u, whence also with respect to z, y_i, $u - A$, and thus satisfy a Lipschitz condition of exponent 1 in these variables, in the closed domain $|z| \leqq F$, $|y_i| \leqq G_i$, $|u - A| \leqq FM_0 + H$. On account of (3) and (4), the

† C. R. Adams and J. A. Clarkson, *On convergence in variation*, Bulletin of the American Mathematical Society, vol. 40 (1934), pp. 413–417.

derivatives $A_{y'}$ and A_u satisfy a Lipschitz condition with respect to z, y_i, u, whence with respect to z, y_i, $u-A$. Therefore we conclude from (16)

$$(17) \qquad |B^*(x) - B^{**}(x)| \leqq K\sum_i \int_0^x |B^*(x) - B^{**}(x)|\,|dg_i(x)|,$$

where K is a certain constant. On iterating (17) m times we easily find

$$(18) \quad \max_{0\leqq x\leqq X} |B^*(x) - B^{**}(x)| \leqq K^m \max_{0\leqq x\leqq X} |B^*(x) - B^{**}(x)| \left(\sum_{i=1}^n G_i\right)^m \Big/ m!$$

which gives $B^*(x)-B^{**}(x)=0$ as $M\to\infty$.

Now the uniqueness of $B^*(x)$ implies, by the remark on page 444, the uniqueness of $u^*(x)$, which in turn justifies defining $u(x)=u^*(x)$ as the solution of (E) for the initial condition $u(0)=0$.

Remark. The assumptions of Theorem 3 state bounds for the functions $a_0, a_1, \cdots, a_n(z, y_1, \cdots, y_n, u)$ holding in a domain that depends on these same bounds. One may ask for a formulation of the theorem such that a statement results for *any* functions $a_0, a_1, \cdots, a_n$, defined in an *arbitrary* neighborhood of the origin. Therefore we observe that there always exists a sub-neighborhood of the form $D_{3,\epsilon}$, provided the constants $F, G_1, \cdots, G_n$, ϵ can be decreased sufficiently. Since the functions $g_{1\mu}(x), \cdots, g_{n\mu}(x)$ are continuous, their total variations are also continuous and may be shown† to converge *uniformly* to those of $g_1(x), \cdots, g_n(x)$. Thus, omitting at most a finite number of values of μ and taking the upper end of the x-interval sufficiently small, makes it possible to choose $G_1, G_2, \cdots, G_n$ arbitrarily small. Whence we conclude the following theorem:

THEOREM 3′. *Suppose that, in a neighborhood of the origin of a* $(z, y_1, \cdots, y_n, u)$*-space the functions* $a_0, a_1, \cdots, a_n$ *are continuously differentiable and the derivatives of* a_0 *satisfy a Lipschitz condition of exponent* 1. *Assume that n continuously differentiable functions* $g_{1\mu}(x), \cdots, g_{n\mu}(x)$ *are defined in an interval I* $(0\leqq x\leqq X)$, *that they tend to continuous functions* $g_1(x), \cdots, g_n(x)$, *and the total variations* $T[g_{i\mu}-g_i]\to 0$ *as* $\mu\to\infty$, *and that* $g_{i\mu}(0)=0$. *Assume furthermore that continuously differentiable functions* $f_\mu(x)$ *converge to a function* $f(x)$ *in I and that for all* μ *we have* $|f_\mu(x)|<F$. *Then the solution* $u_\mu(x)$ *of* (E_μ) *exists in a sufficiently small interval* $0\leqq x\leqq X'$ *which does not depend on* μ, *and converges there to a limit function* $u(x)$ *provided F is sufficiently small.*

The existence Theorems 3 and 3′ can be supplemented by a study of the manner in which the solution $u(x)$ depends on a parameter α on which the known functions in (E) may be supposed to depend. Usual methods of proving

† See Adams and Clarkson, loc. cit.

the continuity of $u(x)$, considered as a function of x and α, from that of the known functions could be carried through with only slight modifications.

Instead of (E) a system of differential equations of the form

$$(19)\quad du_h(x) = a_{0h}(f(x), g_i(x), u_l(x))df(x) + \sum_{i=1}^{n} a_{ih}dg_i(x), \qquad u_h(0) = 0,$$

where $h=1, 2, \cdots, m$, may be studied, and the existence of a solution $u_h(x)$, $h=1, 2, \cdots, m$, can be concluded by a method analogous to that used in the proofs of Theorems 2 and 3. In view of the similarity of the procedure we shall not carry out these generalizations.

3. **Double integrals.** We denote by $T[\alpha, \beta]$ a triangle bounded by the line $\alpha=\beta$ of an (α, β)-plane and the parallels to the axes through (α, β). Similarly $t[f, g]$ designates a triangle of an (f, g)-plane, bounded by $f=g$ and the parallels to the f- and g-axes through (f, g). The elements of area $d\alpha d\beta$ and $dfdg$ are to be counted positive. By $f(\alpha)$ and $g(\beta)$ we understand continuously differentiable mappings of the α-axis on the f-axis and of the β-axis on the g-axis, which are, but for the elements used, identical with each other; $f(\alpha)=g(\beta)$ if $\alpha=\beta$. The range of the four variables α, β, f, g is the domain D with origin as center

$$(D)\qquad |\alpha| \leqq \omega, \quad |\beta| \leqq \omega, \quad |f| \leqq \omega, \quad |g| \leqq \omega,$$

and the function $f(\alpha)$ (and consequently $g(\beta)$) is such that for α and β satisfying (D) the point $(\alpha, f(\alpha), \beta, g(\beta))$ belongs to D. Furthermore there are defined in D three functions a, b, c of α, f, β, g having continuous derivatives up to the fourth order.

We introduce three functions X, Y, Z in D by the relations

$$(20)\qquad X_{fg} = ca_f b_g,$$

$$(21)\qquad Y_f = ca_f b_\beta, \qquad Z_g = ca_\alpha b_g,$$

and the initial conditions

$$\begin{aligned} &(22)\qquad X = X_f = X_g = 0, \\ &(23)\qquad \begin{array}{l} Y = 0, \\ Z = 0, \end{array} \end{aligned} \quad \Bigg\}\ \text{if}\ f = g.$$

We find

$$X(\alpha, f, \beta, g) = \iint_{t[f,g]} ca_f b_g(\alpha, f', \beta, g')df'dg'.$$

Here, as in all integrals that follow, care has been taken to indicate the arguments of the integrand at least in one of the factors of the integrand, to denote the variables of integration by a prime and to denote by subscripts the partial derivatives, while we reserve the symbols $d/d\alpha$ and $d/d\beta$ for total derivatives with respect to these variables.

We are going to study the function

$$
\begin{aligned}
I(\alpha, f, \beta, g) = X(\alpha, f, \beta, g) &- \int_\beta^\alpha (X_\alpha(\alpha', f(\alpha'), \beta, g) - Z)d\alpha' \\
&+ \int_\beta^\alpha (X_\beta(\alpha, f, \beta', g(\beta')) - Y)d\beta' \\
&+ \iint_{T[\alpha,\beta]} (-X_{\alpha\beta}+Z_\beta+Y_\alpha-ca_\alpha b_\beta(\alpha', f(\alpha'), \beta', g(\beta')))d\alpha' d\beta'.
\end{aligned} \tag{24}
$$

In order to abbreviate as much as possible, we write $A\sim B$ if $A-B$ is expressible as a polynomial in a, b, c, their first partial derivatives and their second partial derivatives of the type $\partial^2/\partial\alpha\partial\beta$, $\partial^2/\partial\alpha\partial g$, $\partial^2/\partial\beta\partial f$, $\partial^2/\partial f\partial g$. We write $A\cong B$ if $A-B$ is expressible as an integral over a function which itself is ~ 0.

Thus we find

$$
I_f = X_f + \int_\beta^\alpha (X_{\beta f}(\alpha, f, \beta', g(\beta')) - Y_f)d\beta', \tag{25}
$$

$$
I_g = X_g - \int_\beta^\alpha (X_{\alpha g}(\alpha', f(\alpha'), \beta, g) - Z_g)d\alpha', \tag{26}
$$

$$
\begin{aligned}
I_\alpha = X_\alpha(\alpha, f, \beta, g) &- X_\alpha(\alpha, f(\alpha), \beta, g) + Z(\alpha, f(\alpha), \beta, g) \\
&+ X_\beta(\alpha, f, \alpha, f(\alpha)) - Y(\alpha, f, \alpha, f(\alpha)) \\
&+ \int_\beta^\alpha (- X_{\alpha\beta} + Z_\beta + Y_\alpha - ca_\alpha b_\beta(\alpha, f(\alpha), \beta', g(\beta')))d\beta', \\
&+ \int_\beta^\alpha (X_{\alpha\beta}(\alpha, f, \beta', g(\beta')) - Y_\alpha)d\beta',
\end{aligned} \tag{27}
$$

$$
\begin{aligned}
I_\beta = X_\beta(\alpha, f, \beta, g) &- X_\beta(\alpha, f, \beta, g(\beta)) + Y(\alpha, f, \beta, g(\beta)) \\
&+ X_\alpha(\beta, g(\beta), \beta, g) - Z(\beta, g(\beta), \beta, g) \\
&- \int_\beta^\alpha (X_{\alpha\beta}(\alpha', f(\alpha'), \beta, g) - Z_\beta)d\alpha' \\
&- \int_\beta^\alpha (- X_{\alpha\beta} + Z_\beta + Y_\alpha - ca_\alpha b_\beta(\alpha', f(\alpha'), \beta, g(\beta)))d\alpha',
\end{aligned} \tag{28}
$$

$$
(29)\quad
\begin{aligned}
I_{fg} &= ca_f b_g,\\
I_{f\beta} &= X_{f\beta}(\alpha, f, \beta, g) - X_{f\beta}(\alpha, f, \beta, g(\beta)) + Y_f(\alpha, f, \beta, g(\beta)),\\
I_{g\alpha} &= X_{g\alpha}(\alpha, f, \beta, g) - X_{g\alpha}(\alpha, f(\alpha), \beta, g) + Z_g(\alpha, f(\alpha), \beta, g),\\
I_{\alpha\beta} &= X_{\alpha\beta}(\alpha, f, \beta, g) - X_{\alpha\beta}(\alpha, f(\alpha), \beta, g) + Z_\beta(\alpha, f(\alpha), \beta, g)\\
&\quad - Z_\beta(\alpha, f(\alpha), \beta, g(\beta)) + X_{\alpha\beta}(\alpha, f(\alpha), \beta, g(\beta)) - X_{\alpha\beta}(\alpha, f, \beta, g(\beta))\\
&\quad + Y_\alpha(\alpha, f, \beta, g(\beta)) - Y_\alpha(\alpha, f(\alpha), \beta, g(\beta)) + ca_\alpha b_\beta(\alpha, f(\alpha), \beta, g(\beta)).
\end{aligned}
$$

Hence

$$
\begin{aligned}
I_{f\beta} &= \int_{g(\beta)}^{g} (X_{f\beta g} - Y_{fg}(\alpha, f, \beta, g'))dg' + Y_f(\alpha, f, \beta, g)\\
&= \int_{g(\beta)}^{g} [(ca_f b_g)_\beta - (ca_f b_\beta)_g]dg' + ca_f b_\beta\\
&= \int_{g(\beta)}^{g} [(ca_f)_\beta b_g - (ca_f)_g b_\beta]dg' + ca_f b_\beta,
\end{aligned}
$$

$$
(30)\qquad I_{f\beta} \cong ca_f b_\beta.
$$

Similarly

$$
(31)\qquad I_{g\alpha} \cong ca_\alpha b_g.
$$

Moreover,

$$
\begin{aligned}
&X_{\alpha\beta}(\alpha, f, \beta, g) - X_{\alpha\beta}(\alpha, f(\alpha), \beta, g) + X_{\alpha\beta}(\alpha, f(\alpha), \beta, g(\beta))\\
&\qquad - X_{\alpha\beta}(\alpha, f, \beta, g(\beta)) = \int_{f(\alpha)}^{f}\int_{g(\beta)}^{g} X_{\alpha\beta fg}(\alpha, f', \beta, g')df'dg'.
\end{aligned}
$$

But

$$
\begin{aligned}
X_{\alpha\beta fg} &= (ca_f b_g)_{\alpha\beta} = (c_\alpha a_f b_g + ca_{f\alpha} b_g + ca_f b_{g\alpha})_\beta\\
&\sim c_\alpha a_f b_{g\beta} + c_\beta a_{\alpha f} b_g + ca_{f\alpha\beta} b_g + ca_{f\alpha} b_{g\beta} + ca_f b_{g\alpha\beta}\\
&\sim (c_\alpha a_f b_\beta)_g + (c_\beta a_\alpha b_g)_f + (ca_{\alpha\beta} b_g)_f + (ca_f b_{\alpha\beta})_g + ca_{\alpha f} b_{\beta g},\\
ca_{\alpha f} b_{\beta g} &= (ca_\alpha b_{\beta g})_f - c_f a_\alpha b_{\beta g} - ca_\alpha b_{g\beta f}\\
&\sim (ca_\alpha b_\beta)_{gf} - [(ca_\alpha)_g b_\beta]_f - (c_f a_\alpha b_\beta)_g - (ca_\alpha b_{\beta f})_g.
\end{aligned}
$$

Hence,

$$
\begin{aligned}
\int_{f(\alpha)}^{f}\int_{g(\beta)}^{g} X_{\alpha\beta fg}(\alpha, f', \beta, g')df'dg' &\cong ca_\alpha b_\beta(\alpha, f, \beta, g) - ca_\alpha b_\beta(\alpha, f(\alpha), \beta, g)\\
&\quad + ca_\alpha b_\beta(\alpha, f(\alpha), \beta, g(\beta)) - ca_\alpha b_\beta(\alpha, f, \beta, g(\beta)).
\end{aligned}
$$

Furthermore,

$$Z_\beta(\alpha, f(\alpha), \beta, g) - Z_\beta(\alpha, f(\alpha), \beta, g(\beta)) = \int_{g(\beta)}^{g} Z_{\beta g}(\alpha, f(\alpha), \beta, g')dg'$$

$$= \int_{g(\beta)}^{g} [ca_\alpha b_g(\alpha, f(\alpha), \beta, g')]_\beta dg'$$

$$\cong \int_{g(\beta)}^{g} [ca_\alpha b_\beta(\alpha, f(\alpha), \beta, g')]_g dg'$$

$$\cong ca_\alpha b_\beta(\alpha, f(\alpha), \beta, g) - ca_\alpha b_\beta(\alpha, f(\alpha), \beta, g(\beta)).$$

Similarly,

$$Y_\alpha(\alpha, f, \beta, g(\beta)) - Y_\alpha(\alpha, f(\alpha), \beta, g(\beta)) \cong ca_\alpha b_\beta(\alpha, f, \beta, g(\beta)) - ca_\alpha b_\beta(\alpha, f(\alpha), \beta, g(\beta)).$$

Finally

$$I_{\alpha\beta}(\alpha, f, \beta, g) \cong ca_\alpha b_\beta(\alpha, f, \beta, g). \tag{32}$$

In view of (29), we find

$$\frac{d^2I}{d\alpha d\beta}(\alpha, f(\alpha), \beta, g(\beta)) = ca_\alpha b_\beta + ca_f b_\beta \frac{df(\alpha)}{d\alpha} + ca_\alpha b_g \frac{dg(\beta)}{d\beta} + ca_f b_g \frac{df(\alpha)}{d\alpha}\frac{dg(\beta)}{d\beta}$$

$$= c\,\frac{da(\alpha, f(\alpha), \beta, g(\beta))}{d\alpha}\,\frac{db(\alpha, f(\alpha), \beta, g(\beta))}{d\beta}$$

and

$$I(\alpha, f(\alpha), \beta, g(\beta)) = 0, \qquad \frac{dI}{d\alpha} = 0, \qquad \frac{dI}{d\beta} = 0,$$

for $\alpha = \beta$. Thus

$$I(\alpha, f(\alpha), \beta, g(\beta)) = -\iint_{T[\alpha,\beta]} c\,\frac{da}{d\alpha'}\,\frac{db}{d\beta'}(\alpha', f(\alpha'), \beta', g(\beta'))d\alpha' d\beta'. \tag{33}$$

In order to transform $I_f(\alpha, f, \beta, g)$ we calculate

$$\int_\beta^\alpha X_{\beta f}(\alpha, f, \beta', g(\beta'))d\beta' = \int_\beta^\alpha d\beta'\left[\int_f^{g(\beta')} X_{\beta f g}(\alpha, f, \beta', g')dg' + X_{\beta f}(\alpha, f, \beta', f)\right]$$

$$= \int_\beta^\alpha d\beta' \int_f^{g(\beta')} (ca_f b_g(\alpha, f, \beta', g'))_\beta dg'$$

$$\cong \int_\beta^\alpha d\beta' \int_f^{g(\beta')} (ca_f b_\beta(\alpha, f, \beta', g'))_g dg'$$

$$\cong \int_\beta^\alpha d\beta' [ca_f b_\beta(\alpha, f, \beta', g(\beta')) - ca_f b_\beta(\alpha, f, \beta', f)]$$

since, by (22), $X_{\beta f}(\alpha, f, \beta', f)$ vanishes.

Consequently

$$I_f(\alpha, f, \beta, g) \cong 0. \tag{34}$$

Similarly

$$I_g(\alpha, f, \beta, g) \cong 0. \tag{35}$$

Also we have

$$\begin{aligned} X_\alpha(\alpha, f, \beta, g) &= -\iint_{t[f,g]} (ca_f b_g(\alpha, f', \beta, g'))_\alpha df'dg' \\ &\cong -\iint_{t[f,g]} (ca_\alpha b_g(\alpha, f', \beta, g'))_f df'dg' \\ &\cong -\int ca_\alpha b_g(\alpha, f', \beta, g')dg', \end{aligned}$$

where the integration is to be extended over the boundary of $t[f, g]$; whence

$$X_\alpha(\alpha, f, \beta, g) \cong 0, \qquad X_\beta(\alpha, f, \beta, g) \cong 0.$$

Thus

$$I_\alpha(\alpha, f, \alpha, g) \cong 0, \tag{36}$$

as obviously $Y \cong 0$, $Z \cong 0$.

On writing

$$I_\alpha(\alpha, f, \beta, g) = I_\alpha(\alpha, f, \alpha, g) + \int_\alpha^\beta I_{\alpha\beta}(\alpha, f, \beta', g)d\beta',$$

we find by (36) and (32)

$$I_\alpha(\alpha, f, \beta, g) \cong 0. \tag{37}$$

Similarly,

$$I_\beta(\alpha, f, \beta, g) \cong 0. \tag{38}$$

Finally

$$\begin{aligned} &I(\alpha, f, \beta, g) = I(\alpha, f, \alpha, g) + \int_\alpha^\beta I_\beta(\alpha, f, \beta', g)d\beta', \\ &I(\alpha, f, \beta, g) \cong 0. \end{aligned} \tag{39}$$

The formulas (29), (30), (31), (32), (34), (35), (37), (38), and (39) prove that $I(\alpha, f, \beta, g)$, its first derivatives and its second derivatives of the type

$\partial^2/\partial\alpha\partial\beta$, $\partial^2/\partial\alpha\partial g$, $\partial^2/\partial\beta\partial f$, $\partial^2/\partial f\partial g$ may be expressed in terms of a, b, c, their first and second derivatives of the same type, and integrals over products of such functions.

Henceforth the definition (24) *of I is to be replaced by the explicit formula whose abbreviated equivalent is* (39), and which retains sense in the case that a, b, c admit only of continuous first derivatives and of continuous second derivatives of the indicated type. If we uniformly approximate a, b, c and said derivatives by polynomials in α, f, β, g and their respective derivatives, we may easily see that all of the formulas (29)–(32), (34), (35), (37)–(39) remain valid under the new assumptions and that $I(\alpha, f, \beta, g)$ still retains continuous first and second derivatives of said type.

Moreover a study of the dependence of I on the function $f(\alpha)$ shows that convergence of a sequence of continuously differentiable functions $f_\mu(\alpha)$ to a limit function $f_0(\alpha)$ implies the convergence of the corresponding functionals $I_\mu(\alpha, f, \beta, g)$ to a limit functional $I_0(\alpha, f, \beta, g)$, and uniform convergence of $f_\mu(\alpha)$ to $f_0(\alpha)$ entails uniform convergence of I_μ to I_0. In fact, $f_\mu(\alpha)$ appears in the definition of $I_\mu(\alpha, f, \beta, g)$ only in limits of integration with respect to f' or g', which implies the convergence mentioned of I_μ to I.

From the formulas (29)–(32), (34), (35), (37)–(39) we can derive estimates for $I(\alpha, f, \beta, g)$ and its derivatives. Suppose first that in $|\alpha|, |f|, |\beta|, |g| \leqq \omega$

$$|a|,\ |b|,\ |c| \leqq K, \quad |a_\alpha|, \cdots, |c_g| \leqq K', \quad |a_{\alpha\beta}|, \cdots, |c_{fg}| \leqq K''.$$

The terms suppressed in the above formulas by the use of the symbol $\cong$ are simple, double, triple, and quadruple integrals of polynomials of third degree in $a, b, c, a_\alpha, \cdots, c_{fg}$ with ranges of integration, respectively, $\leqq \rho\omega, \rho\omega^2, \rho\omega^3, \rho\omega^4$, where ρ denotes a sufficiently large number, for instance, 64. On the other hand, any one of the polynomials to be integrated is numerically smaller than a suitable polynomial in K, K', K'' of third degree with positive coefficients. Hence there exists a polynomial p with positive coefficients and of third degree in K, K', K'', such that for any one of formulas (29), $\cdots$, (39) the terms suppressed by the symbol $\cong$ are numerically less than or equal to

$$p(K, K', K'')(\omega + \omega^2 + \omega^3 + \omega^4).$$

We apply these estimates to the case where a, b, c *depend only indirectly* on α, f, β, g and are functions of $\psi_1, \cdots, \psi_n(\alpha, f, \beta, g)$ having third derivatives with respect to $\psi_1, \cdots, \psi_n$. We assume that in $|\alpha|, |f|, |\beta|, |g| \leqq \omega$, the following quantities exist and satisfy

$$(40) \qquad |\psi_1|, \cdots, |\psi_n| \leqq k,$$

$$(41) \qquad |\psi_{1\alpha}|, \cdots, |\psi_{ng}| \leqq k', \quad |\psi_{1\alpha\beta}|, \cdots, |\psi_{ngf}| \leqq k'', \quad (k, k', k'' > 0),$$

and that for ψ_i satisfying (40)

$$|a|, |b|, |c| \leqq L,$$

$$\left|\frac{\partial a}{\partial \psi_i}\right|, \left|\frac{\partial b}{\partial \psi_i}\right|, \left|\frac{\partial c}{\partial \psi_i}\right| \leqq L',$$

$$\left|\frac{\partial^2 a}{\partial \psi_i \partial \psi_j}\right|, \cdots, \left|\frac{\partial^2 c}{\partial \psi_i \partial \psi_j}\right| \leqq L'', \tag{42}$$

$$\left|\frac{\partial^3 a}{\partial \psi_i \partial \psi_j \partial \psi_l}\right|, \cdots, \left|\frac{\partial^3 c}{\partial \psi_i \partial \psi_j \partial \psi_l}\right| \leqq L'''.$$

We then write

$$I(\alpha, f, \beta, g) = I\binom{\alpha, f, \beta, g}{\psi}.$$

In order to utilize the estimates found, it is legitimate to replace K by L, K' by $nL'k'$, K'' by $nLk''+n^2L''k'^2$. Thus from (29), $\cdots$, (39) we obtain

$$\begin{aligned} &\left|I\binom{\alpha, f, \beta, g}{\psi}\right|, |I_\alpha|, \cdots, |I_g| \\ &\qquad \leqq (\omega + \omega^2 + \omega^3 + \omega^4) q(L, L', L'', k, k', k'') \\ &|I_{\alpha\beta}|, |I_{\alpha g}|, |I_{\beta f}|, |I_{fg}| \\ &\qquad \leqq n^2 L L'^2 k'^2 + (\omega + \omega^2 + \omega^3 + \omega^4) q(L, \cdots, k''), \end{aligned} \tag{43}$$

where q is a suitable polynomial with positive coefficients.

We next state bounds for the difference between

$$I\binom{\alpha, f, \beta, g}{\psi} \quad \text{and} \quad I\binom{\alpha, f, \beta, g}{\psi'}$$

and its derivatives, assuming ψ_i' to be another system of continuously differentiable functions of α, f, β, g which satisfy the same inequalities (40), (41) as the ψ_i themselves do. For the sake of simplicity we suppose $\omega < \Omega$, with $\Omega > 0$ and denote by u and v

$$u = \sum_{i=1}^n \max |\psi_i - \psi_i'| + \sum_{i=1}^n \max |\psi_{i\alpha} - \psi_{i\alpha}'| + \cdots + \sum_{i=1}^n \max |\psi_{ig} - \psi_{ig}'|,$$

$$\begin{aligned} v = &\sum_{i=1}^n \max |\psi_{i\alpha\beta} - \psi_{i\alpha\beta}'| + \sum_{i=1}^n \max |\psi_{i\alpha g} - \psi_{i\alpha g}'| \\ &+ \sum_{i=1}^n \max |\psi_{i\beta f} - \psi_{i\beta f}'| + \sum_{i=1}^n \max |\psi_{ifg} - \psi_{ifg}'|, \end{aligned}$$

the maxima to be taken for the domain $|\alpha|, |f|, |\beta|, |g| \leqq \omega$.

We find by a procedure similar to the one previously used,

$$(44.1)\qquad \left|I\binom{\alpha, f, \beta, g}{\psi} - I\binom{\alpha, f, \beta, g}{\psi'}\right| \leqq (u+v)C\omega,$$

$$\left|I_\alpha\binom{\alpha, f, \beta, g}{\psi} - I_\alpha\binom{\alpha, f, \beta, g}{\psi'}\right| \leqq (u+v)C\omega,$$

$$\cdots\cdots\cdots\cdots\cdots\cdots\cdots\cdots$$

$$\left|I_g\binom{\alpha, f, \beta, g}{\psi} - I_g\binom{\alpha, f, \beta, g}{\psi'}\right| \leqq (u+v)C\omega,$$

$$(44.2)\qquad \left|I_{\alpha\beta}\binom{\alpha, f, \beta, g}{\psi} - I_{\alpha\beta}\binom{\alpha, f, \beta, g}{\psi'}\right| \leqq uC + vC\omega,$$

$$\cdots\cdots\cdots\cdots\cdots\cdots\cdots\cdots$$

$$\left|I_{fg}\binom{\alpha, f, \beta, g}{\psi} - I_{fg}\binom{\alpha, f, \beta, g}{\psi'}\right| \leqq uC + vC\omega,$$

with C depending on Ω, k, k', k'', L, L', L'', L'''. In an analogous way the difference between

$$I\binom{\alpha, f, \beta, g}{\psi} \quad \text{and} \quad \bar{I}\binom{\alpha, f, \beta, g}{\psi'}$$

may be estimated under the assumption that the functions used in the formation of $\bar{I}$, namely $\bar{a}$, $\bar{b}$, $\bar{c}(\psi_1, \cdots, \psi_n)$, and their first and second derivatives, differ by less than ϵ from those used for

$$I\binom{\alpha, f, \beta, g}{\psi}.$$

We get

$$\left|I\binom{\alpha, f, \beta, g}{\psi} - \bar{I}\binom{\alpha, f, \beta, g}{\psi'}\right| \leqq (u+v+\epsilon)C\omega,$$

$$(45)\qquad \left|I_\alpha\binom{\alpha, f, \beta, g}{\psi} - \bar{I}_\alpha\binom{\alpha, f, \beta, g}{\psi'}\right|, \cdots, \leqq C(u+v+\epsilon)\omega,$$

$$\left|I_{\alpha\beta}\binom{\alpha, f, \beta, g}{\psi} - \bar{I}_{\alpha\beta}\binom{\alpha, f, \beta, g}{\psi'}\right|, \cdots,$$

$$\left|I_{\beta g}\binom{\alpha, f, \beta, g}{\psi} - \bar{I}_{fg}\binom{\alpha, f, \beta, g}{\psi'}\right| \leqq C(\epsilon+u) + vC\omega.$$

We now are able to formulate the main theorem of this section. Denote by

$$I_{ijl}\begin{pmatrix}\alpha, f, \beta, g\\ \psi\end{pmatrix} \quad \text{the functional} \quad I\begin{pmatrix}\alpha, f, \beta, g\\ \psi\end{pmatrix}$$

formed for

$$a = \psi_j, \qquad b = \psi_l, \qquad c = c_{ijl}.$$

THEOREM 4. *Denote by $\omega > 0$ a number $< \Omega$ and by D the domain $|\alpha|$, $|f|$, $|\beta|$, $|g| \leqq \omega$. Suppose c_{ijl} to be continuously differentiable up to the third order with respect to its arguments $\psi_1, \cdots, \psi_n$, and that for ψ_i in* (40), $a = \psi_j$, $b = \psi_l$, $c = c_{ijl}$ *the relations* (42) *hold. Suppose furthermore that in D the functions $\psi_i^0(\alpha, f, \beta, g)$ are continuously differentiable and admit of continuous second derivatives of the type mentioned such that*

$$(46) \qquad |\psi_i^0(\alpha, f, \beta, g)| \leqq k/2,$$

$$(47) \qquad |\psi^0_\alpha|, \cdots, |\psi^0_g| \leqq k'/2,$$

$$(48) \qquad |\psi^0_{i\alpha\beta}|, \ |\psi^0_{i\alpha g}|, \ |\psi^0_{i\beta f}|, \ |\psi^0_{igf}| \leqq k''/3.$$

Then the system

$$(49) \quad \psi_i(\alpha, f, \beta, g) = \psi_i^0(\alpha, f, \beta, g) + \sum_{j,l=1}^{n} I_{ijl}\begin{pmatrix}\alpha, f, \beta, g\\ \psi\end{pmatrix}, \ (i = 1, 2, \cdots, n),$$

has a solution $\psi_i(\alpha, f, \beta, g)$ existing and uniquely determined in $|\alpha|$, $|f|$, $|\beta|$, $|g| \leqq \omega'$, where $\omega' > 0$ is a number $\leqq \omega$ that may be determined with the aid of Ω, k, k', k'', L, L', L'', L''' only. $\psi_{i\alpha}(\alpha, f, \beta, g), \cdots, \psi_{ig}(\alpha, f, \beta, g)$ exist, are continuous and in absolute value $\leqq k'$, and $\psi_{i\alpha\beta}(\alpha, f, \beta, g), \cdots, \psi_{ifg}(\alpha, f, \beta, g)$ exist, are continuous and in absolute value* $\leqq \max(k'', 12n^5LL'^2k'^2)$.

We start the proof by increasing, if necessary, k'' so as to satisfy the inequality

$$4n^5LL'^2k'^2 < k''/3.$$

We use successive approximations:

$$\psi_i^1(\alpha, f, \beta, g) = \psi_i^0(\alpha, f, \beta, g) + \sum_{k,l=1}^{n} I_{ikl}\begin{pmatrix}\alpha, f, \beta, g\\ \psi^0\end{pmatrix},$$

and generally for $m \geqq 0$

$$(49.1) \qquad \psi_i^{m+1}(\alpha, f, \beta, g) = \psi_i^0(\alpha, f, \beta, g) + \sum_{k,l=1}^{n} I_{ikl}\begin{pmatrix}\alpha, f, \beta, g\\ \psi^m\end{pmatrix}.$$

* In particular ω' does not depend on the function $f(\alpha)$ used in the definition of I_{ijl}.

Denoting, as before, by C a constant depending only on Ω, k, k', k'', L, L', L'', by ∂ a generic first derivative with respect to α, f, β, or g, and by ∂^2 a generic derivative of type $\partial^2/\partial\alpha\partial\beta$, $\partial^2/\partial\alpha\partial g$, $\partial^2/\partial\beta\partial f$, $\partial^2/\partial f\partial g$, we set, for $m>0$,

$$u_m = \sum_{i,j,l=1}^{n} \max \left| I_{ijl}\begin{pmatrix}\alpha, f, \beta, g\\ \psi^m\end{pmatrix} - I_{ijl}\begin{pmatrix}\alpha, f, \beta, g\\ \psi^{m-1}\end{pmatrix}\right|$$

$$+ \sum_{i,j,l=1}^{n} \sum_{\partial} \max \left| \partial(I_{ikl}(\psi^m) - I_{ijl}(\psi^{m-1}))\right|$$

$$v_m = \sum_{i,j,l=1}^{n} \sum_{\partial^2} \max \left| \partial^2\left(I_{ijl}\begin{pmatrix}\alpha, f, \beta, g\\ \psi^m\end{pmatrix} - I_{ijl}\begin{pmatrix}\alpha, f, \beta, g\\ \psi^{m-1}\end{pmatrix}\right)\right|,$$

where the maxima are to be taken in $|\alpha|$, $|f|$, $|\beta|$, $|g| \leqq \omega'$ with $\omega' < \Omega$ to be determined later. On putting

$$u_0 = \sum_{i,j,l=1}^{n} \max \left| I_{ijl}\begin{pmatrix}\alpha, f, \beta, g\\ \psi^0\end{pmatrix}\right| + \sum_{i,j,l=1}^{n} \sum_{\partial} \max \left| \partial I_{ijl}\begin{pmatrix}\alpha, f, \beta, g\\ \psi^0\end{pmatrix}\right|,$$

$$v_0 = \sum_{i,j,l=1}^{n} \sum_{\partial^2} \max \left| \partial^2 I_{ijl}\begin{pmatrix}\alpha, f, \beta, g\\ \psi^0\end{pmatrix}\right|,$$

we find, by (43),

$$u_0 \leqq C\omega', \tag{50}$$

$$v_0 \leqq 4n^5LL'^2k'^2 + C\omega', \tag{51}$$

and, for u_m and v_m the recursion formulas, in view of (44.1) and (44.2),

$$u_{m+1} \leqq C\omega'(u_m + v_m), \tag{52.1}$$

$$v_{m+1} \leqq Cu_m + C\omega' v_m, \tag{52.2}$$

$$(m = 0, 1, 2, \cdots)$$

provided, however, that we can choose $\omega' > 0$ so as to make sure the existence of *all* successive approximations $|\alpha|$, $|f|$, $|\beta|$, $|g| \leqq \omega'$ in the common domain D' ($|\alpha|$, $|f|$, $|\beta|$, $|g| \leqq \omega'$). Now determine $\omega' < \Omega$ so small that

$$1 - C\omega' > 0, \qquad (1 - C\omega')^2 - C^2\omega' > 0,$$

$$U \equiv \frac{C\omega'(1 - C\omega') + (4n^5LL'^2k'^2 + C\omega')C\omega'}{(1 - C\omega')^2 - C^2\omega'} \leqq \frac{\min(k, k')}{2}, \tag{53}$$

$$V \equiv \frac{C^2\omega' + (1 - C\omega')(4n^5LL'^2k'^2 + C\omega')}{(1 - C\omega')^2 - C^2\omega'} \leqq \frac{2k''}{3}.$$

Note that

$$(1 + U + V)C\omega' \leqq U$$

and

$$4n^5LL'^2k'^2 + C\omega' + CU + C\omega' V \leqq V.$$

Now the conditions (46), (47), and (48) permit the construction of

$$I_{ikl}\begin{pmatrix}\alpha, f, \beta, g \\ \psi^0\end{pmatrix}$$

in D, hence a fortiori in D' ($|\alpha|, |f|, |\beta|, |g| \leqq \omega'$), and we certainly have, by (53),

$$u_0 \leqq U,$$
$$v_0 \leqq V.$$

Suppose that we could construct, throughout D', the mth approximation $\psi_i^m(\alpha, f, \beta, g)$ and that

$$\sum_0^m u_j \leqq U, \tag{54}$$

$$\sum_0^m v_j \leqq V. \tag{55}$$

We are then able to prove that we can construct the $(m+1)$st approximation and that

$$\sum_0^{m+1} u_j \leqq U, \qquad \sum_0^{m+1} v_j \leqq V.$$

In fact, we have, in view of (54) and (55), (46), (47), and (48),

$$\begin{aligned} |\psi_i^{m+1}(\alpha, f, \beta, g)| &\leqq |\psi_i^0(\alpha, f, \beta, g)| + \sum_0^m u_j \\ &\leqq |\psi_i^0(\alpha, f, \beta, g)| + U \leqq k, \end{aligned} \tag{56}$$

$$|\partial\psi_i^{m+1}(\alpha, f, \beta, g)| \leqq |\partial\psi_i^0(\alpha, f, \beta, g)| + \sum_0^m u_j \leqq k', \tag{57}$$

$$|\partial^2\psi_i^{m+1}(\alpha, f, \beta, g)| \leqq |\partial^2\psi_i^0(\alpha, f, \beta, g)| + \sum_0^m v_j \leqq k'', \tag{58}$$

and by (52),

$$\sum_0^{m+1} u_j \leqq u_0 + C\omega'\left(\sum_0^m u_j + \sum_0^m v_j\right) \leqq C\omega'(1 + U + V) \leqq U,$$

$$\sum_0^{m+1} v_j \leqq v_0 + C\sum_0^m u_j + C\omega'\sum_0^m v_j \leqq 4n^5LL'^2k'^2 + C\omega' + CU + C\omega' V \leqq V.$$

Thus (54) and (55) hold for all $m \geqq 0$, and we conclude the uniform convergence of $\psi_i^m(\alpha, f, \beta, g)$, $\partial\psi_i^m$, $\partial^2\psi_i^m$ to limit functions $\psi_i(\alpha, f, \beta, g)$ and their corresponding derivatives $\partial\psi_i$, $\partial^2\psi_i$. The continuity relation (45) finally proves

$$I\begin{pmatrix}\alpha, f, \beta, g\\ \psi^m\end{pmatrix} \to I\begin{pmatrix}\alpha, f, \beta, g\\ \psi\end{pmatrix}.$$

Hence by passage to the limit in (49.1) we obtain (49).

The uniqueness follows similarly from the relations analogous to (44.1) and (44.2),

$$u \leqq C\omega'(u + v) \tag{59.1}$$

$$v \leqq Cu + C\omega' v, \tag{59.2}$$

which yield $u(1 - C\omega') \leqq C\omega' v \leqq C\omega' \cdot Cu/(1 - C\omega')$, $u((1 - C\omega')^2 - C^2\omega') \leqq 0$, $u = 0$, $v = 0$, with

$$u = \sum_{i,j,l=1}^{n} \max \left| I_{ijl}\begin{pmatrix}\alpha, f, \beta, g\\ \psi\end{pmatrix} - I_{ijl}\begin{pmatrix}\alpha, f, \beta, g\\ \psi'\end{pmatrix}\right|$$
$$+ \sum_{i,j,l}\sum_{\partial} \max \left| \partial(I_{ijl}(\psi) - I_{ijl}(\psi'))\right|,$$

$$v = \sum_{i,j,l}\sum_{\partial^2} \max \left| \partial^2(I_{ijl}(\psi) - I_{ijl}(\psi'))\right|,$$

$\psi_i(\alpha, f, \beta, g)$ and $\psi_i'(\alpha, f, \beta, g)$ being solutions of (49).

COROLLARY 1. *If, in Theorem* 4, $f(\alpha)$, $\psi_1^0(\alpha, f, \beta, g), \cdots, \psi_n^0(\alpha, f, \beta, g)$, $\partial\psi_1^0(\alpha, f, \beta, g), \cdots, \partial\psi_n^0(\alpha, f, \beta, g)$, $\partial^2\psi_1^0(\alpha, f, \beta, g), \cdots, \partial^2\psi_n^0(\alpha, f, \beta, g)$, *and* $c_{ijl}(\psi_1, \cdots, \psi_n)$ *and its derivatives up to the third order depend on a parameter* μ *and converge uniformly as* $\mu \to \infty$, *then* $\psi_1(\alpha, f, \beta, g), \cdots, \psi_n(\alpha, f, \beta, g)$, $\partial\psi_1, \cdots, \partial\psi_n, \partial^2\psi_1, \cdots, \partial^2\psi_n$ *converge uniformly, as* $\mu \to \infty$, *in* $|\alpha|, |f|, |\beta|, |g| \leqq \omega'' \leqq \omega'$, *where* ω'' *depends only on* $\Omega, k, k', k'', L, L', L''$.

Denote by Δ the operation of taking the difference for two sufficiently large values of μ, and put, in the successive approximations of the proof of Theorem 4,

$$u_m = \sum_{i=1}^{n} \left| \Delta\psi_i^m(\alpha, f, \beta, g)\right| + \sum_{i=1,\partial}^{n} \left| \Delta\partial\psi_i^m\right|,$$

$$v_m = \sum_{i=1,\partial^2}^{n} \left| \Delta\partial^2\psi_i^m(\alpha, f, \beta, g)\right|.$$

Observing that $f(\alpha)$ enters in the functional I only as a limit of integration, as has been remarked earlier, we may use (45) and find, with some

$C=C(\Omega, k, k', k'', L, L', L'', L''')$ and a new and smaller value ω'' of ω', satisfying (53) with the new C:

$$u_m \leqq C(u_{m-1} + v_{m-1} + \epsilon)\omega'' + C\epsilon + u_0,$$
$$v_m \leqq Cu_{m-1} + Cv_{m-1}\omega'' + C\epsilon + v_0,$$
$$u_0 \leqq \epsilon, \qquad v_0 \leqq \epsilon.$$

Hence, for $m\to\infty$, $\lim u_m = u$, $\lim v_m = v$

$$u \leqq C(u + v + \epsilon)\omega'' + C\epsilon + u_0,$$
$$v \leqq Cu + Cv\omega'' + C\epsilon + v_0,$$
$$u(1 - C\omega'') \leqq (C + 1)\epsilon + \epsilon C\omega'' + C\omega''v,$$
$$C\omega''v \leqq \frac{C\omega''}{1 - C\omega''}Cu + \frac{(C + 1)\epsilon C\omega''}{1 - C\omega''}.$$

Hence u and v are $\leqq C'\epsilon$, with $C'(C, \omega'')$, which proves the corollary.

COROLLARY 2. *If, in Theorem* 4, $f(\alpha)$, $\psi_i^0(\alpha, f, \beta, g)$, C_{ijl} *depend on a parameter* μ, *and if* $f(\alpha)$ *converges uniformly as* $\mu\to\mu_0$, *then there exists a subsequence of* μ, *such that* $\psi_i(\alpha, f, \beta, g)$ *and also* $\psi_i(\alpha, f(\alpha), \beta, g(\beta))$ *converge uniformly.*

For by (57), $|\partial\psi_i(\alpha, f, \beta, g)| \leqq k'$ and hence the $\psi_i(\alpha, f, \beta, g)$ are equicontinuous and bounded, in view of (56). Hence there exists a uniformly convergent subsequence, and the corollary follows. From (33) we conclude under the hypothesis of the theorem, that

$$(60) \quad \frac{d^2}{d\alpha d\beta}\psi_i(\alpha, f(\alpha), \beta, g(\beta)) = \frac{d^2}{d\alpha d\beta}\psi_i^0(\alpha, f(\alpha), \beta, g(\beta)) + \sum_{j,l=1}^{n} c_{ijl}(\psi)\frac{d\psi_j}{d\alpha}\frac{d\psi_l}{d\beta},$$

and, for $\alpha=\beta$, $f=f(\alpha)$, $g=f(\alpha)$

$$\psi_i = \psi_i^0, \qquad \partial\psi_i = \partial\psi_i^0.$$

In the application we intend to make of Theorem 4 and its corollaries, the values of the constants such as $\Omega, k, k', k'', L, L', L'', L'''$ are of no importance. What matters, however, is their existence and their interdependence. Therefore, we are led to use the following terminology: we call a function *bounded* if its absolute value is bounded by a positive number irrespective of the values of its arguments and possible other parameters; we call, in a theorem, a quantity *relatively bounded* if its absolute value can be bounded by a positive number which depends only on other bounds *previously* introduced in the theorem; and we use the same term, in a proof, as meaning limitable by bounds, either assumed by the hypotheses of the theorem, or previously introduced in the course of the same proof.

Thus Theorem 4 and Corollary 2 may be formulated as follows:

THEOREM 5. *Suppose c_{ijl} and its derivatives up to the third order with respect to its arguments $\psi_1, \cdots, \psi_n$ bounded for bounded values of ψ_i, and assume that $\psi_i=\psi_i^0(\alpha, f, \beta, g)$, $i=1, \cdots, n$, has derivatives $\partial\psi_i^0$, $\partial^2\psi_i^0$ which are continuous and bounded when α, f, β, g are bounded. Then the system*

$$\psi_i(\alpha, f, \beta, g) = \psi_i^0(\alpha, f, \beta, g) + \sum_{j,l=1}^{n} I_{ijl}\begin{pmatrix}\alpha, f, \beta, g\\ \psi\end{pmatrix}, \qquad (i = 1, 2, \cdots, n),$$

has a solution $\psi_i(\alpha, f, \beta, g)$, continuous together with $\partial\psi_i$ and $\partial^2\psi_i$. This solution exists, is uniquely determined and is relatively bounded together with the derivatives $\partial\psi_i$, $\partial^2\psi_i$ for relatively bounded α, f, β, g. If, in addition, $f(\alpha)$, $\psi_i^0(\alpha, f, \beta, g)$ and c_{ijl} and its derivatives up to the third order depend on a parameter μ, and if $f(\alpha)$ converges uniformly as $\mu\to\mu_0$, then there exists a subsequence of μ such that the corresponding functions $\psi_i(\alpha, f, \beta, g)$ and $\psi_i(\alpha, f(\alpha), \beta, g(\beta))$ converge uniformly for relatively bounded values of α, f, β, g.

4. **Hyperbolic systems.** The results of the preceding section may be used for a study of Cauchy's problem* for the system

$$(61)\qquad \begin{aligned} &\sum_{j=1}^{n} a_{ij}(\phi_1, \cdots, \phi_n)\frac{\partial\phi_j(\alpha, \beta)}{\partial\alpha} = 0, \qquad i = 1, 2, \cdots, m < n,\\ &\sum_{j=1}^{n} a_{ij}(\phi_1, \cdots, \phi_n)\frac{\partial\phi_j(\alpha, \beta)}{\partial\beta} = 0, \qquad i = m + 1, \cdots, n,\end{aligned}$$

in which a_{ij} and its partial derivatives up to the fourth order as well as the reciprocal value of the determinant $|a_{ij}|$ are bounded for bounded values of $\phi_1, \cdots, \phi_n$. The initial line is a bounded neighborhood of the origin on the line $\alpha=\beta$, and on it the unknown functions $\phi_1(\alpha, \beta), \cdots, \phi_n(\alpha, \beta)$ assume relatively bounded values $\xi_1(\alpha), \cdots, \xi_n(\alpha)$ which are continuously differentiable.

In view of the applications we subject the $\xi_1(\alpha), \cdots, \xi_n(\alpha)$ to the following condition:

Condition ϑ. $\xi_1(\alpha), \cdots, \xi_n(\alpha)$ depend continuously on α, and there exists a transformation

$$\zeta_i(\alpha) = \sum_{j=1}^{n} \xi_j(\alpha)\gamma_{ij}, \qquad \xi_j(\alpha) = \sum_{i=1}^{n} \Gamma_{ji}\zeta_i(\alpha),$$

* A study of this Cauchy problem with a view to enlarging the class of admissible initial conditions was undertaken by Margaret Gurney in her dissertation, Brown University, 1935 (unpublished).

with constant γ_{ij} of determinant ± 1 such that the derivatives of $\zeta_2(\alpha)$, $\zeta_3(\alpha), \cdots, \zeta_n(\alpha)$ are bounded.

Then there exists a relatively bounded solution $\phi_1(\alpha, \beta), \cdots, \phi_n(\alpha, \beta)$ of (61) in a relatively bounded (α, β)-neighborhood of the origin, assuming the given initial values, continuous in α, β, and continuously differentiable with respect to α and β.

It should be noticed that the essential content of the above statement lies in the fact that the derivative of $\zeta_1(\alpha)$ has no influence on the determination of the domain of existence.

The idea of the proof is to construct instead of functions $\phi_i(\alpha, \beta)$ other functions ψ_i of four arguments α, f, β, g which reduce to the solution of the initial problem in question for $f=\zeta_1(\alpha)$, $g=\zeta_1(\beta)$. In order to conform with the terminology formerly introduced, we henceforth shall identify $\zeta_1(\alpha)$ with $f(\alpha)$.

We try to satisfy the following conditions for functions $\psi_i^0(\alpha, f, \beta, g)$:

$$\text{(i)} \qquad \frac{d^2\psi_i^0}{d\alpha d\beta}(\alpha, f(\alpha), \beta, g(\beta)) = 0,$$

$$\text{(ii)} \qquad \psi_i^0(\alpha, f(\alpha), \alpha, f(\alpha)) = \xi_i(\alpha),$$

$$\text{(iii)} \qquad \begin{gathered} \sum_{j=1}^n a_{ij}(\xi_1(\alpha), \cdots, \xi_n(\alpha))\Big[\overset{0}{\psi}_{i\alpha}(\alpha, f(\alpha), \alpha, f(\alpha)) \\ + \psi_{if}(\alpha, f(\alpha), \alpha, f(\alpha))\frac{df(\alpha)}{d\alpha}\Big] = 0, \quad i \leqq m, \\ \sum_{j=1}^n a_{ij}(\xi_1(\alpha), \cdots, \xi_n(\alpha))\Big[\overset{0}{\psi}_{i\beta}(\alpha, f(\alpha), \alpha, f(\alpha)) \\ + \psi_{ig}(\alpha, f(\alpha), \alpha, f(\alpha))\frac{df(\alpha)}{d\alpha}\Big] = 0, \quad i > m. \end{gathered}$$

We first introduce $\psi_i^0(\alpha, f, \alpha, f)$ by

$$\text{(62)} \qquad \psi_i^0(\alpha, f, \alpha, f) = \sum_{k=2}^n \Gamma_{ik}\zeta_k(\alpha) + \Gamma_{i1}f, \qquad i = 1, 2, \cdots, n,$$

which yields

$$\text{(63)} \qquad \overset{0}{\psi}_{if}(\alpha, f, \alpha, f) + \overset{0}{\psi}_{ig}(\alpha, f, \alpha, f) = \Gamma_{i1}.$$

To determine ψ_{if}^0 and ψ_{ig}^0 we set up the system

$$\text{(64)} \qquad \begin{aligned} &\sum_{k=1}^n a_{ik}(\psi^0(\alpha, f, \alpha, f))\overset{0}{\psi}_{kf}(\alpha, f, \alpha, f) = 0, \qquad && i = 1, \cdots, m, \\ &\sum_{k=1}^n a_{ik}(\psi^0(\alpha, f, \alpha, f))\overset{0}{\psi}_{kg}(\alpha, f, \alpha, f) = 0, \qquad && i = m+1, \cdots, n, \end{aligned}$$

which together with (63), in view of the boundedness of $|a_{ik}|^{-1}$, determines $\overset{0}{\psi}_{if}(\alpha,f,\alpha,f)$ and $\psi_{ig}^0(\alpha,f,\alpha,f)$ as analytic functions of $a_{ik}(\psi^0(\alpha,f,\alpha,f))$ and thus as relatively bounded functions with relatively bounded and continuous total derivatives with respect to α and f.

From (ii)

$$\overset{0}{\psi}_{i\alpha}(\alpha,f,\alpha,f)+\overset{0}{\psi}_{i\beta}(\alpha,f,\alpha,f)=\sum_{k=2}^{n}\Gamma_{ik}\frac{d\zeta_k(\alpha)}{d\alpha}, \qquad i=1,2,\cdots,n,$$

and by (64) and (iii)

$$\text{(65)}\qquad \begin{aligned}&\sum_{k=1}^{n}a_{ik}(\xi_1(\alpha),\cdots,\xi_n(\alpha))\overset{0}{\psi}_{i\alpha}(\alpha,f(\alpha),\alpha,f(\alpha))=0, && i\leqq m;\\ &\sum_{k=1}^{n}a_{ik}\overset{0}{\psi}_{i\beta}(\alpha,f(\alpha),\alpha,f(\alpha))=0, && i>m,\end{aligned}$$

which determine $\overset{0}{\psi}_{i\alpha}(\alpha,f(\alpha),\alpha,f(\alpha))$ and $\overset{0}{\psi}_{i\beta}(\alpha,f(\alpha),\alpha,f(\alpha))$ as continuous and relatively bounded functions of α for relatively bounded α.

We now put

$$A_i(\alpha,f)=\int_0^f\overset{0}{\psi}_{ig}(\alpha,f',\alpha,f')df'.$$

Obviously, $A_i(\alpha,f)$ has continuous and relatively bounded derivatives with respect to f and α.

Finally put

$$\text{(66)}\qquad \begin{aligned}\overset{0}{\Psi}_i(\alpha,f,\beta,g)&=\overset{0}{\psi}_i(\alpha,f,\alpha,f)+A_i(\beta,g)-A_i(\alpha,f)\\ &\quad-\int_\beta^\alpha[\overset{0}{\psi}_{i\beta}(\alpha',f(\alpha'),\alpha',f(\alpha'))-A_{i\alpha}(\alpha',f(\alpha'))]d\alpha'.\end{aligned}$$

The reader will easily verify that the function $\Psi_i^0(\alpha,f,\beta,g)$, as defined by (66), has the following properties:

$$\begin{aligned}\overset{0}{\Psi}_i(\alpha,f,\alpha,f)&=\overset{0}{\psi}_i(\alpha,f,\alpha,f),\\ \overset{0}{\Psi}_{ig}(\alpha,f,\alpha,f)&=\overset{0}{\psi}_{ig}(\alpha,f,\alpha,f),\\ \overset{0}{\Psi}_{if}(\alpha,f,\alpha,f)&=\frac{d}{df}\overset{0}{\psi}_i(\alpha,f,\alpha,f)-\frac{dA_i(\alpha,f)}{df}=\overset{0}{\psi}_{if}(\alpha,f,\alpha,f),\\ \overset{0}{\Psi}_{i\beta}(\alpha,f(\alpha),\alpha,f(\alpha))&=\overset{0}{\psi}_{i\beta}(\alpha,f(\alpha),\alpha,f(\alpha)),\\ \overset{0}{\Psi}_{i\alpha}(\alpha,f(\alpha),\alpha,f(\alpha))&=\overset{0}{\psi}_{i\alpha}(\alpha,f(\alpha),\alpha,f(\alpha)).\end{aligned}$$

Hence we are justified in considering $\Psi_i^0(\alpha, f, \beta, g)$ as an extension of those elements of the unknown function $\psi_i^0(\alpha, f, \beta, g)$ which were used in the construction of $\Psi_i^0(\alpha, f, \beta, g)$, and we write $\Psi_i^0(\alpha, f, \beta, g) = \psi_i^0(\alpha, f, \beta, g)$. We have, moreover, $\partial^2\psi_i^0 = 0$ so that (i) is true. From (62) we obtain (ii). Formulas (64) and (65) give (iii).

Evidently, $|\psi_i^0(\alpha, f, \beta, g)|$ is less than an arbitrary positive number ϵ if the bounds of the initial data $\xi_1(\alpha), \cdots, \xi_n(\alpha)$ are sufficiently small and if α, f, β, g are relatively bounded.

On differentiating the first m equations of (61) with respect to β, and the last $(n-m)$ equations with respect to α and solving with respect to the mixed derivatives, we obtain a system of the form

$$\frac{\partial^2\phi_i(\alpha, \beta)}{\partial\alpha\partial\beta} = \sum_{j,l} c_{ijl}(\phi_1, \cdots, \phi_n) \frac{\partial\phi_j(\alpha, \beta)}{\partial\alpha} \frac{\partial\phi_l(\alpha, \beta)}{\partial\beta}, \tag{67}$$

where the $c_{ijl}(\phi_1, \cdots, \phi_n)$ have bounded derivatives up to the third order for bounded $\phi_1, \cdots, \phi_n$. Replacing ϕ by ψ, we solve

$$\psi_i(\alpha, f, \beta, g) = \psi_i^0(\alpha, f, \beta, g) + \sum_{j,l=1}^{n} I_{ijl}\begin{pmatrix}\alpha, f, \beta, g\\ \psi\end{pmatrix}$$

with the aid of Theorem 5. By (60) and (i) we have

$$\frac{d^2\psi_i(\alpha, f(\alpha), \beta, g(\beta))}{d\alpha d\beta} = \sum_{i,l} c_{ijl}(\psi(\alpha, f(\alpha), \beta, g(\beta))) \frac{d\psi_j(\alpha, f(\alpha), \beta, g(\beta))}{d\alpha} \frac{d\psi_l(\alpha, f(\alpha), \beta, g(\beta))}{d\beta}.$$

In view of (ii), $\phi_i(\alpha, \beta) = \psi_i(\alpha, f(\alpha), \beta, g(\beta))$ assumes the given initial values, satisfies (61) on $\alpha = \beta$, and has continuous derivatives with respect to α and β and continuous mixed second derivatives with respect to α and β.

A conclusion, familiar in the theory of hyperbolic equations shows that equations (61) are satisfied identically in α and β.

Thus we have established the following theorem:

THEOREM 6. *If in* (61) a_{ij} *and its partial derivatives up to the fourth order and the reciprocal value of the determinant* $|a_{ij}|$ *are bounded for bounded values of* $\phi_1, \phi_2, \cdots, \phi_n$, *and if the initial values of* $\phi_i(\alpha, \beta)$ *on* $\alpha = \beta$ *are relatively bounded in a bounded neighborhood of* $\alpha = 0$ $(=\beta)$ *and satisfy condition* ϑ, *then Cauchy's problem has a solution existing for all relatively bounded* α, β. *This solution has continuous derivatives with respect to* α *and* β. *If the initial values and the* a_{ij} *depend on a parameter* μ *and converge uniformly as* $\mu \to \mu_0$, *then there exists a subsequence of* μ, *for which the corresponding solutions* $\phi_i(\alpha, \beta)$ *converge uniformly.*

COROLLARY 1. *The solution of Theorem 6 is unique.**

Let $|\alpha| \leqq A$, $|\beta| \leqq B$ be the common domain D of existence of two solutions of our initial problem. Denote by u, v, $w(\tau)$

$$u(\tau) = \sum_{i=1}^{n} \max |\Delta\phi_i|, \qquad v(\tau) = \sum_i \max \left| \Delta \frac{\partial\phi_i}{\partial\alpha} \right|,$$

$$w(\tau) = \sum_i \max \left| \Delta \frac{\partial\phi_i}{\partial\beta} \right|,$$

where the operator Δ indicates the difference of the expression following Δ for the two solutions, and the maximum is to be taken on that segment of the line $\tau = |\alpha - \beta|$ which is contained in D. By (67) we have for a suitable constant K

$$u(\tau) \leqq \int_0^\tau (v + w) \, | d\tau |, \qquad u(0) = 0,$$

$$v(\tau) \leqq K \int_0^\tau (u + v + w) \, | d\tau |, \qquad v(0) = 0,$$

$$w(\tau) \leqq K \int_0^\tau (u + v + w) \, | d\tau |, \qquad w(0) = 0.$$

Hence

$$u + v + w \leqq (2K + 1) \int_0^\tau (u + v + w) \, | d\tau |,$$

and by the well known iteration $u \equiv v \equiv w \equiv 0$.

By reasoning very similar to the preceding it may be shown that the dependence of the initial data on a parameter such that the initial data of ϕ_i and those of $\partial\phi_i/\partial\alpha$, $\partial\phi_i/\partial\beta$ satisfy a Lipschitz condition of exponent 1 in the parameter implies a Lipschitz condition of exponent 1 in the solution. Furthermore, passing to the limit from difference quotient to derivative with respect to the parameter we obtain the following corollary:

COROLLARY 2. *If, in Theorem 6, the initial data of ϕ_i and those of $\partial\phi_i/\partial\alpha$, $\partial\phi_i/\partial\beta$ are continuously differentiable with respect to a parameter, the solution and its first derivatives with respect to α and β are also continuously differentiable with respect to the parameter, continuity being understood with respect to the parameter and variables.*

---

* Cf. Hadamard, *Leçons sur le Problème de Cauchy*, Paris, 1932, pp. 488–501.

THE UNIVERSITY OF CALIFORNIA,
BERKELEY, CALIF.

American Journal of Mathematics
Vol. LX, No. 3, July, 1938

A PROPERTY OF SPHERICAL HARMONICS.*

By Hans Lewy.

In this paper will be proved the following theorems:

Theorem 1. *Let $u(x, y, z)$ be a homogeneous polynomial of degree $n > 2$ whose range of values covers positive and negative numbers and whose Hessian $H_{[u]}$ vanishes only at $x = y = z = 0$. Then those points of the unit sphere σ $(x^2 + y^2 + z^2 = 1)$ for which $u = 0$, form two distinct closed convex curves*[1] *without singularities and diametral to each other.*

Theorem 2. *Let $u(x, y, z)$ be a spherical harmonic of degree $n > 2$, i. e. a regular solution of*

$$u_{xx} + u_{yy} + u_{zz} = 0$$

and of

$$xu_x + yu_y + zu_z = nu$$

which does not vanish identically. Then there exists a point distinct from the origin where the Hessian

$$H_{[u]} \equiv \begin{vmatrix} u_{xx} & u_{xy} & u_{xz} \\ u_{xy} & u_{yy} & u_{yz} \\ u_{xz} & u_{yz} & u_{zz} \end{vmatrix}$$

vanishes.

We employ the following

Lemma 1. *If $v(x, y, z)$ be a homogeneous polynomial of degree $m > 1$ and (x_0, y_0, z_0) an arbitrary point where $v(x_0, y_0, z_0) \neq 0$, then the Gaussian curvature of the surface $v(x, y, z) = v(x_0, y_0, z_0)$ has the same sign as $vH_{[v]}$ and vanishes where $H_{[v]}$ vanishes.*

Proof of the Lemma. $|v|^{1/m}$ is of degree 1 and analytic wherever $v \neq 0$. The first derivatives of $|v|^{1/m}$ are of degree 0. On writing down the homogeneity relations for these, we obtain three homogeneous equations whose determinant vanishes. This determinant is $H_{[|v|^{1/m}]}$ and we have for $v \gtrless 0$ resp.

* Received March 14, 1938.

[1] A closed curve without double points is called convex if either its "interior" or its "exterior" is convex, i. e. contains, together with two points P and Q, the smaller arc of the great circle through P and Q.

$$0 = H_{[|v|^{1/m}]}$$

$$= \pm \begin{vmatrix} v_{xx} + (1/m - 1)\dfrac{v_x^2}{v}, \cdots & , \cdots \\ v_{xy} + (1/m - 1)\dfrac{v}{v_x v_y}, \; v_{yy} + (1/m - 1)\dfrac{v_y^2}{v}, \cdots & \\ \cdot \; \cdot \; \cdot \; \cdot \; \cdot \; \cdot \; \cdot \; , \; \cdot \; \cdot \; \cdot \; \cdot \; \cdot \; \cdot \; \cdot & , \cdots \end{vmatrix} \cdot \frac{|v|^{3(1/m-1)}}{m^3}$$

$$= \pm \frac{|v|^{3(1/m-1)}}{m^3} \{H_{[v]} - (1/v)(1 - 1/m)[(v_{xx}v_{yy} - v_{xy}^2)v_z^2 + \cdots]\}.$$

The expression $(v_{xx}v_{yy} - v_{xy}^2)v_z^2 + \cdots$ differs from the Gaussian curvature of $v = \text{const.}$ by a positive factor, and our Lemma follows.

Proof of Theorem I. Replacing, if necessary, u by $-u$, we may assume that $H_{[u]} > 0$ and that the curvature of the surfaces S_+ ($u = \text{const.} > 0$) is positive, the curvature of the surfaces S_- ($u = \text{const.} < 0$) is negative. Furthermore it is legitimate to suppose n even, since for n odd the Hessian is an odd function and the conditions of our Theorem cannot be satisfied. Since $H_{[u]} \neq 0$, the curves $u = 0$ on σ are without singularities and possess a non-vanishing geodetic curvature at each point (i. e. have no contact of higher order with a great circle). For it is a familiar fact of algebraic geometry that $H_{[u]} \neq 0$ implies that the projection of $u = 0$ from the origin O on the tangent plane T of σ at a point P of $u = 0$ has no singularity at P and its curvature at P does not vanish; for the coördinates (x, y, z) on σ can be interpreted as projective coördinates of the projection on T.

Consider that branch of $u = 0$ which passes through P. It must be a closed spherical curve since it does not possess any singularities, and its geodetic curvature does not vanish. Its "interior," therefore, is a convex domain B distinct from a hemisphere.[2]

We show $u > 0$ in B. Assume that a suitable rotation of the system of coördinates transforms the above point P into $(1, 0, 0)$ and the direction of $u = 0$ at P into $dx = dz = 0$. We conclude from the homogeneity of u that at P for $dx = dz = 0$, $dy \neq 0$

$$0 = u = u_x, \quad xu_{xx} = u_{xx} = 0, \quad u_x dx + u_y dy + u_z dz = du = u_y dy = 0,$$
$$u_y = 0, \quad xu_{xy} = u_{xy} = 0.$$

[2] I owe to M. W. Fenchel the following information regarding this theorem about spherical curves of positive geodetic curvature: the theorem, though not explicitly stated in literature, follows immediately from i) a theorem by Möbius that a positively curved arc without singularities nor double points is contained in a hemisphere (Möbius, *Werke II*, pp. 183-187), and ii) the corresponding fact about planar domains bounded by convex curves to which it may be reduced by central projection.

By hypothesis

$$H_{[u]} = \begin{vmatrix} 0 & 0 & u_{xz} \\ 0 & u_{yy} & u_{yz} \\ u_{xz} & u_{yz} & u_{zz} \end{vmatrix} = -u_{xz}{}^2 u_{yy} > 0$$

whence

$$u_{yy} < 0, \quad u_{xz} \neq 0.$$

Now we have along $u = 0$ at P

$$dz = 0, \quad 0 = d^2u/dy^2 = u_{yy} + u_z d^2z/dy^2$$

where y is taken as independent variable. Thus $u_z d^2z/dy^2 > 0$, and u is positive on that side of a neighborhood of $u = 0$ against which $u = 0$ is concave and negative on the other side.

If $u = 0$ at P then $u = 0$ at the diametral point P'. But P and P' cannot lie on the same branch of $u = 0$. For choose in the neighborhood of P $(1, 0, 0)$ two points P_1 and P_2 on $u = 0$ with small $|y_{P_1}|$, $|y_{P_2}|$ and $y_{P_1} < 0$, $y_{P_2} > 0$. B is contained within a loon formed by the two great circles tangent to $u = 0$ at P_1 and P_2. The condition: $d^2z/dy^2 \neq 0$ on $u = 0$ in the neighborhood of P excludes that either circle passes through P. On the other hand P lies within the loon, and consequently P' lies outside of it and of B.[3]

Thus the number of disconnected branches of $u = 0$ is even and each branch bounds a convex domain in which $u > 0$. For suppose $u = 0$ inside of a branch C of $u = 0$, on a certain set of points E. Choose in E a point Q of minimum distance from C. Since $u > 0$ in the neighborhood of C for all points inside of C, Q has a positive distance from C. Thus there would exist a convex curve C', passing through Q and contained within C, in whose exterior necessarily $u < 0$ for a neighborhood of C'. This leads to a contradiction with the assumption that Q is a point of E of minimum distance from C.

Hence the domain $u < 0$ on σ consists of σ minus an even number of disjoint convex domains.

Consider again an arbitrary point of $u = 0$ in the above notation: $P(1, 0, 0)$, $u = u_x = u_{xx} = u_{xy} = u_y = 0$, $u_{xz} \neq 0$. Denote by $Q'(1, 0, z)$ a point for which $|z|$ is small and $u < 0$. The asymptotic directions of the surface $u =$ const. that passes through Q' are determined by

$$(1) \qquad u_x dx + u_y dy + u_z dz = 0$$

$$(2) \qquad u_{xx}dx^2 + 2u_{xy}dxdy + u_{yy}dy^2 + 2u_{xz}dxdz + \cdots = 0.$$

[3] Also cf. Footnote 2.

The second relation expresses that $d^2x = d^2y = d^2z = 0$ imply $d^2u = 0$. On replacing dx by its value in (1) and substituting in (2) we obtain

$$0 = A \equiv u_x(u_{yy}dy^2 + 2u_{yz}dydz + u_{zz}dz^2) \\ + (u_ydy + u_zdz)[(u_{xx}/u_x)(u_ydy + u_zdz) - 2u_{xy}dy - 2u_{xz}dz].$$

As Q' tends to P $(z \to 0)$, this equation reduces to $(dz/dy)^2 = 0$. For

$$\lim_{z\to 0} \frac{u_{xx}u_z - 2u_xu_{xz}}{u_x} = \frac{u_{xxz}u_z + u_{xx}u_{zz} - 2u_{xzz}u_x - 2u_{xz}u_{xz}}{u_{xz}}\Bigg|_{z=0,\,x=1,\,y=0}$$

$$= \frac{u_z^2[(n-1)(n-2) - 2(n-1)^2]}{(n-1)u_z}\Bigg|_{x=1,\,y=0,\,z=0}$$

$$= -nu_z(1,0,0),$$

$$\lim_{z\to 0} \frac{u_{xx}u_y - u_{xy}u_x}{u_x} = 0.$$

Hence the projections of the asymptotic directions at Q' on T tend to the direction of the tangent of $u = 0$ at P as $Q' \to P$. The same result would have been obtained if, instead of $Q'(1, 0, z)$, we had considered the point $\bar{Q}$ $(1/\sqrt{1+z^2}, 0, z/\sqrt{1+z^2})$ which is the intersection of OQ' with σ. For the asymptotic directions are determined by the equations (1) and (2) which are homogeneous in x, y, z.

We next consider the projections from O on the sphere σ of the asymptotic directions at $\bar{Q} = \bar{Q}(\bar{x}, 0, \bar{z})$, given by

$$(d'x, d'y, d'z) = (dx, dy, dz) - (\bar{x}dx + \bar{z}dz)(\bar{x}, 0, \bar{z})$$

with (dx, dy, dz) satisfying (1) and (2). We have

$$\lim_{\bar{z}\to 0} u_x(\bar{x}, 0, \bar{z})/\bar{z} = u_{xz}(1, 0, 0),$$

whence

$$\bar{z}\frac{dx}{dy} = -\bar{z}\frac{u_ydy + u_zdz}{u_xdy} \to \lim_{\bar{z}\to 0} \frac{dz}{dy} \cdot \frac{1}{n-1}.$$

Since $\lim_{\bar{z}\to 0} \frac{dz}{dy} = 0$, we conclude

$$\frac{d'z}{d'y} = \frac{dz - (\bar{x}dx + \bar{z}dz)\bar{z}}{dy} \to 0$$

and

$$\frac{d'x}{d'y} = \frac{dx - (\bar{x}dx + \bar{z}dz)\bar{x}}{dy} \to \lim_{\bar{z}\to 0} \frac{(1-\bar{x}^2)dx}{dy} = \lim_{\bar{z}\to 0} \bar{z}^2\frac{dx}{dy} = 0.$$

Thus is shown that the projections on σ of the asymptotic directions at points $\bar{Q}$ of σ with $u_{\bar{Q}} < 0$ form a field of two vectors which tend to the tangent of C as $\bar{Q}$ approaches C on a normal. It is easily seen that the above limit considerations hold uniformly for all points of C as the higher derivatives of u

on σ are uniformly bounded. The fields of the two vectors is therefore defined and continuous in the domains $u \leqq 0$.

Choose an arbitrary point N of σ for which $u_N > 0$ and project the vector field stereographically from N on a plane. Of the images of the spherical curves $u = 0$, one will contain the image of the domains $u < 0$ while the images of the other curves $u = 0$ will constitute interior boundaries of the domain of definition of the projected field vectors. As the field is without singularities the sum of the rotations of the vectors over the interior boundaries equals that over the exterior boundary. Evidently the contribution of each boundary is 4π. Hence there can be only one interior boundary. In other words, the curves $u = 0$ on σ are precisely two diametrally opposite convex curves without singularities.

The remaining paragraphs will show that for no spherical harmonic u of degree $n > 2$, n even, the curves $u = 0$ on σ can be two distinct diametrally opposite convex curves without singularities.

LEMMA 2. *Consider on σ a closed convex curve C_1 different from a great circle. Denote by B_1 the interior of C_1, and by B_2 the interior of the curve C_2 diametrally opposite to C_1. Then it is possible to locate on σ a circular domain of spherical radius 33° in such a way that its closure lies either in B_1 or in $\sigma - (B_1 + C_1) - (B_2 + C_2)$.*

We omit a proof of this Lemma and refer instead to a note by R. M. Robinson [4] in which is shown that one always can locate in the described way a circular domain of radius $<$ arc tan $\frac{1}{2}\sqrt{3}$ and that this is the best possible upper bound.

LEMMA 3. *Let u be a spherical harmonic of degree $n \geqq 4$. Then each circular domain β of the unit sphere of radius 33° contains a zero of u.*

Suppose there exists a spherical harmonic u of degree $n \geqq 4$ which is positive in a circle β of radius 33° on σ. Introduce polar coördinates (r, θ, ϕ) so that the center of β corresponds to $r = 1$, $\theta = 0$. Consider

$$v(r, \theta) = \int_0^{2\pi} u(r, \theta, \phi)\, d\phi$$

$v(r, \theta)$ is a spherical harmonic of degree n, of the form $cr^n P_n(\cos\theta)$, $c > 0$, studied by Legendre.

We have

$$P_\nu(1) = 1,\ \nu = 0, 1, \cdots; \quad P_0(\xi) \equiv 1, \quad P_1(\xi) \equiv \xi,$$

[4] R. M. Robinson, *Bulletin of the American Mathematical Society*, 1938, pp. 115-116.

and the recursion formula

$$(\nu+1)P_{\nu+1}(\xi) - (2\nu+1)\xi P_\nu(\xi) + \nu P_{\nu-1}(\xi) = 0, \qquad \nu \geqq 1.$$

Denote by $\xi_\nu < 1$ the largest value of ξ for which $P_\nu(\xi) = 0$. By the recursion formula $P_{\nu+1}(\xi_\nu) < 0$ and

$$0 = \xi_1 < \xi_2 < \xi_3 \cdots.$$

Furthermore

$$8P_4(\xi) = 35\xi^4 - 30\xi^2 + 3,$$

$$\xi_4{}^2 = \frac{15 + \sqrt{120}}{35} > \frac{15 + 10.6}{36} \geqq \frac{16^2 \cdot 10}{6^2 \cdot 100} \geqq 0.843^2.$$

Thus we have for $0 < \theta_4 < 90°$ and $\xi_4 = \cos\theta_4$ the inequality

$$\cos\theta_4 > 0.843, \qquad \theta_4 < 33°.$$

Consequently, for $\xi_\nu = \cos\theta_\nu$, $0 < \theta_\nu < 90°$,

$$33° > \theta_4 > \theta_5 > \cdots.$$

But this makes impossible our assumption that u and hence $v(r, \theta)$ are positive in β.

The comparison of Theorem 1 with Lemmas 2 and 3 easily yields Theorem 2.

BERKELEY, CALIFORNIA.

Bulletin of the American Mathematical Society
Vol. 52, No. 9, September, 1946

WATER WAVES ON SLOPING BEACHES

HANS LEWY

1. **Introduction.** Consider the time-dependent velocity potential $\phi(x, y, t)$ of the motion of an incompressible inviscid liquid in two space dimensions x, y. The Newtonian equations and the condition of incompressibility lead to the following mathematical model: Laplace's equation

$$\frac{\partial^2\phi}{\partial x^2} + \frac{\partial^2\phi}{\partial y^2} = 0 \tag{1.1}$$

and Bernoulli's law

$$\frac{\partial\phi}{\partial t} + \left(\left(\frac{\partial\phi}{\partial x}\right)^2 + \left(\frac{\partial\phi}{\partial t}\right)^2\right)\Big/ 2 + gy + p = \text{const.}, \tag{1.2}$$

relating the pressure p and the gravity potential gy per unit mass to the velocity potential; here the x-axis is horizontal, the y-axis vertical and upwards and g is the modulus of gravity acceleration. At the boundary of air and water the pressure is supposed to have a constant value p_0; thus (1.2) relates implicitly the surface elevation y to the velocity potential. A considerable simplification is introduced by assuming the motion to be small of first order so that in (1.2) the quadratic terms may be cancelled. Then the motion becomes a small perturbation of the equilibrium position in which the surface will be thought of as given by $y=0$. For small elevations of the surface one concludes from (1.2) that, but for terms of higher order,

$$y = y(x, t) = -\frac{\partial\phi}{\partial t}(x, 0, t)\Big/ g, \tag{1.3}$$

where the constant on the right hand has been absorbed in a properly modified ϕ. To this equation is added the condition that a particle at the surface remains at the surface; that is (using the "substantial" time derivative of (1.2)),

$$dy(x, t)/dt = -\frac{\partial^2\phi}{\partial t^2}\Big/ g = \partial\phi/\partial y, \tag{1.3.1}$$

An address delivered at the invitation of the Committee to Select Hour Speakers for Far Western Sectional Meetings before the Pasadena meeting of the Society on November 24, 1945; received by the editors May 15, 1946.

where terms of second order, such as $(\partial^2\phi/\partial x\partial t)(\partial\phi/\partial x)$, have been dropped in comparison with $\partial^2\phi/\partial t^2$.

Among these infinitesimal motions the periodic motions are of particular interest. Set

$$\phi(x, y, t) = \phi_1(x, y)\cos \nu t - \phi_2(x, y)\sin \nu t, \tag{1.4}$$

where $\nu > 0$ is the so-called circular frequency. From (1.1) we take

$$(\partial^2/\partial^2 x + \partial^2/\partial^2 y)\phi_1(x, y) = 0, \quad (\partial^2/\partial^2 x + \partial^2/\partial^2 y)\phi_2(x, y) = 0; \tag{1.5}$$

from (1.3)

$$y = y(x, t) = (\nu/g)\phi_1(x, 0)\sin \nu t + (\nu/g)\phi_2(x, 0)\cos \nu t; \tag{1.6}$$

from (1.3.1)

$$\begin{aligned}(\nu^2/g)(\phi_1(x, 0)\cos \nu t - \phi_2(x, 0)\sin \nu t)& \\ = (\partial\phi_1/\partial y)\cos \nu t - (\partial\phi_2/\partial y)\sin \nu t.&\end{aligned} \tag{1.6.1}$$

Special cases occur if one of the two functions ϕ_1, ϕ_2, say ϕ_2, is taken identically zero. Then the crests or troughs and the nodes of the oscillation retain their fixed positions in space. Crests or troughs of the surface elevation occur where $\partial\phi_1(x, 0)/\partial x = 0$, nodes where $\phi_1(x, 0) = 0$. Such a periodic oscillation is called a "standing wave" in contrast to the more general case of a "progressive wave," where the crests and troughs and the nodes change position with time. For the horizontal velocity component v_{nod} of the surface nodes we write (1.6) in the form $\tan \nu t = -\phi_2(x, 0)/\phi_1(x, 0)$ or $\nu t = -\text{arc tan}\,(\phi_2/\phi_1)$; differentiation with respect to t yields

$$v_{\text{nod}} = \nu/((d \text{ arc tan } \phi_1/\phi_2)/dx).$$

Similarly the horizontal component v_{cr} of the crest or trough velocity is found to be

$$v_{\text{cr}} = \nu\bigg/\left(\left(d \text{ arc tan } \frac{d\phi_1}{dx}\bigg/\frac{d\phi_2}{dx}\right)\bigg/dx\right).$$

Both $\phi_1(x, y)$ and $\phi_2(x, y)$ are solutions of the same problem

$$\partial^2\phi/\partial x^2 + \partial^2\phi/\partial y^2 = 0 \tag{1.7}$$

throughout the liquid,

$$\nu^2\phi(x, 0)/g = \frac{\partial\phi}{\partial y}(x, 0) \tag{1.8}$$

at the surface.

One of the fascinating problems concerning waves is the discovery of the influence of the depth of the water on the shape of the wave profile. Take first the classical case of infinite depth. To (1.7) and (1.8) is added the condition that the magnitude of the motion decreases to zero as $y \to -\infty$. Out of the manifold of all harmonic functions the real and imaginary parts of

$$\exp(-i(x+iy)\nu^2/g)$$

are taken as solutions ϕ_1 and ϕ_2 of (1.7) satisfying the boundary condition (1.8). With these is constructed the progressive wave

$$\phi(x, y, t) = R\exp(-i(x+iy)\nu^2/g - it) = \exp(\nu^2 y/g)\cos(\nu^2 x/g + \nu t),$$

which is periodic in x as well as in t; the x-period is $2\pi g/\nu^2$, the t-period $2\pi/\nu$. The wave profile travels unchanged in form in the direction of diminishing x as t increases; the velocity of the profile is $v_{\text{nod}} = v_{\text{cr}} = -g/\nu$ and depends therefore on the frequency.

Airy was the first to give a treatment of the infinitesimal gravity waves in a channel of constant finite depth; his results are the foundation of our knowledge on gravity waves. Again as in the case of infinite depth, the profile of the progressive wave moves, unchanged in form, with constant velocity. But the wave velocity depends on both the frequency and the depth of the channel, and so does the space period of the wave. Airy's formulae show with great clarity the passage from waves in water of finite depth to the well known limit theory of shallow water waves in whose treatment the dependence of the pressure and velocity on the depth is assumed rather than derived.

The problem whose solution forms the object of this paper is the construction of progressive water waves when the depth, instead of being a constant, has a constant slope. This problem was first attacked by E. T. Hanson[1] in 1926. He assumed the angle of the bottom with the surface to be of the form $\pi/(2q)$ with integral q, and gave the expression of one standing wave $\phi_1(x, y)$ in terms of exponential functions, without, however, constructing the progressive wave which requires the derivation of a second standing wave $\phi_2(x, y)$. M. Miche[2] quotes an explicit form of this second function ϕ_2, which he calls "houle," to distinguish it from ϕ_1, called "clapotis." For the derivation

[1] E. T. Hanson, Proc. Roy. Soc. London. Ser. A vol. 111 (1926) p. 491.

[2] M. Miche, *Mouvements ondulatoires de la mer en profondeur constante ou décroissante*, 1945, p. 88. This booklet also contains a most interesting report of various investigations of M. Miche, relating to the breaking of waves. See also M. Miche, *Annales des ponts et chaussées*, 1944, pp. 25–78, 131–164, 270–292, 369–406. Unfortunately, by the time of printing of the present paper, the author had had no opportunity to consult these latter articles.

of his formula he refers the reader to a paper to appear in the Journal de mathématiques pures et appliquées; also his result is restricted to angles of form $\pi/(2q)$.

We shall give in this paper an explicit representation in closed form for a progressive wave for all angles $p\pi/(2q)$ between bottom and surface, provided p and q are integers, p is odd, and $p<2q$. The functions involved in our formulae are exponential functions and exponential integrals of certain special types. (Unfortunately, the number of individual terms occurring in the solution becomes large as p or q become large.) In the course of our investigation we are led to the discovery of a strange and recondite relationship of these waves with the quadratic reciprocity law of number theory. In fact, this law appears as an elementary consequence of the continuous dependence of the finite standing wave on the angle between surface and bottom. We do not go here more deeply into the discussion of certain number-theoretical problems immediately suggested by this relationship, nor do we investigate the continuity of the waves as functions of the angle, for which there appears to exist a more direct approach than through our explicit expressions for these waves. A further topic left untouched is the extension of the principles of construction discussed here to the study of the three-dimensional wave problem where the wave crests are no longer parallel to the shore.

The complexity of our formulae for the waves for large p or q and the scarcity of suitable tables of exponential integrals make the evaluation of these waves a problem of considerable difficulty. It seems important to compare the "exact" theory of infinitesimal waves with other approximations such as the shallow water theory, in order to gauge the differences of the results furnished by the various theories. These problems, for angles $\pi/(2q)$, are attacked by M. Miche (loc. cit.) and by J. J. Stoker in a paper under press.

2. Reformulation of the standing wave problem. The units of length and time will now be chosen so that

$$g = 1 \quad \text{and} \quad \nu = 1.$$

Set

$$x + iy = z = e^{i\theta}$$

and let

$$\beta = \pi p/(2q), \quad p \text{ odd and relatively prime to } q, \text{ and } 0 < p < 2q.$$

Denote by S_β the angular sector $0>\theta>-\beta$. The problem of the standing wave for S_β consists in finding a harmonic function ϕ in S_β satisfying the following boundary and limit conditions, to be stated first for the case $p \leqq q$:

1. $\partial\phi/\partial y - \phi = 0$ on $y=0$, $x>0$;

2. The normal derivative $\partial\phi/\partial n = 0$ on the bottom of S_β;

3. $\phi - A\cos x - B\sin x \to 0$ for suitable constants, not both zero, A and B, as $x\to\infty$, $y=0$;

4. The square of the particle velocity, $(\partial\phi/\partial x)^2 + (\partial\phi/\partial y)^2$, remains bounded as z tends to ∞ in S_β;

5. The flow through any vertical section

$$\int_{-x\tan\beta}^{0} \frac{\partial\phi}{\partial x}(x, y)dy$$

remains bounded as $x\to\infty$.

To these conditions must be added one concerning the behavior of ϕ and its derivatives at the shore (0, 0). We shall see that there is one standing wave where

6a. ϕ and $\partial\phi/\partial x$, $\partial\phi/\partial y$ remain bounded as (x, y) tends to (0, 0), and another where

6b. ϕ behaves like $\log(x^2+y^2)$ near (0, 0).

A solution of the above problem, if it exists, may be considered as the real part of an analytic function $W(z)$ in S_β. It can easily be concluded that $W(z)$ remains bounded as $z\to\infty$. For this purpose consider the conjugate harmonic ψ. Since the normal derivative of ϕ vanishes on the bottom of S_β, so does the tangential derivative of the conjugate ψ, whence ψ is a constant there. But

$$\psi(x, 0) - \psi(x, -x\tan\beta) = \int_{-x\tan\beta}^{0} \frac{\partial\psi}{\partial y}(x, y)dy = \int_{-x\tan\beta}^{0} \frac{\partial\phi}{\partial x}(x, y)dy,$$

so that $\psi(x, 0)$ remains bounded at the surface as $x\to\infty$, by virtue of the 5th condition. From the 4th condition it follows furthermore that $|W(z)|$ can grow at most like const.$|z|$ as $z\to\infty$ within S_β. But then the Phragmén-Lindelöf principle applies and yields the boundedness of $|W(z)|$ for $z\to\infty$ in S_β. Indeed, for the application of this principle it is not at all necessary to demand the condition 4 which states that $|W'(z)|$ remains bounded as $z\to\infty$; but it suffices to demand, besides 1, 2, 3,

4′. $|W(z)| < \exp|z|^{1+\epsilon}$ as $z\to\infty$ in S_β, for all $\epsilon>0$, together with

5′. $|W(z)|$ remains bounded on the surface as $z=x\to\infty$.

For then the principle states that $|W(z)|$ remains bounded as $z\to\infty$ in

S_β. If $p<q$, then 4′ and 5′ imply 4 and 5. For it is then possible to extend $W(z)$ analytically by reflection on the bottom of S_β where condition 2 holds; and the double angle $2\beta=\pi p/q$ is still less than π. In the process of extension any bound for the absolute value of $W(z)$ on the surface of S_β becomes a bound for $|W(z)|$ on the bottom of $S_{2\beta}$. Accordingly, $|W(z)|$ remains bounded on the boundary of $S_{2\beta}$ as $z\to\infty$ and furthermore $|W(z)|<\exp|z|^{1+\epsilon}$ as $z\to\infty$ in $S_{2\beta}$, $\epsilon>0$. Moreover, $W(z)$ can be continued analytically across $y=0$, since this is true for the function $F(z)\equiv idW/(dz)-W$ whose real part vanishes on $y=0$. A bound of $|F(z)|$ for $y\leqq 0$ becomes at the same time a bound of $|F(z)|$ for $y>0$. Let $|W(z)|\leqq M$ for $y\leqq 0$, x sufficiently large; then $|dW/dz|\leqq M/|y|$ by Cauchy's integral theorem. Describe a horizontal-vertical square Q of side 4 about a point x_0 of the surface as center and consider $F_1(z)=F(z)(z-x_0-2)(z-x_0+2)$. On Q we have $|F_1(z)|\leqq 54M$; hence we conclude for $|y|\leqq 2$ that $|F_1(x_0+iy)|\leqq 54M$, where $|F(x_0+iy)|\leqq 54M/(|iy-2|\,|iy+2|)\leqq 14M$, and, finally, from the definition of $F(z)$ and the bound M for $|W(z)|=|W(x_0+iy)|$ that $|dW/dz|\leqq 15M$ for x_0 sufficiently large, $-2\leqq y\leqq 0$. For $y<-2$ Cauchy's estimate gives immediately $|dW/dz|\leqq M$, so that $|dW/dz|$ remains bounded in S_β. Hence 4 and 5 are consequences of 1, 2, 3, 4′, 5′.

If p is no longer less than q, but $p<2q$, we shall still be able to exhibit a solution $W(z)$ of the wave problem satisfying 1, 2, 3, 4′, 5′; but we shall then show explicitly that $|W(z)|$ remains bounded on the bottom of S_β as $z\to\infty$, a fact which, as we have just seen, follows from the Phragmén-Lindelöf principle for $p<q$.

Let λ be an arbitrary complex number of modulus 1, and let $\partial\phi/\partial s$ stand for the derivative in the direction of the vector λ. Then

$$\partial RW/\partial s = \partial\phi/\partial s = R\partial W/\partial s = R\{\lambda dW/(d\lambda s)\} = R\{\lambda dW/dz\}.$$

Accordingly, the condition 1 may be rephrased as

$$(2.1)\qquad R\{(id/dz-1)W(z)\}=0$$

on $y=0$. Similarly for condition 2: set

$$(2.2)\qquad \epsilon^{1/2}=e^{-\beta i},\qquad \epsilon=e^{-2\beta i}=e^{-p\pi i/q};$$

then this condition becomes

$$(2.3)\qquad R\{i\epsilon^{1/2}dW/dz\}=0\quad\text{on}\quad \theta=-\beta=-\pi p/(2q).$$

We furthermore see that reflection on the bottom of S_β yields

$$(2.4)\quad R\{(-i\epsilon d/dz-1)W(z)\}=0\quad\text{on}\quad \theta=-2\beta=-\pi p/q.$$

3. **Derivation of an ordinary differential equation.** Let $f(\xi)$ and $g(\xi)$ be polynomials with real coefficients. Then, by (2.1),

$$R\{f(d/dz)(id/dz - 1)W(z)\} = 0 \tag{3.1}$$

on $y=0$, and by (2.4),

$$\begin{aligned} g(\partial/\partial s)R\{(-&i\epsilon d/dz - 1)W(z)\} \\ &= R\{g(\epsilon d/dz)(-i\epsilon d/dz - 1)W(z)\} = 0 \end{aligned} \tag{3.2}$$

on $\theta = -2\beta$. Suppose we succeed in solving the identical equation

$$f(\xi)(i\xi - 1) \equiv g(\epsilon\xi)(-i\epsilon\xi - 1); \tag{3.3}$$

then

$$f(d/dz)(id/dz - 1)W(z)$$

becomes a function $E(z)$, regular within the sector $0 > \theta > -2\beta$, and whose real part equals zero on the boundary of $S_{2\beta}$ (with the possible exception of the origin). Now all such functions $E(z)$ are known explicitly. Our problem thus becomes that of solving an ordinary differential equation

$$f(d/dz)(id/dz - 1)W(z) = E(z) \tag{3.4}$$

for an appropriate $E(z)$, and of choosing from the manifold of its solutions one, $W(z)$, such as to satisfy the boundary conditions (2.1) and (2.3) as well as certain conditions concerning the behavior of $W(z)$ near $z=0$ and $z=\infty$.

4. **Derivation of $f(\xi)$.**

LEMMA. *There exist polynomials $f(\xi)$ and $g(\xi)$ satisfying* (3.3) *and of degree $q-1$. They are uniquely determined by requiring the degree to be $q-1$ and $f(0)=1$. Furthermore $f(\xi) \equiv g(\xi)$.*

The following proof is valid only if p is an odd number. This is the reason why our explicit construction of waves fails if p is even.

Let

$$\begin{aligned} f(\xi) &= a_0 + a_1\xi + \cdots, \qquad a_0 = 1, \\ g(\xi) &= b_0 + b_1\xi + \cdots. \end{aligned} \tag{4.1}$$

The identity to be solved is

$$(i\xi - 1)(1 + a_1\xi + \cdots) = (-i\epsilon\xi - 1)(b_0 + b_1\xi + \cdots). \tag{4.2}$$

Compare coefficients and find

$$
(4.3)\qquad \begin{aligned}
1 &= b_0,\\
i - a_1 &= -\, i\epsilon b_0 - b_1\epsilon,\\
&\cdots\cdots\cdots\cdots,\\
ia_{n-1} - a_n &= -\, i\epsilon^n b_{n-1} - b_n\epsilon^n,\\
&\cdots\cdots\cdots\cdots.
\end{aligned}
$$

The imaginary part of the left-hand side is a_{n-1} and may be considered known from the preceding equation of the recursion. The imaginary part of the right-hand side involves b_{n-1} and b_n. Now b_{n-1} also may be assumed known from the previous equation. Consequently b_n is determined by equating imaginary parts, provided $I(\epsilon^n)\neq 0$, that is, for all n with $0<n<q$, since p and q have no common divisors. Once b_n is known, we take the value of a_n from the same equation by equating real parts. Thus the recursion furnishes in a unique way all coefficients a_n, b_n for $n<q$. There is a further condition to be satisfied to establish (4.2), namely, (4.3) for $n=q$, which becomes

$$
(4.4)\qquad ia_{q-1} - a_q = -\, i\epsilon^q b_{q-1} - b_q\epsilon^q.
$$

Since $\epsilon^q=-1$ for odd p, this implies

$$
a_{q-1} = b_{q-1};
$$

if this is so, we may assume all further coefficients $a_q, b_q, a_{q+1}, b_{q+1}, \cdots$ to vanish. Now observe that $a_0=b_0=1$, and that the nth recursion formula (4.3) may be satisfied by setting $a_n=b_n$, provided it is true that $a_{n-1}=b_{n-1}$. Indeed, the uniquely determined solution of the recursion formulae can be found from the recursion

$$
a_n(1-\epsilon^n) = a_{n-1}i(1+\epsilon^n).
$$

This gives for a_n the nonvanishing value

$$
(4.5)\qquad a_n = a_{n-1}i(1+\epsilon^n)/(1-\epsilon^n),
$$

which is indeed real since the vector $1-\epsilon^n$ is perpendicular to $1+\epsilon^n$. This holds provided $\epsilon^n\neq 1$, that is, as long as $np/q\neq 2k$ with integral k; hence certainly for odd p and $0<n<q$. Notice that (4.5) can also be written as

$$
(4.5.1)\qquad a_n = a_{n-1}\cot(np\pi/2q) = \prod_{k=1}^{n}\cot(kp\pi/2q), \qquad \text{for } n\geqq 1.
$$

If we had permitted p to be even, then q would necessarily be odd and we would have $\epsilon^q=1$; in this event (4.4) would yield

$$ia_{q-1}(1 + \epsilon^q) = 2a_{q-1}i = a_q - b_q \neq 0,$$

which cannot be solved with real a_{q-1}, a_q, b_q.

Thus we have found

$$f(\xi) \equiv g(\xi) \equiv 1, \qquad \text{for } q = 1,$$

$$f(\xi) \equiv g(\xi) \equiv 1 + \sum_{n=1}^{q-1} i^n \xi^n \prod_{k=1}^{n} \frac{1 + \epsilon^k}{1 - \epsilon^k}, \qquad \text{for } q \geqq 2.$$

LEMMA. *Let $q > 1$. The roots of $f(\xi)$ are*

$$- i\epsilon, \; - i\epsilon^2, \cdots, \; - i\epsilon^{q-1};$$

they are conjugate to and, as a set, identical with

$$i/\epsilon, \; i/\epsilon^2, \cdots, \; i/\epsilon^{q-1}.$$

In fact, since by (3.3)

$$(4.6) \qquad (i\xi - 1)f(\xi) = (- i\epsilon\xi - 1)f(\epsilon\xi),$$

it follows that the root of the right-hand side $\xi = i/\epsilon$ must also be a root of $f(\xi)$ since it is distinct from $-i$. But as all coefficients of $f(\xi)$ are real, the conjugate $-i\epsilon$ is also a root of $f(\xi)$. Substitute $\xi = -i\epsilon$ on the right; then either $\epsilon^2 = -1$ or $f(-i\epsilon^2) = 0$. Hence either $q = 2$, or $-i\epsilon^2$ is a root of $f(\xi)$. In the latter case put $\xi = -i\epsilon^2$ in (4.6) and find either $\epsilon^3 = -1$, $q = 3$, or $f(-i\epsilon^3) = 0$. Repeating this procedure we find that the complete set of roots of $f(\xi)$ is given by $\xi = -i\epsilon^n$, $n = 1, \cdots, q-1$. Hence

$$f(\xi) = c \prod_{n=1}^{q-1} (\xi + i\epsilon^n),$$

where

$$c^{-1} = f(0)/c = \prod_{n=1}^{q-1} (i\epsilon^n) = i^{q-1}\epsilon^{q(q-1)/2} = (- 1)^{(p-1)(q-1)/2},$$

in view of (2.2). Thus

$$(4.7) \qquad f(\xi) = (- 1)^{(p-1)(q-1)/2} \prod_{n=1}^{q-1} (\xi + i\epsilon^n).$$

Upon combining factors which contain conjugate roots of $f(\xi)$, we have

$$f(- i) = (- 1)^{(p-1)(q-1)/2}(- i + i\epsilon)(- i + i\epsilon^{q-1}) \cdots .$$

If q is odd, we have $(q-1)/2$ such pairs of factors; if q is even, we have

$(q-2)/2$ such pairs and the factor $-i+i\epsilon^{q/2}$. Now suppose first that both roots entering such a pair (for example, $-i\epsilon$, $-i\epsilon^{q-1}$) have positive real part. Then the argument of the product of such a pair remains the same if both roots are replaced by $+1$; that is, the argument of the product of such a pair of factors is $\pi/2$. If, on the other hand, both roots involved lie on the left of the imaginary axis, the argument of such a product remains the same if both roots are replaced by -1, which gives precisely $-\pi/2$; and no root $i/\epsilon^n=\pm i$, since $0<n<q$.

Accordingly, for odd q,

$$f(-i) = (-1)^{(p-1)(q-1)/2}(-i)^{\alpha(p,q)}e^{-\pi i(q-1)/4}r,$$

where $r>0$ and $\alpha(p, q)$ is the number of roots of $f(\xi)$ with positive real part. It is easily seen that this formula still holds good for even q. To determine r, we note that

$$r^2 = \left|\prod_1^{q-1}(1-\epsilon^n)\right|^2 = \prod_1^{q-1}(1-\epsilon^n)(1-\epsilon^{-n})$$
$$= \left|\frac{d}{dx}\left(\frac{x^{2q}-1}{x+1}\right)\right|_{x=1} = q,$$

so that

$$f(-i) = (-1)^{(p-1)(q-1)/2}(-i)^{\alpha(p,q)}e^{-\pi i(q-1)/4}q^{1/2}. \tag{4.8}$$

The number of terms with positive real part in the set i/ϵ^j is the same as the number of terms among ϵ^{-j} with negative imaginary parts. Note that

$$e^{\pi ipj/q}\cdot e^{\pi ip(q-j)/q} = -1.$$

Hence, if $e^{\pi ipj/q}$ has a negative imaginary part, so does $e^{\pi ip(q-j)/q}$. Thus, for odd q, α is even and $(-i)^{\alpha(p,q)}=(-1)^{N(p,q)}$, where $N(p, q)$ is defined (for q odd or even) by

$$N(p,q) = \text{number of terms with negative imaginary parts in the set } e^{\pi ipj'/q} \text{ with } j' \text{ integral and } 1 \leqq j' < q/2. \tag{4.9}$$

For even q, setting $j=q/2$ makes $e^{\pi ipj/q}$ equal to i^p, while for $j\neq q/2$ we have the same fact that both $e^{\pi ipj/q}$ and $e^{\pi ip(q-j)/q}$ have negative imaginary parts or both have positive imaginary parts. Accordingly,

$$(-i)^{\alpha(p,q)} = \begin{cases}(-1)^{N(p,q)}, & \text{for } p\equiv 1 \pmod 4,\ q \text{ even},\\ i(-1)^{N(p,q)+1}, & \text{for } p\equiv 3 \pmod 4,\ q \text{ even}.\end{cases}$$

Hence, from (4.8) we can conclude

$$(4.10)\quad f(-i) = \begin{cases} (-1)^{N(p,q)}e^{-\pi i(q-1)/4}q^{1/2}, & \text{for } q \text{ even, } p \equiv 1 \pmod 4,\\ (-1)^{N(p,q)}ie^{-\pi i(q-1)/4}q^{1/2}, & \text{for } q \text{ even, } p \equiv 3 \pmod 4,\\ (-1)^{N(p,q)}e^{-\pi i(q-1)/4}q^{1/2}, & \text{for } q \text{ odd.} \end{cases}$$

Another property of $f(\xi)$ is

$$(4.11)\qquad a_j = a_{q-j-1}(-1)^{(p-1)(q-1)/2}, \qquad j = 1, 2, \cdots, q-1.$$

In fact, the roots of $f(\xi)$ are simple and contain the reciprocal of each of them, whence

$$\xi^{q-1}f(\xi^{-1}) = (-1)^{(p-1)(q-1)/2}f(\xi).$$

5. **Real independence and semi-independence.** The following definitions are variants of the notion of linear independence; they are introduced in order to simplify the formal part of our computations.

DEFINITION. n complex-valued functions $F_1(x), \cdots, F_n(x)$ on a range of the variable x shall be called "really independent" on this range if the relation

$$R\left\{\sum_{\nu=1}^{n} A_\nu F_\nu(x)\right\} = 0$$

with complex coefficients A_ν implies

$$A_1 = A_2 = \cdots = A_n = 0.$$

If and only if the $F_\nu(x)$ are really independent, the $2n$ functions $RF_\nu(x)$ and $IF_\nu(x)$ are linearly independent in the ordinary sense employing real coefficients.

LEMMA. *Let $F_1(x), \cdots, F_n(x)$ be really independent. Then a relation with complex coefficients a_ν, b_ν,*

$$(5.1)\qquad R\left\{\sum_{1}^{n} (a_\nu F_\nu(x) + b_\nu \bar{F}_\nu(x))\right\} = 0,$$

implies

$$(5.2)\qquad a_\nu + \bar{b}_\nu = 0, \qquad \nu = 1, \cdots, n.$$

In fact,

$$0 = R\left\{\sum (a_\nu F_\nu(x) + b_\nu \bar{F}_\nu(x))\right\} \equiv R\left\{\sum (a_\nu + \bar{b}_\nu)F_\nu(x)\right\}.$$

This identity shows that (5.1) follows from (5.2), no matter whether the F_ν are really independent or not.

DEFINITION. If $F_1(x), \cdots, F_n(x)$ are complex, $F(x)$ real, and such that

$$RF_1(x), \cdots, RF_n(x), IF_1(x), \cdots, IF_n(x), F(x)$$

are linearly independent, we shall call $F_1(x), \cdots, F_n(x), F(x)$ "really semi-independent."

LEMMA. *A relation for complex coefficients* a_ν, b_ν, a,

$$(5.3) \qquad \begin{aligned} a_\nu + \bar{b}_\nu &= 0, \qquad \nu = 1, \cdots, n, \\ a + \bar{a} &= 0, \end{aligned}$$

implies

$$(5.4) \qquad R\left\{ \sum_{\nu=1}^{n} (a_\nu F_\nu(x) + b_\nu \bar{F}_\nu(x)) + aF(x) \right\} = 0.$$

Conversely, (5.4) *implies* (5.3) *if the* $F_\nu(x)$ *and* $F(x)$ *are really semi-independent.*

Notation. In order to simplify the printing, we shall often make use of the following notation. Let γ_1 and γ_2 be two complex quantities. We write

$$\gamma_1 \cong \gamma_2,$$

in order to express that γ_1 and γ_2 are complex conjugate.

6. **Integration of** (3.4) **for** $E(z) \equiv 0$. Writing (3.4) in the form

$$(6.1) \qquad \prod_{j=1}^{q} (d/dz - \sigma_j)W(z) = 0,$$

where

$$(6.2) \qquad \sigma_j = i/\epsilon^j = ie^{\pi i p j/q}, \qquad j = 1, \cdots, q,$$

we obtain

$$W(z) = \sum_{j=1}^{q} c_j e^{\sigma_j z}.$$

Note that

$$(6.3) \qquad \sigma_j \cong \sigma_{q-j}, \qquad j = 1, \cdots, q-1,$$

whence

$$e^{\sigma_j z} \cong e^{\sigma_{q-j} z}$$

for real x. For odd q the $e^{\sigma_j x}$ $(j=q, q-1, \cdots, q-(q-1)/2=(q+1)/2)$ are $(q+1)/2$ really independent functions, for even q the $e^{\sigma_j x}$ $(j=q, q-1, \cdots, q/2+1)$ and $e^{\sigma_{q/2} x}$ are $q/2+1$ really semi-independent functions. Both statements follow from the nonvanishing of the Wronskian of $e^{\sigma_1 x}, \cdots, e^{\sigma_q x}$, which in turn is a consequence of the nonvanishing of the Vandermondian of the q distinct numbers $\sigma_1, \cdots, \sigma_q$. In fact a linear form

$$R\left\{\sum_1^n (A_\nu e^{\sigma_\nu x} + B_\nu e^{s_\nu x})\right\},$$

where $s_\nu = \bar{\sigma}_\nu$, can be written as

$$(1/2)\cdot\sum_1^n ((A_\nu + \bar{B}_\nu)e^{\sigma_\nu x} + (\bar{A}_\nu + B_\nu)e^{s_\nu x}).$$

The c_j are determined by the boundary conditions (2.1) and (2.3). We find

$$(6.4)\quad 0 = R\{idW/dz - W\} = R\left\{\sum_1^q (i\sigma_j - 1)c_j e^{\sigma_j z}\right\}, \quad \text{on } y = 0,$$

and

$$(6.5)\quad 0 = R\{i\epsilon^{1/2}dW/dz\} = R\left\{\sum_1^q i\epsilon^{1/2}\sigma_j c_j e^{\sigma_j z}\right\}, \quad \text{on } z = |z|\,\epsilon^{1/2}.$$

Applying the lemmas about real dependence resp. semi-independence in case q is odd resp. even, we obtain as necessary and sufficient conditions for the validity of (6.4)

$$(6.6)\qquad c_j(i\sigma_j - 1) \cong -\,c_{q-j}(i\sigma_{q-j} - 1), \qquad j = 1, \cdots, q-1.$$

Note that, for $j=q$, we have $i\sigma_q - 1 = 0$. In order to exploit (6.5), observe that by (6.2)

$$(6.7)\qquad \sigma_{q-j+1}\epsilon^{1/2} \cong \sigma_j\epsilon^{1/2}, \qquad j = 1, \cdots, q.$$

Hence (6.5) implies

$$(6.8)\qquad i\epsilon^{1/2}\sigma_{q-j+1}c_{q-j+1} \cong -\,i\epsilon^{1/2}\sigma_j c_j,$$

which reduces to

$$\epsilon\sigma_{q-j+1}c_{q-j+1} = \sigma_{q-j}\bar{c}_j$$

or

$$(6.9)\qquad c_{q-j+1} = \bar{c}_j, \qquad j = 1, \cdots, q.$$

The equations (6.6) and (6.9) determine the c_j but for a real factor independent of j. We conclude directly

$$\bar{c}_{q-j} = c_j(i\sigma_j - 1)/(i\sigma_j + 1), \qquad j = 1, \cdots, q-1, \tag{6.10}$$

or

$$c_{j+1} = c_j(i\sigma_j - 1)/(i\sigma_j + 1), \qquad j = 1, \cdots, q-1. \tag{6.11}$$

The equations (6.11) are recursion formulae. They do not exhaust the implications of (6.6) and (6.9). Let q be even, and apply (6.10) with $j=q/2$ and (6.2):

$$\bar{c}_{q/2} = c_{q/2}(1 + \epsilon^{-q/2})/(\epsilon^{-q/2} - 1).$$

Now $\epsilon^{-q/2} = e^{\pi ip/2}$ and p is odd. Hence

$$\bar{c}_{q/2} = c_{q/2}i^{-p}, \qquad \text{for } q \text{ even.} \tag{6.12}$$

Let q be odd, and put $j=(q+1)/2$ in (6.9):

$$\bar{c}_{(q+1)/2} = c_{(q+1)/2}, \quad \text{or} \quad c_{(q+1)/2} \text{ is real for } q \text{ odd.} \tag{6.13}$$

Formulae (6.12) resp. (6.13) together with (6.11) show that the c_j are determined but for a constant real factor.

To compute the c_j, we observe that they are related to the coefficients a_{j-1} of $f(\xi)$ in a simple manner. In fact, we had (4.5),

$$a_j = a_{j-1}i(1 + \epsilon^j)/(1 - \epsilon^j),$$

and (6.11) which can be written

$$c_{j+1} = c_j(1 + \epsilon^j)/(1 - \epsilon^j), \qquad j = 1, \cdots, q-1.$$

Hence

$$c_{j+1}/a_j = -ic_j/a_{j-1}, \qquad j = 1, \cdots, q-1,$$

and

$$c_j = a_{j-1}(-i)^j d, \qquad j = 1, \cdots, q, \tag{6.14}$$

where d does not depend on j. The absolute value of d is irrelevant, but its argument is of importance. Take q even and apply (6.12); we find

$$c_{q/2} = e^{\pi ip/4}r,$$

with a real factor r. Substitute this value of $c_{q/2}$ in the formula (6.14) for $j=q/2$ and remember that all the coefficients a_j are real. We find

$$d = e^{\pi i(p+q)/4}, \qquad \text{for } q \text{ even.} \tag{6.15}$$

Next take q odd. Then, by (6.13), (6.14),

$$(-i)^{(q+1)/2}d = \text{real},$$

and

$$d = i^{(q+1)/2}, \qquad \text{for } q \text{ odd.} \tag{6.16}$$

From (6.14), (4.1), (4.10) we compute

$$\sum_{1}^{q} c_j = -if(-i)d,$$

$$\sum_{1}^{q} c_j = \begin{cases} (-1)^{N(p,q)}q^{1/2}, & \text{for } q \text{ odd,} \\ (-1)^{N(p,q)+(p-1)/4}q^{1/2}, & \text{for } q \text{ even, } p \equiv 1 \pmod 4, \\ (-1)^{N(p,q)+(p+1)/4}q^{1/2}, & \text{for } q \text{ even, } p \equiv 3 \pmod 4. \end{cases} \tag{6.17}$$

Furthermore, by (6.14) and (4.11),

$$c_q = \begin{cases} (-i)^{(q-1)/2}, & \text{for } q \text{ odd,} \\ ie^{-\pi i(p+q)/4}, & \text{for } q \text{ even.} \end{cases} \tag{6.18}$$

Accordingly, our solution $W(z)$, designated henceforth by $W_0(z)$, is

$$W_0(z) = \begin{cases} (-i)^{(q-1)/2}e^{-iz} + i^{(q+1)/2}\sum_{j=1}^{q-1}(-i)^j a_j \exp\left[ize^{i\pi jp/q}\right], \\ \qquad\qquad \text{if } q \text{ odd,} \\ ie^{-\pi i(p+q)/4}e^{-iz} + e^{\pi i(p+q)/4}\sum_{j=1}^{q-1}(-i)^j a_j \exp\left[ize^{i\pi jp/q}\right], \\ \qquad\qquad \text{if } q \text{ even,} \end{cases} \tag{6.19}$$

with

$$a_j = \prod_{n=1}^{j} \cot(np\pi/2q), \qquad a_0 = 1.$$

In order to see that the coefficients c_j actually satisfy (6.6) and (6.9), we notice that these relations contain at most $2q-1$ linearly independent real equations, while on the other hand their consequences (6.11) and (6.12) resp. (6.13) are precisely $2q-1$ linearly independent relations. Thus the number of linearly independent equations (6.6) and (6.9) is $2q-1$, and they follow from (6.11) and (6.12) resp. (6.13).

7. **Behavior of $W_0(z)$ for $z\to\infty$.** For the applicability of (6.19) to the wave problem the behavior of $W_0(z)$ at infinity is of importance.

Now the moduli of the terms under the sum in (6.19) tend to zero or infinity as $z=x\to\infty$, according to whether $R\{ie^{i\pi pj/q}\}<0$ or >0. Furthermore, it is readily seen that the modulus of the sum cannot remain finite unless all terms tend to zero. Now let $1<p<2q$ and consider the angles

$$p\pi/q,\ 2p\pi/q,\ \cdots,\ (q-1)p\pi/q, \tag{7.1}$$

and their conventional representation in the unit circle. Among them there is at least one whose second side lies in the third or fourth quadrant. This is evident if $p>q$, because $p\pi/q$ has this property. For $p<q$ we remember that there are at most $(q-1)$ positions available in the first and second quadrants, namely those of

$$\pi/q,\ 2\pi/q,\ \cdots,\ (q-1)\pi/q.$$

If our contention were wrong, they would have to be filled each by a member of the set (7.1), since no two of them can occupy the same position by virtue of the fact that p and q are relatively prime. In particular there would be one, say $kp\pi/q$, which occupies the same position as $(q-1)\pi/q$. Hence $kp=q-1+nq$ with even n. Here $k\neq q-1$ since otherwise $p-1$ becomes divisible by q in contradiction to $p<q$, $q>1$. But now $(k+1)p\pi/q$ is among the angles (7.1) and its second side is identical with that of

$$(q-1+p)\pi/q = (1+(p-1)/q)\pi,$$

which lies in the third or fourth quadrant.

From the foregoing we gather that, for $1<p<2q$, there exists a j with $1\leqq j\leqq q-1$ for which $R\{ie^{i\pi jp/q}\}>0$. Accordingly $W_0(x)$ does not remain finite as $x\to\infty$ in contradiction of the condition 5′ of §2.

For $p=1$, however, $R\{ie^{i\pi jp/q}\}<0$ for $j=1, 2, \cdots, q-1$, and $W_0(x)$ remains finite as $x\to\infty$. Hence $RW_0(z)$ represents a standing wave for $p=1$, it is free of singularities and behaves for $z\to\infty$ like a simple standing wave in water of infinite depth. More precisely,

$$RW_0(z)\sim R\{e^{-\pi i(q-1)/4}e^{-iz}\}, \qquad \text{for } q \text{ odd or even,}$$

and the complex factor $e^{-\pi i(q-1)/4}$ is decisive in the location of the nodes of this simple wave. These nodes lie at $x=(-(q-1)/4+n)\pi$ with integral n. Accordingly, an increase of one unit in q makes the nodes of the simple wave move one-eighth of a wave length toward shore, if only angles β of the form $\pi/2q$ are considered.

The standing wave $RW_0(z)$ was given by E. T. Hanson.[3] As we have proved, the angles β of the form $\pi/2q$ between surface and floor of the

[3] Loc. cit.

ocean are the only ones for which it will satisfy the condition of finiteness at infinity.

For the construction of a progressive wave, Hanson's standing wave furnishes only one component for $p=1$. For this construction requires the exhibition of a second standing wave of the same period which at large distance from the shore becomes a simple wave whose nodes lie midway between those of the first component.

Plan of the following investigation. The assumption $E(z)=\text{const.}$ underlying the preceding section is a special case of (3.4). More generally we can take

$$E(z) = iz^{2kq/p}, \qquad k = 0, \pm 1, \pm 2, \cdots .$$

All these choices may be expected to lead to a solution $W(z)$ of (2.1) and (2.3), that is, to a $W(z)$ satisfying the boundary conditions. But, generally speaking, these $W(z)$ do not satisfy the condition of finiteness at infinity, and some have too large a singularity at the origin. Our plan is to obtain suitable superpositions of the various solutions which insure finiteness at infinity, indeed two independent such superpositions, so as to enable us to prescribe the location of the nodes of the simple standing wave to which they become asymptotically equal. It is remarkable that this procedure proves successful, in that the number of conditions to be satisfied is larger than the number of available constants. Consequently, it becomes necessary to investigate in detail the asymptotic behavior of the solutions to be constructed, and the algebraic character of the conditions which enforce finiteness at infinity.

The result will be that for all angles $\beta=\pi p/2q$, $1\leq p<2q$, p odd, there exist two standing waves, ϕ_1 and ϕ_2, one which remains finite at the shore (and whose derivatives of first order (the velocity components) remain finite), the other having a logarithmic singularity at $z=0$ (resp. having a velocity which becomes infinite like $1/|z|$).

8. **Properties of exponential integrals.** Unfortunately, the construction of waves runs in a similar, but not identical, pattern when $p\equiv 1 \pmod 4$ and when $p\equiv 3 \pmod 4$. This necessitates two parallel sets of formulae for certain parts of the following investigation.

We begin with $p\equiv 3 \pmod 4$. Let $\alpha \geqq 0$ and consider the integral

$$\int_{\alpha}^{\sigma_j z} e^{-t} t^{\lambda} dt, \tag{8.1}$$

where

$$\lambda = (2kq/p) - 1, \qquad k = 0, \pm 1, \pm 2, \cdots . \tag{8.2}$$

The integral is a multi-valued function of z, but a single-valued function of $\log z$, once $\log \sigma_j$ and the path of integration on the Riemann surface of $\log z$ have been fixed. We shall take $\log z$ real for $z=x>0$ (on the surface) and set, generally,

$$\zeta = \log z \tag{8.3}$$

and

$$\tau_j = \log \sigma_j \tag{8.4}$$

with

$$\tau_j = -p\pi i/2 + jp\pi i/q, \qquad j = 1, \cdots, q. \tag{8.5}$$

This choice of the τ_j has been made with a view to establishing for the τ_j the same relations of conjugation that hold for the σ_j. In fact, the analogues of (6.3) and (6.7) are

$$\tau_{q-j} = p\pi i/2 - jp\pi i/q = \bar{\tau}_j, \qquad 1 \leqq j < q, \tag{8.6}$$

and

$$\tau_j - p\pi i/(2q) \cong p\pi i/2 - p(j - 1/2)\pi i/q = \tau_{q-j+1} - p\pi i/(2q), \qquad 1 \leqq j \leqq q. \tag{8.7}$$

Accordingly, let the path of integration start at that point α which belongs to a real value of $\log \alpha$ and end at that point $\sigma_j z$ whose log is $\zeta + \tau_j$, and define

$$L_{\lambda,\sigma_j}(\zeta) = e^{\sigma_j z}\int_\alpha^{\sigma_j z} e^{-t}t^\lambda dt. \tag{8.8}$$

For real $\log z$, that is, for $z=x>0$ (on the surface), this function satisfies, because of (8.6) and (8.7), (8.3) and (8.4),

$$L_{\lambda,\sigma_j}(\zeta) \cong L_{\lambda,\sigma_{q-j}}(\zeta). \tag{8.9}$$

For $\zeta = \log z = \text{real} - ip\pi/2q$, that is, for $z = r\epsilon^{1/2}$, $r>0$ (on the bottom),

$$L_{\lambda,\sigma_j}(\zeta) \cong L_{\lambda,\sigma_{q-j+1}}(\zeta). \tag{8.10}$$

We find the differential equation

$$(d/dz)L_{\lambda,\sigma_j}(\zeta) - \sigma_j L_{\lambda,\sigma_j}(\zeta) = \sigma_j^{\lambda+1}z^\lambda = z^{-1}e^{(\zeta+\tau_j)2kq/p}, \tag{8.11}$$

valid for all ζ. Thus for real ζ, $z=x>0$,

$$(d/dz)L_{\lambda,\sigma_j}(\zeta) = \sigma_j L_{\lambda,\sigma_j}(\zeta) + x^\lambda(-1)^{kq}, \tag{8.12}$$

and for $\zeta = \text{real} - ip\pi/2q$, $z = r\epsilon^{1/2}$,

(8.13) $$(d/dz)L_{\lambda,\sigma_j}(\zeta) = \sigma_j L_{\lambda,\sigma_j}(\zeta) + r^{\lambda}(-1)^{k(q+1)}\epsilon^{-1/2}.$$

Next take $p\equiv 1 \pmod 4$. Set

(8.4′) $$\tau_j = \log(-\sigma_j),$$

where, again, (8.5) is assumed:

$$\tau_j = -p\pi i/2 + jp\pi i/q, \qquad j = 1, \cdots, q.$$

Then, we have the same relations of conjugations as before, namely (8.6) and (8.7). The definition paralleling (8.8) becomes

(8.8′) $$L_{\lambda,\sigma_j}(\zeta) = e^{\sigma_j z}\int_{\alpha}^{-\sigma_j z} e^t t^{\lambda} dt.$$

It introduces a single-valued function of ζ for $\alpha \geqq 0$, $\log \alpha$ real, and $\lambda = (2kq/p) - 1$ as above. We conclude the same relations (8.9) and (8.10) of conjugation for the new definition (8.8′). Furthermore, the differential equation corresponding to (8.11) becomes

(8.11′) $$(d/dz)L_{\lambda,\sigma_j}(\zeta) = \sigma_j L_{\lambda,\sigma_j}(\zeta) + (-\sigma_j)^{\lambda+1} z^{\lambda},$$

and its special cases corresponding to surface and bottom are again the equations (8.12) and (8.13).

Now consider, no matter what p, the function $W_\lambda(z)$ defined by

(8.14) $$W_\lambda(z) \equiv \sum_{j=1}^{q} c_j L_{\lambda,\sigma_j}(\zeta).$$

We gather from (8.12) that on the surface where $z = x > 0$

$$i(d/dz)W_\lambda(z) - W_\lambda(z) = \sum_1^q c_j(i\sigma_j - 1)L_{\lambda,\sigma_j}(\zeta) + x^{\lambda}(-1)^{kq} i \sum_1^q c_j,$$

and on the bottom where $z = r\epsilon^{1/2}$, $r > 0$,

$$i\epsilon^{1/2}(d/dz)W_\lambda(z) = \sum_{j=1}^{q} i\epsilon^{1/2} c_j \sigma_j L_{\lambda,\sigma_j}(\zeta) + r^{\lambda}(-1)^{k(q+1)} i \sum_1^q c_j.$$

Now, since $\sum_1^q c_j$ is real by (6.17), we have on the surface

$$R\{i(d/dz)W_\lambda(z) - W_\lambda(z)\} = R\sum_1^q c_j(i\sigma_j - 1)L_{\lambda,\sigma_j}(\zeta),$$

and on the bottom

$$R\{i\epsilon^{1/2}(d/dz)W_\lambda(z)\} = R\sum_1^q i\epsilon^{1/2} c_j \sigma_j L_{\lambda,\sigma_j}(\zeta).$$

But then the equations (6.6) and (6.8) are sufficient, because of the conjugation relations (8.9) and (8.10), to insure

$$(8.15)\qquad \begin{aligned} &R\left\{\left(i\frac{d}{dz}-1\right)W_\lambda(z)\right\} = 0 &&\text{on the surface,}\\ &R\{i\epsilon^{1/2}(d/dz)W_\lambda(z)\} = 0 &&\text{on the bottom.}\end{aligned}$$

This shows that $RW_\lambda(z)$ is a harmonic function satisfying the boundary conditions on surface and bottom.

It is of interest that the functions $L_{\lambda,\sigma_j}(\zeta)$ are linearly independent as functions of z for $j=1, \cdots, q$; this implies their real independence or semi-independence (see §5) on the surface as well as on the bottom. But since this fact is of no influence on the explicit formulae which we shall give for the waves, we suppress a proof thereof.

9. **Construction of a second standing wave for $p=1$.** We are now in a position to complete the construction of the progressive wave for $p=1$ by establishing a second standing wave which behaves differently at infinity from Hanson's standing wave given by (6.19).

This second standing wave is

$$(9.1)\qquad RW_{-1}(z) = R\sum_{j=1}^{q} c_j L_{-1,\sigma_j}(\zeta).$$

To show this we study the behavior of $W_{-1}(z)$ for large $|z|$ in S_β.

Notice that a change of the value of the positive lower limit α in the definition (8.8′) of the integral affects $W_{-1}(z)$ merely by adding a real constant times $W_0(z)$ to it. For simplicity's sake take $\alpha=1$. We have

$$L_{-1,\sigma_j}(\zeta) = e^{\sigma_j z}\int_1^{-\sigma_j z} e^t t^{-1}dt = \int_1^{-\sigma_j z} e^{t+\sigma_j z} t^{-1}dt,$$

with the values (8.5)

$$\log - \sigma_j = \tau_j = -\pi i/2 + j\pi i/q, \qquad j = 1, \cdots, q.$$

Hence the arguments of $-\sigma_j z$ lie between $-\pi i/2 + \pi i/2q$ and $\pi i/2$, as z varies in the sector S_β. Thus the variable of integration may be restricted to the half-plane $R(t) \geqq 0$.

With this in mind, let $\gamma=\gamma_1+\gamma_2 i$, $\gamma_1 \geqq 0$, $|\gamma| \geqq 1$. There exists an $M>0$ such that

$$(9.2)\qquad \left|\int_1^{\gamma} e^{t-\gamma}t^{-1}dt\right| \leqq M.$$

For we have

$$(9.3)\qquad \begin{aligned}\int_1^\gamma e^{t-\gamma}t^{-1}dt &= e^{t-\gamma}t^{-1}\Big|_1^\gamma + \int_1^\gamma e^{t-\gamma}t^{-2}dt\\ &= \gamma^{-1} - e^{1-\gamma} + \int_1^\gamma e^{t-\gamma}t^{-2}dt\end{aligned}$$

and

$$\left|\int_1^\gamma e^{t-\gamma}t^{-2}dt\right| \leqq e\int_1^\gamma |t|^{-2}|dt| \leqq e(1 - |\gamma|^{-1} + \pi/2).$$

This latter estimate is obtained by making the path of integration proceed first on the unit circle and then on a ray through the origin and γ. Consequently, (9.2) holds with any M for which

$$M \geqq |\gamma|^{-1} + e^{1-\gamma_1} + e(1 + \pi/2),$$

for example, for $M = 1 + e(2 + \pi/2)$.

Hence $W_{-1}(z)$ remains bounded in the sector S_β as long as $|z| \geqq 1$. Moreover, as $z = x \to \infty$, $L_{-1,\sigma_j}(\zeta) \to 0$ for $j = 1, \cdots, q-1$; for from (9.3) with $-\sigma_j x = \gamma_1 + \gamma_2 i$ we gather $\gamma_1 > 0$ and

$$\begin{aligned}\left|\int_1^\gamma e^{t-\gamma}t^{-1}dt\right| &\leqq \left|\gamma^{-1} - e^{1-\gamma} + \left(\int_1^{\gamma_1/2} + \int_{\gamma_1/2}^\gamma\right)e^{t-\gamma}t^{-2}dt\right|\\ &\leqq |\gamma|^{-1} + e^{1-\gamma_1} + e^{-\gamma_1/2}(1 - 2\gamma_1^{-1})\\ &\quad + |\gamma_1^{-1} - 2\gamma_1^{-1}|.\end{aligned}$$

The terms of the last expression tend individually to zero as $x \to \infty$. On the other hand, consider

$$L_{-1,\sigma_q}(\zeta) = \int_1^{ix} e^{t-ix}t^{-1}dt,$$

as $x \to \infty$. By a deformation of the path of integration in the left half-plane we obtain in a familiar way

$$(9.4)\qquad \int_1^{i\infty} e^t t^{-1}dt = \int_1^{\prime\,-\infty} e^t t^{-1}dt + \pi i = r + \pi i,$$

where the sign $\int'$ denotes the Cauchy principal value of the *real* integral. Accordingly,

$$W_{-1}(x) - c_q e^{-ix}\int_1^{ix} e^t t^{-1}dt \to 0 \quad \text{as} \quad x \to \infty;$$

this leads to

$$W_{-1}(x) - c_q e^{-ix}(r + \pi i) \to 0 \quad \text{as} \quad x \to \infty,$$

since the integral (9.4) converges. Thus $W_{-1}(x) - rW_0(x)$ behaves like $c_q e^{-ix}\pi i$ as $x \to \infty$, while, by (6.18) and (6.19), $W_0(z)$ behaves like $c_q e^{-iz}$. Obviously the nodes of $R\{c_q e^{-iz}\}$ lie midway between those of $R\{c_q e^{-iz}\pi i\}$.

Near the origin we find, since $\sum c_j$ is real, in view of (6.17), that

$$RW_{-1}(z) \sim q^{1/2} \log |z|.$$

10. **Asymptotic behavior of exponential integrals.** We proceed to the study of the asymptotic behavior of the functions $L_{\lambda,\sigma_j}(\zeta)$. For simplicity of notation, we set, for $p \equiv 3 \pmod 4$,

$$\gamma = z\sigma_j, \qquad H_\lambda(\gamma) = e^\gamma \int_\alpha^\gamma e^{-t} t^\lambda dt$$

where now k is supposed not to be divisible by p and

$$\lambda = (2kq/p) - 1, \qquad k = 1, 2, \cdots. \tag{10.1}$$

Furthermore $1 < p < 2q$, $\alpha = 0$. Since $\lambda > 0$, integration by parts yields

$$\int_0^\gamma e^{-t}t^\lambda dt = - e^{-t}t^\lambda \Big|_0^\gamma + \lambda \int_0^\gamma e^{-t}t^{\lambda-1}dt,$$

$$H_\lambda(\gamma) = - \gamma^\lambda - \lambda\gamma^{\lambda-1} - \cdots - \lambda(\lambda - 1) \cdots (\lambda - s + 1)\gamma^{\lambda-s}$$
$$+ \lambda(\lambda - 1) \cdots (\lambda - s)e^\gamma \int_0^\gamma e^{-t}t^{\lambda-s-1}dt,$$

a formula which holds for all integral s with $0 < s < \lambda$. In particular, we shall take the value $s = [\lambda]$, defined as the largest integer less than λ. We find

$$\begin{aligned} W_\lambda(z) &\equiv \sum_1^q c_j L_{\lambda,\sigma_j}(\zeta) \\ &= \sum c_j \sigma_j^\lambda z^\lambda (-1 - \lambda(\sigma_j z)^{-1} - \cdots - \lambda \cdots (\lambda - s + 1)(\sigma_j z)^{-s}) \\ &\quad + \lambda(\lambda - 1) \cdots (\lambda - s) \sum c_j e^{\sigma_j z} \int_0^{\sigma_j z} e^{-t}t^{\lambda-s-1}dt. \end{aligned} \tag{10.2}$$

Now

$$\sum c_j \sigma_j^\lambda = \sum c_j e^{\tau_j \lambda} = \sum c_j e^{\tau_j 2kq/p} \sigma_j^{-1},$$

and by (6.14) and (8.5) this becomes

$$\sum c_j\sigma_j^\lambda = d\sum_1^q a_{j-1}(-i)^j e^{(-p/2+jp/q)\pi i\cdot 2kq/p}\sigma_j^{-1}$$
$$= d\sum a_{j-1}(-i)^{j-1}\epsilon^j(-1)^{kq+1} = f(-i\epsilon)d\epsilon(-1)^{kq+1} = 0,$$

for by the lemma of §4 the roots of $f(\xi)$ are $-i\epsilon, \cdots, -i\epsilon^{q-1}$.

Similarly $\sum_1^q c_j\sigma_j^{\lambda-1} = f(-i\epsilon^2)\cdot\text{const.} = 0$, where the constant denotes a number which does not depend on the index j of summation. Proceeding in a like manner, we get

$$\sum_1^q c_j\sigma_j^{\lambda-s} = f(-i\epsilon^{s+1})\cdot\text{const.} = 0 \tag{10.3}$$

for all $s>0$ with $s\leqq q-2$. This is compatible with $s=[\lambda]=[2kq/p-1]$ for all k with

$$1 \leqq k \leqq (p-1)/2. \tag{10.4}$$

In fact, then

$$[2kq/p] \leqq [(p-1)q/p] \leqq [q-q/p] \leqq q-1,$$
$$[\lambda] = [2kq/p-1] \leqq q-2.$$

For k satisfying (10.4) we conclude from (10.2)

$$\begin{aligned} W_\lambda(z) &= W_{-1+2kq/p}(z) \\ &= \Gamma(2kq/p)(\Gamma(2kq/p-[2kq/p]))^{-1}\sum_1^q c_jL_{-1+2kq/p-[2kq/p]}(\zeta). \end{aligned} \tag{10.5}$$

Set

$$W_k^*(z) = (\Gamma(2kq/p))^{-1}W_\lambda(z). \tag{10.6}$$

By (8.15), W_k^* satisfies the boundary conditions on surface and bottom.

In order to study the behavior of $W_k^*(z)$ for $z=x\to\infty$ and for $z=r\epsilon^{1/2}$, $r\to\infty$, consider one of the integrals on the right of (10.5) (it will suffice to consider only the first of the following alternatives in case $p<q$):

$$e^{\sigma_j x}\int_0^{\sigma_j x} e^{-t}t^{\lambda-[\lambda]-1}dt \text{ resp. } e^{\sigma_j r\epsilon^{1/2}}\int_0^{\sigma_j r\epsilon^{1/2}} e^{-t}t^{\lambda-[\lambda]-1}dt.$$

Suppose first that j is such that

$$\pi/2 \geqq -i\tau_j \geqq -\pi/2 \text{ resp. } \pi/2 \geqq -i\tau_j - p\pi/2q \geqq -\pi/2. \tag{10.7}$$

Then

$$\lim_{x\to\infty}\int_0^{\sigma_j x} e^{-t}t^{\lambda-[\lambda]-1}dt = \Gamma(\lambda-[\lambda])$$

resp.

$$\lim_{r\to\infty}\int_0^{\sigma_j r\epsilon^{1/2}} e^{-t}t^{\lambda-[\lambda]-1}dt = \Gamma(\lambda-[\lambda]),$$

whence

$$e^{\sigma_j x}\int_0^{\sigma_j x} e^{-t}t^{\lambda-[\lambda]-1}dt = \Gamma(\lambda-[\lambda])e^{\sigma_j x} - e^{\sigma_j x}\int_{\sigma_j x}^{\infty} e^{-t}t^{\lambda-[\lambda]-1}dt$$

resp.

$$e^{\sigma_j r\epsilon^{1/2}}\int_0^{\sigma_j r\epsilon^{1/2}} e^{-t}t^{\lambda-[\lambda]-1}dt$$

$$= \Gamma(\lambda-[\lambda])e^{\sigma_j r\epsilon^{1/2}} - e^{\sigma_j r\epsilon^{1/2}}\int_{\sigma_j r\epsilon^{1/2}}^{\infty} e^{-t}t^{\lambda-[\lambda]-1}dt.$$

Here

$$\left|e^{\sigma_j x}\int_{\sigma_j x}^{\infty}\right| = \left|\int_0^{\infty} e^{-u}\left|u+\sigma_j x\right|^{\lambda-[\lambda]-1}du\right|$$

$$\leqq \left|\sigma_j x\right|^{\lambda-[\lambda]-1}\int_0^{\infty} e^{-u}du \to 0,$$

resp.

$$\left|e^{\sigma_j r\epsilon^{1/2}}\int_{\sigma_j r\epsilon^{1/2}}^{\infty}\right| \to 0,$$

since $\lambda-[\lambda]-1<0$. Thus

$$e^{\sigma_j x}\left(\int_0^{\sigma_j x} e^{-t}t^{\lambda-[\lambda]-1}dt - \Gamma(\lambda-[\lambda])\right) \underset{x\to\infty}{\longrightarrow} 0, \text{ resp.}$$

(10.8)

$$e^{\sigma_j r\epsilon^{1/2}}\left(\int_0^{\sigma_j r\epsilon^{1/2}} e^{-t}t^{\lambda-[\lambda]-1} - \Gamma(\lambda-[\lambda])\right) \underset{r\to\infty}{\longrightarrow} 0,$$

if (10.7) holds.

Observe that $\int_0^{\gamma} e^{-t}t^{\lambda-[\lambda]-1}dt$ is multiplied by $e^{2\pi i\lambda}$ as $\log\gamma$ is increased by $2\pi i$. Therefore we can generalize the previous result. Suppose n_j defined equal to n, provided there exists an integral number n such that

$$(10.9)\quad \begin{aligned} &\text{either } n2\pi + \pi/2 \geqq -i\tau_j \geqq n2\pi - \pi/2 \\ &\text{or } \quad n2\pi + \pi/2 \geqq -i\tau_j - p\pi/(2q) \geqq n2\pi - \pi/2. \end{aligned}$$

(Notice that $p<2q$ so that these two conditions never lead to different values of n_j.) Then

$$(10.10)\quad \begin{aligned} &\lim_{x\to\infty}\left\{e^{\sigma_j x}\left(\int_0^{\sigma_j x} e^{-t}t^{\lambda-[\lambda]-1}dt - e^{A\pi i n_j kq/p}\Gamma(\lambda-[\lambda])\right)\right\} = 0 \text{ resp.} \\ &\lim_{r\to\infty}\left\{e^{\sigma_j r\epsilon^{1/2}}\left(\int_0^{\sigma_j r\epsilon^{1/2}} e^{-t}t^{\lambda-[\lambda]-1}dt - e^{A\pi i n_j kq/p}\Gamma(\lambda-[\lambda])\right)\right\} = 0. \end{aligned}$$

On the other hand, if one or both of the conditions (10.9) cannot be satisfied with integral n, but instead we have with some n

$$(10.11)\quad \begin{aligned} &n2\pi + \pi/2 < -i\tau_j < n2\pi + 3\pi/2 \quad \text{resp.} \\ &n2\pi + \pi/2 < -i\tau_j - p\pi/2q < n2\pi + 3\pi/2, \end{aligned}$$

then we must have, for example,

$$\left|e^{\sigma_j x}\int_0^{\sigma_j x}\right| = \left|\int_0^{-\sigma_j x} e^{-u}(u+\sigma_j x)^{\lambda-[\lambda]-1}du\right| \leqq \left|\int_0^{\infty}\right| + \left|\int_{-\sigma_j x}^{\infty}\right|$$
$$\leqq |x|^{\lambda-[\lambda]-1} + |e^{\sigma_j x}|\,\Gamma(\lambda-[\lambda]);$$

hence

$$(10.11.1)\quad \begin{aligned} &\lim_{x\to\infty} e^{\sigma_j x}\int_0^{\sigma_j x} e^{-t}t^{\lambda-[\lambda]-1}dt = 0, \text{ resp.} \\ &\lim_{r\to\infty} e^{\sigma_j r\epsilon^{1/2}}\int_0^{\sigma_j r\epsilon^{1/2}} e^{-t}t^{\lambda-[\lambda]-1}dt = 0. \end{aligned}$$

These estimates show that

$$(10.12)\quad \begin{aligned} &\lim_{x\to\infty}\left\{W_k^*(x) - \sum_{j=1}^{q} c_j\delta_{jk}e^{\sigma_j x}\right\} = 0, \\ &\lim_{r\to\infty}\left\{W_k^*(r\epsilon^{1/2}) - \sum_{1}^{q} c_j\delta_{jk}e^{\sigma_j r\epsilon^{1/2}}\right\} = 0, \end{aligned}$$

where δ_{jk} stands for

$$(10.13)\quad \delta_{jk} = \begin{cases} e^{A\pi i n_j kq/p} & \text{if (10.9) holds,} \\ 0 & \text{if (10.11) holds} \end{cases}$$

(the value of n_j in the second case is immaterial for the validity of

(10.12) because of the vanishing of the exponential factor). The possible values of n_j in (10.9) are

$$(10.14)\quad n_j = (p+1)/4;\ (p-3)/4, \cdots, 0, -1, \cdots, -(p-3)/4.$$

The first of these occurs only for $j=q$, that is, $n_q=(p+1)/4$. All the other possible values of n_j must be successive integers since two successive terms $-i\tau_j-p\pi/2q$, $-i\tau_j$ or $-i\tau_j$, $-i\tau_{j+1}-p\pi/2q$ differ by $p\pi/2q<\pi$. On the other hand, the smallest value of n_j occurs either in $n_j2\pi-\pi/2\leqq -i\tau_j\leqq n_j2\pi+\pi/2$ or in $n_j2\pi-\pi/2\leqq -i\tau_j-p\pi/2q \leqq n_j2\pi+\pi/2$ when this n_j is the smallest integer following that one, n', which would occur if j could become equal to zero; this would give

$$2n'\pi+\pi/2 = -p\pi/2, \qquad n' = -(p+1)/4.$$

The alternative in the relations (10.9) to (10.14) is dictated by the possibility that $p>q$. Had we demanded that $p<q$, then the Phragmén-Lindelöf principle would have permitted us to fix our attention solely on the surface, as was explained in §2.

Next we give the parallel developments for $p\equiv 1 \pmod 4$. Integrate by parts for $\lambda>0$:

$$\int_0^{-\gamma} e^t t^\lambda dt = e^t t^\lambda\Big|_0^{-\gamma} - \lambda\int_0^{-\gamma} e^t t^{\lambda-1}dt = \cdots,$$

$$\begin{aligned} e^\gamma\int_0^{-\gamma} e^t t^\lambda dt &= (-\gamma)^\lambda - \lambda(-\gamma)^{\lambda-1} + \cdots \\ &\quad + (-1)^s(-\gamma)^{\lambda-s}\lambda(\lambda-1)\cdots(\lambda-s+1) \\ &\quad + (-1)^{s+1}e^\gamma\int_0^{-\gamma} e^t t^{\lambda-s-1}dt\lambda(\lambda-1)\cdots(\lambda-s), \end{aligned}$$

where $s=[\lambda]$. Then, by (8.8′),

$$\begin{aligned} W_\lambda(z) &\equiv \sum_{j=1}^{q} c_j L_{\lambda,\sigma_j}(\zeta) \\ &= \sum_1^q \Big[c_j(-\sigma_j z)^\lambda (1+\lambda(\sigma_j z)^{-1} + \cdots \\ &\quad + \lambda\cdots(\lambda-s+1)(\sigma_j z)^{-s}) \\ &\quad + c_j(-1)^{s+1}e^{\sigma_j z}\int_0^{-\sigma_j z} e^t t^{\lambda-s-1}dt\lambda\cdots(\lambda-s)\Big]. \end{aligned} \tag{10.2'}$$

Now

$$\sum_1^q c_j(-\sigma_j)^\lambda = \sum_1^q c_j e^{\tau_j\lambda} = d\sum_1^q a_{j-1}(-i)^j e^{(-p/2+jp/q)2kq\pi i/p}(-\sigma_j)^{-1}$$
$$= \text{const.} \sum a_{j-1}(-i\epsilon)^{j-1} = \text{const.}\ f(-i\epsilon) = 0,$$
$$\cdots\cdots\cdots\cdots\cdots\cdots,$$
$$\sum_1^q c_j(-\sigma_j)^{\lambda-s} = \text{const.}\sum_1^q a_{j-1}(-i)^{j-1}\epsilon^{(j-1)(s+1)}$$
$$= \text{const.}\ f(-i\epsilon^{s+1}) = 0,$$

since (10.3) and (10.4) apply again. Hence, for $k=1, 2, \cdots, (p-1)/2$, $\lambda=(2kq/p)-1$,

$$\begin{aligned} W_k^*(z) &\equiv (\Gamma(\lambda+1))^{-1}W_\lambda(z) \\ &= (-1)^{[2kq/p]}(\Gamma(\lambda-[\lambda]))^{-1}\sum_{j=1}^q c_j L_{\lambda-[\lambda]-1,\sigma_j}(\zeta) \end{aligned} \tag{10.6$'$}$$

satisfies the boundary conditions on surface and bottom.

For large x, resp. large r, consider

$$e^{\sigma_j x}\int_0^{-\sigma_j x} e^t t^{\lambda-[\lambda]-1}dt \quad \text{resp.} \quad e^{\sigma_j r\epsilon^{1/2}}\int_0^{-\sigma_j r\epsilon^{1/2}} e^t t^{\lambda-[\lambda]-1}dt.$$

For such j as make

$$-\pi/2 < -i\tau_j < \pi/2 \quad \text{resp.} \quad -\pi/2 < -i\tau_j - p\pi/2q < \pi/2,$$

we have $R(-\sigma_j)>0$, and, with $\gamma=-\sigma_j x$,

$$e^{-\gamma}\int_0^\gamma e^t t^{\lambda-[\lambda]-1}dt = \int_0^\gamma e^{t-\gamma}t^{\lambda-[\lambda]-1}dt \to 0, \tag{10.11.1$'$}$$

as $x\to\infty$; or resp. (10.11.1′) holds with $\gamma=-\sigma_j r\epsilon^{1/2}$ and $R(-\sigma_j\epsilon^{1/2})>0$ as $r\to\infty$. Similarly, (10.11.1′) still holds if there is an integral n for which

$$\begin{aligned} &-\pi/2+2n\pi < -i\tau_j < \pi/2+2n\pi \quad \text{resp.} \\ &-\pi/2+2n\pi < -i\tau_j - p\pi/2q < \pi/2+2n\pi. \end{aligned} \tag{10.11$'$}$$

On the other hand, if there is an integral n such that

$$\begin{aligned} &\pi/2+2n\pi \leqq -i\tau_j \leqq 3\pi/2+2n\pi \quad \text{resp.} \\ &\pi/2+2n\pi \leqq -i\tau_j - p\pi/2q \leqq 3\pi/2+2n\pi, \end{aligned} \tag{10.9$'$}$$

then let n_j be defined by $n_j=n$ in (10.9′). We have $R\gamma\leqq 0$, where

$$\gamma = -\sigma_j x \quad \text{resp.} \quad \gamma = -\sigma_j r e^{1/2},$$

$$e^{-\gamma}\int_0^{\gamma} e^t t^{\lambda-[\lambda]-1}dt = e^{-\gamma}\left(\int_0^{-\infty}+\int_{-\infty}^{\gamma}\right),$$

$$\int_0^{-\infty} e^t t^{\lambda-[\lambda]-1}dt = -\int_0^{\infty} e^{-|t|}\,|t|^{\lambda-[\lambda]-1}d\,|t|\,e^{\pi i(1+2n_j)(\lambda-[\lambda]-1)}$$

$$= e^{(2n_j+1)\pi i 2kq/p}(-1)^{[2kq/p]}\Gamma(\lambda-[\lambda]),$$

$$\left|e^{-\gamma}\int_{-\infty}^{\gamma} e^t t^{\lambda-[\lambda]-1}dt\right| = \left|\int_{-\infty}^{\gamma} e^{t-\gamma}t^{\lambda-[\lambda]-1}dt\right|$$

$$= \left|\int_{-\infty}^{0} e^u(u+\gamma)^{\lambda-[\lambda]-1}du\right|$$

$$\leqq |\gamma|^{\lambda-[\lambda]-1}\to 0,$$

as $x\to\infty$ resp. $r\to\infty$.

Hence

$$\lim_{x\to\infty}\left[e^{\sigma_j x}\int_0^{-\sigma_j x} e^t t^{\lambda-[\lambda]-1}dt - (-1)^{[2kq/p]}\Gamma(\lambda-[\lambda])e^{\sigma_j x}e^{(2n_j+1)\pi i 2kq/p}\right] = 0, \tag{10.10'}$$

resp. a similar limit equation for $r\to\infty$, if $x\sigma_j$ is replaced by $r\sigma_j\epsilon^{1/2}$. Accordingly, by (10.6')

$$\lim_{x\to\infty}\left\{W_k^*(x) - \sum_1^q c_j\delta_{jk}e^{\sigma_j x}\right\} = 0, \text{ resp.}$$

$$\lim_{r\to\infty}\left\{W_k^*(r\epsilon^{1/2}) - \sum_1^q c_j\delta_{jk}e^{\sigma_j r\epsilon^{1/2}}\right\} = 0, \tag{10.12'}$$

were now

$$\delta_{jk} = \begin{cases} e^{2\pi i kq(2n_j+1)/p} & \text{if (10.11') holds,}\\ 0 & \text{otherwise.}\end{cases} \tag{10.13'}$$

Here the possible values of n_j range from $(p-1)/4$, occurring for $j=q$ when $-\sigma_q=i$, $\tau_q=p\pi/2$, to $-(p-1)/4$, occurring for the smallest j for which (10.9') holds. For this value exceeds by 1 the value $-(p+3)/4$ which would occur if $j=0$ were admissible; for then

$$3\pi/2 + 2n'\pi = -p\pi/2, \qquad n' = -(p+3)\pi/4.$$

11. **Extension to the case $\lambda=-1$, $p\equiv 3$ (mod 4).** If k in (8.2) is set equal to zero, α can no longer be chosen as zero if the integral (8.1) is to converge. We set $\alpha=1$ and study the asymptotic behavior

of $L_{-1,\sigma_j}(\zeta)$ defined by

$$(11.1)\qquad L_{-1,\sigma_j}(\zeta) = \begin{cases} e^{\sigma_j z}\displaystyle\int_1^{\sigma_j z} e^{-t}t^{-1}dt, & \text{for } p \equiv 3 \pmod 4, \\ e^{\sigma_j z}\displaystyle\int_1^{-\sigma_j z} e^{-t}t^{-1}dt, & \text{for } p \equiv 1 \pmod 4. \end{cases}$$

Here again the generalities about the multivalued character of the integral apply. If ζ increases by $2\pi i$, $L_{-1,\sigma_j}(\zeta)$ changes in the following manner:

$$(11.2)\qquad L_{-1,\sigma_j}(\zeta + 2\pi i) = L_{-1,\sigma_j}(\zeta) + 2\pi i e^{\sigma_j z}.$$

Furthermore, there exists the conjugation

$$(11.3)\qquad L_{-1,\sigma_j}(\zeta) \cong L_{-1,s_j}(\bar{\zeta}),$$

where $s_j = \bar{\sigma}_j$. Let $\gamma \neq 0$, $-\pi/2 \leqq \arg \gamma \leqq \pi/2$, $\log \gamma = \log |\gamma| + i \arg \gamma$. Then

$$e^{\gamma}\int_1^{\gamma} e^{-t}t^{-1}dt = e^{\gamma}\left(\int_1^{\infty} - \int_{\gamma}^{\infty}\right)$$

$$= e^{\gamma}\int_1^{\infty} e^{-t}t^{-1}dt - \int_0^{\infty} e^{-t}(t+\gamma)^{-1}dt,$$

$$(11.4)\qquad \left| e^{\gamma}\int_1^{\gamma} e^{-t}t^{-1}dt - e^{\gamma}\int_1^{\infty} e^{-t}t^{-1}dt \right| \leqq |\gamma|^{-1},$$

$$(11.5)\qquad e^{\gamma}\int_1^{\gamma} e^{-t}t^{-1}dt - e^{\gamma}\int_1^{\infty} e^{-t}t^{-1}dt \to 0$$

as $\gamma \to \infty$, $|\arg \gamma| \leqq \pi/2$. But if $\pi/2 + \delta < \arg \gamma < 3\pi/2 - \delta$, $\delta > 0$, consider $e^{\gamma}\int_1^{\gamma} e^{-t}t^{-1}dt$ where the integral is extended along a path leading in the upper half-plane from 1 to $\gamma/|\gamma|$ and thence to γ on a straight line. We have, with $\kappa = \gamma/|\gamma| = -\kappa_1 + i\kappa_2$, $\kappa_1 > \sin \delta > 0$,

$$\left| e^{\gamma}\int_1^{\gamma} e^{-t}t^{-1}dt \right| \leqq \left| e^{\gamma}\int_1^{\kappa} e^{-t}t^{-1}dt \right| + \left| \int_{\kappa}^{\gamma} e^{-(t-\gamma)}t^{-1}dt \right|,$$

$$\left| \int_{\kappa}^{\gamma} e^{-(t-\gamma)}t^{-1}dt \right| \leqq \left| \int_1^{|\gamma|} e^{-\kappa(t-|\gamma|)}t^{-1}dt \right|$$

$$\leqq \left| \int_1^{|\gamma|/2} \right| + \left| \int_{|\gamma|/2}^{|\gamma|} \right|$$

$$\leqq e^{-\kappa_1|\gamma|/2} \log (|\gamma|/2) + 2(\kappa_1|\gamma|)^{-1}.$$

Hence $e^{\gamma}\int_1^{\gamma} e^{-t}t^{-1}dt \to 0$ as $\pi/2 + \delta \leqq \arg \gamma \leqq 3\pi/2 - \delta$ and $|\gamma| \to \infty$. Consequently, for n_j as defined by (10.9) and $\zeta=\log x$ real resp. $\zeta=\log r - ip\pi/2q$, $\log r$ real,

$$L_{-1,\sigma_j}(\zeta) - 2\pi i n_j e^{\sigma_j x} - e^{\sigma_j x}\int_1^\infty e^{-t}t^{-1}dt \to 0, \qquad \text{if } R\sigma_j \geqq 0, \tag{11.6}$$

resp.

$$L_{-1,\sigma_j}(\zeta) - 2\pi i n_j e^{\sigma_j r\epsilon^{1/2}} - e^{\sigma_j r\epsilon^{1/2}}\int_1^\infty e^{-t}t^{-1}dt \to 0, \qquad \text{if } R(\sigma_j\epsilon^{1/2}) \geqq 0, \tag{11.7}$$

while always

$$\lim_{x\to\infty} L_{-1,\sigma_j}(\zeta) = 0, \qquad \text{if } R\sigma_j < 0,\ \zeta = \log x, \tag{11.8}$$

resp.

$$\lim_{x\to\infty} L_{-1,\sigma_j}(\zeta) = 0, \qquad \text{if } R(\sigma_j\epsilon^{1/2}) < 0,\ \zeta = \log (r\epsilon^{1/2}). \tag{11.9}$$

The formulae (11.6) and (11.7) lead to corresponding relations for

$$W_{-1}(\zeta) \equiv \sum_1^q c_j L_{-1,\sigma_j}(\zeta). \tag{11.10}$$

12. **Extension to the case $\lambda=-1$, $p\equiv 1$ (mod 4).** Let $R-\sigma_j>0$. Then with some integral n and $\gamma=-\sigma_j x$, $R\gamma>0$,

$$e^{-\gamma}\int_1^\gamma e^t t^{-1}dt = 2\pi i n e^{-\gamma} + \int_1^\gamma e^{t-\gamma}t^{-1}dt,$$

where the last integral can be taken along a straight line and

$$\left|\int_1^\gamma e^{t-\gamma}t^{-1}dt\right| \leqq \left|\int_1^{\gamma/|\gamma|}\right| + \left|\int_{\gamma/|\gamma|}^\gamma\right|$$
$$\leqq e^{-R\gamma}\cdot\text{const.} + e^{-R\gamma}\int_1^{R\gamma} e^t t^{-1}dt.$$

Consequently, if γ tends to ∞ on a ray through the origin which lies in $R\gamma>0$, we have $e^{-\gamma}\int_1^\gamma e^t t^{-1}dt \to 0$. Now suppose n_j defined by (10.9′), that is,

$$\pi/2 + 2n_j\pi \leqq -\tau_j i \leqq 3\pi/2 + 2n_j\pi$$

resp.

$$\pi/2 + 2n_j\pi \leqq -\tau_j i - p\pi i/2q \leqq 3\pi/2 + 2n_j\pi.$$

We have in the first alternative

$$e^{\sigma_j x}\int_1^{-\sigma_j x} e^t t^{-1}dt = e^{\sigma_j x}\int_1^{-\infty} e^t t^{-1}dt + e^{\sigma_j x}\int_{-\infty}^{-\sigma_j x} e^t t^{-1}dt,$$

where

$$\int_1^{-\infty} e^t t^{-1}dt = \int_1^{\prime -\infty} e^t t^{-1}dt + (2n_j + 1)\pi i,$$

with $\int'$ understood as Cauchy's principal value, and where the path of $\int_{-\infty}^{-\sigma_j x}$ remains in the negative half-plane. Now as $x \to \infty$,

$$\left| e^{\sigma_j x}\int_{-\infty}^{-\sigma_j x} e^t t^{-1}dt \right| = \left| \int_{-\infty}^{-\sigma_j x} e^{t+\sigma_j x} t^{-1}dt \right|$$

$$= \left| \int_{-\infty}^{0} e^t (t - \sigma_j x)^{-1}dt \right| \leqq x^{-1} \to 0,$$

so that

$$(12.1)\qquad e^{\sigma_j x}\left[\int_1^{-\sigma_j x} e^t t^{-1}dt - (2n_j + 1)\pi i - \int_1^{\prime -\infty} e^t t^{-1}dt\right] \to 0,$$

$$\text{for } x \to \infty,\ R\sigma_j \geqq 0.$$

For the second alternative of (10.9′) we have similarly

$$(12.2)\qquad e^{\sigma_j r \epsilon^{1/2}}\left[\int_1^{-\sigma_j r \epsilon^{1/2}} e^t t^{-1}dt - (2n_j + 1)\pi i - \int_1^{\prime -\infty} e^t t^{-1}dt\right] \to 0,$$

$$\text{for } r \to \infty,\ R\sigma_j \epsilon^{1/2} \geqq 0.$$

13. **Construction of the finite standing wave.** It is now possible to exhibit a solution of our boundary problem in the form

$$(13.1)\qquad \Omega_0(z) = W_0(z) - \sum_{k=1}^{(p-1)/2} A_k W_k^*(z);$$

$\Omega_0(z)$ remains bounded as z tends to infinity on the surface and bottom of S_β, and remains bounded as $z \to 0$. Moreover, the derivative $d\Omega_0(z)/dz$ will also remain bounded near $z=0$ as is obvious from the definition (6.19) of $W_0(z)$ and the definitions (10.6) resp. (10.6′) and (10.2) resp. (10.2′). Our problem is to determine the A_k in such a way as to eliminate by superposition the various infinities introduced for large $|z|$ by each of the summands in (13.1). It suffices if this can be done on the surface for $p<q$, and on the surface and bottom for $q<p<2q$. Let $p\equiv 3 \pmod 4$. Fix the attention first on the surface. The estimates (10.10) and (10.12) lead to the equation

$$\lim_{x\to\infty}\left[\Omega_0(x) - \sum_j{}' c_j e^{\sigma_j x}\left(1 - \sum_{k=1}^{(p-1)/2} \delta_{jk}A_k\right)\right] = 0, \tag{13.2}$$

where it is necessary to extend the sum $\sum'$ only over those values j for which the first alternative of (10.9) applies. Now suppose that for those values of n_j for which the first alternative (10.9) applies, *but with the exception of* $j=q$, we can solve the system of linear equations

$$\sum_{k=1}^{(p-1)/2} \delta_{jk}A_k = 1, \qquad j \neq q, \tag{13.3}$$

in the unknowns A_k. Then (13.2) reduces to

$$\lim_{x\to\infty}\left[\Omega_0(x) - c_q e^{-ix}\left(1 - \sum_{k=1}^{(p-1)/2} \delta_{qk}A_k\right)\right] = 0. \tag{13.4}$$

In order that $\Omega_0(x)$ should behave like a simple wave at infinity, we shall have to show in addition that

$$-1 + \sum_{k=1}^{(p-1)/2} \delta_{qk}A_k \neq 0. \tag{13.5}$$

It follows directly that $\Omega_0(x)$ stays bounded also on the bottom if the A_k satisfy the system of equations (13.3).

Now (10.14) shows that (13.3) represents exactly $(p-1)/2$ distinct equations, namely

$$\sum_{k=1}^{(p-1)/2} A_k e^{4\pi iknq/p} = 1, \quad n = -(p-3)/4, \cdots, +(p-3)/4. \tag{13.6}$$

A similar analysis in the case $p\equiv 1 \pmod 4$ shows the corresponding equations to be

$$\sum_{k=1}^{(p-1)/2} A_k e^{2\pi ikq(2n+1)/p} = 1, \tag{13.6'}$$

$$n = -(p-1)/4, \cdots, (p-5)/4.$$

We proceed to give the values of the A_k explicitly. There occurs in this deduction a remarkable analogy to the computation of the coefficients a_j of the polynomial $f(\xi)$ introduced in §4, with the role of p and q interchanged. Consider the polynomial of degree $(p-1)/2$,

$$F(\xi) = \sum_{0}^{(p-1)/2} A_k \xi^k, \qquad \text{with } A_0 = -1,$$

which vanishes at the $(p-1)/2$ distinct points

$$(13.7) \qquad \xi = e^{4\pi i n q/p}, \quad n = -(p-3)/4, \cdots, (p-3)/4, \text{ for } p \equiv 3,$$

resp.

$$(13.7') \qquad \xi = e^{2\pi i q(2n+1)/p}, \quad n = -(p-1)/4, \cdots, (p-5)/4, \text{ for } p \equiv 1 \pmod 4).$$

We have

$$(13.8) \qquad \begin{aligned} F(\xi) &= \prod_{|n| \leq (p-3)/4} (\xi - \eta^n), \qquad \eta = e^{4\pi i q/p}, \text{ for } p \equiv 3 \pmod 4), \\ F(\xi) &= -\prod_{n \geq -(p-1)/4}^{(p-5)/4} (\xi - \eta^{2n+1}), \qquad \eta = e^{2\pi i q/p}, \text{ for } p \equiv 1 \pmod 4). \end{aligned}$$

Evidently (13.6) resp. (13.6′) hold. To test (13.5) we write for $p \equiv 3 \pmod 4$)

$$(13.9) \qquad -1 + \sum_{k=1}^{(p-1)/2} \delta_{qk} A_k = F(e^{4\pi i q(p+1)/(4p)}) = F(e^{(p+1)q\pi i/p}) \neq 0$$

since $e^{(p+1)q\pi i/p}$ coincides with none of the roots (13.7). Similarly for $p \equiv 1 \pmod 4$)

$$(13.9') \qquad \begin{aligned} -1 + \sum_{1}^{(p-1)/2} \delta_{qk} A_k &= F(e^{2\pi i (q/p)(2(p-1)/4+1)}) \\ &= F(e^{(p+1)q\pi i/p}) \end{aligned}$$

which does not vanish for a similar reason. It is easy to evaluate the A_k explicitly. We have for $p \equiv 3 \pmod 4$)

$$(\xi - \eta^{(p-3)/4}) F(\eta\xi) = \eta^{(p-1)/2} F(\xi)(\xi - \eta^{-(p+1)/4}),$$

whence

$$\sum_{0}^{(p-1)/2} A_k \xi^{k+1} (\eta^k - \eta^{(p-1)/2}) = \sum_{0}^{(p-1)/2} A_k \xi^k (\eta^k - 1) \eta^{(p-3)/4}.$$

But

$$\eta^{p/2} = 1, \qquad \eta^{p/4} = (-1)^q$$

and

$$A_{k-1} = (-1)^q A_k(\eta^{k-3/4} - \eta^{-3/4})/(\eta^{k-1} - \eta^{-1/2})$$

so that

$$A_k = (-1)^q A_{k-1}(\sin (2k-1)q\pi/p)/\sin 2kq\pi/p$$

(13.10)
$$= (-1)^{kq+1}\prod_{\nu=1}^{k} \frac{\sin (2\nu-1)q\pi/p}{\sin 2\nu q\pi/p}; \qquad A_0 = -1.$$

Similarly for $p \equiv 1 \pmod 4$

$$(\xi - \eta^{(p-1)/2})F(\eta^2\xi) = \eta^{p-1}(\xi - \eta^{-(p-1)/2})F(\xi),$$

$$\sum A_k\xi^{k+1}(\eta^{2k} - \eta^{p-1}) = \sum A_k\xi^k(\eta^{(p-1)/2}\eta^{2k} - \eta^{(p-1)/2}).$$

But here

$$\eta^p = 1, \quad \eta^{p/2} = (-1)^q, \quad A_k = (-1)^q A_{k-1}\eta^{-1/2}(\eta^{2k-1} - 1)/(\eta^{2k} - 1)$$

$$A_k = (-1)^q A_{k-1}(\sin (2k-1)q\pi/p)/\sin 2kq\pi/p,$$

which again leads to the result (13.10).

We find jointly for $p \equiv 3$ and $p \equiv 1 \pmod 4$ from (13.8)

$$(-1)^{(p+1)/2}F(e^{\pi iq(p+1)/p}) = (e^{\pi iq(p+1)/p} - e^{\pi iq(p-3)/p}) \cdots,$$

and by elementary simplification

$$(-1)^{(p+1)/2}F(e^{\pi iq(p+1)/p}) = e^{\pi iq(p^2-1)/(4p)} \prod_{1}^{(p-1)/2} (e^{2\pi iqk/p} - e^{-2\pi iqk/p}),$$

and by a reasoning similar to that employed in §4,

(13.11)
$$(-1)^{(p+1)/2}F(e^{\pi iq(p+1)/p}) = e^{\pi iq(p^2-1)/4p} i^{(p-1)/2}(-1)^{N(2q,p)}p^{1/2}.$$

Here $N(2q, p)$ represents again the number of terms in the set $e^{2\pi iq/p}, e^{4\pi iq/p}, \cdots, e^{\pi i(p-1)q/p}$ whose imaginary part is negative. Accordingly, by (13.4) and (13.9) resp. (13.9′),

(13.12)
$$\lim_{x\to\infty} [\Omega_0(x) + c_q e^{-ix}F(e^{(p+1)q\pi i/p})] = 0,$$

or

(13.13)
$$0 = \lim_{x\to\infty} [\Omega_0(x) + c_q e^{-ix}e^{\pi iq(p^2-1)/4p} \cdot i^{(p-1)/2}(-1)^{N(2q,p)+(p+1)/2}p^{1/2}].$$

14. **Construction of a second standing wave for** $p \equiv 3 \pmod 4$. **Second interpolating function** $G(\xi)$. Consider the polynomial

(14.1)
$$G(\xi) = \sum_{1}^{(p-1)/2} B_k\xi^k$$

where

$$B_k = -i(A_k/2)(\eta^k - \eta^{-k})/(\eta^k + \eta^{-k} - 2) \tag{14.2}$$
$$= -(A_k/2)\cot 2\pi qk/p$$

is real. We have

$$B_k = -i(A_k/2)((1-\eta^k)^{-1} - (1-\eta^{-k})^{-1}),$$

$$G(\xi\eta) - G(\xi) = \sum B_k\xi^k(\eta^k - 1) = (i/2)\left(\sum_1^{(p-1)/2} \xi^k A_k(1+\eta^k)\right) \tag{14.3}$$
$$= (i/2)(F(\xi) + F(\xi\eta) + 2),$$

since $A_0 = -1$. Also

$$G(1) = \sum_1^{(p-1)/2} B_k,$$

whence, by (13.8) and (14.3),

$$G(\eta^n) = in + \sum_1^{(p-1)/2} B_k, \text{ for } n = -(p-3)/4, \cdots, (p-3)/4, \tag{14.4}$$

and

$$G(\eta^{(p+1)/4}) = i(p+1)/4 + (i/2)F(\eta^{(p+1)/4}) + \sum B_k. \tag{14.5}$$

Thus the B_k are real numbers for which, with δ_{jk} from (10.13),

$$\sum B_k\delta_{jk} = in_j + \sum B_k, \qquad \text{if (10.9) with } j \neq q, \tag{14.6}$$

and

$$\sum B_k\delta_{qk} = i(p+1)/4 + (i/2)F(\eta^{(p+1)/4}) + \sum B_k.$$

Set

$$\pi\Omega_{-1}(z) \equiv W_{-1}(z) - 2\pi\sum_1^{(p-1)/2} W_k^*(z)B_k \tag{14.7}$$
$$+ W_0(z)\left(2\pi\sum B_k - \int_1^\infty e^t t^{-1}dt\right).$$

By (14.7), (14.6), (11.6), (11.10) we have the limit relation

$$\lim_{x\to\infty}\Bigg[-\pi\Omega_{-1}(x) + 2\pi i\sum_j{}' c_j n_j e^{\sigma_j x} - 2\pi\sum_j{}'\sum_{k=1}^{(p-1)/2} B_k\delta_{jk}c_je^{\sigma_j x}$$
$$+ \sum_j{}' 2\pi c_j\sum B_k e^{\sigma_j x}\Bigg] = 0,$$

where the sum $\sum'$ extends over all j with (10.9). This immediately yields

(14.8) $$\lim_{x\to\infty} \left[-\pi\Omega_{-1}(x) - \pi i c_q F(\eta^{(p+1)/4})e^{-ix}\right] = 0.$$

Compare (14.8) with the limit formula (13.12) for $\Omega_0(x)$. Clearly the nodes of $\Omega_{-1}(x)$ lie asymptotically midway between those of $\Omega_0(x)$. In a similar manner it is seen that $\Omega_{-1}(z)\to 0$ on the bottom of S_β.

For $z\to 0$, $\Omega_{-1}(z)$ has a logarithmic singularity while $\Omega_0(z)$ remains finite. Furthermore, $\Omega_{-1}(z)$ satisfies the differential equations on surface and bottom since it is a linear combination with real coefficients of functions having this property. Thus $R\Omega_{-1}(z)$ is the velocity potential of a second standing wave and

$$\cos tR\Omega_0(z) - \sin tR\Omega_{-1}(z)$$

is the time-depending velocity potential of a progressive wave, advancing toward the shore $z=0$, which at large distances from shore is a simple progressive wave of wave length 2π. We observe that it has a logarithmic singularity at $z=0$.

15. **Same for $p\equiv 1$ (mod 4).** Set, as before

$$G(\xi) = \sum_{1}^{(p-1)/2} B_k\xi^k,$$

with

$$\begin{aligned} B_k &= -(i/2)A_k((1-\eta^{2k})^{-1} - (1-\eta^{-2k})^{-1}) \\ &= -(A_k/2)\cot(2\pi kq/p), \end{aligned}$$

where now

$$\eta = e^{2\pi iq/p}.$$

We find

(15.1) $$G(\xi\eta^2) - G(\xi) = (i/2)(F(\xi) + F(\xi\eta^2) + 2),$$

since $A_0 = -1$. Also

(15.2) $$G(\eta^{2n+1}) = G(\eta) + in,\quad n = -(p-1)/4, \cdots, +(p-5)/4,$$

(15.3) $$G(\eta^{(p+1)/2}) = G(\eta) + i(p-1)/4 + (i/2)F(\eta^{(p+1)/2}).$$

Now, by (15.1),

$$G(\eta) - G(\eta^{-1}) = i,$$

since $F(\eta) = F(\eta^{-1}) = 0$. Hence, because B_k is real,

$$G(\eta) = i/2 + \text{real}.$$

Thus (15.2), (15.3) become

$$G(\eta^{2n+1}) = i(n + 1/2) + RG(\eta),$$

$$n = -(p-1)/4, \cdots, (p-5)/4,$$

$$G(\eta^{(p+1)/2}) = i(p + 1)/4 + (i/2)F(\eta^{(p+1)/2}) + RG(\eta).$$

Accordingly

$$\sum B_k\delta_{jk} = i(n_j + 1/2) + RG(\eta),$$

if (10.9′) holds with $j \neq q$,

$$\sum B_k\delta_{qk} = i(p + 1)/4 + (i/2)F(\eta^{(p+1)/2}) + RG(\eta).$$

Set

$$\pi\Omega_{-1}(z) = W_{-1}(z) - 2\pi \sum W_k^*(z)B_k + W_0(z)\left(2\pi RG(\eta) - \int_1^{l-\infty} e^{-t}t^{-1}dt\right).$$

Then

$$0 = \lim_{x\to\infty}\left[-\pi\Omega_{-1}(x) + 2\pi i \sum_j{}' c_j(2n_j + 1)e^{\sigma_j x} - 2\pi \sum_j{}' c_j e^{\sigma_j x} \sum_k B_k\delta_{jk} + \left(2\pi RG(\eta) - \int_1^{l-\infty} e^t t^{-1}dt\right)\sum{}' c_j e^{\sigma_j x}\right].$$

Consequently

$$0 = \lim_{x\to\infty}\left[-\pi\Omega_{-1}(x) - \pi i c_q F(e^{\pi i(p+1)q/p})e^{-ix}\right].$$

The same conclusions about the behavior of $\Omega_{-1}(z)$ on the bottom and near $z=0$, and about the asymptotic distribution of the nodes of $R\Omega_0(x)$ and $R\Omega_{-1}(x)$, hold as did for $p\equiv 3 \pmod 4$. Again

$$\cos tR\Omega_0(z) - \sin tR\Omega_{-1}(z)$$

is the velocity potential of a progressive wave advancing toward shore which at large distance from shore becomes a simple wave of length 2π and which has a logarithmic singularity at the shore.

16. **About the continuity of the standing wave as function of the angle $p\pi/2q$ and the connection with the quadratic reciprocity law of number theory.** Let us inspect the asymptotic location of the surface nodes of the finite standing wave $\Omega_0(z)$ as $z=x\to\infty$. On entering the

value of c_q from (6.18) in (13.13) we find that $\Omega_0(x)$ differs infinitely little from

$$(-i)^{(q-1)/2+(p-1)/2}(-1)^{N(2q/p)}e^{\pi i(p^2-1)q/4p}e^{-ix}p^{1/2}, \qquad \text{if } q \text{ odd},$$
$$(-i)^{(p-1)/2}ie^{-\pi i(p+q)/4+\pi i(p^2-1)q/4p}(-1)^{N(2q,p)}e^{-ix}p^{1/2}, \qquad \text{if } q \text{ even}.$$

Thus the nodes of $R\Omega_0(x)$ come to lie asymptotically at the points $x=\pi/2+m\pi+qp\pi/4-q\pi/4p-\pi(q+p-2)/4$, with m integral for q odd, $x=m\pi+qp\pi/4-q\pi/4p-\pi(1+q)/4$, with m integral for q even. If the oddness of p is utilized, it is seen that the nodes lie asymptotically at

$$(16.1) \qquad x = m\pi - \pi/4 - q\pi/4p,$$

no matter whether q is odd or even; for if q is odd,

$$qp - q - p \equiv -1 \pmod 4,$$

and if q is even

$$qp - 1 - q \equiv -1 \pmod 4.$$

The result (16.1) is the generalization of a property announced at the end of §7 for the case $p=1$.

Next let us study the ratio of $\lim_{x\to\infty}(e^{ix}\Omega_0(x))$ and $\Omega_0(0)$. By (13.1), (10.6)

$$\Omega_0(0) = W_0(0) = \sum_1^q c_j,$$

as given by (6.17). Accordingly, by elementary simplification of the exponents, one obtains, for example, for odd p and q the formula

$$\lim_{x\to\infty} (e^{ix}\Omega_0(x))/\Omega_0(0)$$
$$= (-1)^{N(2q,p)+N(p,q)+(qp-p-q+1)/4}e^{\pi i(1-q/p)/4}(p/q)^{1/2}.$$

Now suppose that the left-hand side depends continuously on the angle $p\pi/2q$ between surface and bottom. Then the same applies to the right-hand side, whence in particular the expression

$$(-1)^{N(2q,p)+N(p,q)+(qp-p-q+1)/4}$$

must be continuous in p/q for p and q odd. But since it equals ± 1, this expression must have a constant value, which is determined as $+1$ by taking $p=q=1$ or $p=1$, $q=3$. Furthermore it is easily seen from the definition of $N(p, q)$ that, for odd p and q, $N(p, q)=N(2p, q)$. Thus the assumption of a continuous dependence of the finite standing wave on p/q implies

$$(-1)^{N(2q,p)+N(2p,q)} = (-1)^{(p-1)(q-1)/4}.$$

This is, of course, the quadratic reciprocity law of number theory.

We shall not investigate here the above continuity question, nor shall we go here into a discussion of those problems which are immediately suggested by the above embedding of a number theoretical fact concerning two integers p and q into a continuous law concerning the quantity p/q.

UNIVERSITY OF CALIFORNIA

Reprinted from *Proceedings of the ICM*
Vol. I, ©1948, John Wiley & Sons, Inc.

The Dock Problem

By K. O. FRIEDRICHS and HANS LEWY

Suppose one half-plane of an infinite water surface is covered with a rigid plate, the "dock." How does this dock influence waves standing or traveling perpendicularly to the edge of the dock on the free water surface? This problem may be considered as a special case of the problem of waves on a sloping beach, viz. for a beach with a slope angle of 180°, see the introductory paper [4]. For beach angles which are an integer fraction of a right angle, $\omega = \pi/2q$, the problem was solved by Miche [1], Stoker, and Lewy [2, 3]; for beach angles of the form $\omega = p\pi/2q$ with relatively prime integers p and q, the problem was solved by Lewy in case p is odd. The present problem corresponds to the case $q = 1$, $p = 2$. Although it may be possible to make $p/2q$ tend to the limit 1 and follow through this passage to the limit in the explicit expressions for the waves, we attack the problem differently, using a method related to the Laplace transformation. It turns out that the problem permits a simple explicit solution which offers the possibility of discussing the nature of the solution and of determining it numerically. We shall in particular discuss the behavior of the wave motion near the edge of the dock, i.e. near the line along which the water surface and the dock meet. In addition, we shall show graphically the amplitudes of the waves on the free surface and of the pressure under the dock.

Aside from its interest as a flow problem, the dock problem is of significance as a new example of a potential problem with different linear boundary conditions on different parts of the boundary.[1]

Since we are interested in two-dimensional flow, all quantities are assumed constant along any parallel to the edge. We place an (x, y)-coordinate system such that the water fills the lower half-plane $y < 0$, the free water surface is the positive x-axis, $y = 0$, $x > 0$, and the dock is the negative x-axis, $y = 0$, $x < 0$. We set $x + iy = z = re^{i\theta}$. Further we choose the units of length and time such that the frequency of the oscillations of the water surface and the acceleration of gravity have the value one. The water flow is then described by a complex potential function

$$\chi(z) \cos t$$

from which the velocity components u and v are derived:

$$u - iv = \frac{d\chi(z)}{dz} \cos t.$$

[1]A solution of the problem of two- and three-dimensional waves over a channel of finite depth with a dock has been announced by A. E. Heins [5].

The boundary condition that the pressure be constant on the free surface leads, by the well known process of linearization, see e.g. Stoker [3, p. 6], to the condition

$$\text{(I)} \qquad \mathcal{I}m\left(\frac{d}{dz}\chi + i\chi\right) = 0, \qquad y = 0, \qquad x > 0,$$

while the condition that the water slides under the dock surface leads to the condition

$$\text{(II)} \qquad \mathcal{I}m\left(\frac{d}{dz}\chi\right) = 0, \qquad y = 0, \qquad x < 0.$$

At infinity, i.e. as $|z| \to \infty$, we further require

$$\text{(III)} \qquad \chi(z) \sim \chi_\infty e^{-iz},$$

$$\text{i.e.} \qquad \chi(z)e^{iz} \to \chi_\infty, \qquad \theta_* \leq \theta \leq 0,$$

$$\chi(z) \to 0, \qquad -\pi \leq \theta \leq \theta_{**},$$

for any angles θ_*, θ_{**} with $-\pi < \theta_* < \theta_{**} < 0$ and with an appropriate complex constant $\chi_\infty \neq 0$.

These two conditions, which are evidently not contradictory, assure that, at large distances from the dock, the water executes sinusoidal motions near the surface and quiets down in other directions.

At the edge we permit a logarithmic singularity, i.e. we require that, with an appropriate constant k,

$$\text{(IV)} \qquad \chi(z) - k \log z \quad \text{is bounded at} \quad z = 0.$$

For the significance of this logarithmic singularity see Stoker [3, p. 11].

We shall show two solutions of the problem: one, $\chi^R(z)$, bounded at the edge $z = 0$, the other, given by

$$(1) \qquad \chi^S(z) = -\frac{d}{dz}\chi^R(z),$$

and possessing a logarithmic singularity at the edge. Specifically, the behavior of the solutions at $z = \infty$ is given by

$$(2) \qquad 2^{1/2}\chi^R(z) - e^{i\pi/8}e^{-iz} \to 0, \qquad 2^{1/2}\chi^S(z) - ie^{i\pi/8}e^{-iz} \to 0, \qquad z \to \infty,$$

if z approaches infinity within a sector which excludes the negative real axis, $-\pi < \theta_* \leq \theta \leq 0$, while if z approaches infinity in a sector which excludes the positive real axis, $-\pi \leq \theta \leq \theta_{**} < 0$,

$$(3) \qquad \chi^R(z) \to 0, \qquad \chi^S(z) \to 0, \qquad z \to \infty.$$

Both statements, (2) and (3), are in agreement in the overlapping parts of the sectors. The behavior at the origin is given by

(4) $$\chi^R(z) \to \frac{1}{2} \quad \text{as} \quad z \to 0,$$

(5) $$\chi^S(z) - \frac{1}{2\pi} \log(-z) \to -\frac{1 - C}{2\pi} \quad \text{as} \quad z \to 0,$$

C being the Euler constant.

The relation

(6) $$\chi^R(z)e^{iz}|^{\infty}/\chi^R(0) = 2^{1/2}e^{i\pi/8}$$

is particularly remarkable, since it is in agreement with the value $\eta(p/q)^{1/2}$ found for the same ratio for beaches of slope angle $p\pi/2q$, see [2, p. 744], η being an appropriate number of absolute value one. Note that in our case, $p = 2$, $q = 1$, p is not odd. Refined expressions for the behavior of the functions $\chi(z)$ as $z \to 0$ and $z \to \infty$ will be given in the Appendix.

The solutions $\chi^R(z)$, $\chi^S(z)$ are given by the integrals

(7) $$\chi^R(z) = \frac{1}{2\pi} \int_{\mathcal{P}} e^{\zeta z}\psi(\zeta)\, d\zeta,$$

(8) $$\chi^S(z) = -\frac{1}{2\pi} \int_{\mathcal{P}} e^{\zeta z}\zeta\psi(\zeta)\, d\zeta,$$

in which $\mathcal{P}$ is any path running in the lower half-plane $\eta \leq 0$ from $\zeta = 0$ to $\zeta = \infty$ entering $\zeta = \infty$ in a direction along which ζz has a negative real part and skirting the point $\zeta = -i$ so as to leave it to the right. Such a path $\mathcal{P}$ will be called "admissible" for the value z.

The property of the function $\psi(\zeta)$ which insures that the boundary conditions are satisfied is that $\psi(\zeta)$ is real on the positive ξ-axis and $(\zeta + i)\psi(\zeta)$ is real on the negative ξ-axis. The second of these properties follows from the first one if $\psi(\zeta)$ is defined as a one-valued function of $\log \zeta$ which satisfies the relation

$$\psi(\zeta e^{2\pi i}) = \frac{\zeta + i}{\zeta - i}\psi(\zeta)$$

in obvious notation. A solution of this equation can be given in terms of an infinite product. It happens that this infinite product can be evaluated explicitly. For this reason we confine ourselves to describing the correct function $\psi(\zeta)$ explicitly and to verifying that it has those properties which insure that (7) and (8) are the solutions of our problem.

The function $\psi(\zeta)$ is defined in the lower half-plane, $\eta \leq 0$, ($\zeta \neq 0$), except for the point $\zeta = -i$, at which it has a pole. It is given by the two equivalent representations

$$(9)_0 \qquad \psi(\zeta) = \zeta^{-1/2}(1 - i\zeta)^{-1}(1 + i\zeta)^{-1/2} \exp \frac{1}{\pi} \int_0^\zeta \log (i\zeta_*)^{-1} \frac{d\zeta_*}{\zeta_*^2 + 1},$$

$$(9)_\infty \qquad \psi(\zeta) = \zeta^{-2}(1 + i\zeta^{-1})^{-1}(1 - i\zeta^{-1})^{-1/2} \exp \frac{1}{\pi} \int_\zeta^{-i\infty} \log (i\zeta_*) \frac{d\zeta_*}{\zeta_*^2 + 1}.$$

Note that the function $\int_\zeta^{-i\infty} \log (i\zeta_*) \, d\zeta_*/(\zeta_*^2 + 1)$ is regular in the lower half-plane. The equivalence of the two representations follows from the relation

$$(10) \qquad \begin{aligned} \int_0^{-i\infty} \log (i\zeta_*) \frac{d\zeta_*}{\zeta_*^2 + 1} &= \int_0^\infty \log (i\xi_*) \frac{d\xi_*}{\xi_*^2 + 1} \\ &= \frac{i\pi}{2} \int_0^\infty \frac{d\xi_*}{\xi_*^2 + 1} = \frac{i\pi^2}{4}. \end{aligned}$$

From the relation

$$\int_0^{-1} \log (i\zeta_*) \frac{d\zeta_*}{\zeta_*^2 + 1} = \int_{-i}^{-i\infty} \log (\tau i) \frac{d\tau}{\tau^2 + 1}$$

obtained by substituting $- 1/\tau$ for ζ_* and from (10), we conclude

$$(11) \qquad \int_0^{-i} \log (i\zeta_*) \frac{d\zeta_*}{\zeta_*^2 + 1} = \frac{i\pi^2}{8}$$

and we see that $(\zeta + i)\psi(\zeta)$ is regular at $\zeta = -i$ and assumes there the value

$$(12) \qquad (\zeta + i)\psi(\zeta) \,|^{-i} = ie^{i\pi/8}.$$

The path of integration $\mathcal{P}$ is any curve coming from the origin, skirting the point $\zeta = -i$ leaving it to the right, and then going to infinity in a direction in which the real part of ζz is negative. Thus from the expressions $(9)_0$ and $(9)_\infty$ it is clear the $\psi(\zeta)$ behaves at $\zeta = 0$ and at $\zeta = \infty$ in such a way that the two integrals (7) and (8) represent analytic functions of z which are regular in the lower half-plane, $y \leq 0$, $(z \neq 0)$.

In order to *verify that the boundary conditions* I *and* II *are satisfied* we use different representations of the function $\psi(\zeta)$. From the relations

$$(13)_0 \qquad i \int_0^\zeta \frac{d\zeta_*}{\zeta_*^2 + 1} = \frac{1}{2} \log \frac{1 + i\zeta}{1 - i\zeta}, \qquad \text{except for} \qquad \xi = 0, \ |\, \zeta \,| \geq 1,$$

$$(13)_\infty \qquad i \int_\zeta^\infty \frac{d\zeta_*}{\zeta_*^2 + 1} = \frac{1}{2} \log \frac{1 + i\zeta^{-1}}{1 - i\zeta^{-1}}, \qquad \text{except for} \qquad \xi = 0, \ |\, \zeta \,| \leq 1,$$

we find

$$(14)_0 \qquad \psi(\zeta) = \zeta^{-1/2}(1 + \zeta^2)^{-3/4} \exp \frac{1}{\pi} \int_0^\zeta \log \zeta_*^{-1} \frac{d\zeta_*}{\zeta_*^2 + 1}$$

$$(15)_0 \quad \psi(\zeta) = \zeta^{-1/2}(1 - i\zeta)^{-1}(1 + \zeta^2)^{-1/4} \exp \frac{1}{\pi} \int_0^{\zeta} \log(-\zeta_*)^{-1} \frac{d\zeta_*}{\zeta_*^2 + 1},$$

valid for $\eta \leq 0$ except for $\xi = 0, \eta \leq -1$,

$$(14)_\infty \quad \psi(\zeta) = \zeta^{-2}(1 + \zeta^{-2})^{-3/4} \exp \frac{1}{\pi} \int_{\zeta}^{\infty} \log \zeta_* \frac{d\zeta_*}{\zeta_*^2 + 1},$$

valid for $\eta \leq 0$ except for $\xi = 0, \eta \geq -1$. These representations make it evident that:

$\psi(\zeta)$ is real positive on the positive real axis,

$$\zeta = \xi = 0, \qquad \eta = 0,$$

$(\zeta + i)\psi(\zeta)$ is real positive on the negative real axis,

$$\zeta = \xi < 0, \qquad \eta = 0.$$

For values $z = x < 0$, $y = 0$, on the negative real z-axis we may take as the path $\mathcal{P}$ the positive real ζ-axis, $\zeta = \xi > 0$, $\eta = 0$. From the representations (7) and (8) we can then see that $\chi^R(z)$ and hence also $d\chi^R(z)/dz$ and $d\chi^S(z)/dz = -d^2\chi^R(z)/dz^2$ are real for $z = x < 0$, $y = 0$. *The boundary condition* II *is therefore satisfied.* In order to *verify that the boundary condition* I *is satisfied* we observe that the integrand in

$$\left(\frac{d}{dz} + i\right)\chi^R(z) = \int_{\mathcal{P}} e^{\zeta z}(\zeta + i)\psi(\zeta)\, d\zeta$$

is regular at $\zeta = -i$. Hence we can deform the path $\mathcal{P}$ into the negative real ζ-axis, $\zeta = \xi < 0$, $\eta = 0$, if z lies on the positive real z-axis, $z = x > 0$, $y = 0$. Since $(\zeta + i)\psi(\zeta)$ is positive real for $\zeta = \xi < 0$, $\eta = 0$, we see that $(d/dz + i)\chi^R(z)$ is real for $z = x > 0$, $y = 0$. The same statement follows similarly for $\chi^S(z)$. The boundary condition I is therefore satisfied.

To derive *the behavior* of χ^R and χ^S *at infinity* we make use of the pole of $\psi(\zeta)$ at $\zeta = -i$. Suppose the point $z = re^{i\theta}$ lies in any sector excluding the positive real axis, $-\pi \leq \theta \leq \theta_{**} < 0$. Then ζz has a negative real part on any ray $\zeta = \rho e^{i\lambda}$, $\lambda =$ constant whose angle λ lies in the interval $-\pi/2 < \lambda < -\pi/2 - \theta_{**}$. Since any such ray in the lower half-plane, run through from 0 to ∞, leaves the point $\zeta = -i$ to the right, it is admitted as a path $\mathcal{P}$. If we now let z approach ∞ it is clear that only the neighborhood of $\zeta = 0$ will contribute asymptotically to the integrals (7) and (8). By $(14)_0$ the function $\psi(\zeta)$ behaves like $\zeta^{-1/2}$ at the origin; consequently, $\chi^R(z)$ behaves like $1/(2(-\pi z)^{1/2})$ for large values of z and hence vanishes as $z \to \infty$. If, on the other hand, $z = re^{i\theta}$ lies in the sector $-\pi \leq \theta_* \leq \theta \leq 0$, we shift our path to any ray $\zeta = \rho e^{i\lambda}$, $\lambda =$ constant, with $-3\pi/2 - \theta_* < \lambda < -\pi/2$, and then argue as before. In doing so, however, we cross the pole at $\zeta = -i$. The residue contributed by the pole is just $1/2^{1/2}e^{i\pi/8}e^{-iz}$ for $\chi^R(z)$ and $1/2^{1/2}ie^{i\pi/8}e^{-iz}$ for $\chi^S(z)$ as seen from (12). Thus the formulas (2) and (3) are justified. Refined expressions for the behavior of the functions $\chi(z)$ as $z \to \infty$ will be given in the Appendix.

To determine *the behavior* of the functions $\chi(z)$ *at the edge* $z = 0$ we must establish an expansion of $\psi(\zeta)$ for large values of ζ. Since

$$(16) \qquad \int_\zeta^\infty \log \zeta_* \frac{d\zeta_*}{\zeta_*^2 + 1} = \zeta^{-1}(\log \zeta + 1) - \frac{1}{3}\zeta^{-3}\left(\log \zeta + \frac{1}{3}\right) + \cdots$$

we have from $(14)_\infty$

$$(17) \qquad \psi(\zeta) = \zeta^{-2} + \frac{1}{\pi}\zeta^{-3}(\log \zeta + 1) + 0(\zeta^{-4}\log^2 \zeta).$$

Thus $\psi(\zeta)$ vanishes at infinity like ζ^{-2} and the integral

$$(18) \qquad P = \frac{1}{2\pi}\int_{\mathcal{P}} \psi(\zeta)\, d\zeta$$

is absolutely integrable and does not depend on which path $\mathcal{P}$ is chosen. Hence, $\chi^R(z)$ approaches P if z approaches 0 through values for which the path $\mathcal{P}$ is admissible, see page 5. Consequently, if the function $\chi^R(z)$ is defined for $z = 0$ by $\chi^R(0) = P$, it is continuous there.

To determine the value of P we consider the integral

$$(19) \qquad Q = \frac{1}{2\pi}\int_{\mathcal{P}} \{\zeta\psi(\zeta) - \zeta(\zeta^2 + 1)^{-1}\}\, d\zeta,$$

which is also independent of the choice of $\mathcal{P}$. The term subtracted from $\zeta\psi(\zeta)$ is so chosen that the integrand vanishes at infinity like $1/\pi\, \zeta^{-2} \log \zeta$, as seen from (17), and remains integrable at $\zeta = 0$.

If one takes the path $\mathcal{P}$ in (18) and (19) as the positive ξ-axis, it becomes clear that both P and Q are real.

We now combine the two integrals to

$$(20) \qquad Q + iP = \frac{1}{2\pi}\int_{\mathcal{P}} \{(\zeta + i)\psi(\zeta) - \zeta(\zeta^2 + 1)^{-1}\}\, d\zeta.$$

Since $(\zeta + i)\psi(\zeta)$ is regular in the whole lower half-plane ($\zeta \neq 0$), we may swing the path of integration into the negative ξ-axis, taking the residue due to the second contribution into account. We obtain

$$(21) \qquad Q + iP = \frac{1}{2\pi}\int_0^{-\infty} \{(\zeta + i)\psi(\zeta) - \zeta(\zeta^2 + 1)^{-1}\}\, d\zeta + \frac{i}{2}.$$

Now, $(\zeta + i)\psi(\zeta)$ is real on the negative ξ-axis. Hence we have

$$(22) \qquad P = \frac{1}{2},$$

$$(23) \qquad Q = \frac{1}{2\pi}\int_0^{-\infty} \{(\zeta + i)\psi(\zeta) - \zeta(\zeta^2 + 1)^{-1}\}\, d\zeta.$$

Relation (22) yields formula (4) since

$$\chi^R(z) \to P \quad \text{as} \quad z \to 0, \tag{24}$$

as was shown above.

We now apply the same process in the opposite direction. We consider the integral

$$R = \frac{1}{2\pi}\int_0^{-\infty}\left\{\zeta(\zeta + i)\psi(\zeta) - \left[1 + \frac{1}{\pi}\zeta(\zeta^2 + 1)^{-1}(\log \zeta + i\pi + 1)\right]\right\} d\zeta \tag{25}$$

which, by (17), converges absolutely and permits swinging the path of integration back to the positive real axis, except for a residue due to the second contribution. We note that R is real since $(\zeta + i)\psi(\zeta)$ and $\log \zeta + i\pi$ are real on the negative ξ-axis. We now combine R with Q, as given by (25) and (23), to

$$R - iQ = \frac{1}{2\pi}\int_0^{-\infty}\left\{(\zeta^2 + 1)\psi(\zeta) - \left[1 + \frac{1}{\pi}\zeta(\zeta^2 + 1)^{-1}(\log \zeta + 1)\right]\right\} d\zeta \tag{26}$$

and swing the path back into a path $\mathcal{P}$. Since $(\zeta^2 + 1)\psi(\zeta)$ is regular in the lower half-plane, we obtain

$$R - iQ = \frac{1}{2\pi}\int_{\mathcal{P}}\left\{(\zeta^2 + 1)\psi(\zeta) - \left[1 + \frac{1}{\pi}\zeta(\zeta^2 + 1)^{-1}(\log \zeta + 1)\right]\right\} d\zeta - \frac{1}{4} - \frac{i}{2\pi}. \tag{27}$$

The integral in (27) is real if $\mathcal{P}$ is taken as the positive ξ-axis. Consequently, we have

$$Q = \frac{1}{2\pi}. \tag{28}$$

From this result we can derive formula (5), which gives the behavior of $\chi^S(z)$ at $z = 0$. We first observe that, by the definition (19) of Q, relation (28) implies

$$\frac{1}{2\pi}\int_{\mathcal{P}} e^{\zeta z}\{\zeta\psi(\zeta) - \zeta(\zeta^2 + 1)^{-1}\}\, d\zeta \to \frac{1}{2\pi} \tag{29}$$

if z approaches 0 through values for which $\mathcal{P}$ is admissible. Thus, by (8), the behavior of $\chi^S(z)$ at $z = 0$ is reduced to that of

$$\int_{\mathcal{P}} e^{\zeta z}\zeta(\zeta^2 + 1)^{-1}\, d\zeta.$$

In order to discuss this integral we may lead the path $\mathcal{P}$ first along the positive ξ-axis from 0 to any point $\zeta = \alpha$, $\alpha > 0$, and then to infinity in such a way that it eventually runs along a straight line on which ζz is real and negative. Consequently, we may write

$$\int_{\mathcal{P}} e^{\zeta z}(\zeta^2 + 1)^{-1}\zeta\, d\zeta = \int_0^\alpha e^{\zeta z}(\zeta^2 + 1)^{-1}\zeta\, d\zeta + \int_{-\alpha z}^\infty e^{-\sigma}(\sigma^2 + z^2)^{-1}\sigma\, d\sigma$$

$$(30) \qquad = \int_0^\alpha (e^{\zeta z} - 1)(\zeta^2 + 1)^{-1}\zeta\, d\zeta$$

$$- \log(-z) + \frac{1}{2}(1 - e^{\alpha z}) \log(z^2 + \alpha^2 z^2) + \frac{1}{2}\int_{-\alpha z}^\infty e^{-\sigma} \log(\sigma^2 + z^2)\, d\sigma.$$

The path in the last integral, which begins at the point $-\alpha z$ in the upper half-plane, enters infinity along a section of the positive ξ-axis; if $-\alpha z$ lies in the left half-plane the path skirts the point $\sigma = -iz$ so as to leave it to the right. Letting now z approach 0 we obtain

$$(31) \qquad \int_{\mathcal{P}} e^{\zeta z}(\zeta^2 + 1)^{-1}\zeta\, d\zeta + \log(-z) \to \int_0^\infty e^{-\sigma} \log \sigma\, d\sigma = -C,$$

C being the Euler constant. Consequently we have from (8), (29), and (31)

$$(32) \qquad \chi^S(z) - \frac{1}{2\pi} \log(-z) \to -\frac{1 - C}{2\pi}$$

as $z \to 0$. Thus formula (5) is established.

Since the function $-\chi^S(z)$ is the derivative of $\chi^R(z)$, the formula (5) gives at the same time a refined behavior of $\chi^R(z)$ at $z = 0$,

$$(33) \qquad \chi^R(z) = \frac{1}{2} - \frac{1}{2\pi}\{z \log(-z) - (2 - C)z\} + 0(z),$$

the symbol $0(z)$ representing a function for which $z^{-1}0(z) \to 0$ as $z \to 0$.

Refined asymptotic representations of the functions $\chi^R(z)$ and $\chi^S(z)$ at $z = 0$ could be derived by continuing the process of swinging the path of integration of appropriate integrals forward and backward. One step in this process is carried out in the Appendix by Cathleen S. Morawetz. It will be seen that this procedure becomes very cumbersome. Two different methods of deriving asymptotic expansions of analytic functions satisfying the two boundary conditions I and II on the two rays of the real axis will be given in a later publication.

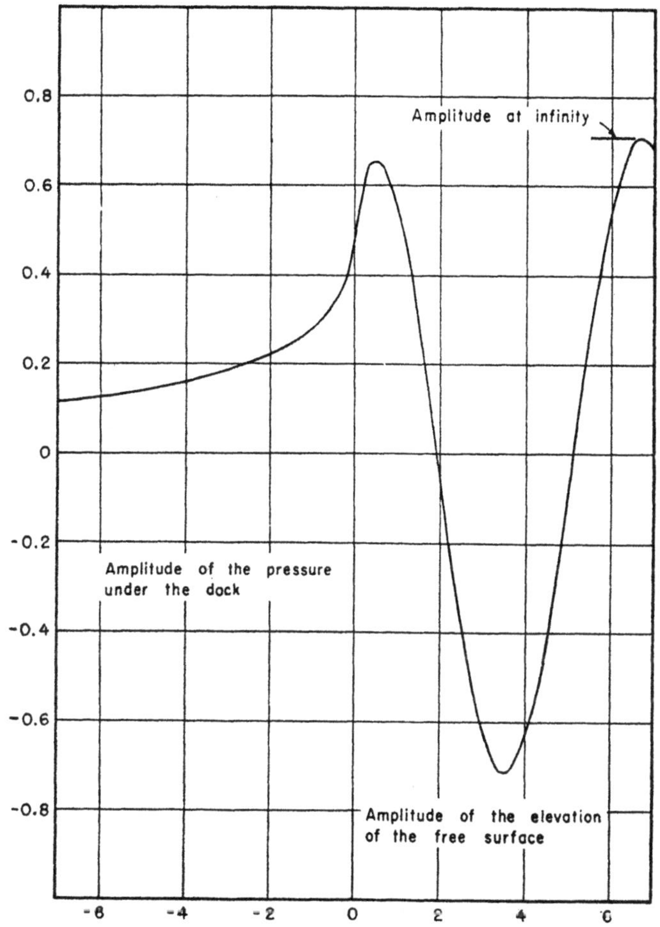

FIGURE 1. Regular standing wave.

Explanation of the Figures

The figures show the values of the velocity potential $\varphi(x, 0) = \Re e\, \chi(x)$ along the dock, $x < 0$, and along the free surface, $x > 0$. The elevation of the free surface is given by

$$\eta = g^{-1}\sigma\varphi(x, 0) \sin \sigma t, \qquad x > 0;$$

the pressure under the dock surface—in excess over the atmospheric pressure—is given by

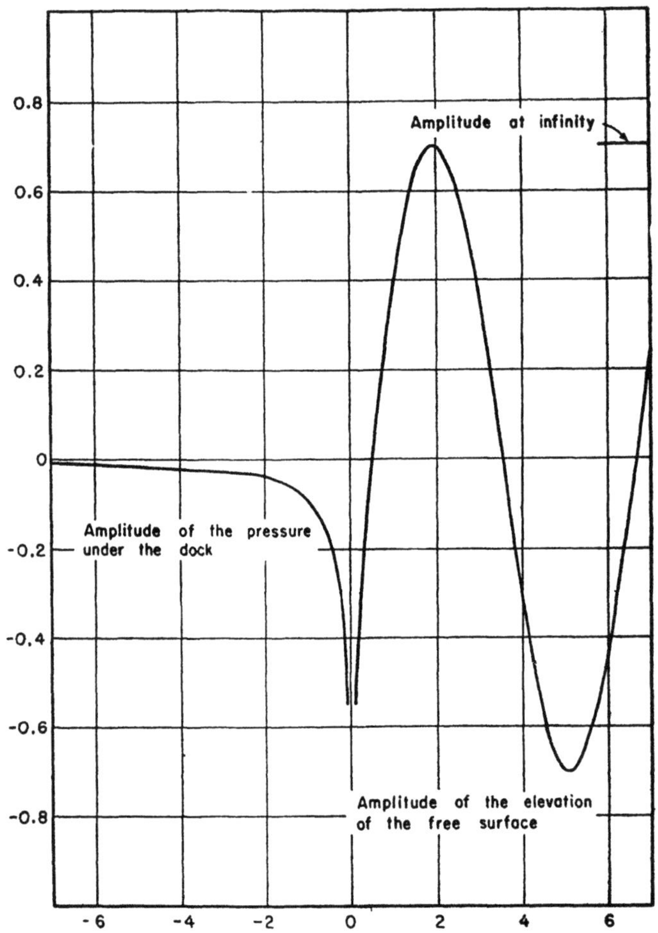

FIGURE 2. Singular standing wave.

$$p = \rho\sigma\varphi(x, 0) \sin \sigma t, \qquad x < 0.$$

Here g is the acceleration of gravity, ρ is the density of water, and σ is the circular frequency of the motion.

The unit of length is chosen as $g\sigma^{-2}$. Figures 1 and 2 refer to the wave motion described by the function $\chi^R(z)$ and $\chi^S(z)$ respectively.

Appendix. *Refined Asymptotic Expressions for the Complex Potentials of the Dock Problem*

To find terms of higher order in *the expansions of* $\chi^R(z)$ *and* $\chi^S(z)$ *about the point at infinity* we need terms of higher order in the expansion of $\psi(\zeta)$ about the origin. This expansion as derived from $(14)_0$ is

$$\psi(\zeta) = \zeta^{-1/2}\left\{\left[1 + \frac{\zeta}{\pi} + \zeta^2\left(-\frac{3}{4} + \frac{1}{2\pi^2}\right)\cdots\right] + \frac{\zeta \log \zeta^{-1}}{\pi}\left[1 + \frac{\zeta}{\pi} + \cdots\right]\right.$$

$$\left. + \frac{\zeta^2(\log \zeta^{-1})^2}{2\pi^2}[1 + \cdots] + \cdots\right\}. \tag{34}$$

Using (34) the integrand $e^{\zeta z}\psi(\zeta)$ in the expression (7) for $\chi^R(z)$ can be written as

$$e^{\zeta z}\psi(\zeta) = \left\{\left[1 + \frac{1}{\pi}\frac{d}{dz} + \left(-\frac{3}{4} + \frac{1}{2\pi^2}\right)\frac{d^2}{dz^2} + \cdots\right]e^{\zeta z}\zeta^{-1/2}\right.$$

$$+ \left[\frac{1}{\pi}\frac{d}{dz} + \frac{1}{\pi^2}\frac{d^2}{dz^2} + \cdots\right]e^{\zeta z}\zeta^{-1/2}\log \zeta^{-1} \tag{35}$$

$$\left. + \left[\frac{1}{2\pi^2}\frac{d^2}{dz^2} + \cdots\right]e^{\zeta z}\zeta^{-1/2}(\log \zeta^{-1})^2\right\}$$

The path $\mathcal{P}$ over which $e^{\zeta z}\psi(\zeta)$ is to be integrated may, as was argued previously, be taken as the positive real axis if z approaches infinity in a sector which excludes the positive real axis. Then as $z \to \infty$

$$2\pi\chi^R(z) \sim \left[1 + \frac{1}{\pi}\frac{d}{dz} + \left(-\frac{3}{4} + \frac{1}{2\pi^2}\right)\frac{d^2}{dz^2}\cdots\right]\int_0^\infty e^{\zeta z}\zeta^{-1/2}\,d\zeta$$

$$+ \left[\frac{1}{\pi}\frac{d}{dz} + \frac{1}{\pi^2}\frac{d^2}{dz^2}\cdots\right]\int_0^\infty e^{\zeta z}\zeta^{-1/2}\log \zeta^{-1}\,d\zeta$$

$$+ \left[\frac{1}{2\pi^2}\frac{d^2}{dz^2} + \cdots\right]\int_0^\infty e^{\zeta z}\zeta^{-1/2}(\log \zeta^{-1})^2\,d\zeta$$

or by substituting $\sigma = -\zeta z$ and taking the positive real axis as the path of integration for σ,

$$2\pi\chi^R(z) \sim \left[1 + \frac{1}{\pi}\frac{d}{dz} + \left(-\frac{3}{4} + \frac{1}{2\pi^2}\right)\frac{d^2}{dz^2}\cdots\right]\frac{1}{(-z)^{1/2}}\int_0^\infty e^{-\sigma}\sigma^{-1/2}\,d\sigma$$

$$+ \left[\frac{1}{\pi}\frac{d}{dz} + \frac{1}{\pi^2}\frac{d^2}{dz^2}\right]\frac{1}{(-z)^{1/2}}\int_0^\infty e^{-\sigma}\sigma^{-1/2}(\log(-z) - \log \sigma)\,d\sigma \tag{36}$$

$$+ \left[\frac{1}{2\pi^2}\frac{d^2}{dz^2}\right]\frac{1}{(-z)^{1/2}}\int_0^\infty e^{-\sigma}\sigma^{-1/2}(\log(-z) - \log \sigma)^2\,d\sigma.$$

The integrals in (36) can be evaluated immediately in terms of Γ-functions. Thus we find that, *if z approaches infinity within a sector which excludes the positive real axis, the asymptotic expansions of $\chi^R(z)$ and $\chi^S(z) = -d\chi^R(z)/dz$ are given by*

$$(37)\quad \begin{aligned} 2^{1/2}\chi^R(z) \to -\frac{ia_0}{\pi(2z)^{1/2}}\Big\{1 &- \frac{1}{2\pi z}[\log(-z) - 1 - a_1] \\ &+ \frac{1}{4\pi^2 z^2}\Big[\frac{3}{2}(\log(-z))^2 - 1 + 3a_1)\log(-z) - \frac{9}{4}\pi^2 + \frac{3}{2} + a_1 - \frac{3}{2}a_2\Big]\Big\}, \end{aligned}$$

$$(38)\quad 2^{1/2}\chi^S(z) \to -\frac{ia_0}{(2z)^{1/2}}\Big\{\frac{1}{2\pi z} - \frac{1}{4\pi^2 z^2}[3\log(-z) - 5 - 3a_1]\Big\},$$

corresponding to relation (3). The constants a_1 and a_2 are given by

$$a_0 = [\Gamma(x)]_{x=1/2} = 1.772$$

$$a_1 = \Big[\frac{d\Gamma(x)}{dx}/\Gamma(x)\Big]_{x=1/2} = 4.934$$

$$a_2 = \Big[\frac{d^2\Gamma(x)}{dx^2}/\Gamma(x)\Big]_{x=1/2} = 6.136.$$

On the other hand *if z approaches infinity within a sector which excludes the negative real axis* we must, as was explained earlier, cross the pole of $\psi(\zeta)$ at $\zeta = -i$ in order to shift the path $\mathcal{P}$ to the positive real axis. In this case we must include the residue at $\zeta = -i$ and we find, corresponding to relation (2),

$$(39)\quad \begin{aligned} 2^{1/2}\chi^R(z) \sim e^{i\pi/8}e^{-iz} - \frac{ia_0}{\pi(2z)^{1/2}}\Big\{1 &- \frac{1}{2\pi z}[\log(-z) - 1 - a_1] \\ &+ \frac{1}{4\pi^2 z^2}\Big[\frac{3}{2}(\log(-z))^2 - (1 + 3a_1)\log(-z) - \frac{9}{4}\pi^2 + \frac{3}{2} + a_1 - \frac{3}{2}a_2\Big]\Big\} \end{aligned}$$

$$(40)\quad 2^{1/2}\chi^S(z) \sim ie^{i\pi/8}e^{-iz} - \frac{ia_0}{(2z)^{1/2}}\Big\{\frac{1}{2\pi z} - \frac{1}{4\pi^2 z^2}[3\log(-z) - 5 - 3a_1]\Big\}.$$

To obtain a better description of *the behavior of χ^R and χ^S at the edge $z = 0$* the process of swinging the path of integration of appropriate integrals through the lower half plane can be continued.

The appropriate integral to introduce for the next step in this process is

$$(41)\quad \begin{aligned} S = \frac{1}{2\pi}\int_0^\infty \Big\{(\zeta^2 + 1)\zeta\psi(\zeta) - \zeta\Big[1 &+ \frac{\zeta}{\pi}(\zeta^2 + 1)^{-1}(\log\zeta + 1) \\ &+ \frac{1}{2\pi^2}(\zeta^2 + 1)^{-1}(\log\zeta + 1)^2 + \frac{1}{4}(\zeta^2 + 1)^{-1}\Big]\Big\}\, d\zeta \end{aligned}$$

The behavior of ψ as $\zeta \to \infty$ is given, according to (14_∞), by

$$\psi \to (\zeta^2+1)^{-1} + \frac{\zeta}{\pi}(\zeta^2+1)^{-2}(\log\zeta+1) + \frac{1}{2\pi^2}(\zeta^2+1)^{-2}(\log\zeta+1)^2$$
$$+\frac{1}{4}(\zeta^2+1)^{-2} + M(\zeta)$$

where $\int_0^\infty \zeta^3 M(\zeta)d\zeta$ exists, see equation (17). Hence the integral S converges absolutely.

Combining S with iR we form from (27) and (41)

$$S + iR = \frac{1}{2\pi}\int_{\mathcal{P}}\left\{(\zeta^2+1)(\zeta+i)\psi(\zeta)\right.$$
$$(42) \qquad -(\zeta+i)\left[1+\frac{\zeta}{\pi}(\zeta^2+1)^{-1}(\log\zeta+1)\right]$$
$$\left.-\left[\frac{1}{2\pi^2}(\zeta^2+1)^{-1}(\log\zeta+1)^2+\frac{i}{4}(\zeta^2+1)^{-1}\right]\right\}d\zeta - \frac{i}{4}$$

Since $(\zeta^2+1)(\zeta+i)\psi(\zeta)$ is regular in the whole lower half plane, $\zeta \neq 0$, we may swing the path $\mathcal{P}$ back again to the negative ξ-axis taking into account the pole at $\zeta = -i$. Since $(\zeta+i)\psi(\zeta)$ is real on this axis, we find from (42) and (27)

$$(43) \qquad R = \frac{1}{2\pi}\int_0^\infty\left\{(\zeta^2+1)\psi(\zeta) - \left[1+\frac{\zeta}{\pi}(\zeta^2+1)^{-1}(\log\zeta+1)\right]\right\}d\zeta - \frac{1}{4} = \frac{1}{4\pi^2}+\frac{1}{16}.$$

Using relation (43) we determine in much the same way as (32) was determined, that, as $z \to 0$,

$$(44) \qquad \frac{d^2\chi^R}{dz^2} = -\frac{d\chi^S}{dz} = \int_0^\infty e^{\zeta z}\zeta^2\psi(\zeta)\,d\zeta \sim \frac{1}{4\pi^2} - \frac{3}{16} - \frac{1}{2\pi z}$$
$$-\frac{1}{2\pi^2}[\log(-z)+C] + \frac{1}{2\pi^2}\left[\frac{1}{2}(\log(-z))^2 + C\log(-z) - \frac{\pi^2}{24}+\frac{1}{2}D\right],$$

where C is again the Euler constant and

$$D = \left[\frac{d^2}{dx^2}\Gamma(x)\right]_{x=1} = 1.978$$

By integrating relation (44) twice and using relations (32) and (33) to determine the integration constants we find the following *expressions for the asymptotic expansions of* $\chi^R(z)$ *and* $\chi^S(z) = -d\chi^R/dz$ *at* $z = 0$,

$$\begin{aligned}\chi^S(z) \sim \frac{1}{2\pi} \log(-z) - \frac{1}{2\pi}(1 - C) - \frac{z}{4\pi^2}\Big[(\log(-z))^2 \\ + (2C - 4)\log(-z) + 5 - \frac{5}{6}\pi^2 - 4C + D\Big],\end{aligned} \tag{45}$$

$$\begin{aligned}\chi^R(z) \sim \frac{1}{2} - \frac{z}{2\pi}[\log(-z) - (2 - C)] + \frac{z^2}{8\pi^2}\Big[(\log(-z))^2 \\ + (2C - 5)\log(-z) + \frac{15}{2} - \frac{5}{6}\pi^2 - 5C + D\Big]\end{aligned} \tag{46}$$

Bibliography

1. Miche, A., *Mouvements ondulatoires de la mer en profondeur constante ou décroissante.* Annales des ponts et chaussées, Tome 114, pp. 25-78, 131-164, 270-292, 369-406, 1944.
2. Lewy, H., *Water Waves on Sloping Beaches.* Bulletin of the American Mathematical Society, Volume 52, No. 9, pp. 737-775, 1946.
3. Stoker, J. J., *Surface Waves in Water of Variable Depth.* Applied Mathematics Group—New York University No. 153; Quarterly of Applied Mathematics, Volume 5, No. 1, pp. 1-54, April 1947.
4. *Waves over Beaches of Small Slope, under a Dock, under an Overhanging Cliff, and past Plane Barriers.* Communications on Applied Mathematics, Volume 1, No. 2, 1948, p. 101.
5. Heins, A. E., *Water Waves over a Channel with a Dock.* Bulletin of the American Mathematical Society, Volume 53, No. 5, p. 497, 1947.

COURANT
Anniversary Volume
1948

ON THE CONVERGENCE OF SOLUTIONS OF DIFFERENCE EQUATIONS

Hans Lewy

The convergence of the solutions of difference equations as the length of the elementary mesh size tends to zero, has been studied in extenso for the case of hyperbolic equations in a paper by R. Courant, K. Friedrichs and the present author.[1)] The methods used there to establish convergence employ certain energy sums and are of sufficient generality to yield convergence for a wide class of equations. Nevertheless, it appears to be of interest to follow up in more explicit terms the passage to the limit, and this paper intends to show how this can be done in the example of the equation of cylindrical waves. The equation in question is

$$u_{zz} = u_{xx} + u_{yy} , \tag{1}$$

in which the suffixes stand for the taking of standard difference quotients in the directions of the three axes of independent variables x,y,z on which the function u depends. The particular initial problem to be treated here is

$$u = f(x,y), \qquad u_z = 0 \qquad \text{for } z = 0, \tag{2}$$

where f(x,y) is supposed to vanish of sufficiently high order as x or y tend to infinity, in order to permit the following manipulations of its Fourier integral.

One can first consider an exponential solution $e^{i(\alpha x+\beta y+\gamma z)}$

1) Ueber die partiellen Differenzengleichungen der mathematischen Physik, Math. Ann. 1928, vol. 100, pp. 32-74.

of (1). Let the mesh size in the x,y,z-directions be h, k, ℓ. The equation (1) becomes a condition for α, β, γ

$$(3) \qquad -\frac{e^{i\ell\gamma}+e^{-i\ell\gamma}-2}{\ell^2} = \frac{4}{\ell^2}\sin^2\frac{\gamma\ell}{2} = \frac{4}{h^2}\sin^2\frac{\alpha h}{2} + \frac{4}{k^2}\sin^2\frac{\beta k}{2} \; .$$

In this equation we consider $\alpha, \beta, h, k, \ell$ as given real quantities and γ as function of these. It is easily seen that there always exists a real γ solving (3), if and only if

$$(4) \qquad \frac{1}{\ell^2} \geqq \frac{1}{h^2} + \frac{1}{k^2} \; .$$

For in that case the expression

$$\frac{\ell^2}{h^2}\sin^2\frac{\alpha h}{2} + \frac{\ell^2}{k^2}\sin^2\frac{\beta k}{2}$$

will be between the limits 0 and 1 and accordingly equal to $\sin^2\frac{\gamma\ell}{2}$ for some real value of γ. Conversely, if the condition (4) does not hold, there is no real value of γ satisfying (3) whenever $\sin^2\frac{\beta k}{2}$ and $\sin^2\frac{\alpha h}{2}$ equal 1.

Consequently, if (4) holds, the two solutions

$$(5) \qquad e^{i(\alpha x + \beta y + \gamma z)} \quad \text{and} \quad e^{i(\alpha x + \beta y - \gamma z)},$$

with γ from (3), are absolutely bounded above by 1 as α, β, x, y, z range over all real values.

The fact that (3) determines γ only but for the addition of an integral multiple of $2\pi/\ell$ does not introduce an ambiguity in the functions (5) as solutions of the difference equation (1) because these are needed only for values of z which are integral multiples of ℓ; at such points the values (5) are unaffected by the mentioned ambiguity of γ.

If α, β, x, y, z are restricted to an arbitrary bounded region, it becomes possible to give an asymptotic evaluation of γ as function of $\alpha, \beta, h, k, \ell$ and accordingly of (5) as h, k, ℓ tend to zero under observance of the inequality (4). We have

$$\gamma = \frac{1}{\ell} \text{ arc sin} \left(\frac{4\ell^2}{h^2} \sin^2 \frac{\alpha h}{2} + \frac{4\ell^2}{k^2} \sin^2 \frac{\beta k}{2} \right)^{1/2}$$

and by an application of Taylor's theorem

$$\gamma = \sqrt{\alpha^2 + \beta^2} \left(1 - \frac{\alpha^4 h^2 + \beta^4 k^2 - (\alpha^2 + \beta^2)^2 \ell^2}{6(\alpha^2 + \beta^2)} + \dots\right).$$

The solution of the initial problem (2) is now written in the form

$$(6) \qquad u = \iint_{-\infty}^{\infty} g(\alpha, \beta)\left(e^{i(\alpha x + \beta y + \gamma z)} + e^{i(\alpha x + \beta y - \gamma z)}\right) d\alpha d\beta,$$

where 2g is the Fourier transform of f(x,y), i.e.,

$$f(x,y) = \iint_{-\infty}^{\infty} 2g(\alpha, \beta) e^{i(\alpha x + \beta y)} d\alpha \, d\beta .$$

We notice that the boundedness of (5) for all real α, β, x, y, z together with the continuity of γ as function of $\alpha, \beta, h, k, \ell$ in any bounded region of the α, β plane as h, k, ℓ tend to zero, imply the convergence of (6) as h, k, ℓ tend to zero, to the solution

$$\iint_{-\infty}^{\infty} g(\alpha, \beta)\left(e^{i(\alpha x + \beta y + \sqrt{\alpha^2 + \beta^2}\, z)} + e^{i(\alpha x + \beta y - \sqrt{\alpha^2 + \beta^2}\, z)}\right) d\alpha d\beta$$

of the differential equation (1).

It remains to give a geometrical interpretation of the condition (4). The difference equation (1) states a relation between the values of u at a point (x_0, y_0, z_0) and at the points

$$(x_0 \pm h, y_0, z_0), \ (x_0, y_0 \pm k, z_0) \text{ and } (x_0, y_0, z_0 \pm \ell).$$

These six points are the corners of an actahedron whose faces are given by one of the following eight equations

$$\pm (x - x_0)/h \pm (y - y_0)/k \pm (z - z_0)/\ell = 1.$$

They intersect the plane $z = z_0$ in the lines

(7) $$\pm (x - x_0)/h \pm (y - y_0)/k = 1.$$

Now pass a cone (the characteristic cone of the differential equation (1)) through $(x_0, y_0, z + \ell)$ of equation

$$(x - x_0)^2 + (y - y_0)^2 = (z - z_0 - \ell)^2.$$

This cone intersects with $z = z_0$ in the circle

$$(x - x_0)^2 + (y - y_0)^2 = \ell^2.$$

Thus the lines (7) avoid the interior of the circle if and only if (4) holds, since their distance from (x_0, y_0, z_0) is

$$\left(\frac{1}{h^2} + \frac{1}{k^2}\right)^{-1/2};$$

and (4) states that the faces of the octahedron intercept the cone only at its vertex.

The above considerations remain applicable if the equation of the cylindrical waves is replaced by any other hyperbolic or parabolic equation with constant coefficients.

Reprinted from Vol. I, PROCEEDINGS OF THE
INTERNATIONAL CONGRESS OF MATHEMATICIANS, 1950
Printed in U.S.A.

DEVELOPMENTS AT THE CONFLUENCE OF ANALYTIC BOUNDARY CONDITIONS[1]

HANS LEWY

1. I wish to discuss two prototypes of a class of problems which are frequently encountered in the applications of analysis to elasticity and hydrodynamics as well as in the study of complex functions defined by geometric properties of the induced conformal mapping. The choice of the two prototypes is personal; I was led to the one example by an investigation of the nature of water waves near a horizontal obstacle floating on the surface of the liquid, but I noticed that the essence of the treatment of this problem applies to a larger group of problems. Roughly speaking, we are dealing with functions of two real variables which satisfy an elliptic analytic differential equation (such as Laplace's) and which, on the boundary of their domain of definition, satisfy two analytic boundary conditions C_1 and C_2, with C_1 holding on a portion of the boundary and C_2 holding on an adjacent portion. The problem is to describe the behaviour of the solution of the differential equation near the point of confluence of the two boundary conditions, according to rising orders of vanishing of functions usable in the development, in a way similar to the development of a function, regular for $z = 0$, with the aid of ascending powers of z.

More precisely the two problems here to be discussed are:

Problem A: The conformal map of a portion of the upper half of a complex $z = x + iy$ plane on a domain of the upper $\zeta = u + iv$ plane, bounded (partially) by a portion of the negative real axis and a portion of a parabola Γ of form $v = 2u^2$, the two meeting at the origin under an angle π so that the tangent of the boundary remains continuous at the origin.

Problem B: The harmonic function $u(x, y)$ is continuous in a neighborhood of the origin for $y \leqq 0$ and satisfies on the negative x-axis the condition $\partial u/\partial y = 0$ and on the positive x-axis the condition $\partial u/\partial y = u$.

Problem B is a linear problem while A is nonlinear. Let $F(z)$ be the complex function which yields the desired conformal mapping of A, resp. the analytic function of the complex argument $z = x + iy$ whose real part is the function u occurring in B. An immediate simple observation tells us that $F(z)$ can in both cases be analytically extended into a neighborhood of the origin of an infinitely leaved Riemann surface with a logarithmic branch point at the origin. It can be proved that $F(z)$ remains continuous as $z \to 0$ provided this is interpreted to mean that $|\operatorname{Im} \log z|$ remains bounded as $z \to 0$.

Denote by z and z^+ points of the Riemann surface such that

$$\log z^+ = \log z + 2\pi i,$$

[1] For proofs and further results consult the author's memoir of the same title, University of California Publications in Mathematics, vol. 1 (1950) pp. 247–280.

and set, for brevity, $F(z) = F$, $F(z^+) = F^+$. For real positive z the defining property of $F(z)$ in A leads to the equation

$$F^+ - F + i(F^+ + F)^2 = 0, \tag{A1}$$

while in case B we find

$$(F^+)' - (F)' + i(F^+ + F) = 0. \tag{B1}$$

These equations must remain valid everywhere on the Riemann surface since they hold for positive z. (B1) is a difference-differential equation connecting the values of the same function (and its derivative) on successive leaves of the Riemann surface. (A1) is a difference equation of the same kind; since z does not enter explicitly, the relation between F, F^+, F^{++}, $\cdots$ is a simple functional iteration in which z plays the rôle of a uniformizing parameter.

2. To formulate briefly the answers to the problems stated above, consider power products of z and $z \log z$. Denote by (m, n) the degree of the product $z^m(z \log z)^n$; write $(m, n) = (m', n')$ if $m = m'$, $n = n'$, and write $(m, n) < (m', n')$ if $z^{m'}(z \log z)^{n'}/[z^m(z \log z)^n]$ tends to 0 as $z \to 0$. As m and n range over the non-negative integers, the degrees (m, n) are thus lined up in the order of their increase. It makes sense to talk of a function $F(z)$ as asymptotically equal to a polynomial of degree (m, n),

$$F(z) \sim \sum_{(0,0)}^{(m,n)} A_{\mu\nu} z^\mu (z \log z)^\nu,$$

if the difference between the left and right sides, divided by $z^m(z \log z)^n$, tends to zero as $z \to 0$. Obviously the last equality implies all those asymptotic relations which are obtained by striking successively from the right side the terms of highest degree. Accordingly a relation $F(z) \sim \sum_{(0,0)}^{\infty} A_{\mu\nu} z^\mu (z \log z)^\nu$ with infinite sum can be interpreted as the infinitude of relations obtained by equating the left side to any section of finite degree on the right, and no statement about convergence of the infinite sum is necessary to make this asymptotic relation meaningful.

Now with this terminology there holds the following theorem:

THEOREM. *Every solution $F(z)$ of problem* A *or* B *is asymptotically equal to an infinite double series in z and z* log *z, whose coefficients are uniquely determined by $F(z)$.*

While I cannot reproduce here the proof of this theorem, I state at least the essential lemma from which it follows.

LEMMA. *Suppose that for positive t and real* log *t a function $F_1(t) \sim M_1(t, t \log t)$, where $M_1(t, t \log t)$ is a polynomial of degree (m, n). Let a be positive and $F_1(t)$*

be bounded and measurable in $0 < t < a$. *Set*

$$F_2(z) = \int_0^a F_1(t) \log (t - z)\, dt$$

with $0 \leqq |z| < a$, $0 \leqq \operatorname{Im} \log z < 2\pi i$, *and with* $\log (t - z)$ *that branch of the function which is real for* $t > z$. *Then there exists a polynomial* $M_2(z, z \log z)$ *of degree* $(m, n + 1)$ *such that*

$$F_2(z) \sim M_2(z, z \log z).$$

3. Returning to the double power series in z and $z \log z$ which is the asymptotic representation of $F(z)$, we shall call it $P(z, z \log z)$ and indicate a simple way of finding its coefficients. First let $F(z)$ be a solution of B. Set formally

$$P(z, z \log z) = \varphi_0(z \log z) + z\varphi_1(z \log z) + z^2\varphi_2(z \log z) + \cdots .$$

Then evidently

$$P(z^+, z^+ \log z^+) = \varphi_0(z \log z + 2\pi i z) + z\varphi_1(z \log z + 2\pi i z) + \cdots .$$

Applying Taylor's series to $\varphi_0(z \log z + 2\pi i z)$, $\varphi_1(z \log z + 2\pi i z)$, $\cdots$, we get

$$P(z^+, z^+ \log z^+) = P(z, z \log z) + 2\pi i z\varphi_0'(z \log z) + 2\pi i z^2\varphi_1'(z \log z) + \cdots ,$$

$$\frac{dP}{dz}(z, z \log z) = (\log z + 1)\varphi_0'(z \log z) + \varphi_1(z \log z)$$

$$+ z(\log z + 1)\varphi_1'(z \log z) + \cdots ,$$

$$\frac{dP}{dz}(z^+, z^+ \log z^+) = (2\pi i + \log z + 1)(\varphi_0'(z \log z) + 2\pi i z\varphi_0''(z \log z) + \cdots)$$

$$+ \varphi_1(z \log z) + \cdots .$$

(B1) begins formally with the terms free of z (i.e., depending only on $z \log z$),

(B1.0) $$2\pi i\varphi_0'(z \log z) + 2\pi i z \log z\varphi_0''(z \log z) + 2i\varphi_0(z \log z).$$

Next come the terms divisible by z with coefficients that are functions of $z \log z$; these terms involve only φ_0 and φ_1 and their derivatives, etc. Setting (B1.0) equal to zero, we see that φ_0, being a power series of $z \log z$, must be proportional to the Bessel function $J_0[((z \log z)/\pi)^{1/2}]$. Similarly φ_1 can be determined as a solution of a linear differential equation whose known members depend on φ_0, found previously. The recursive evaluations of $\varphi_0, \varphi_1, \varphi_2, \cdots$ introduce a constant of integration at each step, which agrees with the fact that there are infinitely many solutions of the proposed problems.

It is of interest that a less formal and more structural relation between $F(z)$ and $\varphi_0 = J_0[((z \log z)/\pi)^{1/2}]$ can be given. Consider a fixed value $\alpha \neq 0$ and the infinitely many solutions of the equation

(S) $$z \log z = \alpha.$$

They may be arranged according to increasing values of $|\operatorname{Im}\log z|$. As $\operatorname{Im}\log z$ tends to ∞ or to $-\infty$, the value z tends necessarily to 0. Let $z_1, z_2, \cdots$ be such a sequence of solutions of (S) with $z \to 0$ and $\operatorname{Im}\log z \to \infty$. Then

$$\lim_{n\to\infty} F(z_n) = \varphi_0(\alpha),$$

$$\lim_{n\to\infty} (F(z_n) - \varphi_0(\alpha))/z = \varphi_1(\alpha),$$

$$\cdots\cdots\cdots\cdots\cdots\cdots\cdots$$

In view of this relationship, I propose to name $\varphi_0, \varphi_1, \cdots$ the "shadows" of $F(z)$ at $z = 0$ of order $0, 1, \cdots$.

Analogous considerations for A lead to the recursive determination of the shadows of $F(z)$ in A. We find without difficulty

$$\varphi_0 = 0,$$

$$\varphi_1 = (c + 2\pi^{-1}z\log z)^{-1}, \qquad c \neq 0 \text{ a constant.}$$

$$\cdots\cdots\cdots\cdots\cdots$$

4. Since with each $F(z)$ is associated a unique double power series $P(z, z\log z)$, the natural question is whether or not there is a way of regaining $F(z)$ when $P(z, z\log z)$ is known. The answer is positive in case B and is covered by the following theorem.

THEOREM. *A solution $F(z)$ of* (B1) *which is regular for $z \neq 0$, $|z| \leqq \rho$, $\rho > 0$, and continuous for $z \to 0$, can be developed in a double power series in z and $z\log z$ which converges uniformly and absolutely to $F(z)$ for all z with $|z| \leqq \rho$ and bounded $|\operatorname{Im}\log z|$.*

The double power series involved is of course the same one which yields the asymptotic development.

A further development of $F(z)$ is based on the observation that if $G_0(z)$ is a solution of (B1) which is continuous for $z \to 0$, then $G_1(z) = \int_0^z G_0(t)\,dt$ is another such solution, and so are $G_2(z) = \int_0^z G_1(t)\,dt, \cdots$. Assume $G_0(0) \neq 0$. The following theorem holds:

THEOREM. *Any $F(z)$, solution of* (B1) *and continuous at the origin, can be developed in a series*

$$F(z) = \sum_{\nu=0}^{\infty} c_\nu G_\nu(z)$$

which converges uniformly and absolutely in a proper concentric subcircle of the common domain of existence of $F(z)$ and $G_0(z)$.

This theorem illustrates the perfect analogy with the developability of a regular function of z in a power series in z, where likewise the functions used are the iterated integrals of the first of them, namely the constant 1.

To prove the quoted theorems I had to devise some new analytic tools which fit the special singularity of the functions $F(z)$. It would take more than the allotted time to describe them in this short address. Unfortunately, these tools depend strongly on the linearity of the equation (B1), and I have no hope of extending them to the case A. I do not know whether $P(z, z \log z)$ converges in case A, and I am inclined to doubt it.

UNIVERSITY OF CALIFORNIA,
BERKELEY, CALIF., U. S. A.

DEVELOPMENTS AT THE CONFLUENCE OF ANALYTIC BOUNDARY CONDITIONS

BY

HANS LEWY

UNIVERSITY OF CALIFORNIA PRESS
BERKELEY AND LOS ANGELES
Vol. 1, No. 7, 1950

DEVELOPMENTS AT THE CONFLUENCE OF ANALYTIC BOUNDARY CONDITIONS

By

HANS LEWY

IN THE THEORY of partial differential equations and the disciplines which borrow their tools from it, it has been customary to approach the determination of the solution of boundary-value problems from two opposed points of view: either the data at hand are supposed to possess a maximum of analyticity, or the conditions of regularity of the data are weakened as far as possible and compatible with existence or uniqueness of the solution. The first approach offers the advantage of the availability of power-series developments which not only lend themselves to the investigation of local properties of the solution, but often reduce the problem of the construction of the solution to the determination of coefficients from recursion formulae. The second approach, however, often lays bare the deeper topological principles on which existence and uniqueness of a solution must rest, once the help of formal developments is no longer available.

The first approach is open to the criticism that the data of the problem and the nature of its solution are restricted not by the structure of the problem, but by that of the method used for its solution. The second approach, on the other hand, can frequently be criticized for the very generality of the results, the descent from which to the properties of special cases often remains a problem comparable to, or greater than, that of the establishment of the general solution.

There is, however, in the opinion of the author, a large class of problems, and one of increasing importance in the applications, which escapes the first treatment because the data are not totally analytic, but for which the second treatment is too general because the data can be termed "piecewise analytic." An example from the theory of conformal mapping of the upper half plane on a domain bounded by a curve Γ will illustrate the situation. Suppose Γ to be composed of two curves Γ_1 and Γ_2, both analytic, and meeting under an angle π at a point P which in the mapping corresponds to the origin of the first plane. How does the mapping function behave in the neighborhood of the origin? The important work of O. D. Kellogg and Stefan Warschawski† gives a partial answer by asserting that the mapping function will have a continuous derivative at the origin, because the assumption implies a Lipschitz condition for the tangent of Γ near P. But the assumption that Γ_1 and Γ_2 are both analytic up to the point P does not yield improvement of the result.

In this paper we shall prove, however, that it is possible to obtain an infinite asymptotic development for the mapping function near the origin. This development is a formal double power series in the two arguments z and $z \log z$. This result is proved in § **3**.

The example just given is a special case of what we shall call development at the

† O. D. Kellogg, *Harmonic functions and Green's integral*, Trans. Am. Math. Soc., 13 (1912), 109–132. A theorem with a localized condition is found in Stefan Warschawski, *Ueber das Randverhalten der Ableitung der Abbildungsfunktion bei conformer Abbildung*, Math. Zeitschr., 35 (1932), 321–456. The theorem is contained in a "Zusatz" on page 447.

confluence of two analytic boundary conditions. This paper deals in extenso also with a second boundary problem in which a harmonic function $u(x, y)$ of the following kind is considered: $u(x, y)$ is supposed to exist for $y < 0$ in the neighborhood of the origin and to assume continuous boundary values on the x-axis, such that on the negative axis the normal derivative is zero while on the positive axis the condition holds: $\partial u/\partial y = au$ with analytic $a(x)$. This special problem, of interest in the theory of surface waves, is taken up because it is the simplest type of a linear problem with a confluence of analytic boundary conditions. Our result in this case is stronger than in the first example, for now a development for the analytic function F of $z = x + iy$ whose real part is $u(x, y)$, can again be obtained as a power series in z and $z \log z$, but this power series is not only an asymptotic estimate but is shown to converge and equal the function $F(z)$.

While the asymptotic representations in both examples mentioned are consequences of theorem 1 which deals with a simple singular integral equation, the sharper result valid for the second example made necessary a detailed study of the behavior of $F(z)$ in the logarithmic Riemann surface with the origin as branch point. It is shown that $F(z)$ satisfies a certain differential-difference equation which connects function and derivative at points z having the same base point in the ordinary plane, but belonging to successive sheets of the Riemann surface. A fundamental tool is the establishment of an integral invariant of any two solutions of this equation.

The two examples treated here are special cases of a very general situation. Although the investigation has been pushed considerably further, it would have been unwieldy to present all the material in one paper. Instead an attempt has been made to point up those aspects of the problem which are capable of generalization to other cases. A further communication on the subject is intended.

§ 1. **Notations.** We consider a complex variable z ranging over a Riemann surface on which the origin is a logarithmic branchpoint. By z^+, respectively z^-, we denote the same point as z, located on the next respectively preceding sheet of the Riemann surface, i.e. such that

$$\log z^+ = \log z + 2\pi i, \qquad \text{resp.} \qquad \log z^- = \log z - 2\pi i\,.$$

Consider power products of z and $z \log z$, say $z^m(z \log z)^n$, where m and n are nonnegative integers. It is possible to arrange these power products in a natural order analogous to the usual ordering of the powers of z according to increasing degree. Call (m, n) the degree of $z^m(z \log z)^n$. We have

$$z^{m'}(z \log z)^{n'}/[z^m(z \log z)^n] \to 0 \quad \text{as} \quad z \to 0\dagger)\,,$$

provided either

$$m + n < m' + n'$$

or

$$m + n = m' + n' \quad \text{and} \quad n' < n\,.$$

† $z \to 0$ is always understood with the restriction: $|\operatorname{Im} \log z|$ is bounded as $z \to 0$.

If one of these two conditions holds, we shall say that the degree (m, n) is less than (m', n'). We can now speak of polynomials in z and $z \log z$ of degree (m, n), as meaning a sum of power products of the above form, of degree inferior or equal to (m, n), with constant coefficients; observe that there are finitely many power products of degree less than a given one; they can be arranged in the order of increasing degree. Similarly there is an obvious interpretation of an asymptotic relation

$$F(z) \sim \sum A_{mn} z^m (z \log z)^n,$$

whether the sum is finite or not, as signifying the several limit relations arranged according to increasing degree:

$$\begin{aligned} &F(z) - A_{00} \to 0, \\ &[F(z) - A_{00} - A_{01} z \log z]/(z \log z) \to 0, \\ &[F(z) - A_{00} - A_{01} z \log z - A_{10} z]/z \to 0, \\ &\cdots\cdots\cdots\cdots\cdots\cdots, \end{aligned}$$

as $z \to 0$.

The purpose of this section is the proof of the following:

THEOREM 1. Suppose that for positive t, $F(t) \sim M(t, t \log t)$, where M is a polynomial of degree (m, n). Let $A > 0$ be a constant and $F(t)$ bounded and measurable in $0 < t < A$, and set

$$F_1(z) = \int_0^A F(t) \log (t - z)\, dt.$$

Then we have, for arbitrary z,

$$F_1(z) \sim M_1(z, z \log z),$$

where M_1 is a polynomial of degree $(m, n + 1)$.

In this theorem the integration extends over positive values, $\log (t - z)$ is understood as that branch of the function whose imaginary part is zero for negative z and positive t, and the range of z to be considered is $0 \leqq |z| < A$, $0 \leqq \operatorname{Im} \log z \leqq 2\pi i$.

The proof of this theorem requires several lemmas.

LEMMA 1. Let $F(t) = t^k (\log t)^n$, $F_1(z) = \int_0^A F(t) \log (t - z) dt$, $0 < A$, $0 \leqq k$, $0 \leqq n$. There exists a polynomial $p(z)$ of degree $\leqq k$ such that, for $z \to 0$,

$$F_1(z) - p(z) - z^{k+1} (\log z)^{n+1}/[(k + 1)(n + 1)] = o(z^{k+1} (\log z)^{n+1}).\dagger$$

Proof: For positive z with $\log z$ real, $0 < z < A$, we conclude

$$F_1(z^+) - F_1(z) = 2\pi i \int_0^z t^k (\log t)^n\, dt$$

† k, n are integers; $o(g(z))$ denotes, as usual, a quantity, which, divided by $g(z)$, tends to zero as $z \to 0$.

$$= [2\pi i/(k+1)]\{z^{k+1}(\log z)^n - n\int_0^z t^k(\log t)^{n-1}\,dt\}$$

$$= z^{k+1}\,Q(\log z)\,,$$

where Q is a polynomial of degree n. Now consider the expression

$$G(z) = z^{k+1}Q_1(\log z)\,,$$

where Q_1 is a polynomial of degree $n+1$, satisfying the difference equation

$$Q_1(\log z + 2\pi i) - Q_1(\log z) = Q(\log z)\,.$$

The coefficients of Q_1 can be found as solution of linear recursion formulae which determine Q_1 uniquely except for an additional constant. The function $\gamma(z) = F_1(z) - G(z)$ satisfies accordingly the condition $\gamma(z^+) = \gamma(z)$, for positive z in $0 < z < A$. Furthermore, $\gamma(z)$ is regular for $0 < |z|, 0 < \operatorname{Im}\log z < 2\pi i$, and continuous as z approaches a point of the positive real axis. Therefore $\gamma(z)$ is a single-valued regular function, and it remains bounded as $z \to 0$. Hence $\gamma(z)$ is a power series in z and we can write

$$\gamma(z) = p(z) + z^{k+1}\gamma_1(z)$$

with $p(z)$ a polynomial of degree $\leqq k$ and $\gamma_1(z)$ a power series. On the other hand, we gather from the difference equation for Q_1 that the highest coefficient of Q_1 is $[(k+1)(n+1)]^{-1}$; thus

(1.1) $$F_1(z) = p(z) + z^{k+1}(\log z)^{n+1}/[(k+1)(n+1)] + z^{k+1}\gamma_1(z) + z^{k+1}Q_2(\log z)\,,$$

where Q_2 is a polynomial of degree $\leqq n$. This equation implies the lemma to be proved.

Lemma 2. $\int_0^A (\log t)^n \log(t-z)dt = a + R(z)$ where

$$R(z)/[z(\log z)^{n+1}] \to (n+1)^{-1}$$

as $z \to 0$, and

$$a = \int_0^A (\log t)^{n+1}dt\,.$$

This is a special case of lemma 1.

Lemma 3. There exists a number N, such that

$$q(\alpha) \equiv |\log(1-\alpha)|/|\log(1-|\alpha|)| < N$$

for all complex α with $|\alpha| \neq 1$. Here it is assumed that Im (log) is between the limits $-\pi$ and $+\pi$ inclusive.

Proof: $q(\alpha)$ could tend to ∞ only if for a subsequence of α's one of the following three alternatives held: $|\alpha| \to \infty$ or $\alpha \to 1$ or $|\alpha| \to 0$. In the first two cases the numerator of q would tend to ∞, in the third case the denominator would tend to zero. We show that in no case does q tend to ∞. If $|\alpha| \to 1$, then

$$\begin{aligned}\log(1-\alpha) &= \log\alpha + \log(1/\alpha - 1)\\ &= \log|\alpha| + i\theta \pm i\pi - 1/\alpha - \cdots,\\ \log(1-|\alpha|) &= \log|\alpha| \pm i\pi - 1/\alpha - \cdots;\end{aligned}$$

but $\log|\alpha| \to \infty$, so that $q \to 1$ as $|\alpha| \to \infty$. Next, if $\alpha \to 1$, then with $|\theta| \leqq 2\pi$,

$$|1-\alpha| \geqq |1-|\alpha||, \ |\log(1-|\alpha|)| \geqq |\log|1-\alpha||$$

and

$$|\log|1-\alpha|| \to \infty.$$

But then

$$q = |(\log|1-\alpha| + i\theta)|/|\log(1-|\alpha|)| \leqq 1 + 1 = 2,$$

if α is close enough to 1. Finally if $\alpha \to 0$, we have

$$q = |-\alpha - \cdots|/|-|\alpha| - \cdots| \to 1.$$

Thus q cannot tend to ∞ and the existence of N has been proved.

LEMMA 4. Let $\beta(t)$ be measurable and absolutely bounded by $\epsilon > 0$ for $0 < t < A'$. Let $0 < z < A'$ and $A' < 1$. There exists a constant C depending only on the non-negative integer n such that

$$(1.2)\qquad \left|\int_0^{A'} \beta(t)(\log t)^n(\log(t-z) - \log t)dt\right| < C\epsilon|z(\log z)^{n+1}|$$

for z sufficiently small, e.g. $z < 1/e$.

From lemma 2 we have for small positive z

$$\left|\int_0^{A'} (\log t)^n(\log(t-z) - \log t)dt\right| \leqq 2z|\log z|^{n+1}.$$

Thus for small positive z in view of $0 < A' < 1$,

$$\left|\int_z^{A'} \beta(t)(\log t)^n(\log(t-z) - \log t)dt\right|$$

$$\leqq \epsilon \left| \int_z^{A'} (\log t)^n (\log (t - z) - \log t) dt \right|$$

$$\leqq \epsilon \left(\left| \int_0^{A'} \right| + \left| \int_0^z \right| \right)$$

$$< 2\epsilon z \, |\log z|^{n+1} + \epsilon \left| \int_0^z \right| .$$

Furthermore,

$$\left| \int_0^z \beta(t) (\log t)^n (\log (t - z) - \log t) dt \right|$$

$$\leqq \epsilon \left| \int_0^z (\log t)^n |\log (t - z) - \log t| dt \right|$$

$$\leqq \epsilon z \left| \int_0^1 (\log z + \log t)^n | \log (t - 1) - \log t| dt \right|$$

$$\leqq C_1 \epsilon z \, |\log z|^{n+1},$$

where C_1 depends only on n. Consequently

$$\left| \int_0^{A'} \beta(t) (\log t)^n (\log (t - z) - \log t) dt \right| \leqq (2\epsilon + 2C_1\epsilon) z \, |\log z|^{n+1},$$

q.e.d.

LEMMA 5. The inequality (1.2) of lemma 4 holds for complex z if $|z|$ is small and $0 \leqq \arg z \leqq 2\pi$.

Proof: By lemma 3 for $\alpha = z/t$,

$$\left| \int_{|z|}^{A'} \beta(t) (\log t)^n (\log (t - z) - \log t) dt \right|$$

$$\leqq \epsilon \left| \int_{|z|}^{A'} (\log t)^n (\log (t - |z|) - \log t) dt \right| N$$

$$\leqq \left(2C\epsilon |z| \, |\log |z||^{n+1} + \epsilon \left| \int_0^{|z|} \right| \right) N ;$$

$$\left| \int_0^{|z|} \beta(t) (\log t)^n (\log (t - z) - \log t) dt \right|$$

$$\leqq \epsilon \left| \int_0^{|z|} (\log t)^n N (\log (t - |z|) - \log t) dt \right| \leqq C_1 N \epsilon |z| \, |\log |z||^{n+1} .$$

Hence, as before,

$$\left| \int_0^{A'} \beta(t)(\log t)^n(\log (t - z) - \log t)dt \right|$$

$$\leqq N(2C\epsilon + 2C_1\epsilon)|z| \, |\log |z| \, |^{n+1}$$

$$\leqq 2N\epsilon(C + C_1)|z(\log z)^{n+1}|,$$

and lemma 5 follows.

LEMMA 6: Let $\Phi(z)$ be a complex-valued function defined on a ray $\arg z = h$, $0 \leqq h \leqq 2\pi$, for small $|z|$, and suppose that $\Phi(z)$ admits continuous derivatives up to the s-th order. Then there is a θ with $0 < \theta < 1$ such that

$$|\Phi(z) - \Phi(0) - z\Phi'(0) - \cdots - z^s\Phi^{(s)}(0)(s\,!)^{-1}| \leqq |z^s[\Phi^{(s)}(\theta z) - \Phi^{(s)}(0)](s\,!)^{-1}| .$$

Proof: Taylor's formula with the integral remainder yields

$$\Phi(z) - \Phi(0) - z\Phi'(0) - \cdots - z^{s-1}\Phi^{(s-1)}(0)\,((s-1)\,!)^{-1}$$

$$= \int_0^z (z - t)^{s-1}\Phi^{(s)}(t)dt\,((s-1)\,!)^{-1}.$$

Now

$$\left| \int_0^z (z - t)^{s-1}(\Phi^{(s)}(t) - \Phi^{(s)}(0))dt \right| \leqq |z|^s|\Phi^{(s)}(\theta z) - \Phi^{(s)}(0)|/s,$$

whence the lemma.

LEMMA 7. Let $0 < A' < 1$, $\beta(t)$ measurable and absolutely $<\epsilon$ for $0 < t < A'$, and set, with non-negative integers s and n,

$$\Phi_s(z) = \int_0^{A'} \beta(t)(\log t)^n(z - t)^s \log (t - z)dt .$$

There exists a polynomial $p_s(z)$ of degree $\leqq s$, such that for $0 \leqq \arg z \leqq 2\pi$, $|z|$ small,

$$|\Phi_s(z) - p_s(z)| \leqq \epsilon C|z^{s+1}(\log z)^{n+1}|,$$

where the constant C depends only on s and n.

Proof: We have

$$\Phi_s^{(s)}(z) = s\,! \int_0^{A'} \beta(t)(\log t)^n \log (t - z)dt + \text{const},$$

whence, by the preceding lemma,

$$|\Phi_s(z) - p_s(z)| \leqq |z|^s \left| \int_0^{A'} \beta(t)(\log t)^n(\log(t - z\theta) - \log t)dt \right| ,$$

where $p_s(z)$ is a polynomial of degree $\leqq s$ and $0 < \theta < 1$. (Observe that the coefficients of p_s do not depend on arg z).
We now apply lemma 5 and obtain

$$|\Phi_s(z) - p_s(z)| \leqq |z|^{s+1}\theta|\log \theta z|^{n+1}C\epsilon .$$

Next observe that the ratio $\theta|\log \theta z|^{n+1}/|\log z|^{n+1}$ is bounded for $|z| < e^{-1}$ and $0 < \theta < 1$ by some constant $N'(n)$ since this is also true for $\theta(\log \theta/\log|z|)^k$ with arbitrary non-negative $k \leqq n$. The lemma follows.

We are now in a position to prove the main theorem of this section. By hypothesis we have $F(z) - M(z, z\log z) = o(z^m(z\log z)^n)$, or $F(z) - M(z, z\log z) = \beta(z)z^m(z\log z)^n$ with $|\beta(z)| < \epsilon$, provided $|z| < A'(\epsilon) < 1$. We can estimate as follows:

$$t^{m+n} = (z - (z - t))^{m+n},$$

$$\int_0^{A'} \beta(t)t^m(t\log t)^n \log(t - z)dt = \sum_{s=0}^{m+n} c_s z^{m+n-s}\Phi_s(z) ,$$

$$\left| \int_0^{A'} \beta(t)t^m(t\log t)^n \log(t - z)\,dt - \sum_{s=0}^{m+n} c_s z^{m+n-s}p_s(z) \right| \leqq \sum_0^{m+n} |c_s|\epsilon C|z^{m+n+1}(\log z)^{n+1}| ,$$

or

$$\left| \int_0^{A'} \beta(t)t^m(t\log t)^n \log(t - z)dt - p(z) \right| \leqq \epsilon C'|z^{m+n+1}(\log z)^{n+1}| ,$$

where $C' = C'(m, n)$ and $p(z)$ is a polynomial of degree $\leqq m + n$.

From (1.1) we gather that

$$\int_0^{A'} M(t, t\log t) \log(t - z)dt$$

is a polynomial in z and $z\log z$, plus a power series in z. Finally, in case $A > A'(\epsilon)$, we know that for $|z| < A'$,

$$\int_{A'}^{A} F(t) \log(t - z)dt$$

is a power series in z. Combining the last three facts we see that $F_1(z)$ equals a polynomial $M_1(z, z\log z)$ of degree $(m, n + 1)$ with an error of at most $2\epsilon C'|z^{m+n+1}(\log z)^{n+1}|$, provided $|z|$ is small enough. Here ϵ is an arbitrary positive number. Let us make sure that the coefficients of M_1 are independent of ϵ.

Put

$$M_1(z, z\log z) = B_{00} + B_{01}z\log z + B_{10}z + \cdots + B_{mn+1}z^m(z\log z)^{n+1}$$

where the B are permitted to depend on ϵ, and consider the inequality

$$(1.3) \qquad |F_1(z) - M_1(z, z\log z)| \leqq 2\epsilon C' |z^m(z\log z)^{n+1}| .$$

We find, upon letting $z \to 0$, that $B_{00} = F_1(0), B_{01} = \lim (F_1(z) - B_{00})/(z\log z), \cdots$. Thus all coefficients except B_{mn+1} are uniquely determined by the function $F_1(z)$ alone. In order to see that also B_{mn+1} can be uniquely determined by $F_1(z)$, divide the inequality (1.3) by $|z^m(z\log z)^{n+1}|$ and get

$$(1.4) \qquad |F_2(z)/[z^m(z\log z)^{n+1}] - B_{mn+1}(\epsilon)| \leqq 2\epsilon C'$$

where† $F_2(z) = F_1(z) - B_{00} - B_{01}z\log z - \cdots - B_{m'n'}z^{m'}(z\log z)^{n'}$, and B_{mn+1} may conceivably depend on ϵ. Now take another value of z, say $\tilde{z}$, with sufficiently small $|\tilde{z}|$ and apply to it the same inequality (1.4). Addition yields

$$|F_2(z)/[z^m(z\log z)^{n+1}] - F_2(\tilde{z})/[\tilde{z}^m(\tilde{z}\log \tilde{z})^{n+1}]| \leqq 4\epsilon C',$$

provided $|z|$ and $|\tilde{z}|$ are sufficiently small. Hence Cauchy's criterium yields the existence of $\lim F_2(z)/[z^m(z\log z)^{n+1}]$. If we define B_{mn+1} as this limit, we easily gather from (1.4) that $|B_{mn+1} - B_{mn+1}(\epsilon)| \leqq 2\epsilon C'$. Thus we have found for a certain polynomial M_1 whose coefficients are independent of ϵ, that for all $\epsilon > 0$,

$$|F_1(z) - M_1(z, z\log z)| \leqq 4\epsilon C' |z^m(z\log z)^{n+1}| ,$$

provided $|z|$ is small enough. This is the statement of the theorem.

§ 2. **Rules of computation.** In order to apply the result of the preceding section we state a few rules of computation with polynomials in z and $z\log z$.

The degree of the product of a polynomial of degree (r, s) and a polynomial of degree (m, n) is $(r + m, s + n)$.

Proof: This follows once it has been shown that the term of highest degree in the product is obtained by multiplication of the highest terms of the two polynomials. Hence it suffices to show that the degree of any term of the product remains below that of the highest two terms if at least one of the factors has a degree inferior to the highest. In other words it suffices to show:

$$(2.1) \qquad (r', s') < (r, s) \text{ and } (m', n') \leqq (m, n)$$

implies $(r' + m', s' + n') < (r + m, s + n)$. Now we either must have both inequalities $r' + s' < r + s$, $m' + n' \leqq m + n$, or we have $r' + s' = r + s$, $s' > s$ and $m' + n' < m + n$, or we have $r' + s' = r + s$, $s' > s$, and $m' + n' = m + n$, $n' \geqq n$. In the first two cases we find

$$r' + m' + s' + n' < r + m + s + n ;$$

in the third case

$$r' + m' + s' + n' = r + m + s + n, \qquad s' + n' > s + n. \quad \text{q.e.d.}$$

† (m', n') here stands for the highest degree $< (m, n + 1)$.

Notation: If in a polynomial $M(z, z \log z)$ all terms of degree $< (k, l)$ have a vanishing coefficient, then (k, l) is called a grade of M.

Rule: Let $F_1(z) \sim M(z, z \log z)$ and $F_2(z) \sim M'(z, z \log z)$ where M has the grade (k, l), the degree (m, n) and where M' has the grade (k', l') and the degree (m', n'). Then there is a polynomial P such that

$$F_1(z)F_2(z) \sim P(z, z \log z),$$

where grade of P is the sum of the grades, i.e. $(k + k', l + l')$, and degree of P is the minimum of $(k + m', l + n')$ and $(k' + m, l' + n)$.

The proof of this rule is obvious.

From this rule derives the following one by induction: If $\zeta \equiv F(z) \sim M(z, z \log z)$ where M is of grade (k, l) and degree (m, n), then the positive integral powers of ζ have similar asymptotic relations

$$\zeta^j \sim M'(z, z \log z)$$

where M' is of grade (jk, jl) and degree $(m + (j - 1)k, n + (j - 1)l)$.

THEOREM 2. If $\eta = G(\zeta)$ where $G(\zeta)$ is a convergent power series for small $|\zeta|$, $G(0) = 0$, and if $\zeta \sim M(z, z \log z)$ with M of grade (k, l) and degree (m, n) and if $(k, l) > (0, 0)$, then η likewise has an asymptotic expansion $\eta \sim M''(z, z \log z)$, where grade of M'' is $\geqq (k, l)$ and the degree is (m, n).

Proof: In view of the preceding rule, there is a smallest positive integral j_0 such that $\zeta^{j_0} \sim M'(z, z \log z)$ and M' is of grade $> (m, n)$. Write $\eta = a_1 \zeta + a_2\zeta^2 + \cdots + (a_{j_0} + \alpha)\zeta^{j_0}$ where $\alpha \to 0$ as $\zeta \to 0$. Now each of the terms $a_1\zeta, \cdots, a_{j_0-1}\zeta^{j_0-1}$ permits an asymptotic representation of degree (m, n) or higher, hence their sum certainly does, too. Since $(k, l) > (0, 0)$, we have $\zeta \to 0$ as $z \to 0$. But the last term is absolutely less than a constant times $|z^{j_0k}(z \log z)^{j_0l}|$ as $z \to 0$, and this is by assumption of degree $> (m, n)$, hence certainly $o(z^m (z \log z)^n)$. Consequently, η has a representation of degree (m, n) or higher, and grade of M'' is $\geqq$ the minimum of the grades of the individual terms, thus $\geqq (k, l)$.

COROLLARY. Let $\zeta \sim M(z, z \log z)$ where M is an infinite power series in z and $z \log z$ whose absolute term is 0, and let $\eta = G(\zeta)$ be a convergent power series of ζ with $G(0) = 0$. Then η likewise has an infinite asymptotic representation $\eta \sim M' (z, z \log z)$.

Proof: A relation $\zeta \sim M(z, z \log z)$ with M an infinite series is equivalent to infinitely many relations $\zeta \sim M_{ij}(z, z \log z)$ where the M_{ij} are the successive polynomial sections of the infinite series M. From the previous theorem we deduce corresponding infinitely many polynomial representations $\eta \sim M_{ij}'(z, z \log z)$; we order them according to increasing degree. These M_{ij}' are necessarily sections of an infinite power series in z and $z \log z$ since the successive coefficients B_{mn} of these polynomials are obtainable, as we have seen (p. 249), from $\eta = G(\zeta(z))$ by passage to the limit $z \to 0$, hence are uniquely determined by the degree (m, n) of the power product which they multiply. This completes the proof of the corollary.

If one adds a convergent power series $\delta(z)$ to a function having an asymptotic representation of degree (m, n), one obtains a sum which likewise has an asymptotic representation of degree (m, n). For $\delta(z)$ has an asymptotic representation of arbitrarily high degree, hence certainly of degree (m, n), and the sum of two functions having a representation of same degree (m, n) is a function with the same property.

If $z = cZ(1 + Zp(Z))$, where $c \neq 0$ and $p(Z)$ is a convergent power series for small $|Z|$, then $\log z = \log Z + P(Z)$ where $P(Z)$ has the same description as $p(Z)$. We have $\lim z^m(z \log z)^n/[Z^m(Z \log Z)^n] = c^{m+n}$ as z or Z tend to zero. A polynomial $M(z, z \log z)$ becomes a convergent double power series in Z and $Z \log Z$, and $o(z^m(z \log z)^n) = o(Z^m(Z \log Z)^n)$. Hence an asymptotic estimate of degree (m, n) in z and $z \log z$ yields an estimate of same degree in Z and $Z \log Z$. The converse holds since a similar relation $Z = cz(1 + zp(z))$ expresses Z in terms of z.

If a function $F(z)$ has an asymptotic representation by an infinite power series in z and $z \log z$, then the integral $\int_0^z F(t)dt$ has likewise an infinite representation whose coefficients are obtainable from the termwise integration of the series of $F(z)$. This is immediately concluded from the fact that integration of $o(z^k)$ yields $o(z^{k+1})$.

§ 3. A theorem on conformal mapping. Consider a univalent conformal mapping $F(z)$ of a semineighborhood $y > 0$ of the origin of a $z = x + iy$ plane into a semineighborhood of the origin of a $\zeta = u + iv$ plane, such that $z = 0$ goes into $\zeta = 0$, a portion $0 \leqq x \leqq A$ of the positive x-axis into a curve Γ_1, a portion $0 \geqq x \geqq -A$ of the negative x-axis into a curve Γ_2, where Γ_1 and Γ_2 are respectively given by equations

$$\Gamma_1 : \quad v = u^2 f_1(u), \qquad u \geqq 0,$$
$$\Gamma_2 : \quad v = u^2 f_2(u), \qquad u \leqq 0,$$

in which f_1 and f_2 are convergent power series in u for sufficiently small $|u|$.

THEOREM 3.

$$F(z) \sim \sum A_{mn} z^m (z \log z)^n, \tag{3.1}$$

where the sum on right hand is infinite.

Proof: There is no loss of generality in assuming $f_2(u) \equiv 0$. In fact, let $U > 0$ be a real number such that $f_2(u)$ converges for $-U \leqq u \leqq U$ and let $\lambda = G(\zeta)$ be a conformal mapping for which the curve $v = u^2 f_2(u)$ goes into the real λ-axis so that origin corresponds to origin, $0 < u < U$ goes into a portion of the positive axis and $0 > u > -U$ into a portion of the negative axis of the λ-plane. There exists such a mapping in which G is an analytic function of ζ for sufficiently small $|\zeta|$. The image in G of the analytic curve Γ_1 is again an analytic curve of the λ-plane of analogous description relative to the λ-plane as Γ_1 has relative to the ζ-plane. The map from the z-plane into the ζ-plane can be obtained by first mapping onto the λ-plane by $\lambda = H(z)$ and superposing afterwards the inverse $G^{-1}(\lambda)$ of $G(\zeta)$. $G^{-1}(\lambda)$ is a power series in λ. Suppose we know

$$H(z) \sim \sum B_{mn} z^m (z \log z)^n,$$

then we find for $F(z)$

$$F(z) \sim G^{-1}(H(z)) .$$

From the corollary of theorem 2 follows that $F(z)$ likewise has an asymptotic representation (3.1).

We shall suppose, therefore, that Γ_2 is a segment of the negative real ζ-axis and drop the suffix from $f_1(u)$ and Γ_1. Restricting ourselves to a sufficiently small neighborhood of $z = 0$, we see that the slope of the image in $F(z)$ of the real z-axis satisfies a Lipschitz condition in terms of the arc length of the image. A theorem by Kellog and Warschawski† states that the mapping function $F(z)$ has, on the axis, a continuous derivative $F'(z)$.

Consider, with $F(z) = \zeta = u(x, y) + iv(x, y)$, the analytic function of z for $y > 0$,

$$\psi(z) = -\pi^{-1}\int_0^A \log(t - z)\partial v(t, 0)/\partial t\, dt$$

where $\log(t - z)$ is an analytic continuation with respect to z of that branch which is real for $z < 0$. We extend $\psi(z)$ analytically through the negative axis and obtain easily for positive z by passage to the limit

$$\psi(z^+) - \psi(z) = -\pi^{-1}\int_0^z 2\pi i\partial v(t, 0)/\partial t\, dt$$

$$= -2i\, v(z, 0)\, .$$

But we similarly have

$$F(z^+) - F(z) = -2i\, v(z, 0)$$

so that the difference $\delta(z) = F(z) - \psi(z)$ satisfies for positive z the condition $\delta(z^+) = \delta(z)$. This difference, like its two terms, is bounded on approach to $z \to 0$. It is therefore an analytic function of z and developable in a power series in z.

We now make use of relation between u and v on $t > 0$ and find

$$\psi(z) = -\pi^{-1}\int_0^A \log(t - z)(2f(u) + uf'(u))u\, \partial u/\partial t\, dt\, . \tag{3.2}$$

Here $\partial u/\partial t$ possesses, by the theorem of Kellog-Warschawski, a limit as $z > 0$ tends to zero. We express this by saying that, for positive t, $\partial u/\partial t$ has an asymptotic representation of degree (0, 0). Similarly, $u(t, 0)$ has a representation of degree (1, 0) and grade (1, 0) and $2f(u) + uf'(u)$ has a representation of degree (1, 0) and grade (0, 0). By the multiplication rule of section 2 the expression $(2f(u) + uf'(u))u\, \partial u/\partial t$ is representable of grade (1, 0) and degree (1, 0). By theorem 1, $\psi(z)$ has an asymptotic representation of degree (1, 1), and, since $\delta(z)$ is regular, the same holds for $F(z) = \psi(z) + \delta(z)$, and a fortiori for $u(t, 0)$ as $t \to 0$. In order to utilize this result by reinserting it in (3.2), we must first obtain a better asymptotic estimate for $\partial u/\partial t$.

LEMMA. $\partial u/\partial t$ has a representation of degree (0, 1). To show this we prove that $F'(t)$ whose real part is $\partial u/\partial t$, has such representation.

Consider an analytic reflection on the curve $v = u^2f(u)$ of the part of the ζ-plane for which $v > u^2f(u)$. Call ζ^* the image of ζ and observe that the complex conjugate

† Warschawski, *loc. cit.*

of ζ^* is a function $H(\zeta)$, regular for sufficiently small $|\zeta|$. We now extend the function $F(z)$ analytically across the positive real axis by setting $F(\bar{z}) = (F(z))^* = \overline{H}(F(z))$. Observe that this extension defines $F(z)$ for all z of small modulus and arg z between 0 and $-\pi$. We have seen that for arg z between 0 and π,

$$F(z) \sim \cdots + A_{11}z^2 \log z \tag{3.3}$$

hence, because of the analyticity of $H(\zeta)$ near $\zeta = 0$ and $H(0) = 0$,

$$H(F(z)) \sim \cdots + B_{11}\, z^2 \log z\,,$$

$$F(\bar{z}) = \overline{H(F(z))} \sim \cdots + B_{11}\, \bar{z}^2 \log \bar{z}\,.\dagger$$

Thus the asymptotic relation for $F(z)$ holds for arg z between $-\pi$ and π. Now let $\epsilon > 0$ be given. If $t > 0$ is sufficiently small then we know that for a circle C_t of radius t about the center t we have

$$F(z) = F_{11}(z) + R(z)$$

where $|R(z)| < \epsilon\, |z^2 \log z|$ and $F_{11}(z)$ is a polynomial in z and $z \log z$ of degree (1, 1). Thus, by Cauchy's formula,

$$2\pi i\, F'(t) = \int_{C_t} (F_{11}(z) + R(z))(z - t)^{-2}\, dz\,.$$

Here evidently

$$2\pi i\, F_{11}'(t) = \int_{C_t} F_{11}(z)(z - t)^{-2} dz\,,$$

and

$$\left| \int_{C_t} R(z)(z - t)^{-2}\, dz \right| < \int_{C_t} \epsilon |\log z|\, |z/(z - t)|^2\, |dz|\,.$$

Now on C_t we have $z = t + te^{i\theta}$, $\quad |dz| = t|d\theta|\,,\quad |z - t| = t\,, \quad |z| \leqq 2t$,

$$\left| \int_{C_t} R(z)(z - t)^{-2} dz \right| < 4\epsilon t \int_{-\pi}^{\pi} (|\log t| + |\log (e^{i\theta} + 1)|) d\theta < N\epsilon t\, |\log t|$$

where N is a constant. But this means since $\epsilon > 0$ was arbitrary, that

$$F'(t) \sim F_{11}'(t)$$

where $F_{11}'(t)$ is a polynomial of degree (0, 1). This is the lemma.

We now reinsert the asymptotic estimates for u, $2f(u) + uf'(u)$, and $\partial u/\partial t$ in (3.2). Their degrees are resp. (1, 1), (1, 1), (0, 1), and the grades (1, 0), (0, 0), (0, 0). It follows that the product $(2f(u) + uf'(u))\partial u/\partial t$ has an estimate of degree (0, 1) and

† This polynomial in $\bar{z}$, $\bar{z} \log \bar{z}$ has the same coefficients as in (3.3) since both agree for $z > 0$.

grade (0, 0). If we now multiply by u, we obtain an estimate of degree $\min[(1 + 0, 0 + 1), (0 + 1, 0 + 1)] = (1, 1)$.

We now apply this estimate to (3.2). We find by the theorem 1 that $\psi(z)$ possesses an estimate of degree (1, 2) and the same holds for $F(z) = \psi(z) + \delta(z)$. An argument similar to the one just given will now show that $\partial u/\partial t$ possesses an estimate of degree (0, 2), while of grade (0, 0). On the other hand $(2f(u) + uf'(u))$ has an estimate of degree identical with that of u, and of grade (0, 0). Thus

$$E = u(2f + uf')\partial u/\partial t$$

has an estimate of degree (1, 2) and grade (1, 0). Hence $u(t)$ and $F(z)$ have estimates of degree (1, 3). Generally let F and u be estimated of degree $(1, n)$. Then $\partial u/\partial t$ has an estimate of degree $(0, n)$ and $(2f + uf')$ has one of degree $(1, n)$. The respective grades can be assumed to be (1, 0), (0, 0), (0, 0). It can easily be verified that $(2f+uf')\ \partial u/\partial t$ has an estimate of degree $(0, n)$ and grade (0, 0). Thus the product E of the three quantities has an estimate of degree $(1, n)$, and by insertion in **(3.2)** we conclude that $F(z)$ and thereby $u(t, 0)$ has an estimate of degree $(1, n + 1)$. By mathematical induction it follows that $F(z)$ has an estimate of degree $(1, n)$ with n arbitrarily large, or expressed differently, $F(z)$ is asymptotically represented by an infinite power series in z and $z \log z$. Observe that simultaneously it is proved that $F'(z)$ is asymptotically representable by the termwise differentiated power series.

§ **4. Singularities at the confluence of linear boundary conditions.** Consider a harmonic function u defined and continuous in a semicircle $y < 0$ about the origin, assuming continuous boundary values on the x-axis and satisfying the following boundary conditions there:

$$\partial u/\partial y = 0\,, \qquad \text{for} \quad y = 0\,,\ x < 0\,,$$

$$\partial u/\partial y - a(x)u = 0\,, \quad \text{for} \quad y = 0\,,\ x > 0\,;$$

here $a(x)$ is a (real) power series in x, convergent for the values in question of x, with $a(0) \neq 0$. Let $v(x, y)$ be the conjugate harmonic function of u, which must be constant on $y = 0$, $x < 0$, and assumes the value 0 there. Denote by $F(z)$ the function $u + iv$ which is regular in $z = x + iy$ within the semicircle.

THEOREM 4:1. $F(z) \sim M(z, z \log z)$, where M is a double power series in z and $z \log z$.

Proof: **a.** We show first the continuity of v—hence of F—as z approaches the x-axis. This is trivial for points where $x < 0$ since v can immediately be extended analytically across the negative real axis. It is likewise obvious for points where $x > 0$. For we have, for $x > 0$, $\partial v/\partial x = -\partial u/\partial y = -a(x)u$. Hence the conjugate harmonic function v has continuously differentiable boundary values on the positive axis. Thus $F(z)$ is continuous for positive z. The proof of continuity at the origin, however, requires further work. Observe first that there is a limit of $v(x, 0)$ as x tends to zero through positive values, because of the ordinary differential equation $dv(x, 0)/dx = -a(x)\, u(x, 0)$ where a and u are continuous. This $\lim v(x, 0) = v(0+, 0)$

is necessarily zero. In fact let $v(0+, 0) = \alpha$ and $\alpha \neq 0$. Consider, with polar coördinates r and θ, the double integral

$$I = \iint \partial u/\partial r \, dr \, d\theta$$

extended over the portion $y < 0$ of an annulus $R/2 < r < R$ where R is some small positive number. Let σ and σ' be the two bounding semicircles. Then

$$I = \int_\sigma u d\theta - \int_{\sigma'} u d\theta ,$$

consequently $I \to 0$ as $R \to 0$ because u is continuous near the origin. On the other hand, by the Cauchy-Riemann equations, $\partial u/\partial r = r^{-1} \partial v/\partial \theta$, so that the double integral

$$I = \int_{R/2}^{R} v(r, 0) - v(-r, 0))r^{-1} \, dr$$

$$= \int_{R/2}^{R} v(r, 0) \, r^{-1} \, dr ,$$

since $v = 0$ on the negative real axis. We have, accordingly, the estimate

$$|I| \geqq (|\alpha|/2) \int_{R/2}^{R} r^{-1} \, dr = (|\alpha| \log 2)/2 ,$$

since for sufficiently small positive values of x, $|v(x, 0)| > |\alpha|/2$. This contradicts $I \to 0$ as $R \to 0$. Hence $v(0+, 0) = 0$. This completes the proof that $v(x, 0)$ is continuous at the origin, considered as function of x alone. In order to show the continuity of v at the origin as function of x and y, we prove that, given $\epsilon > 0$, and taken (x_0, y_0) sufficiently close to the origin, the double integral of v over a circle Γ of radius $|y_0|$ about (x_0, y_0) is absolutely $\leqq 6\epsilon y_0^2$. By the average property of harmonic functions it will follow that $|v(x_0, y_0)| \leqq 6\epsilon\pi^{-1}$, which implies the continuity at the origin. To establish the inequality, we apply Cauchy's integral theorem to a rectangle with sides parallel to the axis and obtain, with positive p and $0 < \xi < p$

$$\int_\xi^p v(x, y) dx = \int_\xi^p v(x, 0) dx + \int_0^y u(p, \eta) d\eta - \int_0^y u(\xi, \eta) d\eta ,$$

whence for $\xi \to 0$ in view of the supposed continuity of u,

$$\int_0^p v(x, y) dx = \int_0^p v(x, 0) dx + \int_0^y (u(p, \eta) - u(0, \eta)) d\eta .$$

Similarly

$$\int_0^{-p} v(x, y)dx = \int_0^{-p} v(x, 0)dx + \int_0^{y} (u(-p, \eta) - u(0, \eta))d\eta .$$

Thus, if p_1 and p_2 are any two real numbers of small modulus,

$$\left| \int_{p_1}^{p_2} v(x, y)dx \right| \leqq \epsilon |p_2 - p_1| + \epsilon |y| .$$

Consequently by integration of this inequality

$$\left| \iint_\Gamma v(x, y)dxdy \right| \leqq \epsilon 2|y_0| \int_0^{2|y_0|} |dy| + \epsilon \int_0^{2|y_0|} |y|\,|dy| \leqq 6\epsilon |y_0|^2,$$

q.e.d.

b. Difference-differential equation. The analytic extension of $F(z)$ across the negative axis is given by

$$F(\bar{z}) = \overline{F(z)} .$$

Accordingly, we find for points z of the positive axis for which $\operatorname{Im} \log z = 0$,

$$2u(x, 0) = F(z) + F(z^-), \quad 2\partial u/\partial y = i(F'(z) - F'(z^-)) ,$$

so that the boundary condition becomes, for $z > 0$,

$$F'(z) - F'(z^-) + i(F(z) + F(z^-))a(z) = 0 . \tag{4.1}$$

But $F(z)$ and $F(z^-)$ are analytic functions of z in the neighborhood of a point $z > 0$, and so is $a(z)$. Hence (4.1) holds for arbitrary z by analytic extension; it can be used to effect this extension, since (4.1) is an ordinary differential equation for $F(z^-)$ if $F(z)$ is known and conversely. Thus $F(z)$ must exist on the entire Riemann logarithmic surface which has the origin as branch point and lies over the original circle of existence of $u(x, y)$. This function $F(z)$ tends to a (real) constant c as $z \to 0$ through any sequence of values z, belonging to arbitrary sheets of the Riemann surface, provided $|\arg z|$ remains bounded. To see this write (4.1) in the form

$$(\Delta F(z))' - ia(z)\Delta F(z) = -2ia(z)F(z), \qquad \Delta F(z) = F(z) - F(z^-) ,$$

which is solved by

$$\Delta F(z) = \exp\left\{ i \int_0^z a(t)dt \right\} \int_0^z -2ia(Z)F(Z) \exp\left\{ -i \int_0^Z adt \right\} dZ \tag{4.2}$$

for $z > 0$. But right hand of (4.2) represents an analytic function for arbitrary $z \neq 0$, and it tends to zero for $z \to 0$, $0 \geqq \arg z \geqq -2\pi$. Hence

$$\Delta F(z) = F(z) - F(z^-) \to 0$$

at the same time, and thus

$$\lim_{z\to 0} F(z^-) = \lim_{z\to 0} F(z) = c\,.$$

Thus continuing we find $\lim_{z\to 0} F(z) = c$ whenever $0 \geqq \arg z \geqq -2\pi n$ for all fixed integral $n > 0$. A corresponding conclusion can be found for $0 \leqq \arg z \leqq 2\pi n$ from the equation

$$(\Delta F(z))' + ia(z)\Delta F(z) = 2ia(z)F(z^-)\,.$$

Equation (4.1) can be simplified by introduction of the new variable

$$Z = \int_0^z a(t)dt\,.$$

If we set $a(0) = a$, then $a \neq 0$ by hypothesis and

$$Z = az(1 + p(z)) = Z(z)$$

where $p(z)$ is a convergent power series. We find

$$\log Z(z^-) = -2\pi i + \log Z(z) = \log Z^-\,.$$

Thus (4.1) becomes, upon writing $F(z) = \Phi(Z)$,

$$\Phi'(Z) - \Phi'(Z^-) + i(\Phi(Z) + \Phi(Z^-)) = 0\,. \tag{4.3}$$

Since an asymptotic estimate of degree (m, n) in z and $z \log z$ yields an asymptotic estimate of same degree in Z and $Z \log Z$, it is therefore legitimate to assume henceforth that the function $a(z)$ in (4.2) is identically equal to 1. [Notice that the rôle of the real axis is preserved as far as the boundary conditions for u are concerned, if $a(0) > 0$.]

c. In order to conclude† the proof of theorem 4, consider a point $z \neq 0$, $-2\pi < \arg z < 0$, and the integral along the positive axis

$$\psi(z) = \int_0^R \Delta F(t)\,(t - z)^{-1}dt\,,$$

where $\arg t = 0$, $R > |z| > 0$ a small number, and $\Delta F(t) = F(t) - F(t^-)$. We have, on passing to the limit $z \to$ real >0, $\arg z = 0$,

$$\psi(z) - \psi(z^-) = -2\pi i(F(z) - F(z^-))\,. \tag{4.4}$$

On the other hand, an integration by parts yields in view of the continuity of F at the origin

$$\psi(z) = \log(R - z)\Delta F(R) - \int_0^R \Delta F'(t)\log(t - z)dt\,.$$

† In case $a(0) > 0$; the general case in which $a(0) \neq 0$ is concluded at the beginning of § 5.

Apply (4.1) and find

$$\psi(z) = \log (R - z)\Delta F(R) + i \int_0^R (F(t) + F(t^-)) \log (t - z)dt \tag{4.5}$$

$$= \log (R - z)\Delta F(R) + 2i \int_0^R u(t, 0) \log (t - z)dt$$

On the other hand, because of (4.4) the sum $F(z) + (2\pi i)^{-1}\psi(z)$ stays single valued; because of (4.5) it remains continuous near $z = 0$; it is therefore a power series in z. Hence, finally,

$$F(z) = -\pi^{-1} \int_0^R u(t, 0) \log (t - z)dt + p(z) ,$$

where $p(z)$ is a convergent power series. Now we apply theorem 1. We know that $u(t, 0) \sim u(0, 0)$; this being an estimate of degree (0, 0). $F(z)$ has an estimate of degree (0, 1). Thus $u(t, 0) = \mathrm{Re}(t)$ has an estimate of degree (0, 1), hence by theorem 1, $F(z)$ has an estimate of degree (0, 2) etc. Thus, ultimately,

$$F(z) \sim M(z, z \log z)$$

where M is an infinite power series: q.e.d.

It is of interest to observe that the asymptotic representation of $F(z)$ can be obtained even in the case that $a(x)$, instead of being a power series in x, is a real continuous function of x for $x > 0$ which admits of an asymptotic representation by a double power series in x and $x \log x$. The substitution $Z = \int_0^z a(t)dt$ no longer leads to the simple equation (4.3) because it is no longer necessarily true that

$$\log Z(z^-) = -2\pi i + \log Z(z) = \log Z^- .$$

We still can conclude that $F(z)$ is continuous at $z = 0$. Proceeding as before, we obtain the equation

$$F(z) = -\pi^{-1} \int_0^R a(t)u(t, 0) \log (t - z)dt + P(z)$$

where $P(z)$ is a convergent power series in z. We again apply theorem 1. We know that $u(t, 0) \sim u(0, 0)$. Thus $a(t)u(t, 0)$ possesses an estimate of degree (0, 0), whence $F(z)$ has an estimate of degree (0, 1). Thus $u(t, 0) = \mathrm{Re}\, F(t)$ has an estimate of degree (0, 1). The same now holds for $a(t)u(t, 0)$, whence $F(z)$ has an estimate of degree (0, 2), etc. Thus ultimately, for $0 \geqq \arg z \geqq -2\pi$

$$F(z) \sim M(z, z \log z) \tag{4.6}$$

where M is an infinite power series. Thus we have the

THEOREM 4:2. Let u be a (real) harmonic function of x and y, continuous and defined in a semicircle $y < 0$ about the origin, assuming continuous boundary values on the x-axis and satisfying the following boundary conditions there:

$$\partial u/\partial y = 0, \qquad \text{for} \quad y = 0, x < 0,$$

$$\partial u/\partial y = a(x)u, \quad \text{for} \quad y = 0, x > 0$$

Here $a(x)$ is a real continuous function of x for $x > 0$ which possesses there an asymptotic estimate by an infinite power series in x and $x \log x$. Denote by $F(z)$ the analytic function of $z = x + iy$ on the plane slit along the positive x-axis which is continuous at $z = 0$, regular for $z \neq 0$ in a neighborhood of the origin and whose real part is $u(x, y)$. Then $F(z)$ has an asymptotic estimate (4.6).

§ **5. Unique determination by asymptotic estimate.** A solution of the difference–differential equation

$$F'(z) - F'(z^-) + i(F(z) + F(z^-)) = 0 \tag{5.1}$$

is not necessarily real on the negative real axis and derives not always from a boundary problem of harmonic functions in the way considered in the preceding section. But the asymptotic development which was obtained there applies to all solutions of (5.1) which are continuous at the origin. To see this, it is only necessary to observe that the function $G(z) = e^{-iz}F(z)$ satisfies the equation

$$G'(z) - G'(z^-) = +2i\,G(z)\,, \tag{5.2}$$

whence

$$G(z) + (2\pi i)^{-1}\int_0^R (G(t) - G(t^-))(t - z)^{-1}dt$$

is a convergent power series $P(z)$. Applying an integration by parts and the relation (5.2) we have for small $|z| < R$,

$$P(z) = G(z) + (2\pi i)^{-1}(G(R) - G(R^-))\log(R - z) - \pi^{-1}\int_0^R G(t)\log(t - z)dt\,,$$

where $G(t)$ has an estimate of degree (0, 0). By virtue of theorem 1, $G(z)$ and thus $G(t)$ have an estimate of degree (0, 1); it follows by induction as before that $G(z)$ has an estimate of form $G(z) \sim M(z, z \log z)$ where M is an infinite power series. The same holds for $F(z) = e^{iz}G(z)$, since e^{iz}, like $G(z)$, admits of an estimate of arbitrarily high degree in z and $z \log z$, whence the same is true for the product $e^{iz}G(z)$.

THEOREM 5. Let $F(z)$ be a solution of (5.1) and continuous at the origin. Suppose that all the coefficients of the infinite asymptotic estimate of F are zero. Then $F(z) \equiv 0$ identically.

Proof: We show that $G(z)$ which has an asymptotic estimate of the same description as $F(z)$ is identically zero. $G(z)$, $z \neq 0$, has derivatives of all orders; these derivatives tend to zero as $z \to 0$, as is seen by the following argument. Let $\epsilon > 0$ be given and let $R < 1$ be small enough so that for $|z| \leqq R$, $-\pi \leqq \arg z \leqq \pi$,

$$|G(z)| < \epsilon |z|^{n+1},$$

where n is a given positive integral number. A point z, $\operatorname{Re} z \geqq 0$, $|z| = t < R/2$, is interior to a domain D whose boundary consists of the negative axis from $-R$ to $-t/2$, a circle of radius $t/2$ described clockwise about the origin, the negative axis from $-t/2$ to $-R$ and a circle of radius R described counterclockwise about the origin. $G(z)$ is regular within D. Denoting by D' the boundary of D, we have for all z in D with $|z| = t$, $\operatorname{Re} z \geqq 0$,

$$G^{(n)}(z) = n!\,(2\pi i)^{-1} \int_{D'} G(Z)(Z - z)^{-(n+1)} dZ$$

whence, estimating separately the integrals over the two circles and along the portion of the negative axis,

$$|G^{(n)}(z)| \leqq n!\,\epsilon \left\{ (t/2)^{n+2}/(t/2)^{n+1} + R^{n+2}/(R/2)^{n+1} + \pi^{-1} \int_{-t/2}^{-R} |dZ| \right\}$$

$$\leqq n!\,\epsilon \,\{t/2 + R \cdot 2^{n+1} + \pi^{-1} R\} .$$

Since $t < R < 1$, this becomes arbitrarily small with $\epsilon > 0$. Hence $G^{(n)}(z) \to 0$ as $z \to 0$, $\operatorname{Re} z \geqq 0$. Observe that the estimate just given for $\operatorname{Re} z \geqq 0$ can be repeated in analogous manner for $\operatorname{Re} z \leqq 0$ by choosing another domain D_1, similar to D, and applying to it similar estimates. Hence $G^{(n)}(z)$ tends to zero as z tends to zero. Furthermore, since equation (5.2) has constant coefficients, $G^{(n)}(z)$ is solution of the same equation (5.2). This will enable us to derive an upper bound for the n-th derivative of $G(z)$. Let ρ with $0 < \rho < 1/e$ be a number such that $G(z)$ exists and is regular for $z \neq 0$, $|z| \leqq 4\rho$, $-3\pi \leqq \arg z \leqq 3\pi$. Multiplying $G(z)$ by a constant does not change the hypotheses nor the conclusion of the theorem, hence we may assume that for the values of z in question $|G(z)| < 1$ and $G^{(n)}(z) \to 0$ as $z \to 0$ for all non-negative integral n. Apply Cauchy's integral formula to a circle of radius ρ about any point Z with $2\rho \geqq |Z| \geqq \rho$ for which $|\arg Z| \leqq 2\pi$ and obtain the bound

$$|G^{(n)}(Z)| \leqq n!\,\rho^{-n} . \tag{5.3}$$

Now let z be a point with $z \neq 0$, $|z| \leqq \rho$, $\pi/2 \geqq \arg z \geqq -\pi/2$. We slit the plane along the negative real axis, describe a circle $C_{2\rho}$ of radious 2ρ about the origin and a second circle of smaller radius which we later make tend to 0, and express $G^{(n)}(z)$ by a Cauchy integral over the boundary of the included domain. Observing the vanishing of $G^{(n)}$ at the origin, we find

$$G^{(n)}(z) = (2\pi i)^{-1} \int_{C_{2\rho}} G^{(n)}(Z)(Z - z)^{-1} dZ + (2\pi i)^{-1} \int_{-2\rho}^{0} (G^{(n)}(Z) - G^{(n)}(Z^-))\,(Z - z)^{-1} dZ , \tag{5.4}$$

where in the second integral we take $\arg Z = \pi$. In view of (5.3) and (5.2),

$$|G^{(n)}(z)| \leqq n!\,\rho^{-n} + \pi^{-1} \left| \int_{-2\rho}^{0} G^{(n-1)}(Z)(Z-z)^{-1} dZ \right| . \tag{5.5}$$

But

$$\int_{-2\rho}^{0} G^{(n-1)}(Z)(Z-z)^{-1} dZ = G^{(n-1)}(Z) \log (Z-z) \Big|_{-2\rho}^{0} - \int_{-2\rho}^{0} G^{(n)}(Z) \log (Z-z) dZ .$$

Thus, since $G^{(n-1)}(0) = 0$,

$$\left| \int_{-2\rho}^{0} G^{(n-1)}(Z)(Z-z)^{-1} dZ \right| \leqq (n-1)!\,\rho^{-n+1} |\log(-2\rho - z)| + \left| \int_{-2\rho}^{0} G^{(n)}(Z) \log (Z-z) dZ \right| ,$$

and

$$|G^{(n)}(z)| \leqq n!\,\rho^{-n} \cdot + \pi^{-1} (n-1)!\,\rho^{-n+1} |\log(-2\rho - z)| + \pi^{-1} \left| \int_{-2\rho}^{0} G^{(n)}(Z) \log (Z-z) dZ \right| .$$

The integral on the right is further split into $\int_{-2\rho}^{-\rho}$ and $\int_{-\rho}^{0}$. For the first integral the estimate (5.3) still holds. Furthermore, the expressions $\int_{-2\rho}^{-\rho} |\log (Z - z)| dZ$, $\rho |\log(-2\rho - z)|$ are bounded by a constant which depends, because of the limitations to which z and ρ are subject on neither ρ, z, nor n. Thus, with some such constant h, for $\pi/2 \geqq \arg z \geqq -\pi/2$, $\rho \geqq |z|$

$$\left| G^{(n)}(z) \right| \leqq n!\,\rho^{-n} h + \pi^{-1} \left| \int_{-\rho}^{0} G^{(n)}(z) \log (Z-z) dZ \right| . \tag{5.6}$$

In a similar way we obtain for $|z| \leqq \rho$, $\pi/2 \leqq \arg z \leqq 3\pi/2$, that

$$|G^{(n)}(z)| \leqq n!\,\rho^{-n} h + \left| \int_{0}^{\rho} G^{(n)}(Z) \log (Z-z) dZ \right| , \tag{5.7}$$

where the argument of Z under the integral sign is 2π.

Denote by $\gamma_n, \gamma_n', \gamma_n''$ respectively the max $|G^{(n)}(z)|$ for $0 \leqq z \leqq \rho$, $\arg z = 0$, for $0 \leqq z \leqq \rho$, $\arg z = \pi$, and for $0 \leqq z \leqq \rho$, $\arg z = 2\pi$. Then we can estimate the last two integrals

$$\left| \int_{-\rho}^{0} G^{(n)}(Z) \log (Z-z) dZ \right| \leqq \gamma_n' \int_{-\rho}^{0} |\log (Z-z)| dZ = \gamma_n' I_1 ,$$

$$\left| \int_{0}^{\rho} G^{(n)}(Z) \log (Z-z) dZ \right| \leqq \gamma_n'' \int_{0}^{\rho} |\log (Z-z)| dZ = \gamma_n'' I_2 ,$$

and here $|\log(Z-z)| \leqq \pi + |\log|Z-z|| \leqq \pi + |\log|z||$; so that both integrals satisfy the same inequality

$$I_1 \text{ and } I_2 \leqq \int_{-\rho}^{0} (\pi + |\log|Z||)dZ = \int_{0}^{\rho} (\pi + |\log Z|)dZ \leqq \pi\rho + \rho(1 - \log\rho) ;$$

since $\rho < 1/e$, this is $\leqq e^{-1}(\pi + 2)$. Hence (5.6) and (5.7) become

$$\gamma_n \leqq n!\,\rho^{-n} h + s\gamma_n' , \tag{5.8}$$

$$\gamma_n' \leqq n!\,\rho^{-n} h + s\gamma_n'' , \tag{5.9}$$

with $s = (\pi + 2)/(e\pi) < 2/e < 1$. Furthermore, (5.2) yields

$$\gamma_n'' \leqq \gamma_n + 2\gamma_{n-1}'' , \tag{5.10}$$

and, by induction,

$$\gamma_n'' \leqq \gamma_n + 2\gamma_{n-1} + \cdots + 2^{n-1}\gamma_1 + 2^n\gamma_0'' ,$$

where, incidentally, $\gamma_0'' \leqq 1$. Set $\mu_n = \max[\gamma_n, \gamma_n']$. From (5.8), (5.9) and (5.10) we gather the inequality for $n \geqq 1$:

$$\mu_n \leqq n!\,\rho^{-n} h + s(\mu_n + 2\mu_{n-1} + \cdots + 2^{n-1}\mu_1 + 2^n\mu_0) , \tag{5.11}$$

and

$$\mu_0 \leqq 1 .$$

There is a positive number σ which can be chosen small enough so that for all n

$$\mu_n \leqq n!\,\sigma^{-n}.$$

This inequality certainly applies to $n = 0$, $n = 1$. We only have to show, that if the above inequality holds for $1, 2, \cdots, n$, it will also hold for $n + 1$, provided

$$\sigma < 4^{-1}, \qquad \sigma < \rho , \qquad (h/\rho + 4s)\sigma \leqq 1 - s .$$

We have from (5.11)

$$(1 - s)\mu_{n+1} \leqq (n + 1)!\,\rho^{-n-1}h + 2s\sigma^{-n}n!\,(1 + 2\sigma + \cdots + 2^n\sigma^n) .$$

Since $\sigma < 4^{-1}$, $2\sigma < 1/2$, and $\sigma < \rho$,

$$(1 - s)\mu_{n+1} \leqq (n + 1)!\left(\rho^{-n-1}h + 2s\sigma^{-n}\cdot\frac{1 - (2\sigma)^{n+1}}{1 - 2\sigma}\right)$$

$$\leqq (n+1)!\,(\rho^{-n-1}h + 4s\sigma^{-n})$$
$$\leqq (n+1)!\,\sigma^{-n}\,(h/\rho + 4s)\,.$$

Thus

$$\mu_{n+1} \leqq (n+1)!\,\sigma^{-n}\,(h/\rho + 4s)(1-s)^{-1} \leqq (n+1)!\,\sigma^{-n-1}\,,$$

which proves the induction.

For the function of the real argument t, $G(t)$, we know therefore that its n-th derivative $G^{(n)}(t)$ is absolutely $\leqq n!\,\sigma^{-n}$, provided $0 < t \leqq \rho$. Consider the remainder of the Taylor development about the origin of $\operatorname{Re} G(t)$ or $\operatorname{Im} G(t)$ for $0 < t < \sigma/2$. This remainder must then be $\leqq \max_{0 \leqq t \leqq \rho} |G^{(n)}(t)|\,\sigma^n 2^{-n}\,(n!)^{-1}$, hence $\leqq 2^{-n}$; thus it tends to zero as $n \to \infty$. But this means that both $\operatorname{Re} G(t)$ and $\operatorname{Im} G(t)$ are represented by convergent Taylor series. Now the coefficients of these series are all zero since all derivatives of G as well as G itself vanish at the origin, as was shown previously. Consequently $G(t)$ is identically zero for $0 < t < \sigma/2$ and thus $G(z)$ vanishes identically for all z, q.e.d.

§ **6. An integral invariant.** The fact that the asymptotic development about the origin determines uniquely the function $F(z)$, solution of (5.1) and continuous near $z = 0$, makes it desirable to devise a method by which this development can be summed so as to yield back the function to which it belongs. In order to obtain this goal, we introduce the following integral invariant.

Let $F(z)$ and $G(z)$ be any two solutions of (5.1), regular for $z \neq 0$ and $|z| < \rho$ on the logarithmic Riemann surface.† Denote by $Q[F, G]$ the expression

$$Q[F,G] = 2\int_{z^+}^{z} F(t)G(t)dt + i\Delta F(z)\Delta G(z)\,, \tag{6.1}$$

where the integration extends over a circle about the origin leading from z^+ to z and the symbol $\Delta F(z)$ is an abbreviation of $F(z^+) - F(z)$. Then the theorem holds:

THEOREM 6:1. $Q[F', G] = -Q[F, G']$ and $Q[F, G]$ is independent of z.

Proof: We have

$$Q[F',G] + Q[F,G'] = 2\int_{z^+}^{z} (F'G + G'F)dt + i\Delta F'(z)\Delta G(z) + i\Delta F(z)\Delta G(z)\,.$$

Now

$$2\int_{z^+}^{z} (F'G + G'F)dt = 2FG\Big|_{z^+}^{z} = 2(F(z)G(z) - F(z^+)G(z^+))\,;$$

on the other hand, by (5.1),

$$\begin{aligned} i(\Delta F'(z)\Delta G(z) + \Delta F(z)\Delta G'(z)) &= (F(z) + F(z^+))\,(G(z^+) - G(z)) \\ &\quad + (F(z^+) - F(z))\,(G(z^+) + G(z)) \\ &= 2(F(z^+)G(z^+) - F(z)G(z)) \\ &= 2\Delta(F(z)G(z))\,, \end{aligned}$$

whence the first part of the theorem.

† Actually it suffices to demand the equation (5.1) and regularity of F and G only for $0 < |z| < \rho$ $0 \leqq \arg z \leqq 2\pi$, because the equation (5.1) provides a means of extension into the whole of $0 < |z| < \rho$ on the logarithmic Riemann surface. Notice that continuity at $z = 0$ is not required in theorem 6:1.

To prove the second part, consider the domain D bounded by a circle about the origin leading from z^+ to z, a path P leading from z to Z, a circle about the origin from Z to Z^+, and a path congruent to P from Z^+ to z^+. Within D both F and G are regular. Hence by Cauchy's integral theorem, the integral of $2FG$, extended over the boundary of D, vanishes. Thus

$$(6.2)\quad \begin{cases} 0 = 2\int FG\,dt = 2\left(\int_{z^+}^{z} + \int_{z}^{Z} - \int_{Z^+}^{Z} - \int_{z^+}^{Z^+}\right) \\ \quad = 2\int_{z^+}^{z} FG\,dt - 2\int_{Z^+}^{Z} FG\,dt - 2\int_{z}^{Z} \Delta(F(t)G(t))dt\,. \end{cases}$$

But, as above,

$$2\Delta(FG) = i(\Delta F \Delta G)'\,,$$

whence

$$2\int_{z}^{Z} \Delta(F(t)G(t))dt = i\Delta F \Delta G \Big|_{z}^{Z}$$

If we substitute this into (6.2) we obtain

$$Q[F, G, z] - Q[F, G, Z] = 0\,,$$

where, for distinction, the notation

$$Q[F, G, Z] = 2\int_{Z^+}^{Z} FG\,dt + i\Delta F(Z)\Delta G(Z)$$

has been employed. Hence $Q[F, G, z]$ is independent of z, which is the second part of the theorem.

THEOREM 6:2. If $F(z)$ is a solution of (5.1) for $z \neq 0$ and $|z| < \rho$, then the same holds for $F'(z)$ and, if $F(z)$ is continuous for $z \to 0$, also for $\int_0^z F(t)dt = \Phi(z)$.

Proof: The first part is obvious. The second part follows from

$$\begin{aligned} \Phi'(z) = F(z),\ \Delta\Phi'(z) &= \Delta F(z) = \Delta(F(z) - F(0)) \\ &= \int_0^z \Delta F'(t)dt \\ &= -i\int_0^z (F(t) + F(t^+))dt \\ &= -i(\Phi(z) + \Phi(z^+))\,. \end{aligned}$$

THEOREM 6:3. If $F(z)$ is a solution of (5.1) for $z \neq 0$ and $|z| < \rho$, then

$$Q[F, F'] = 0\,.$$

Proof:

$$Q[F, F'] = 2\int_{z^+}^{z} FF'dt + i\Delta F'(z)\Delta F(z)$$

$$= F^2(z) - F^2(z^+) + (F(z^+) + F(z))\,(F(z^+) - F(z))$$

$$= 0\,.$$

THEOREM 6:4. If F and G are solutions of (5.1) and continuous at $z = 0$, then

$$Q[F', G'] = -Q[F, G''] = -4iF(0)G(0)\,.$$

The proof follows from the formula (6.8) to be derived below.

THEOREM 6:5. Let $F(z)$ be a solution of (5.1), continuous at the origin, and regular for $z \neq 0$, $|z| \leqq \rho$. Let $\Phi_0(z)$ be a solution of (5.1), continuous at the origin with $\Phi_0(0) \neq 0$, and regular for $|z| \leqq 2\rho$. Denote by $\Phi_n(z)$, $n = 1, 2, \cdots$ the integral $\int_0^z \Phi_{n-1}(t)dt$. Then $F(z)$ can be developed in a series

$$F(z) = \sum_{\nu=0}^{\infty} c_\nu \Phi_\nu(z) \tag{6.3}$$

which converges absolutely and uniformly for $|z| \leqq \rho$ and where the constants c_ν are defined by recurrence relations (6.14) and (6.15).

Proof: We start the proof by remarking that the repeated integrals and derivatives of $\Phi_0(z)$ can be simply estimated. In fact, let, for $|\arg z| \leqq 3\pi$ and $|z| \leqq 2\rho$, $|\Phi_0(z)| \leqq M$; then obviously

$$|\Phi_1(z)| \leqq M|z|, \cdots, |\Phi_n(z)| \leqq M|z|^n/n\,! \tag{6.4'}$$

On the other hand, if $|z| = \rho$, $|\arg z| \leqq 2\pi$,

$$|\Phi_0^{(n)}(z)| \leqq M\rho^{-n}, n = 1, 2, \cdots.$$

Accordingly, by (6.1), computed for $z = \rho$, if $\beta_n = Q[\Phi_0^{(n)}, \Phi_0^{(n)}]$,

$$|\beta_n| \leqq 4\pi M^2\rho^{1-2n} + 4M^2\rho^{-2n} = \rho^{-2n}M^2(4\pi\rho + 4), \quad n = 1, 2, \cdots. \tag{6.4}$$

On the other hand, we find

$$Q[\Phi_0, \Phi_0] = 0\,,$$

if the invariant Q is computed for $z \to 0$ where Φ_0 is continuous. If we differentiate termwise the asymptotic development of $\Phi_0(z)$ and collect the terms of the same form, we obtain a certain development which need no longer be a power series in z and $z \log z$, but it becomes one if multiplied by z. Thus the terms of the formally differentiated power series for Φ_0 can be arranged in a natural order, i.e., the one obtainable after multiplication by z. It is readily seen by an application of Cauchy's integral formula to a circle of radius $|z|/2$ about the point z that the derivative of a function $o(z^k)$ equals $o(z^{k-1})$. Hence the formally differentiated power series for Φ_0 equals the derivative of Φ_0 asymptotically in the sense that $\Phi_0'(z)$ minus those terms where z appears in a power $\leqq k$, $k \geqq 0$, is $o(z^k)$. Similar facts apply to the relation between the higher derivatives of Φ_0 and those of the associated power series in z and $z \log z$.

Now put

$$\Phi_0(z) \sim \Phi_0(0)\,(1 + a_{01} z \log z + a_{10} z + \cdots)\,; \tag{6.5}$$

then

$$\Phi_0'(z) \sim \Phi_0(0)\,(a_{01} \log z + (a_{10} + a_{01}) + \cdots)\,, \tag{6.6}$$

and (5.1) yields

$$0 = \Delta\Phi_0'(z) + i(\Phi_0(z) + \Phi_0(z^+)) \sim \Phi_0(0)\,(a_{01} 2\pi i + 2i + \cdots)$$

or, since $\Phi_0(0) \neq 0$,

$$a_{01} = -\pi^{-1}. \tag{6.7}$$

Thus

$$Q[\Phi_0', \Phi_0'] = 2\int_{z^+}^{z} \Phi_0'\Phi_0' dt + i\Delta\Phi_0'(z)\Delta\Phi_0'(z)$$

$$= 2\int_{z^+}^{z} \Phi_0'\Phi_0' dt + i(-i)^2\,(2\Phi_0(0))^2 + \cdots$$

can be evaluated for $\lim z = 0$ and we find

$$\beta_1 = Q[\Phi_0', \Phi_0'] = -4i\Phi_0^2(0)\,. \tag{6.8}$$

We have, furthermore,

$$Q[\Phi_\nu, \Phi_0^{(\mu)}] = 0, \qquad \mu < \nu + 2\,. \tag{6.9}$$

For $Q[\Phi_\nu, \Phi_0^{(\mu)}] = (-1)^\mu\, Q[\Phi_{\nu-\mu}, \Phi_0]$, by theorem 6:1; if $\nu - \mu \geqq 0$, the invariant vanishes because of the continuity near the origin of $\Phi_{\nu-\mu}(z)$ and $\Phi_0(z)$; and if $\mu = \nu + 1$, it vanishes by virtue of theorem 6:3, as $Q[\Phi_\nu, \Phi_0^{(\nu+1)}] = \pm Q[\Phi_0, \Phi_0']$. Now let s be a non-negative integer, $\mu = \nu + 2s + 1$. Then $Q[\Phi_\nu, \Phi_0^{(\nu+2s+1)}] = \pm Q[\Phi_0^{(s)}, \Phi_0^{(s+1)}] = 0$ by theorem 6:3; thus

$$Q[\Phi_\nu, \Phi_0^{(\nu+2s+1)}] = 0\,, \qquad s \geqq 0\,. \tag{6.10}$$

Similarly we have

$$Q[\Phi_\nu, \Phi_0^{(\nu+2s)}] = (-1)^{\nu+s}\, Q[\Phi_0^{(s)}, \Phi_0^{(s)}] = (-1)^{\nu+s}\,\beta_s\,, \quad s \geqq 0\,. \tag{6.11}$$

It is now evident what the relations are by which we compute the successive coefficients c_ν. If (6.3) were to hold and the termwise application of the following functional could be justified, we would have, by virtue of the relations (6.9), (6.10), (6.11),

$$(6.12)\qquad Q[F, \Phi_0^{(\mu)}] = Q[\sum_0^\infty c_\nu \Phi_\nu, \Phi_0^{(\mu)}] = Q[\sum_0^{\mu-2} c_\nu\Phi_\nu, \Phi_0^{(\mu)}], \quad \mu = 2, 3, \cdots,$$

$$(6.13)\qquad Q[F, \Phi_0^{(\mu)}] = c_{\mu-2}\cdot(-)^{\mu-1}\beta_1 + c_{\mu-4}\cdot(-)^{\mu-2}\beta_2 + \cdots, \quad \mu = 2, 3, \cdots.$$

The sum on right hand ends with the last term in which the index of c is $\geqq 0$. Thus (6.13) can be decomposed in two systems, the first containing only even indices, the second only odd indices, namely,

$$(6.14)\qquad \begin{cases} Q[F, \Phi_0^{(2)}] = -c_0\beta_1, \\ Q[F, \Phi_0^{(4)}] = -c_2\beta_1 + c_0\beta_2, \\ Q[F, \Phi_0^{(6)}] = -c_4\beta_1 + c_2\beta_2 - c_0\beta_3, \\ \cdots\cdots\cdots, \end{cases}$$

and

$$(6.15)\qquad \begin{cases} Q[F, \Phi_0^{(3)}] = c_1\beta_1, \\ Q[F, \Phi_0^{(5)}] = c_3\beta_1 - c_1\beta_2, \\ Q[F, \Phi_0^{(7)}] = c_5\beta_1 - c_3\beta_2 + c_1\beta_3, \\ \cdots\cdots\cdots\cdots. \end{cases}$$

Relations (6.14) and (6.15) determine recursively the coefficients c_j, since $\beta_1 \neq 0$. In order to shorten the following estimation of the c_j, we shall assume that $|\Phi_0(0)| = 1/2$ so that

$$(6.16)\qquad |\beta_1| = 1.$$

This is no loss of generality, since a multiplication of Φ_0 by a constant $\neq 0$ does not alter the hypotheses or the conclusion of the theorem.† Let γ be an upper bound of $|F|$ on the circle $|z| = \rho$, $0 \leqq \arg z \leqq 2\pi$. We find in view of (6.4)

$$\alpha_\nu = |Q[F, \Phi_0^{(\nu)}]| \leqq \gamma M(4\pi\rho + 4)\rho^{-\nu},$$

$$|\beta_\nu| \leqq M^2(4\pi\rho + 4)\rho^{-2\nu}.$$

† The following estimates (6.17) and (6.18) remain generally true if the constant h is suitably reinterpreted.

Thus there exists a constant h independent of ν, such that

$$\alpha_\nu \leqq h\rho^{-\nu}, \qquad |\beta_\nu| \leqq h\rho^{-2\nu}.$$

We now show that

$$(6.17) \qquad |c_{2j}| \leqq h\rho^{-2}[(1 + h\rho^{-2})\rho^{-2}]^j.$$

Indeed, for $j = 0$, we find

$$|c_0| = \alpha_2 \leqq h\rho^{-2}.$$

Suppose (6.17) to be true for $j = 0, 1, \cdots, n$; then by (6.14)

$$\begin{aligned}|c_{2n+2}| &\leqq \alpha_{2n+4} + |c_0||\beta_{n+2}| + \cdots + |c_{2n}||\beta_2|, \\ &\leqq h\rho^{-2n-4} + h\rho^{-2}h\rho^{-2n-4} + h\rho^{-2}(1 + h\rho^{-2})\rho^{-2}h\rho^{-2n-2} + \cdots \\ &\qquad + h\rho^{-2}[(1 + h\rho^{-2})\rho^{-2}]^n h\rho^{-4} \\ &\leqq h\rho^{-2n-4}(1 + h\rho^{-2} + h\rho^{-2}(1 + h\rho^{-2}) + \cdots + h\rho^{-2}(1 + h\rho^{-2})^n) \\ &= h\rho^{-2n-4}(1 + h\rho^{-2}((1 + h\rho^{-2})^{n+1} - 1)/(1 + h\rho^{-2} - 1)) \\ &= h\rho^{-2}((1 + h\rho^{-2})\rho^{-2})^{n+1}.\end{aligned}$$

Hence (6.17) holds for all $j \geqq 0$.

Similarly, from (6.15),

$$(6.18) \qquad |c_{2j+1}| \leqq h\rho^{-3}((1 + h\rho^{-2})\rho^{-2})^j.$$

Combining (6.17) and (6.18) we can say that there is a constant C such that

$$|c_n| < C^{n+1}, \qquad n = 0, 1, \cdots.$$

Since, by (6.4′), $|\Phi_n(z)| \leqq M|z|^n/n!$, we have

$$\sum_0^\infty |c_n\Phi_n(z)| \leqq M\sum_0^\infty C^{n+1}|z|^n/n! = MCe^{C|z|},$$

whence the absolute and uniform convergence of $\sum_0^\infty c_n\Phi_n(z)$ for $|z| \leqq \rho$. It remains to prove that $\sum_0^\infty c_n\Phi_n(z) = F(z)$. For this purpose we consider the difference $\delta(z) = F(z) - \sum_0^\infty c_n\Phi_n(z)$. $\delta(z)$ is itself a solution of (5.1) since $\sum_0^\infty c_n\Phi_n(z)$ converges uniformly in a sufficiently small neighborhood of any $z \neq 0$ with $|z| < \rho$, hence is termwise differentiable there and the differentiated series converges uniformly. Furthermore, $\delta(z)$ is continuous at the origin since $F(z)$ and all $\Phi_n(z)$ are. Finally,

$$Q[\delta, \Phi_0^{(\mu)}] = 0, \qquad \mu = 2, 3, \cdots,$$

in view of (6.12). By theorem 5 it suffices to show that all coefficients of the asymptotic development for δ are zero, in order to deduce the identical vanishing of $\delta(z)$. Let then $az^m(z \log z)^n$ be the term of smallest degree in the development for $\delta(z)$ for which $a \neq 0$. If $n = 0$, consider $\delta^{(m)}(z)$ which is solution of (5.1) and continuous at the origin; its term of lowest degree is necessarily $am\,!$, since the higher terms in the development of $\delta(z)$ have at least degree $(m, 1)$, hence their m-th derivative has grade $(0, 1)$. By theorem 6:4, $Q[\delta, \Phi_0^{(m+2)}]$ would not vanish, since

$$Q[\delta, \Phi_0^{(m+2)}] = \pm Q[\delta^{(m+1)}, \Phi_0'] = \pm 4i\delta^{(m)}(0)\,\Phi_0(0) = \pm 4m\,!\,ai\Phi_0(0)$$

and $\Phi_0(0) \neq 0$ by hypothesis. Consequently, $n > 0$. Consider, then, $Q[\delta, \Phi_0^{(m+n+1)}] = \pm Q[\delta^{(m+n)}, \Phi_0']$. We find easily that

$$\lim_{z \to 0} (\delta^{(m+n)}(z) - a(m+n)\,!\,(\log z)^n - \cdots) = 0,$$

where the omitted terms are a polynomial in $\log z$ of degree $< n$. We also know from (6.6) and (6.7) that

$$\lim_{z \to 0} (\Phi_0'(z) + \pi^{-1}\Phi_0(0) \log z) = \text{const.}$$

We now compute the invariant $Q[\delta^{(m+n)}, \Phi_0']$ by forming it for a point z and letting z tend to zero. The integral part, $2\int_{z^+}^{z} \delta^{(m+n)}(t)\,\Phi_0'(t)dt$ necessarily tends to zero as the radius of the circle of integration converges to zero. On the other hand

$$\lim_{z \to 0} \Delta\Phi_0'(z) = -\pi^{-1}\,\Phi_0(0)2\pi i \neq 0,$$

but $\lim \Delta\delta^{(m+n)}(z) = \lim \Delta(a(m+n)\,!\,(\log z)^n + \cdots)$ does not exist, if $n > 1$, and equals $a(m+n)\,!\,2\pi i$ for $n = 1$. In neither case do we obtain $Q[\delta, \Phi_0^{(m+n+1)}] = \pm Q[\delta^{(m+n)}, \Phi_0'] = 0$. This is in contradiction with the assumption concerning $\delta(z)$. Hence $a \neq 0$ always leads to a contradiction. This proves $\delta(z) \equiv 0$, and theorem 6:5 follows.

§ 7. **Explicit solutions.** We obtain explicit solutions $F(z)$ of the equation

$$F'(z^+) - F'(z) + i(F(z^+) + F(z)) = 0, \tag{7.1}$$

by putting

$$F(z) = (2\pi i)^{-1} \int z^\alpha g(z, \alpha) d\alpha, \tag{7.2}$$

where $g(z, \alpha)$ is a single valued analytic function of the two complex variables z and α, regular for a neighborhood of $z = 0$, $\alpha = 0$, with exception of the points

(z, α) for which $\alpha = 0$. We suppose the exponential $z^\alpha = \exp(\alpha \log z)$ to be that branch of the function which is real and positive for positive real z with $\arg z = 0$ and real α. The integral is to be taken over a closed simple contour surrounding the origin of the α-plane. We have

$$(7.3) \qquad F'(z) = (2\pi i)^{-1} \int z^\alpha (dg/dz + \alpha z^{-1} g) d\alpha ,$$

$$F(z^+) = (2\pi i)^{-1} \int z^\alpha g e^{2\pi i \alpha} d\alpha .$$

Thus the left hand of (7.1) becomes

$$(7.4) \qquad (2\pi i)^{-1} \int z^\alpha H(z, \alpha) d\alpha$$

with

$$H(z, \alpha) = (e^{2\pi i \alpha} - 1)(dg/dz + \alpha z^{-1} g) + i(e^{2\pi i \alpha} + 1) g ,$$

and (7.4) vanishes if $H(z, \alpha)$ is regular as function of α in a circle about $\alpha = 0$, with the origin a point of regularity. Hence let $g(z, \alpha)$ satisfy the ordinary differential equation in z in which α appears as parameter:

$$(7.5) \qquad dg/dz + (\alpha z^{-1} + \cot \pi\alpha) g = H(z, \alpha)(e^{2\pi i \alpha} - 1)^{-1} .$$

This equation has the explicit solution

$$(7.6) \qquad g(z, \alpha) = e^{-z \cot \pi\alpha} z^{-\alpha} \int H(z, \alpha)(e^{2\pi i \alpha} - 1)^{-1} e^{z \cot \pi\alpha} z^\alpha dz ,$$

in which the constant of integration must be adjusted so as to make $g(z, \alpha)$ regular in z and α for $\alpha \neq 0$ in a neighborhood of the origin of the z, α-plane. A simple way to achieve this is to consider the function

$$(7.7) \qquad g(z, \alpha) = e^{-z \cot \pi\alpha} z \int t^\alpha e^{tz \cot \pi\alpha} H(tz, \alpha) dt (e^{2\pi i \alpha} - 1)^{-2} ,$$

where the integral is extended over a path beginning at $t = 1$, surrounding the origin of the t-plane in the positive sense and ending at $t = 1^+$. This function $g(z, \alpha)$ is obviously regular for small $|\alpha|$, provided $\alpha \neq 0$. Furthermore, for $\operatorname{Re} \alpha < 0$, $\alpha \neq 1, 2, \cdots$, and $z \neq 0$,

$$g(z, \alpha) = e^{-z \cot \pi\alpha} z \int_0^1 t^\alpha e^{tz \cot \pi\alpha} H(tz, \alpha)\, dt\, (e^{2\pi i \alpha} - 1)^{-1}$$

or

$$(7.8) \qquad g(z, \alpha) = e^{-z \cot \pi\alpha} z^{-\alpha} \int_0^z t^\alpha e^{t \cot \pi\alpha} H(t, \alpha)\, dt\, (e^{2\pi i \alpha} - 1)^{-1} ,$$

and the last form is a special case of (7.6). Since (7.7) is regular in α for $\alpha \neq 0$, (7.7) satisfies the differential equation (7.1) for all values of $\alpha \neq 0$ in question, by analytic extension from real positive α. Observe that (7.7) defines $g(z, \alpha)$ as a single valued function of z and α, regular in a neighborhood of the origin $z = \alpha = 0$ for $\alpha \neq 0$, provided $H(z, \alpha)$ satisfies the condition:

(7.9) $zH(z, \alpha)$ is regular in z and α for small $|\alpha|$ and $|z|$.

The aim of this section is the proof of the following theorem.

THEOREM 7. All solutions of (7.1) which are continuous at $z = 0$ can be obtained from (7.2) where $g(z, \alpha)$ is given by (7.7) and $H(z, \alpha)$ is a suitable function satisfying the condition (7.9).

Proof: We begin by studying that solution $\Phi_0(z)$ which is obtained by setting $(e^{2\pi i\alpha} - 1)^{-1} H(z, \alpha) = z^{-1}$. We know that $g(z, \alpha)$ is, for $\alpha \neq 0$, a regular function of z near $z = 0$ and may thus be written as

$$g(z, \alpha) = k_0 + k_1 z + \cdots$$

where the coefficients of the power series are functions of α, regular in α in a neighborhood of the origin $\alpha = 0$ but excluding possibly the point $\alpha = 0$. In order to compute k_0, we take an α with $\operatorname{Re} \alpha > 0$ and find from (7.8)

$$k_0 = \lim_{z\to 0} g(z, \alpha) = \lim_{z\to 0} \{z^{-\alpha}\int_0^z t^{\alpha-1} e^{t \cot \pi\alpha} dt \}$$

$$= \alpha^{-1}.$$

Hence

$$\Phi_0(0) = (2\pi i)^{-1} \lim_{z\to 0} \int z^\alpha (k_0 + k_1 z + \cdots) d\alpha$$

$$= 1.$$

Let us agree to denote the function H and g just employed in the construction of $\Phi_0(z)$, by H_0 and g_0. We understand as before that $\Phi_\nu(z)$, $\nu = 1, 2, \cdots$ are the repeated integrals from zero to z of $\Phi_0(z)$. We can easily establish the relations which yield the corresponding functions H_ν, g_ν. In fact, (7.3) shows that if g belongs to F, then $z^{-\alpha}d(z^\alpha g)/dz$ belongs to $F'(z)$. Similarly, (7.5) can be written

$$H(z, \alpha) (e^{2\pi i\alpha} - 1)^{-1} = z^{-\alpha}d(z^\alpha g)/dz + \cot \pi\alpha\, g.$$

Replace g by $z^{-\alpha}d(z^\alpha g)/dz$, then H is replaced by $z^{-\alpha}d(z^\alpha H)/dz$. Conversely, integration of $F(z)$ from zero to z effects a replacement of g by $z^{-\alpha}\int_0^z t^\alpha g(t, \alpha)dt$ and of H by $z^{-\alpha} \int_0^z tH(t, \alpha)dt$; these formulae are at first defined for $\operatorname{Re} \alpha > 0$ and are for other α to be understood as analytic extensions thereof. We find in particular

$$(e^{2\pi i\alpha} - 1)^{-1} H_\nu = z^{\nu-1} (\alpha(\alpha + 1) (\alpha + 2) \cdots (\alpha + \nu - 1))^{-1}, \quad \nu = 1, 2, \cdots,$$

$$g_\nu = e^{-z\cot\pi\alpha} z^{-\alpha}\int_0^z t^{\alpha+\nu-1} e^{t\cot\pi\alpha} dt\,(\alpha(\alpha+1)\cdots(\alpha+\nu-1))^{-1},$$

$$\Phi_\nu = (2\pi i)^{-1}\int z^\alpha g_\nu(z,\alpha)d\alpha .$$

Now we know from theorem 6:5 that an arbitrary solution $F(z)$ of (7.1) which is continuous at $z = 0$, can be written in the form

$$F(z) = \sum_0^\infty c_\nu \Phi_\nu(z) .$$

If $F(z)$ is regular for $|z| \leqq \rho$, $z \neq 0$, the above series converges and there is a constant $C > 0$ so that $|c_j| < C^{j+1}$ for $j = 0, 1, 2, \cdots$. But if we consider the sum

$$(7.10)\quad H(z,\alpha) = \sum_0^\infty c_\nu H_\nu = (e^{2\pi i\alpha} - 1)\sum_0^\infty c_\nu z^{\nu-1}(\alpha(\alpha+1)\cdots(\alpha+\nu-1))^{-1},$$

it, too, must converge absolutely, and uniformly in α for a neighborhood of $\alpha = 0$ from which the point $\alpha = 0$ has been excluded by a small circle about it. It follows from the representation (7.7) that $\sum_0^\infty c_\nu g_\nu$ converges uniformly in α in the same domain, and finally we obtain by a further interchange of summation and integration that

$$F(z) = (2\pi i)^{-1}\int z^\alpha \sum_0^\infty c_\nu g_\nu d\alpha = (2\pi i)^{-1}\int z^\alpha g d\alpha,$$

where g is given by (7.7) with H as in (7.10). H obviously satisfies the condition (7.9) and theorem 7 is proved.

§ **8. Convergence of double power series.** Formula (7.7) shows that, taken a small circle about $\alpha = 0$, on its circumference the function $g(z, \alpha)$ is a power series in z, convergent uniformly and absolutely in z and α for all values of z in $|z| \leqq \rho$. Here ρ denotes the radius of a circle on whose closure $F(z)$ is regular for $z \neq 0$. Accordingly, the expression $z^\alpha g(z, \alpha)$ is an absolutely convergent power series in z and log z with coefficients depending on α, which converges uniformly and absolutely for α on the circumference mentioned and for z in $0 < \epsilon \leqq |z| \leqq \rho$ as long as $|\arg z|$ remains bounded. A further integration with respect to α leaves a uniformly and absolutely convergent power series in z and log z. It is easy to see that this power series is actually a power series in z and z log z. In fact, inspection of (7.8) and (7.10) shows that $g(z, \alpha)$ is of the form

$$\alpha^{-1}\sum_0^\infty z^n\alpha^{-n}p_n(\alpha)$$

where $p_n(\alpha)$ are certain power series in α. Thus $(\log z)^n$ appears in the development of $F(z)$ multiplied by z^n or a higher power of z.

Furthermore this development must converge uniformly for $0 \leqq |z| \leqq \rho$ and bounded $|\arg z|$. In fact, if $|z| < \epsilon$, $|\arg z| \leqq A$, and $\epsilon > 0$ is chosen small enough, then $|z \log z|/|\epsilon(Ai + \log \epsilon)| \leqq 1$ so that the remainder of the development at the point z is majorized by that at the point ϵe^{iA} with argument A. Hence:

THEOREM 8. Every solution of (7.1) which is continuous at $z = 0$ and regular for $z \neq 0$ and $|z| \leqq \rho$, can be developed there in a double power series in z and $z \log z$ which converges absolutely and converges uniformly for $|z| \leqq \rho$ and bounded $|\arg z|$.

Remark: The double power series in z and $z \log z$ is identical with the asymptotic development of $F(z)$, obtained above. Since there can be only one asymptotic development, it suffices to show that the remainder after a term of degree (m, n) in the series is $o(z^m(z \log z)^n)$. This in turn need be established only for $n = 0$ as the remainder in any case is a finite polynomial in z and $z \log z$ plus the remainder after a term of the form $(m, 0)$. Now set, in the above notations, $R_m(\epsilon e^{iA})$ equal to this remainder in the series of absolute values formed for $z = \epsilon e^{iA}$. All the terms of the corresponding remainder at an arbitrary z with $|z| \leqq \epsilon$, $|\arg z| \leqq A$, are of the type $|z^{m+s} (\log z)^k|$, $0 \leqq k \leqq m + s$, $s \geqq 1$. We have for small $|z|$

$$|z/\epsilon| \leqq |z \log z|/[\epsilon|\log \epsilon + iA|] \leqq 1,$$

$$|z^{m+s} (\log z)^k| \leqq (|z \log z|^{m+1}/(\epsilon\, |\log \epsilon + iA|)^{m+1})\epsilon^{m+s}|\log \epsilon + iA|^k,$$

so that for the remainder at z

$$R_m(z) \leqq \{|z \log z|^{m+1}/(\epsilon\, |\log \epsilon + iA|)^{m+1}\}\, R_m(\epsilon e^{iA}).$$

Thus

$$R_m(z) = o(|z|^m),$$

which yields the desired estimate.

§ 9. A theorem on surface waves.

THEOREM 9. Let $F(z)$ be a solution of (7.1), regular except at the origin for all z and continuous at the origin. Assume that for $0 \geqq \arg z \geqq -2\pi$, $|F(z)|$ is bounded and that $F(x)$ and $F(x^-)$ tend to zero as the real variable x tends to ∞. Then $F(0) \equiv 0$.

Proof: In a paper written in collaboration with K. Friedrichs† we found an explicit solution of (7.1), hereafter called $\Phi_0(z)$, which is regular in the whole plane with the exception of $z = 0$, and which is continuous at the origin where $\Phi_0(0) \neq 0$; $\Phi_0^{(n)}(z)$, $n = 2, 3, \cdots$ has the following asymptotic behavior for $|z| \to \infty$: For $0 \geqq \arg z \geqq -\pi/2$,

$$|\Phi_0^{(n)}(z) - A_0(-i)^n e^{-iz}| < C_n\, |z|^{-3/2};$$

for $-\pi/2 \geqq \arg z \geqq -3\pi/2$, $|\Phi_0^{(n)}(z)| < C_n|z|^{-3/2}$; and for $-3\pi/2 \geqq \arg z \geqq -2\pi$, $|\Phi_0^{(n)}(z) - \bar{A}_0 i^n e^{iz}| < C_n|z|^{-3/2}$. Here $A_0 \neq 0$ is a certain constant, $\bar{A}_0$ its conjugate, and C_n is a positive constant. The estimates hold uniformly as $|z| \to \infty$, i.e. for $|z| > \rho(C_n)$. We prove that $Q[F, \Phi_0^{(n)}] = 0$ for $n = 2, 3, \cdots$. Observe first that F tends uniformly to zero as $|z| \to \infty$ and $0 \geqq \arg z \geqq -2\pi$. [In fact, if the plane

† K. O. Friedrichs and Hans Lewy, *The Dock Problem*, Communications Appl. Math., 1 (1948), 135–148.

slitted along the positive axis is mapped conformally into the unit circle, F becomes a bounded function inside the unit circle with continuous boundary values which tend to zero as the point on the boundary moves into the image of $x = \infty$, $x^- = \infty$. Since F remains bounded, f is continuous at the image of ∞, which yields the statement.] Observe next that $Q[F, \Phi_0^{(n)}]$ may be evalued with the aid of a circle of large radius R about the origin, leading from R to R^-. We have, with arbitrary $\epsilon > 0$,

$$Q[F, \Phi_0^{(n)}] \equiv 2\int_R^{R^-} F(t)\Phi_0^{(n)}(t)dt + i(F(R) - F(R^-))\,(\Phi_0^{(n)}(R) - \Phi_0^{(n)}(R^-)),$$

$$|Q[F, \Phi_0^{(n)}]| \leqq 2\epsilon\left|\int_R^{R^-} \Phi_0^{(n)}(t)dt\right| + 2\epsilon|\Phi_0^{(n)}(R) - \Phi_0^{(n)}(R^-)|$$

$$\leqq 4\epsilon\,|A_0| + 4\epsilon\, C_n R^{-1/2}\cdot 2\pi + 4\epsilon\,(C_n R^{-3/2} + |A_0|)$$

whence

$$Q[F, \Phi_0^{(n)}] = 0, \qquad n = 2, 3, \cdots.$$

If we enter this result into the equations (6.14) and (6.15) which determine the coefficients of F in the development $F = \sum_0^\infty c_\nu\Phi_\nu$ we find that all c_ν vanish, whence $F(z) \equiv 0$, q.e.d.

COROLLARY: There is, but for a constant factor, only one solution $F(z)$ of (7.1) which is continuous at the origin, regular in the whole plane except at the origin and remains absolutely bounded for $|z| \to \infty$, $0 \geqq \arg z \geqq -2\pi$, and for which there is a constant A such on the positive axis $F(x) - Ae^{-ix} \to 0$, $F(x^-) - \bar{A}e^{ix} \to 0$, as x resp. $x^- \to \infty$.

Proof: The function constructed in the paper quoted, and called $\Phi_0(z)$ here, has the properties demanded of $F(z)$. If $\psi(z)$ were a similar function with $\psi(x) - Be^{-ix} \to 0$, $\psi(x^-) - \bar{B}e^{ix} \to 0$ as x resp. $x^- \to \infty$, then consider (notice $A_0 \neq 0$)

$$F(z) = \psi(z) - (B/A_0)\Phi_0(z).$$

$F(z)$ satisfies the conditions of theorem 9; hence $F(z) \equiv 0$.

Commun. in Pure and Appl. Mathematics
Vol. 4, 1951

On Minimal Surfaces with Partially Free Boundary

By HANS LEWY
Berkeley, California

Introduction

In his book [1] on Dirichlet's principle, R. Courant discusses in the sixth chapter the problem of minimal surfaces with boundaries partially restricted to curves and partially to surfaces. His results contain in particular the existence of a minimal surface bounded by a rectifiable Jordan arc γ whose endpoints lie on a closed, connected, and bounded surface S which otherwise has no point in common with γ, and on which the endpoints of γ can be joined by rectifiable arcs. This minimal surface is a surface μ of least area among all those which are continuous images of a semicircle $u^2 + v^2 \leq 1$, $v > 0$ of a u,v-plane, in such a way that the circumference is mapped topologically on γ minus endpoints, and that, if the point (u, v) tends to a point of the diameter of the semicircle, the distance of the corresponding point of μ from S tends to zero. The proof consists in minimizing the Dirichlet integral of the point vector of the surface among the surfaces of this type and invoking a theorem on conformal mapping of polygonal domains in order to prove the minimum area property.

While the existence problem can thus be considered solved, the nature of the point set γ' to which the point of μ tends when the parameter point (u, v) converges to a point of the u-axis, is treated in the book only if S is a plane. For this case, Riemann had furnished the argument showing that γ' is necessarily an analytic curve and that μ can be extended analytically across S. There exists, however, a short note by Courant [2] which sketches a proof that γ' is a continuous image of the diameter of the semicircle provided S is convex, and which asserts the rectifiability of γ' under further assumptions of global nature relative to γ and S.

In the present paper we are essentially concerned with the structure of γ' when the surface S is an orientable closed bounded surface which is locally analytic at each of its points. We show in our first part that this assumption insures rectifiability and continuity of γ' in its dependence on the point of the diameter.

In our second part we prove that γ' is analytic. The basic idea of this proof is similar to the one [3] by which we showed that a minimal surface can be analytically extended across an analytic boundary arc: a system of differential equations is formed which establishes a relation between the regular functions of a complex variable whose real parts generate μ, at points $u + iv$

and $u - iv$. The integration of this system furnishes the analytic extension of μ across S.

1. *Admissibility*

We shall call a surface S of the three-dimensional Euclidean space admissible at a point P of S if for some sphere Σ_ρ of radius ρ with P as center, the portion of S contained in Σ_ρ can, upon introduction of proper Cartesian coordinates, be written as, and is identical with,

$$z - z_0 = \psi(x - x_0, y - y_0), \qquad (x - x_0)^2 + (y - y_0)^2 + (z - z_0)^2 < \rho^2,$$

where ψ is a convergent power series in $x - x_0$, $y - y_0$, and x_0, y_0, z_0 are the coordinates of the point P. We call a surface S admissible if it is admissible at each of its points and if it is orientable, closed, connected, and bounded.

For example, the points $P(x, y, z)$ which satisfy an analytic relation $G(x, y, z) = 0$ form such an admissible surface S if

(1) there are points satisfying $G = 0$,

(2) for all such points not all three derivatives $\partial G/\partial x$, $\partial G/\partial y$, $\partial G/\partial z$ vanish simultaneously.

(3) these points are all connected and lie within a fixed sphere.

If S is admissible, a positive number p_S can be assigned such that all points Q of the space which lie within distance p_S from S have the property that one and only one point P, "the foot of Q", exists on S such that $PQ =$ distance from Q to S. For these points Q, the vector from the origin to Q can uniquely be written as

$$\mathbf{OQ} = \mathfrak{x} + w\xi$$

where $\mathfrak{x}$ is the vector from the origin to P and ξ is a unit normal depending continuously on P on S. Thus the metric of the portion of space in question becomes

$$dQ^2 = (d\mathfrak{x})^2 + 2w\, d\mathfrak{x}\, d\xi + w^2\, (d\xi)^2 + (dw)^2.$$

In this formula $(d\mathfrak{x})^2$ represents the metric on S while $(d\xi)^2$ and $d\mathfrak{x}\, d\xi$ are two other quadratic forms on S for which we have the estimates

$$(1) \qquad |\, d\mathfrak{x}\, d\xi \,| \leq M(d\mathfrak{x})^2, \qquad (d\xi)^2 \leq M(d\mathfrak{x})^2,$$

with fixed constant M. Thus, if p_S is chosen sufficiently small, and if $|\, w \,| \leq p_S$,

$$(d\mathfrak{x})^2 + 2w\, d\mathfrak{x}\, d\xi + w^2(d\xi)^2 \geq (d\mathfrak{x})^2(1 - 2\,|\, w \,|\, M - w^2 M) \geq 0$$

and accordingly

$$dQ^2 \geq (dw)^2.$$

Remark: When we speak of area of a surface or of length of a curve, we shall always understand these notions in terms of their integral expressions referred to the parameters through which surface or curve are given. This convention may in some cases lead to a larger value than a purely geometrical definition; e.g. the portion of the x-axis given in parametric form by $x = t - t^2$ from $t = 0$ to $t = 1$ has, according to this convention, the length

$$\int_0^1 | dx/dt | \, dt = \int_0^1 | 1 - 2t | \, dt = 1/2$$

while the "geometrical" length is 1/4; etc.

Let S be admissible and let σ be a surface which is the continuous image of a semicircle of the u,v-plane. Denote by $l(w)$ the length (zero, finite or infinite) of the intersection of σ with the surface parallel to, and at distance w from S. Then the area of σ satisfies the obvious inequality

$$\text{area of } \sigma \geq \int_{-p_S}^{p_S} l(w) \, dw. \tag{2}$$

For the portion of σ contained within distance p_S of S is equal to a double integral $\iint ds_w \, ds_p$ where ds_w measures the element of arc on the sets of distance w from S while ds_p is the arc length on perpendicular trajectories. But as we have seen, $ds_p \geq | dw |$, whence (2).

2. *Theorem*

We now formulate the problem of least area which we are discussing. Let S be admissible. Let γ be a rectifiable Jordan arc whose distinct endpoints lie on S but which contains no other point of S. Let γ have the additional property that *near its endpoints* the distance $w = w(Q)$ of the point Q of γ from S is a strictly monotonic function of the arc length s and remains smaller there than on the remainder of γ. Let the class K of surfaces consist of all surfaces which are continuous images of the semi-circle D: $u^2 + v^2 \leq 1$, $v > 0$ in such a way that the semi-circumference is mapped topologically on γ, while the distance of the point of the surface from S tends to zero if the parameter point (u, v) tends to a point of the diameter of the semicircle.

According to Courant [1] there is in K a surface of least area. We designate it by μ. The coordinates of the point of μ are harmonic functions in the interior of D. Moreover, the vector $X(u, v)$ to the point of μ satisfies in the interior of D the two scalar equations expressing conformality of the map from D to μ:

$$(\partial X/\partial u)^2 = (\partial X/\partial v)^2, \qquad (\partial X/\partial u)(\partial X/\partial v) = 0.$$

We wish to prove the

Theorem: μ contains a portion μ^* which has an area equal to that of μ, is itself a surface of K and is such that there is a conformal map from D to μ^* which remains continuous on the diameter $-1 \leq u \leq +1$, $v = 0$. On μ^* the image of the diameter is a rectifiable curve γ' of S which joins the endpoints of γ.

3. *A Topological Lemma*

Let α be a point of $u^2 + v^2 = 1$ such that the image A of α on μ has maximum distance from S among the points of γ. Suppose the orientation of the normals S such that the distance of a point from S is positive if the point is on the same side of S as A, and non-positive elsewhere. Choose a value $w' > 0$ sufficiently small so that $w' < w(\alpha) = w(A)$, $w' < p_S$ and that the distance of Q from S assumes the value w' at points Q_1 and Q_2 of γ within the two monotonicity intervals near the endpoints. Consider the set $i(w')$ of points P of D for which $w(P) = w'$ and which can be joined to α by an open set of points R such that $w(R) > w'$. Denote by P_1 and P_2 the originals of Q_1 and Q_2 on the circumference of the unit circle.

Lemma: $i(w')$ contains a Jordan arc joining P_1 with P_2. This arc $J(w')$ is uniquely determined.

Proof: $i(w')$ is a closed set, has (in view of the assumption on γ and the smallness of w') exactly the two points P_1 and P_2 in common with $u^2 + v^2 = 1$, and has positive distance from the u-axis because the function w tends to zero upon approach to the axis. Let $0 < \rho < 1$, and denote the set in common to D and $u^2 + v^2 \leq \rho^2$ by D_ρ. If ρ is sufficiently close to 1, $i(w')$ must intersect with $C_\rho : u^2 + v^2 = \rho^2$ near P_1 and near P_2 since we find on this circumference points near P_1 and near P_2 for which w is larger and for which it is smaller than w'. The intersection $i(w')\cdot C_\rho$ consists therefore of two closed and distinct sets $\pi_{1\rho}$ and $\pi_{2\rho}$, with $\pi_{1\rho}$ close to P_1 and $\pi_{2\rho}$ close to P_2. The intersection $i(w')\cdot D_\rho$ consists of a finite number of analytic arcs. Since, in D_ρ, $i(w')\cdot D_\rho$ separates points near α from points near the origin, $i(w')\cdot(D_\rho - C_\rho)$ must contain a connected subset τ connecting a point of $\pi_{1\rho}$ with a point of $\pi_{2\rho}$. This subset consists of analytic arcs. It must contain a Jordan curve $J_\rho(w')$ connecting a point of $\pi_{1\rho}$ with a point of $\pi_{2\rho}$ as is seen if we detach successively from τ all arcs both sides of which can be joined in D to α by open sets with $w > w'$, and all complete loops of $i(w')\cdot(D_\rho - C_\rho)$. Moreover, there are in $D_\rho - C_\rho$ at most finitely many Jordan arcs connecting a point of $\pi_{1\rho}$ with a point of $\pi_{2\rho}$, since we have only a finite set of analytic arcs of which these Jordan arcs must be composed.

Let $\bar{\rho} > \rho$. The intersection of a $J_{\bar{\rho}}(w')$ with $D_\rho - C_\rho$ contains necessarily a $J_\rho(w')$. Now consider an increasing sequence ρ_k, $k = 1, 2, \cdots$, of numbers close to and tending to 1. Among the finitely many $J_{\rho_1}(w')$ there is one contained in infinitely many $J_{\rho_k}(w')$ for $k > 1$; we denote it by J_1. Moving from a point of J_1 on $J_{\rho_k}(w')$ in either direction until we first hit C_{ρ_2}, we obtain a $J_{\rho_2}(w')$ which contains J_1. Infinitely many of the $J_{\rho_k}(w')$ must now contain the same $J_{\rho_2}(w')$, denoted by J_2. Continuing in this way we obtain a sequence J_1, J_2, $\cdots$ of Jordan arcs, each contained in the following. Since J_k joins a point of $\pi_{1\rho_k}$ to a point of $\pi_{2\rho_k}$, and since the points $\pi_{1\rho_k}$ tend to P_1 and those of $\pi_{2\rho_k}$ tend to P_2, the set limit of J_k as $k \to \infty$ is a Jordan arc joining P_1 to P_2. Let this limit be denoted by $J(w')$. This is the first part of the Lemma. There can be no other Jordan arc $J'(w')$ in $i(w')$ which joins P_1 and P_2. Otherwise $J(w')$ would pass through points separated in D from α by $J'(w')$ or conversely; but all points of $i(w')$ can be connected with α by open sets on which $w > w'$. The Lemma follows.

Denote by $D(w')$ the simply connected region of D bounded by the circumference $u^2 + v^2 = 1$ between P_1 and P_2 and by $J(w')$; denote the image on μ of $J(w')$ by g and the small parts of γ from its endpoints to Q_1 and Q_2 by γ_1 and γ_2.

It is possible to assume that g is rectifiable for proper choice w_0 of w'. This follows from the finiteness of the area of μ and the estimate (2) applied to it. Consider that portion of μ which is the image of $D - D(w_0)$. It must be a surface of least area, which is the image of a simply connected parameter domain; it is bounded by the curve $\gamma_1 + g + \gamma_2$ and by S, in the sense that the distance from S tends to zero as the parameter point tends to the u-axis. For if there were another surface of less area of the same description, it could be pieced together with the portion of μ corresponding to $D(w_0)$. The so-composed surface would lie in K, but have less area than μ, contradicting the least area property of μ.

This fact makes it possible to assume in the discussion of the intersection of μ with S that the surface of least area μ is itself bounded by S and by a rectifiable curve γ consisting of $\gamma_1 + g + \gamma_2$; these are prescribed as the image of the semicircle circumference in such a way that the circumference can be divided into three successive parts: on the first part, w rises strictly monotonically from 0 to w_0, on the second w equals w_0, and on the third w descends strictly monotonically to zero. Furthermore $w_0 < p_S$. We shall nowhere have to assume that γ is a Jordan arc. This hypothesis, necessary in order to insure existence of μ at the outset, is dropped as soon as the existence of a surface of least area bounded by γ and S can be taken for granted.

The domains $D(w)$, $0 < w < w_0$, are increasing monotonically as w decreases. For let $w_1 > w_2$. $J(w_1)$ and $J(w_2)$ have no points in common, and they join respectively points on the circumference with $w = w_1$ and $w = w_2$. Since the latter are closer to the diameter than the former, $D(w_2)$ contains $D(w_1)$.

The Jordan curves $J(w)$ possess the property of lower semi-continuity from the right in w, in the following sense. Let $w_1, w_2, \cdots$ be a monotonically decreasing sequence tending to some value $w' > 0$. Consider the set J' of points which are limits of points $P_1, P_2, \cdots$, lying respectively on $J(w_1), J(w_2), \cdots$. Then J' contains $J(w')$.

For let P' be on $J(w')$, $\bar{P}$ a point close to P' and belonging to an open set on which $w > w'$ and through which P' can be joined with α, and let $w(\bar{P})$ equal $\bar{w}$. Join $\bar{P}$ and α by a path on which $w > w'$. On this path the minimum of w is still greater than w', whence greater than some $w'' > w'$. It follows that for $w'' > w_i$, $\bar{P}$ lies in $D(w_i)$ as $D(w_i)$ certainly contains all points joinable to α on a set where $w > w_i$. P' however is not in $D(w_i)$ as it lies on $J(w')$ and $w' < w_i$. Hence there must be on the straight segment $\bar{P}P'$ a point P_i of $J(w_i)$. This shows that P' is a limit point of points of $J(w_i)$ for $i \to \infty$, q.e.d.

Denote by $L^0(w)$ the length of the image of $J(w)$ on μ. In view of the semi-continuity from the right of $J(w)$ and the semi-continuity of the arc length we conclude the lower semi-continuity from the right of $L^0(w)$. Denote by $L(w)$ the length (possibly infinite) of the set intersection of μ with the surface parallel to S at distance w. From (2) we gather, in view of $L(w) \geq L^0(w)$, that

$$\text{(3)} \qquad \text{area of } \mu \geq \int_{-ps}^{ps} L(w)\, dw \geq \int_0^{w_0} L(w)\, dw \geq \int_0^{w_0} L^0(w)\, dw.$$

4. *Comparison Surface*

With γ we associate a surface σ_γ bounded by γ and S, by considering the totality of segments normal to S that end at a point Q of γ. Denote by s the arc length on γ. The distance of a point Q of γ from S becomes a continuous function $f(s)$ which starts out by rising monotonically from 0 to w_0 and after an interval of constancy, decreases monotonically to 0. The points of σ_γ are defined as the continuous images of the points of the region of an s, w-diagram, contained between the s-axis and the curve $w = f(s)$. About the area of this σ_γ we can make the statement that

$$\text{area of } \sigma_\gamma \leq \int_0^{w_0} l(w)\, dw$$

where $l(w)$ has the meaning defined above. For this $l(w)$ we shall establish an estimate. $l(w)$ is the length of that curve of σ_γ which is the image of the straight segment of the s,w-diagram from a point s_1, $f(s_1) = w$ to the point s_2, $f(s_2) = w$. Write $s_1 = s_1(w)$, $s_2 = s_2(w)$. We have with

$$R\, ds = ((d\mathfrak{x})^2 + 2w\, d\mathfrak{x}\, d\xi + w^2(d\xi)^2)^{1/2}$$

that

$$l(w) = \int_{s_1(w)}^{s_2(w)} R\, ds.$$

Here $\mathfrak{x}$, ξ are always the same functions of s (no matter what w) to wit, the point vector and normal vector of the foot of γ on S. For $w < w_0$ sufficiently small, we have by (1)

$$dl(w) = dw \int_{s_1(w)}^{s_2(w)} R^{-1}(d\mathfrak{x}\, d\xi + w(d\xi)^2)/ds + R(s_2\,,\, w)\, ds_2 - R(s_1\,,\, w)\, ds_1\,;$$

$$|\, dl(w) + R(s_1\,,\, w)\, ds_1 - R(s_2\,,\, w)\, ds_2\,|$$

$$\leq | dw | \int_{s_1}^{s_2} R^{-1}(s, w)M(1 + w)(d\mathfrak{x}/ds)^2 \, ds$$

$$\leq | dw | \int_{s_1}^{s_2} R^{-1}M(1 + w)R^2(1 - 2wM - w^2M)^{-1} \, ds$$

$$\leq | dw | \, l(w)M(1 + w)(1 - 2wM - w^2M)^{-1}.$$

Abbreviating, set

$$\int_{w_0}^{w} - R(s_1 , w) \, ds_1 + R(s_2 , w) \, ds_2 = A(w),$$

$$M(1 + w)(1 - 2wM - w^2M)^{-1} = a(w),$$

$$| dl(w) - dA(w) | \leq | dw | \, a(w)l(w),$$

$$dA(w) \geq 0 \qquad \text{for} \qquad dw < 0.$$

Therefore $l(w)$ satisfies for $dw < 0$ the inequality

$$| dl(w) | \leq -dw \, a(w)l(w) + dA(w), \qquad l(w) \geq 0.$$

Now if $Y(w)$ solves the corresponding equality

$$dY(w) = -dw \, a(w)Y(w) + dA(w), \qquad Y(w_0) = l(w_0),$$

i.e. if

$$(4) \qquad Y(w) = \exp\left\{-\int_{w_0}^{w} a(t) \, dt\right\}\left(l(w_0) + \int_{w_0}^{w} \exp\left\{\int_{w_0}^{t} a(t') \, dt'\right\} dA(t)\right)$$

then, certainly, by a familiar conclusion,

$$l(w) \leq Y(w)$$

for $0 \leq w \leq w_0$. We verify that for $0 \leq w \leq \hat{w} \leq w_0$ also

$$Y(w) = \exp\left\{-\int_{\hat{w}}^{w} a(t) \, dt\right\}\left(Y(\hat{w}) + \int_{\hat{w}}^{w} \exp\left\{\int_{\hat{w}}^{t} a(t') \, dt'\right\} dA(t)\right).$$

We now compare the $Y(w)$ with $L^0(w)$ defined above. Since

$$\text{area of } \mu \leq \text{area of } \sigma_\gamma$$

by the minimal property of μ, and since

$$\text{area of } \sigma_\gamma \leq \int_0^{w_0} l(w) \, dw \leq \int_0^{w_0} Y(w) \, dw,$$

we take from (3) the inequality

$$\int_0^{w_0} L^0(w) \, dw \leq \int_0^{w_0} Y(w) \, dw$$

which makes it impossible that $L^0(w) > Y(w)$ for all w in $0 < w < w_0$. Now, either there is a $\hat{w} > 0$ such that for $0 < w < \hat{w}$ always $L^0(w) > Y(w)$ or there exists a sequence of positive values $w_1, w_2, \cdots$ tending monotonically to zero, for which $L^0(w_i) \leq Y(w_i)$. The first alternative is impossible. Otherwise we could find a maximum value $\hat{w}$ of w so that for $0 < w < \hat{w}$, $L^0(w) > Y(w)$ while arbitrarily near and above $\hat{w}$ there are values of w for which $L^0(w) \leq Y(w)$. The semi-continuity of $L^0(w)$ and the continuity of $Y(w)$ show that then $L^0(\hat{w}) \leq Y(\hat{w})$. Now consider the surface $\sigma_{\hat{\gamma}}$ generated by the normal segments of S passing through the image $\hat{\gamma}$ of $J(\hat{w})$ on μ. We form a surface in K bounded by γ and S and being a continuous image of a simply connected domain by piecing together this new $\sigma_{\hat{\gamma}}$ with the portion of μ which is the image of the domain $D(\hat{w})$ defined above. The portion π of μ to be altered is replaced by one of smaller area, for

$$\text{area of } \pi \geq \int_0^{\hat{w}} L^0(w)\, dw > \int_0^{\hat{w}} Y(w)\, dw \geq \text{area of } \sigma_{\hat{\gamma}} .$$

But this contradicts the minimum property of μ. Thus there exists a set of values $w_1, w_2, \cdots$ descending monotonically to zero for which $L^0(w_i) \leq Y(w_i)$. The curves on μ which are the images of $J(w_i)$ contain, because of the compactness of bounded curves of bounded length, a subsequence which converges to a limit curve γ' whose length cannot exceed $Y(0)$. γ' lies necessarily on S and joins the two endpoints of γ.

If we consider the portion μ^* of μ which is the image of the sum of all $D(w_i)$ for $w_i \to 0$, then μ^* is indeed bounded by γ and S, and has area neither greater than that of μ itself nor smaller, because of the minimum property of μ. We shall therefore consider μ^* as equivalent with the surface μ of least area bounded by S and γ.

We have not demonstrated that as w_i descends to zero the sum of the sets $D(w_i)$ is the whole semicircle $u^2 + v^2 \leq 1$, $v > 0$. We know, however, that this becomes true if the u,v-plane is subjected to a conformal mapping, since the sum of the $D(w_i)$ for $w_i \to 0$ is, like every one of the summands, a simply connected domain. We shall therefore consider that such conformal mapping has been performed, if necessary, with the added precaution that the circumference of the semicircle is mapped into itself and the point α remains fixed. The coordinates of the point of μ remain harmonic functions in D, having the property that w as function of (u, v) tends to 0 as (u, v) approaches a point of the diameter. More is true: the Jordan arcs $J(w)$ can now be assumed to have no other limit points than those of the diameter, all points of which are limits of points of $J(w)$ as w descends to zero. We shall show that the map of the semicircle on μ remains continuous on the diameter which is mapped into γ'.

To this end we perform a conformal mapping of $D(w_i)$ into the upper half x,y-plane in such a way that α goes into the infinite point of the x-axis, and the two portions of the semicircle circumference which belong to $D(w_i)$, with endpoint α go into the rays from 1 to ∞ and from $-\infty$ to -1, while $J(w_i)$ goes into the segment from -1 to 1 of the x-axis. Taking, if necessary, a proper subsequence of the w_i, it can be supposed that fixed interior points of the upper x,y-plane go into interior points P_i of the domain $D(w_i)$ with P_i tending to an interior limit point P of D as $w_i \to 0$.

We now consider a Cartesian coordinate of the point Q of μ, which is first of all a harmonic function of u, v in D and becomes, through the diverse mappings from the u,v-plane into the x,y-plane, a harmonic function $U_i(x, y)$. The following properties hold for $U_i(x, y)$: $U_i(x, y) \to U(x, y)$ as $i \to \infty$; the Dirichlet integral of $U_i(x, y)$ over the half plane remains, as $i \to \infty$, bounded by the area of μ; the function $U_i(x, y)$ assumes continuous boundary values $f_i(x)$ on $y = 0$, these being the values of the coordinate on a portion of γ and on $J(w_i)$; the total variation $T[f_i(x)]$ of $f_i(x)$ over the x-axis is bounded by the sum m of the length of γ and the bound $Y(0)$ of the length of the image of $J(w_i)$ on μ. We now apply a reasoning similar to that employed in [3]. Designating by $\theta(P, x)$ the angle from the negative y-axis to the vector from P to a point x of the x-axis, we have

$$U_i(P) = -\pi^{-1} \int_{-\infty}^{\infty} \theta(P, x)\, df_i(x) + \text{const}.$$

By Lemma 2 of the quoted paper, the total variation T_p of $U_i(x, y)$ over segments perpendicular (or parallel) to the x-axis is bounded by $T[f_i(x)]$. Because of the semi-continuity of the total variation, the total variation T_p of $U(x, y)$ is likewise bounded by m. Thus $U(x, y)$ assumes limit values when $y \to 0$, to be denoted by $f(x)$. Now, again by the semicontinuity of the total variation we have

$$\int_{-\infty}^{\infty} | df(x) | \leq \liminf_{y \to 0} \int_{-\infty}^{\infty} | d_x U(x, y) |$$

$$\leqq \liminf_{y \to 0} \liminf_{i \to \infty} \int_{-\infty}^{\infty} | d_x U_i(x, y) | \leq m.$$

Representing $U(x, y)$ by an integral over its values on a parallel to the x-axis and passing to the limit as the parallel approaches the axis, we obtain $U(P) = \pi^{-1} \int_{-\infty}^{\infty} f(x)\, d\theta\,(P, x)$ which shows that $U(x, y)$ assumes the boundary value $f(x)$ upon approach to any point x of continuity of $f(x)$. If there existed a point x_0 of discontinuity of $f(x)$, then the boundedness of $T[f(x)]$ would require the existence of $f(x_0-)$ and $f(x_0+)$ with $| f(x_0+) - f(x_0-) | = \epsilon > 0$. According to the proof of a well-known Lemma (cf. [1], p. 18), an arbitrarily small neighborhood of x_0 contains circumferences of semicircles about x_0, whose radii form a set of positive measure, such that the total variation of $U(x, y)$ remains $< \epsilon/2$ on these circumferences provided the Dirichlet integral of U is bounded. Since there are no more than enumerably many discontinuities of $f(x)$, the end-points of such circular arcs may be taken as points x_1 and x_2 of continuity of $f(x)$ where $U(x, y)$ assumes the boundary values $f(x_1)$ and $f(x_2)$. It follows that $| f(x_1) - f(x_2) | < \epsilon/2$, as well as $> \epsilon/2$, if x_1 and x_2 are sufficiently close to x_0. This contradiction proves continuity of $f(x)$ for all x.

5. *Transversality Condition*

We now establish in a three-dimensional neighborhood N of the curve of S a vector field η which is continuously differentiable and becomes tangential to S when the carrier point is a point of S. Denoting by X_1, X_2, X_3 the coordinates of a point and by η_1, η_2, η_3 the components of the vector of the field at this point, we can construct a continuous deformation of a subneighborhood N' of N with the aid of the differential equations

$$dX_1 : \eta_1 = dX_2 : \eta_2 = dX_3 : \eta_3 = dt$$

where t is a deformation parameter, whose value 0 corresponds to the undeformed state. The deformation obviously transforms points of S into points of S because there the vector η is tangential. Take two points a and b, $-1 < a < b < 1$, of the x-axis and a small rectangle R in the x,y-plane with the segment $a \leq x \leq b$ as base. In R we define a continuously differentiable function $h(x, y)$ which assumes the value 0 on the other three sides of R, is equal to 1 on all points of a subrectangle R' on the same base, and lies between zero and 1 on $R - R'$; we complete the definition of h by setting it equal to zero in the upper half plane minus R. Now we subject μ to the deformation $t = h(x, y)\beta$ where we make the number β tend to zero. The stationary character of the Dirichlet integral over the half plane

$$\iint [(\partial X/\partial x)^2 + (\partial X/\partial y)^2]\, dx\, dy$$

where X_1, X_2, X_3, are the coordinates of the points of μ as functions of (x, y), yields in the familiar manner of the calculus of variations that for the vectors X and η

$$\iint_R [X_x(h\eta)_x + X_y(h\eta)_y]\, dx\, dy = 0. \tag{5}$$

Here we interpret the integral over R as the limit of integrals over rectangles $\hat{R}$ obtained from R by slicing off a rectangular portion containing the segment $a \leq x \leq b$ of the axis and letting $\hat{R}$ approach R. We denote by Y_1, Y_2, Y_3 the coordinates of a conjugate minimal surface. We find from (5) first

$$\lim_{y \to 0} \int_a^b h(x, y)\eta(X)X_y\, dx = 0, \tag{6}$$

or, in terms of the vector Y of the conjugate surface

$$\lim_{y \to 0} \int_a^b h(x, y)\eta(X)d_xY(x, y) = 0. \tag{7}$$

Now, upon repeating the proof of Lemma 3 in [3], we find that the rectifiability of the boundary $\gamma + \gamma'$ of μ and the continuity of $X(x, 0)$ imply the continuity and rectifiability of $Y(x, 0)$, and the equation

$$T_a^b[Y_j(x, 0)] = \lim_{y\to 0} T_a^b[Y_j(x, y)], \qquad j = 1, 2, 3, \text{ for all } a, b.$$

Hence (7) becomes

$$\int_a^b h(x, 0)\bar{\eta}(x)\, dY(x, 0) = 0, \tag{7'}$$

where $\bar{\eta}(x) = \eta(X(x, 0))$. Since $h(x, 0)$ is an arbitrary function in C^1, rising on $[a, b]$ from 0 to 1 and descending to 0, and since $\bar{\eta}(x)$ is continuous and $Y(x, 0)$ continuous of bounded variation, we can let $h(x, 0)$ tend to the function 1 in $a < x < b$ and obtain

$$\int_a^b \bar{\eta}(x)\, dY(x, 0) = 0. \tag{8}$$

Here $\bar{\eta}(x)$ is an arbitrary vector tangential to S at $X(x, 0)$, embeddable in a continuously differentiable vector field.

6. *Analyticity*

Since the admissible surface S is analytic, the neighborhood of each of its points possesses a development of one of the coordinates in terms of the two others which after a movement assumes the form

$$X_3 = \psi(X_1, X_2) = \sum_{j+k\geq 2} a_{jk}X_1^jX_2^k \tag{9}$$

and represents S in that neighborhood.

Theorem: γ' is analytic and μ can be continued analytically across S.

The proof establishes a system of differential equations connecting the regular functions $\xi_j = X_j + iY_j$, $j = 1, 2, 3$, of the complex variable $z = x + iy$ with the complex conjugate $X_j - iY_j$ at the conjugate point $\bar{z} = x - iy$. Pick a point of the x-axis between -1 and 1. To it corresponds a point of γ'. Applying a translation along the x-axis in the x,y-plane and a movement of S in the three-dimensional space we shall suppose that the point in question on the x-axis is its origin and that to it corresponds the origin of the three-dimensional space in whose neighborhood S assumes the form (9). We introduce three auxiliary functions Ξ_1, Ξ_2, Ξ_3 by the following differential equations

$$\begin{aligned} d\Xi_1 + \psi_1(\Xi_1, \Xi_2)\, d\Xi_3 &= d\xi_1 + \psi_1(\Xi_1, \Xi_2)\, d\xi_3 \\ d\Xi_2 + \psi_2(\Xi_1, \Xi_2)\, d\Xi_3 &= d\xi_2 + \psi_2(\Xi_1, \Xi_2)\, d\xi_3 \\ -\psi_1(\Xi_1, \Xi_2)\, d\Xi_1 - \psi_2(\Xi_1, \Xi_2)\, d\Xi_2 + d\Xi_3 &= 0 \end{aligned} \tag{10}$$

where

$$\psi_1(\Xi_1, \Xi_2) = \partial\psi(\Xi_1, \Xi_2)/\partial\Xi_1, \qquad \psi_2(\Xi_1, \Xi_2) = \partial\psi(\Xi_1, \Xi_2)/\partial\Xi_2,$$

with the initial conditions $\Xi_1 = \Xi_2 = \Xi_3 = 0$ at $z = 0$. The three functions ξ_1, ξ_2, ξ_3, regular in the upper half plane, are continuous upon approach to the real axis and their total variations along segments parallel to the x-axis tend to their values for these segments on the x-axis, by virtue of Lemma 1 of [3]. The determinant of the coefficients of the differentials on the left hand of (10) is 1 at the origin and remains therefore $\neq 0$ in a complex neighborhood of the Ξ_1, Ξ_2-space. It becomes therefore possible to replace (10) by the algebraically equivalent system

$$(11) \qquad \begin{aligned} d\Xi_1 &= \Delta^{-1}\{(1 + \psi_2^2)\, d\xi_1 - \psi_1\psi_2\, d\xi_2 + \psi_1\, d\xi_3\} \\ d\Xi_2 &= \Delta^{-1}\{-\psi_1\psi_2\, d\xi_1 + (1 + \psi_1^2)\, d\xi_2 + \psi_2\, d\xi_3\} \\ d\Xi_3 &= \Delta^{-1}\{\psi_1\, d\xi_1 + \psi_2\, d\xi_2 + (\psi_1^2 + \psi_2^2)\, d\xi_3\} \end{aligned}$$

with $\Delta = 1 + \psi_1^2 + \psi_2^2$. This system is employed in its integrated form for the construction of the functions Ξ_1, Ξ_2, Ξ_3 in the part $y \geq 0$ of a neighborhood of the origin of the x,y-plane. As in [3], we solve the integral equations

$$(12) \qquad \begin{aligned} \Xi_1(z) &= \int \Delta^{-1}\{(1 + \psi_2^2)\, d\xi_1 - \psi_1\psi_2\, d\xi_2 + \psi_1\, d\xi_3\} \\ \Xi_2(z) &= \int \Delta^{-1}\{-\psi_1\psi_2\, d\xi_1 + (1 + \psi_1^2)\, d\xi_2 + \psi_2\, d\xi_3\} \\ \Xi_3(z) &= \int \Delta^{-1}\{\psi_1\, d\xi_1 + \psi_2\, d\xi_2 + (\psi_1^2 + \psi_2^2)\, d\xi_3\} \end{aligned}$$

by successive approximations. We first integrate over a portion of the y-axis and, with the values determined there as initial values, we integrate over segments parallel to the x-axis. The crux of the matter is that

(i) all of the successive approximations exist and are continuous in a fixed neighborhood of the origin for $y \geq 0$;

(ii) they converge uniformly;

(iii) the functions Ξ_1, Ξ_2, Ξ_3 thus found are uniquely determined by (12) and regular for $y > 0$;

(iv) on $y = 0$, Ξ_1, Ξ_2, Ξ_3 have bounded derivatives with respect to $s = \int [(dX_1(x, 0))^2 + (dX_2)^2 + (dX_3)^2]^{1/2}$, almost everywhere in s, because the same holds for ξ_1, ξ_2, ξ_3.

The proof of these facts runs precisely as the proof for the analogous statements concerning an analogous differential equation in [3]. The only difference is that here we have three simultaneous differential equations for three functions instead of one equation for one function. We therefore omit a repetition of the argument.

From (10) we gather that $\Xi_1 = X_1, \Xi_2 = X_2, \Xi_3 = X_3$ satisfies equations (12) *on the x-axis*. This will follow if (10) or (11) can be proved to be satisfied by these substitutions on $y = 0$, almost everywhere in the variable s. Now as far as the third equation in (10) is concerned, it expresses, upon this substitution, the fact that the vector dX is perpendicular to the vector of components $-\psi_1(X_1, X_2), -\psi_2, 1$; this is a consequence of γ' being a curve of S. As to the first equation of (10), consider a vector field $\eta(X) = (1, 0, \psi_1(X_1, X_2))$. This field is tangential to S when the carrier point is on S, and it is analytic in the neighborhood of the origin. If we consider η along γ' and set, for a neighborhood of the origin of the x-axis, $\eta = \bar{\eta}(x)$, then (8′) yields by differentiation with respect to the value s at the upper limit b,

$$dY_1 + \psi_1(X_1, X_2)\, dY_3 = 0$$

almost everywhere in s. Then after substituting X_1, X_2, X_3 for Ξ_1, Ξ_2, Ξ_3 the first equation (10) becomes an identity. The second equation in (10) follows similarly from the consideration of the field $\eta = (0, 1, \psi_2(X_1, X_2))$ which on S is tangential to S. We therefore have established (10) almost everywhere in s for $y = 0$, whence (12) is fulfilled there for $\Xi_1 = X_1, \Xi_2 = X_2, \Xi_3 = X_3$. The solution of (12) being unique these relations must hold for $y = 0$.

Consider the points $z = x + iy$, with $y < 0$, such that $\bar{z} = x - iy$ falls in the neighborhood where the functions ξ_i and Ξ_i exist and are regular. Set

$$\bar{\xi}_i(z) = -\xi_i(\bar{z}) + 2\Xi_i(\bar{z}).$$

The function $\xi_i(z)$ is a conjugate of a regular function, taken for the conjugate value of the argument, hence is necessarily regular for $y < 0$. Upon approach to the axis from the lower half plane ξ_i remains continuous and assumes the boundary values

$$-\bar{\xi}_i(x) + 2\bar{\Xi}(x) = X_i(x, 0) + iY_i(x, 0),$$

i.e. the same boundary values as upon approach from the upper half plane. Hence $\xi_i(z)$ is analytic in a full neighborhood of the origin and the analytic extensibility of X across the x-axis is proved.

Bibliography

1. Richard Courant, *Dirichlet's Principle, Conformal Mapping, and Minimal Surfaces*, Interscience Publishers, New York, 1950.
2. Richard Courant, *On Plateau's problem with free boundaries*, Proceedings of the National Academy of Science, Volume 31, No. 8, 1945, pp. 242–246.
3. Hans Lewy, *On the boundary behavior of minimal surfaces*, Proceedings of the National Academy of Science, Volume 37, No. 2, 1951, pp. 103–110.

Reprinted from the Proceedings of the NATIONAL ACADEMY OF SCIENCES,
Vol. 37, No. 2, pp. 103–110. February, 1951

ON THE BOUNDARY BEHAVIOR OF MINIMAL SURFACES

BY HANS LEWY

UNIVERSITY OF CALIFORNIA

Communicated by G. C. Evans, December 16, 1950

This paper deals with the analytic extension of a minimal surface across its boundary.

LEMMA 1: *Let $U(x, y)$ be harmonic and absolutely bounded in the half plane $y > 0$. Assume $U(x, y)$ to possess*[1] *continuous boundary values $U(x, 0)$ on $y = 0$ which admit limits $U(\infty, 0)$ and $U(-\infty, 0)$. Assume furthermore that the total variation from $x = -\infty$ to $x = \infty$ of $U(x, 0)$, denoted by $T_{-\infty}^{\infty}[U(x, 0)]$, is bounded. Denote similarly by $T_a^b[U(x, y)]$ the total variation of $U(x, y)$ as x ranges from a to b where a and $b > a$ are any two numbers including $\pm\infty$, and y is kept constant. Then, for $y > 0$,*

$$T_{-\infty}^{\infty}[U(x, y)] \leq T_{-\infty}^{\infty}[U(x, 0)],$$

and

$$\lim_{y \to 0} T_a^b[U(x, y)] = T_a^b[U(x, 0)].$$

Proof: $U(x, y) = U(P)$ is for $y > 0$ expressed by the formula

$$U(P) = \pi^{-1}\int_{Q=-\infty}^{\infty} f(Q)d\theta(PQ)$$

where $\theta(PQ)$ is the angle from the negative y-axis to the vector from P to Q, and $f(Q) = U(Q)$ for Q on the x-axis. Integration by parts yields

$$U(P) = -\pi^{-1}\int \theta(PQ)df(Q) + (f(\infty) - f(-\infty))/2. \qquad (1)$$

Differentiating with respect to x we find

$$(\partial/\partial x)U(x, y) = \pi^{-1}\int -\frac{\partial}{\partial x}\theta(PQ)df(Q), \qquad (2)$$

which we also write with θ_x instead of $(\partial/\partial x)\theta(PQ)$ and where we note that $-\theta_x$ is positive. Thus

$$T_{-\infty}^{\infty}[U(x, y)] = \int_{-\infty}^{\infty}|dU(x, y)| \leq \int\int_{-\infty}^{\infty} - \theta_x\, dx|df(Q)|/\pi. \quad (3)$$

Now $\int_{-\infty}^{\infty} - \theta_x\, dx = \pi/2 - (-\pi/2) = \pi$. Hence the desired result

$$T_{-\infty}^{\infty}[U(x, y)] \leq T_{-\infty}^{\infty}[U(x, 0)].$$

Similarly $T_a^b[U(x, y)] \leq \pi^{-1}\int_{-\infty}^{\infty}[\theta(P_1Q) - \theta(P_2Q)]|df(Q)|$, $P_1 = P(a, y)$, $P_2 = P(b, y)$. The bracketed expression is bounded by π and tends uniformly to π as $y \to 0$, for $a + \epsilon \leq Q \leq b - \epsilon$, and to 0 for $Q \geq b + \epsilon$ or for $Q \leq a - \epsilon$ if $\epsilon > 0$ is assigned sufficiently small. For the two missing intervals from $a - \epsilon$ to $a + \epsilon$ and from $b - \epsilon$ to $b + \epsilon$ we gather from the continuity of $f(Q)$ which entails (as is familiar) that of $T_a^x[f(Q)]$, that their contribution to the last integral becomes for all $y > 0$ less than an arbitrary small $\epsilon' > 0$ provided ϵ is sufficiently small. Thus $\overline{\lim}_{y\to 0} \int_{-\infty}^{\infty}[\theta(P_1Q) - \theta(P_2Q)]|df(Q)| \leq \pi \int_{a+\epsilon}^{b-\epsilon}|df(Q)| + \pi\epsilon' \leq \pi\int_a^b|df(Q)| + \pi\epsilon'$, whence

$$\overline{\lim_{y\to 0}}\, T_a^b[U(x, y)] \leqq T_a^b[U(x, 0].$$

But since also lower semicontinuity holds, i.e.,

$$\underline{\lim_{y\to 0}}\, T_a^b[U(x, y)] \geq T_{a+\epsilon}^{b-\epsilon}[U(x, 0)]$$

no matter how small ϵ, hence also for $\epsilon = 0$, we have the desired continuity of $T_a^b[U(x, y)]$ as $y \to 0$.

A theorem for the circle analogous to Lemma 1 was proved by Evans.[2]

LEMMA 1′: *If $U_1(x, y)$, $U_2(x, y)$, $U_3(x, y)$ are harmonic and absolutely bounded in $y > 0$ and assume on $y = 0$ continuous boundary values of bounded total variation, then the length $L(y)$ of the curve of coordinates $U_1(x, y)$, $U_2(x, y)$, $U_3(x, y)$ for y fixed and $-\infty < x < \infty$ tends to the length L_0 of the curve $U_1(x, 0)$, $U_2(x, 0)$, $U_3(x, 0) = f_1(Q), f_2(Q), f_3(Q)$, as $y \to 0$.*

Proof: $\pi L(y) = \pi \int_{-\infty}^{\infty} dx\,[(\partial U_1(x, y)/\partial x)^2 + (\partial U_2/\partial x)^2 + (\partial U_3/\partial x)^3]^{1/2}$

$$= \int_{-\infty}^{\infty} dx[(\int_{-\infty}^{\infty} - \theta_x\, df_1(Q))^2 + \ldots]^{1/2}$$
$$\leq \int_{-\infty}^{\infty} dx[(\int_{-\infty}^{\infty} - \theta_x|df_1(Q)|)^2 + \ldots]^{1/2}$$

since $-\theta_x > 0$.

Now

$$[(\int_{-\infty}^{\infty} - \theta_x|df_1(Q)|)^2 + \ldots]^{1/2} \leq \int_{-\infty}^{\infty} - \theta_x[|df_1(Q)|^2 + \ldots]^{1/2}$$

because this formula expresses the fact that a straight line joining the origin to the point with coordinates $\int_{-\infty}^{\infty} - \theta_x|df_1(Q)|$, $\int_{-\infty}^{\infty} - \theta_x|df_2(Q)|$, $\int_{-\infty}^{\infty} - \theta_x|df_3(Q)|$ is the shortest line of this property. Hence

$$\pi L(y) \leq \int_{-\infty}^{\infty} dx \int_{-\infty}^{\infty} - \theta_x(|df_1|^2 + \ldots)^{1/2}$$

and by Lemma 1,

$$\overline{\lim_{y\to 0}}\ \pi L(y) \leq \lim_{y\to 0} \int_{-\infty}^{\infty} dx \int_{-\infty}^{\infty} - \theta_x(|df_1(Q)|^2 + \ldots)^{1/2} =$$
$$\pi \int_{-\infty}^{\infty} [|df_1(Q)|^2 + \ldots]^{1/2} = \pi L_0.$$

But the semicontinuity of the length requires

$$\underline{\lim_{y\to 0}}\ L(y) \geq L_0.$$

Thus $L(y)$ is continuous as $y \to 0$, q. e. d.

LEMMA 2: *Under the hypotheses of Lemma 1, the total variation of U on segments perpendicular to the x-axis is at most* $T_{-\infty}^{\infty}\ [f(Q)]$.

Proof: It suffices to prove the case when the segment ends on the x-axis, at the origin. Write $U = U_+ + U_-$, where

$$U_+(P) = \pi^{-1}\int_0^{\infty} - \theta(PQ)df(Q) + \text{const.},$$
$$U_- = \pi^{-1}\int_{-\infty}^{0} - \theta(PQ)df(Q) + \text{const.}$$

We have, with $dP = dy > 0$, $d_P\theta(PQ) = \theta_y\, dy < 0$, if $Q > 0$; $\theta_y\, dy > 0$, if $Q < 0$; $|dU_+(P)| \leqq \pi^{-1} \int_0^{\infty} - d_P\theta(PQ)|df(Q)|$, $|dU_-(P)| \leqq \pi^{-1} \int_{-\infty}^{0} d_P\theta(PQ)|df(Q)|$

Thus, integrating over the segment in question,

$$\int |dU_+(P)| \leq \pi^{-1} \int_0^{\infty} \theta(OQ)|df(Q)| = (1/2) \int_0^{\infty} |df(Q)|$$

and similarly

$$\int |dU_-(P)| \leq 1/2\int_{-\infty}^{0} |df(Q)|.$$

Hence

$$\int |dU(P)| \leq \int |dU_+(P)| + \int |dU_-(P)| \leq (1/2)T_{-\infty}^{\infty}[f(Q)].$$

LEMMA 3: *Let S be a minimal surface of the three-dimensional* ξ, η, ζ *space, and let S be bounded by a rectifiable Jordan curve C of length L. Take the conformal mapping of S on the upper half plane* $y > 0$ *of an x, y-plane which establishes a continuous one-to-one correspondence* α *between C and the x-axis and for which* ξ, η, ζ *become bounded harmonic functions of* x, y *in* $y > 0$ *with boundary values according to* α, *for which in* $y > 0$ *the two relations hold:*

$$\xi_x^2 + \eta_x^2 + \zeta_x^2 = \xi_y^2 + \eta_y^2 + \zeta_y^2,$$

$$\xi_x\xi_y + \eta_x\eta_y + \zeta_x\zeta_y = 0$$

with $\xi_x = (\partial/\partial x)\xi, \ldots.$

Denote by $\xi^*(x, y)$, $\eta^*(x, y)$, $\zeta^*(x, y)$ *conjugate harmonics of* $\xi(x, y)$, η, ζ *and call the surface* S^*, *formed with these functions as coordinates of its points, the conjugate minimal surface. Denote by* $L(y)$ *and* $L^*(y)$ *the length of the curves on* S *and* S^*, *respectively, corresponding to fixed* y *and* $-\infty < x < \infty$. *Then* $L(y) = L^*(y) \to L(0) = L^*(0)$ *as* $y \to 0$.

Proof: If ξ^*, η^*, ζ^* are taken so as to vanish at (0, 0), they become bounded harmonic functions assuming continuous boundary values of bounded total variation on the x-axis. In fact, let $0 < h' < h$, then

$$\begin{aligned}|\xi^*(a, h) - \xi^*(a, h')| \leq \int_{h'}^{h} |\xi_y^*(a, y)|dy &\leq \int_{h'}^{h} (\xi_y^{*2} + \eta_y^{*2} + \zeta_y^{*2})^{1/2} dy \\ &\leq \int_{h'}^{h} (\xi_x^2 + \eta_x^2 + \zeta_x^2)^{1/2} dy \\ &\leq \int_{h'}^{h} (\xi_y^2 + \eta_y^2 + \zeta_y^2)^{1/2} dy.\end{aligned}$$

Now by Lemma 2, $\int_0^\infty |\xi_y(a, y)|dy \leq T_{-\infty}^{\infty}[\xi(x, 0)] \leq L(0)$.

Hence $\int^\infty |\xi_y^*(a, y)|dy$ converges and is $\leq 3L(0)$, thereby insuring that ξ^* tends to a limit as $y \to 0$. Similar facts hold for η^*, ζ^*. That these boundary values of ξ^*, η^*, ζ^* are continuous follows easily from

$$\begin{aligned}|\xi^*(a, 0) - \xi^*(b, 0)| \leq T_a^b [\xi^*(x, 0)] &\leq \liminf_{y\to 0} T_a^b [\xi^*(x, y)] \\ &\leq \lim_{y\to 0} \int_a^b (\xi_x^{*2} + \eta_x^{*2} + \zeta_x^{*2})^{1/2} dx \\ &= \lim_{y\to 0} \int_a^b (\xi_x^2 + \eta_x^2 + \zeta_x^2)^{1/2} dx \\ &\leq \text{length of } C \text{ from } x = a \text{ to } x = b,\end{aligned}$$

which also shows that the total variation of the boundary values of ξ^* is continuous and remains bounded. Hence Lemma 1′ applies to ξ^*, η^*, ζ^* as well as to ξ, η, ζ; Lemma 3 follows.

LEMMA 3′: *Under the hypotheses of Lemma 3, we have more generally*

$$L_a^b(y) = L^{*b}_a(y) \to L_a^b(0) = L^{*b}_a(0).$$

Proof: Because of semicontinuity the only other possibility could be $L_a^b(0) < \liminf L_a^b(y)$. But then

$$L(0) = L_{-\infty}^a(0) + L_a^b(0) + L_b^\infty(0) < \liminf L_{-\infty}^{\infty}(y),$$

which contradicts Lemma 1′. Analogous considerations exclude $L^{*b}_a(0) < \liminf L^{*b}_a(y)$, as $y \to 0$ through an arbitrary sequence.

LEMMA 4: *Denote by* $L(0, \gamma)$ *the length of the boundary curve of the surface with coordinates* $\xi \cos \gamma + \xi^* \sin \gamma, \eta \cos \gamma + \eta^* \sin \gamma, \zeta \cos \gamma + \zeta^* \sin \gamma$. *Then* $L_a^b(0, \gamma) = L_a^b(0, 0) = L_a^b(0)$.

Proof: An elementary computation shows $L(y, \gamma) = L(y, 0)$, and the passage $y \to 0$ gives, in view of Lemma 3′, the desired relation.

LEMMA 5: *Denote by* s *the length on* C. *Excepting a set of measure zero with respect to* s, *we have on* C

$$0 = (d\xi/ds)(d\xi^*/ds) + (d\eta/ds)(d\eta^*/ds) + (d\zeta/ds)(d\zeta^*/ds). \quad (4)$$

Proof: Since

$$L_a^b(0, \gamma) = L_a^b(0, 0) = \int_{x=a}^{b} [(d(\xi \cos \gamma + \xi^* \sin \gamma)/ds)^2 + \ldots]^{1/2} ds$$

we find by differentiation with respect to γ, setting $\gamma = 0$,

$$0 = \int_{x=a}^{b} [(d\xi/ds)(d\xi^*/ds) + \ldots]ds$$

which is the integrated form of Lemma 5, from which it follows by differentiation with respect to $s = s(b)$.

THEOREM: *Under the assumptions of Lemma 3 about* S *and its coordinates* $\xi, \eta, \zeta(x, y)$, suppose furthermore that the boundary curve contains an analytic arc A which, without loss of generality, may be taken in the form $\eta = \varphi(\xi), \zeta = \psi(\xi)$ with $\varphi(0) = \psi(0) = \varphi'(0) = \psi'(0) = 0$, and with φ and ψ analytic functions of the argument ξ for sufficiently small values of $|\xi|$; and let the origin of the ξ, η, ζ-space correspond to the origin of the x, y-plane. Then $\xi(x, 0), \eta(x, 0), \zeta(x, 0)$ are analytic functions of x at $x = 0$ and can be continued analytically near the origin across $y = 0$.

With the exception, noted by Riemann, of straight line arcs, no other example of an analytic curve has heretofore been shown to possess the property of automatically making the minimal surface analytically extensible across it.[3]

Proof: Set

$$\lambda = (\xi + i\xi^*)/2, \mu = (\eta + i\eta^*)/2, \nu = (\zeta + i\zeta^*)/2, z = x + iy.$$

$\lambda, \mu, \nu(z)$ are analytic functions for $y > 0$, continuous in $y \geqq 0$. If we take the integral $\int |d\lambda| + |d\mu| + |d\nu|$, extended over the x-axis, a parallel to the x-axis, or over the y-axis, it remains by Lemmas 1, 2, 3 bounded above by $6L(0)$. We now define the continuous function $\sigma(x, y) = \int |d\lambda| + |d\mu| + |d\nu|$ by extending the integral from $(0, 0)$ to $(0, y)$ to (x, y) over straight segments. For sufficiently small $h > 0$, the square q of the corners $-h, +h, h + 2ih, -h + 2ih$ has the property that within q, $\sigma(x, y) < \epsilon$ for given $\epsilon > 0$.

Denote by ρ, M two positive numbers such that for the complex quantity Ξ satisfying $|\Xi| \leqq \rho$, we have $\varphi(\Xi)$ and $\psi(\Xi)$ regular and

$$|\varphi'(\Xi)| \leq M|\Xi|, \psi'(\Xi) \leq M|\Xi|.$$

Take ϵ so small that

$$M\epsilon < 1/8 \text{ and } 4\epsilon < \rho.$$

We determine a function $\Xi(z)$ as solution of the differential equation

$$d\Xi = 2[1 + \varphi'^2(\Xi) + \psi'^2(\Xi)]^{-1}\{d\lambda + \varphi'(\Xi)d\mu + \psi'(\Xi)d\nu\} \quad (5)$$

with $\Xi(0) = 0$.[4] Successive approximations give in the usual manner $\Xi(z)$ by integrating over straight segments from 0 to iy to $x + iy$ for all $x + iy = z$ of q. To show this, we observe, calling Ξ_k the kth approximation, that $\Xi_0 \equiv 0$, and

$$|d\Xi_1| \leq 2d\sigma, \ |\Xi_1| \leq 2\sigma.$$

Furthermore, suppose that $|\Xi_k| \leq 4\sigma$; then, by (5),

$$\begin{aligned} |d\Xi_{k+1}| &\leq 2[1 - 2M^2|\Xi_k|^2]^{-1}\, d\sigma \\ &\leq 2[1 - 2M^2.16\sigma^2]^{-1}\, d\sigma \\ &\leq 2[1 - 32(M\epsilon)^2]^{-1}\, d\sigma \\ &\leq 4d\sigma, \end{aligned}$$

and $|\Xi_{k+1}| \leq 4\sigma$.

Thus Ξ_k exists in q for all integral $k \geq 0$. For a certain $M^\wedge$, to be described presently, we find

$$|d|\Xi_{k+2} - \Xi_{k+1}|| \leq |d(\Xi_{k+2} - \Xi_{k+1})| \leq M^\wedge|\Xi_{k+1} - \Xi_k|d\sigma,$$

whence

$$|\Xi_1 - \Xi_0| \leq 2\sigma, \ |\Xi_2 - \Xi_1| \leq 2M^\wedge\sigma^2/2, \ |\Xi_{k+1} - \Xi_k| \leq 2\sigma(M^\wedge\sigma)^k/(k+1)!$$

Thus $\Xi_0 + (\Xi_1 - \Xi_0) + (\Xi_2 - \Xi_1) + \ldots$ has the majorant $2(e^{M^\wedge\sigma} - 1)/M^\wedge$ and converges uniformly in q. Here $M^\wedge$ is the maximum of $|G(\kappa) - G(\kappa')| : |\kappa - \kappa'|$ for the following possibilities

$$\begin{aligned} |\kappa|, |\kappa'| \leq \rho, \kappa' \neq \kappa, G(\kappa) &= 2[1 + \varphi'(\kappa)^2 + \psi'(\kappa)^2]^{-1} = G_1(\kappa), \\ G(\kappa) &= G_1(\kappa)\varphi'(\kappa), \\ G(\kappa) &= G_1(\kappa)\psi'(\kappa). \end{aligned}$$

Now set $\Xi = \lim_{k \to \infty} \Xi_k$. We have in q the integrated form of (5),

$$\Xi = 2\int [1 + \varphi'^2(\Xi) + \psi'^2(\Xi)]^{-1}\{d\lambda + \varphi'(\Xi)d\mu + \psi'(\Xi)d\nu\} \quad (6)$$

since the Stieltjes integral on the right hand is the limit as $k \to \infty$ of the analogous integral with Ξ replaced by Ξ_k.

That the construction yields an analytic function of z near all $z = iy$, $y > 0$, follows from the fact that (i) there is for arbitrary initial $x_0 + iy_0$, $y_0 > 0$, a solution of (5) which is analytic, according to the classical existence theorem and (ii) that there is no more than one solution of (5). For call $\delta\Xi$ the difference of two solutions of (5). We have

$$|\delta\Xi| \leq M^{\wedge}\int|\delta\Xi|d\sigma$$

with $\delta\Xi$ vanishing initially. This inequality yields $\delta\Xi \equiv 0$ in familiar manner.

That the solution Ξ is continuous in the whole of q follows from the same property of the Ξ_k. The latter fact follows from the convergence of the total variations of λ, μ, ν over parallel segments which approach the x-axis normally. It is true that *a priori* (5) will hold on $y = 0$ only except for a set of measure zero relative to the variable s, but the integrated form holds everywhere.

I say that Ξ coincides with $\xi(x, 0)$ for $y = 0$. In fact, we have on $y = 0$ because $\eta = \varphi(\xi)$, $\zeta = \psi(\xi)$, that almost everywhere in the sense mentioned, in view of (4),

$$2[1 + \varphi'^2(\xi) + \psi'^2(\xi)]^{-1}\{d\xi/2 + \varphi'(\xi d\eta/2 + \psi'(\xi)\, d\zeta/2 + i(d\xi^*/2 + \varphi'(\xi)d\eta^*/2 + \psi'(\xi)d\zeta^*/2)\} = d\xi.$$

Moreover $\xi(0, 0) = 0$. Hence ξ is a possible solution of (5) on $y = 0$ and the solution is unique as proved.

Let $z = x + iy$ be a point such that $\bar{z} = x - iy$ lies in q. Put $\lambda(z) = \overline{\Xi(\bar{z}) - \lambda(\bar{z})}$.
On $y = 0$, this new λ coincides with the former definition. In fact,

$$\begin{aligned}\lambda(x) = \overline{\Xi(x) - \lambda(x)} &= \xi(x, 0) - (\xi(x, 0) - i\xi^*(x, 0))/2 \\ &= (\xi(x, 0) + i\xi^*(x, 0))/2.\end{aligned}$$

This new $\lambda(z)$ is regular, being the result of a succession of the two anticonformal mappings $z \rightarrow \bar{z}$ and $\bar{z} \rightarrow \overline{\Xi(\bar{z}) - \lambda(\bar{z})}$. We can therefore apply the well-known theorem that λ is regular within q plus its image by reflection on $y = 0$, in particular on $y = 0$.

μ and ν can equally be continued across $y = 0$, thus establishing their analyticity at $z = 0$. We effect the extension by putting for $y < 0$, $z = x + iy$,

$$\mu(z) = -\overline{\mu(\bar{z})} + \overline{\varphi(\Xi(\bar{z}))},\ \nu(z) = -\overline{\nu(\bar{z})} + \overline{\psi(\Xi(\bar{z}))}.$$

First of all, these new definitions agree with the previous ones for μ, ν on $y = 0$, since there

$$\begin{aligned}-\overline{\mu(x)} + \overline{\varphi(\Xi(x))} &= -(\eta - i\eta^*)/2 + \varphi(\xi) \\ &= (\eta + i\eta^*)/2 = \mu(x)\end{aligned}$$

and similarly

$$-\bar{\nu}(x) + \overline{\psi(\Xi(x))} = \nu(x).$$

Secondly, the same argument as before yields the fact that for $y < 0$, μ and ν are regular functions of z, thus becoming regular also on the x-axis.

The Theorem is proved.

The idea underlying the analytic extension, of forming a differential equation connecting the extension with the given function, is capable of application in many other situations. We have treated similarly the problem of the "free boundary surface" for minimal surfaces. Other problems concerning analytic boundary conditions come under the same heading. We intend to return to the subject elsewhere.

[1] It suffices to demand assumption of boundary values upon normal approach to the boundary.

[2] Evans, G. C., *The Logarithmic Potential*, Am. Math. Soc. Colloquium Publications VI, New York, 1927, pp. 35ff.

[3] Courant, R., *Dirichlet's Principle, Conformal Mapping and Minimal Surfaces*, Interscience Publishers, 1950, esp. p. 118.

[4] More precisely a solution of the integrated equation (6).

Permissions

Birkhäuser Boston thanks the original publishers of the papers of Hans Lewy for granting permission to reprint specific papers in this collection.

[21] Reprinted from *American Journal of Mathematics*, **60**, ©1938 Johns Hopkins University Press

[23] Reprinted from *Bulletin of the American Mathematical Society*, **52**, ©1946 the American Mathematical Society

[24] Reprinted from *Communications in Pure and Applied Mathematics*, **1**, ©1948 John Wiley and Sons

[26] Reprinted from *International Congress of Mathematicians, Cambridge, MA*, **1**, ©1950 the American Mathematical Society

[27] Reprinted from *University of California Publications in Mathematics, New Series*, ©1950 the University of California Press

[28] Reprinted from *Communications in Pure and Applied Mathematics*, **4**, ©1951 John Wiley and Sons

[29] Reprinted from *Proceedings of the National Academy of Sciences*, **37**, ©1951 the National Academy of Sciences

[30] Reprinted from *Proceedings of the American Mathematical Society*, **3**, ©1952 the American Mathematical Society

[32] Reprinted from *National Bureau of Standards*, **18**, ©1952 the National Institute of Standards and Technology

[33] Reprinted from *Annals of Mathematics*, **56**, ©1952 Johns Hopkins University Press

[34] Reprinted from *Communications in Pure and Applied Mathematics*, **5**, ©1952 John Wiley and Sons

[36] Reprinted from *Annals of Mathematics*, **64**, ©1935 Johns Hopkins University Press

[37] Reprinted from *Communications in Pure and Applied Mathematics*, **9**, ©1956 John Wiley and Sons

[38] Reprinted from *Annals of Mathematics*, **66**, ©1957 Johns Hopkins University Press

[39] Reprinted from *Journal of Mathematics and Mechanics*, **6**, ©1957 the Graduate Institute for Mathematics and Mechanics, Indiana University

[40] Reprinted from *Journal of Mathematics and Mechanics*, **8**, ©1959 the Graduate Institute for Mathematics and Mechanics, Indiana University

[41] Reprinted from *Bulletin of the American Mathematical Society*, **65**, ©1959 the American Mathematical Society

[42] Reprinted from *Communications in Pure and Applied Mathematics*, **13**, ©1960 John Wiley and Sons

[43] Reprinted from *Partial Differential Equations and Continuum Mechanics*,©1961 the University of Wisconsin Press

[44] Reprinted from *Communications in Pure and Applied Mathematics*, **14**, ©1961 John Wiley and Sons

[47] Reprinted from *Annali della Scuola Normale Superiore di Pisa*, **19**, ©1965 Scuola Normale Superiore di Pisa

[48] Reprinted from *Journal of Mathematics and Mechanics*, **14**, ©1965 the Graduate Institute for Mathematics and Mechanics, Indiana University

[49] Reprinted from *Communications in Pure and Applied Mathematics*, **17**, ©1965 John Wiley and Sons

[51] Reprinted from *Journal of Mathematics and Mechanics*, **17**, ©1968 the Graduate Institute for Mathematics and Mechanics, Indiana University

[52] Reprinted from *Annals of Mathematics*, **88**, ©1968 Johns Hopkins University Press
[53] Reprinted from *American Journal of Mathematics*, **91**, ©1969 Johns Hopkins University Press
[54] Reprinted from *Communications in Pure and Applied Mathematics*, **22**, ©1969 John Wiley and Sons
[56] Reprinted from *Rendiconti della Accademia Nazionale dei Lincei*, **8** , ©1970 the Accademia Nazionale dei Lincei
[58] Reprinted from *Journal d'Analyse Mathématique*, **23**, ©1970
[60] Reprinted from *Archive for Rational Mechanics and Analysis*, **41**, ©1971 Springer-Verlag
[61] Reprinted from *Journal of Differential Geometry*, **6**, ©1972 Lehigh University
[62] Reprinted from *Rendiconti della Accademia Nazionale dei Lincei*, **217**, ©1975 the Accademia Nazionale dei Lincei
[63] Reprinted from *Annali della Scuola Normale Superiore di Pisa*, **4**, ©1976 Scuola Normale Superiore di Pisa
[64] Reprinted from *Contributi del Centro Linceo Interdisciplinare di Scienze Matematiche e Loro Applicazione*, No. 35, ©1977 Accademia Nazionale dei Linzei
[65] Reprinted from *Communications in Partial Differential Equations*, **2**, ©1977 Marcel Dekker Inc.
[67] Reprinted from *Annali della Scuola Normale Superiore di Pisa*, **6**, ©1979 Scuola Normale Superiore di Pisa
[68] Reprinted from *Manuscripta Mathematica*, **26**, ©1979 Springer-Verlag
[69] Reprinted from *Communications in Pure and Applied Mathematics*, **33**, ©1980 John Wiley and Sons
[70] Reprinted from *Nachrichten der Gesellschaft der Wissenschaften Göttingen*, **4**, ©1982 Georg-August-Universitt Göttingen
[72] Reprinted from *Journal für die reine und Angewandte Mathematik*, **379**, ©1987 Walter de Gruyter
[73] Reprinted from *Partial Differential Equations and the Calculus of Variations*, Essays in Honor of Ennio DeGiorgi, ©1989 Birkhäuser

MIX
Papier aus verantwortungsvollen Quellen
Paper from responsible sources
FSC® C105338

If you have any concerns about our products,
you can contact us on
ProductSafety@springernature.com

In case Publisher is established outside the EU,
the EU authorized representative is:
Springer Nature Customer Service Center GmbH
Europaplatz 3, 69115 Heidelberg, Germany

Printed by Libri Plureos GmbH
in Hamburg, Germany